Polyacetylene

Chemistry, Physics, and Material Science

Polyacetylene

Chemistry, Physics, and Material Science

James C. W. Chien

Department of Polymer Science and Engineering
University of Massachusetts
Amherst, Massachusetts

 1984

ACADEMIC PRESS
(Harcourt Brace Jovanovich, Publishers)

Orlando   San Diego   San Francisco   New York   London
Toronto   Montreal   Sydney   Tokyo   São Paulo

CHEMISTRY

7189 - 7434

ACADEMIC PRESS, INC.
Orlando, Florida 32887

United Kingdom Edition published by
ACADEMIC PRESS, INC. (LONDON) LTD.
24/28 Oval Road, London NW1 7DX

Library of Congress Cataloging in Publication Data

Chien, James C. W., Date
    Polyacetylene: chemistry, physics, and material
science.

    Includes index.
    1. Polyacetylenes. I. Title.
TP1180.P53C48 1984   668.4'23   83-7237
ISBN 0-12-172460-3

PRINTED IN THE UNITED STATES OF AMERICA

84 85 86 87   9 8 7 6 5 4 3 2 1

To my wife, Stella,
for her loving devotion

Contents

Preface

Acetylene was first polymerized to a linear conjugated polymer by Natta, Mazzanti, and Corradini in 1958 using a $Ti(OBu)_4/AlEt_3$ catalyst. Subsequently, a number of other catalysts were shown to polymerize acetylene with varying degrees of efficiency. The polymer formed was usually a gray or black semicrystalline powder that was insoluble in any solvent and decomposed before melting. In the early 1960s Hatano, Ikeda, and co-workers began investigating the properties of polyacetylene. In 1974, Shirakawa, working with Ito and Ikeda, found a way to prepare a mechanically strong, free-standing film of the polymer. During a lecture tour in Japan, MacDiarmid learned of the finding and invited Shirakawa to Philadelphia to work with him and Heeger. They, together with other co-workers, demonstrated in 1977 that the conductivity of polyacetylene can be increased by 13 orders of magnitude upon doping with various electron-accepting and electron-donating substances. This generated a great deal of interest and created virtually a new field of research.

There are two prime forces urging ahead researchers on polyacetylene. The first is the recognition by Heeger and Schrieffer that novel physics is emerging from such studies. Many properties can be interpreted by the concepts of soliton and polaron. It is no wonder that condensed-matter physicists constitute the largest single group of researchers of the field. The second is the appreciation by MacDiarmid of the novel chemistry of the unusual carbonium ions and carbanions of polyacetylene. But relatively few chemists have entered the field. Most surprisingly, very little activity has been generated among the polymer scientists.

The basic purpose of this book is to survey and appraise the present state of polyacetylene research. For such a rapidly expanding interdisciplinary field, it would seem logical to edit a book with contributions from experts in various areas. Undoubtedly there probably will be many such books forthcoming. However, each chapter would quite likely be appreciated mainly by readers of that discipline. It is my feeling that if polyacetylene research is to grow further, the physicists need to know the materials they are working with, and the chemists should have an understanding of the physics that differentiates polyacetylene from many other substances. Accordingly, one goal of this book is to serve different types of audiences, with the hope that it is at a level such that many readers can benefit.

Polyacetylene is a very reactive macromolecule, extremely sensitive to heat, oxygen, and other, unknown, aging processes with consequences of chemical and physical transformations difficult to characterize quantitatively. There are controversies in both observations and interpretations. Some of these are unavoidable as one tries to understand a new material. Most others are attributable to sample differences in preparation and history. Another goal of this book is to comment critically on the sample-related controversial results based on my own experiences with the material.

A synthetic polymer often comprises a family of substances even though they are produced from the same monomer. The members may differ in molecular weights, molecular weight distributions, crystallinity, morphologies, isomeric and stereoisomeric structures, etc. Take a simple example of polybutadiene; the polymers can have different *cis*-1,4, *trans*-1,4, and 1,2 contents. Polyacetylene is similar in this respect. Therefore, each type of polyacetylene obtained with different catalysts under various conditions should be treated as a dissimilar material and requires full characterization. The problems are compounded with doped polyacetylenes. The nature of dopant species and the homogeneity of doping are very important matters. Consequently, the properties of polyacetylenes given in this book are subject to changes as better methods of synthesis, characterization, handling, and doping are developed. The book is written to anticipate such developments and wherever possible indicates the probable changes in properties.

I feel that a book addressing the goals described above should be written by one person in the interest of cohesiveness. There are several risks involved in such an endeavor. The presentations may be at a highly technical level in areas in which the author is knowledgeable and at low levels in other areas. The coverage of literature may also be uneven. I hope the readers will be understanding of minor errors and unintentional omissions of published works.

After the introduction, Chapter 2 gives a detailed treatment of polymerization. This is done so that the interested investigators can prepare polyacetylene comparable in quality to the best available material. Among the various classes of polymerization initiators, Ziegler–Natta catalysts are the most complicated. The mechanisms and kinetics of acetylene polymerization are given here. The direct determination of polyacetylene molecular weight by radioquenching is presented. Finally, a number of other catalyst systems for acetylene polymerizations are described.

The crystal structures and morphology of undoped polyacetylenes are discussed in Chapters 3 and 4, respectively. In particular, the controversy concerning fibrillar or lamellar morphology is resolved. There are also differences in the crystal structures for polyacetylene. They are all included in Chapter 3. Whether the differences are due to polymorphism or other causes will require further investigation especially with better oriented specimens. The all-important degree of bond length alternation seems to have been settled.

Polyacetylene can be made with predominantly cis or nearly pure trans isomeric structures. Since they differ markedly in certain properties, the methods of isomerization are treated in Chapter 5 with emphasis on optimum conditions. Isomerization creates neutral defects or solitons which can be best studied by magnetic resonance techniques. The principal results are described. The other spectroscopic, physical, and mechanical properties of undoped polyacetylene are given in Chapter 6.

Chapter 7 discusses the various chemical reactions of polyacetylene and polymethylacetylene. At elevated temperatures there are facile electron–proton and electron–methyl exchange processes not known for other polymers. Autoxidation and stabilization of the polymers are treated here. The often invoked cross-linking reaction is scrutinized at the end of this chapter.

Doping is the key for imparting metallic conductivity to polyacetylene. The various methods of doping are given in Chapter 8. The nature of iodine dopant is generally agreed to be I_3^- and I_5^-. However, there is considerable disagreement about the nature of AsF_5 dopant in polyacetylene. Probably several different species can be formed depending on experimental conditions. The rate of diffusion of dopant is a subject of some concern. The probable mechanisms of doping are proposed.

Much controversy exists concerning the properties of doped polyacetylene, especially regarding the magnetic susceptibility. IBM investigators reported that only Pauli susceptibility was found in AsF_5 doped *cis*-polyacetylene. Scientists at the University of Pennsylvania and Xerox found Curie susceptibility in lightly doped polymer, no susceptibility between $y \approx 10^{-3}$ and $y \approx 5 \times 10^{-2}$ (y is moles of dopant per mole of CH), and Pauli susceptibility in heavily doped polyacetylene. There are other reports of no

Pauli susceptibility in "metallic" polyacetylenes. I believe the discordant results originate from sample differences. Doping introduces new infrared active vibration modes, which are independent of the dopant. Different explanations have been proposed for their origins. These and other properties of doped polyacetylenes are the topics of Chapter 9.

Theoretical models for polyacetylene are presented in Chapter 10. Chapter 11 treats the transport properties. First, the various possible mechanisms for carrier transport are offered and then the factors influencing conductivity are discussed. The role of sample homogeneity is closely examined as it has significant influence on polyacetylene research.

The book concludes with the description of a few technical applications of polyacetylene.

Chapters 2, 3, 4, and 7 are based substantially on the recent published and unpublished results of our laboratories at the University of Massachusetts. Our work also contributes to the contents of Chapters 5, 8, 9, and 11. The efforts of my co-workers, J. Capistran, L. C. Dickinson, J. L. Fan, R. Gable, R. Gooding, K. D. Gourley, J. A. Hirsch, J. R. Reynolds, M. A. Schen, K. Shimamura, J. M. Warakomski, G. E. Wnek, Y. Yamashita, and X. Yang, and the collaboration of my colleagues, Professor F. E. Karasz and Professor C. P. Lillya, are deeply appreciated. Our work received invaluable support from DARPA, the Office of Naval Research, and the National Science Foundation. Expenses for the preparation of the book were partially paid for by these grants.

I acknowledge the excellent efforts of P. Barschenski in preparing the manuscript with assistance from M. Putnam, J. A. Hirsch, and N. Finkenaur, who drew most of the figures, and of Dr. L. C. Dickinson in proofreading. I am especially grateful to Professor A. G. MacDiarmid and Professor A. J. Heeger, who had encouraged and stimulated this work. The kindness of researchers to allow me to reprint figures and tables of their published and unpublished works is deeply appreciated.

Chapter 1

Introduction

Organic polymers have been traditionally associated with dielectric materials. One of the earlier applications of organic polymers was the use of phenolic resins in electrical outlets and light switches. Polyethylene is widely used for wire coating. Polymers differ from other inorganic and organic substances in many basic ways that may not be fully appreciated by physicists dealing with polymers for the first time. On the other hand, concepts that describe semiconductors and metals and their electrical properties may be unfamiliar to many polymer scientists. Therefore, it seems desirable to present here brief introductions to a few basic principles of polymer chemistry and of solid-state physics in order to furnish some background to the main subjects of this book on polyacetylene. This is followed by brief surveys of charge-transfer salts and conducting polymers other than polyacetylene to show that there is a broad interest in organic conducting materials. Finally, a historical discussion on polyacetylene and a general description of its properties are given to set the stage for this book.

1.1 SYNTHETIC POLYMERS

Polymers are synthesized from monomers through reactions of functional groups of the monomers. One large class is the vinyl polymer, of which polyacetylene is a member. The monomers contain unsaturated

carbon–carbon bonds; a monomer is converted to polymer by stimulating
the opening of the unsaturated carbon–carbon bond with a free radical,
ionic, or metal coordination initiator. The polymerization of acetylene can
be simply written as

$$n\text{CH}\equiv\text{CH} \xrightarrow{\text{initiator}} \left(\text{CH}=\text{CH}\right)_n \qquad (1.1)$$

The detailed mechanism of this addition polymerization is very complicated and not yet entirely elucidated, as discussed in Section 2.5. The
resulting polymer is named by just adding the prefix *poly* to the name of the
monomer, e.q., polyacetylene. In this particular case a simplified notation,
$[\text{CH}]_x$, is widely used and generally accepted.

   Other polymerizations proceed via stepwise reactions of functional
groups. An example is the synthesis of poly(phenylene vinylene) by the
Wittig reaction:

$$(1.2)$$

In this instance, the name of the structural unit is placed in parentheses and
prefixed with poly. This nomenclature system is used for all polymers
derived from two or more monomers.

   The value of *n* is usually large; it corresponds to the degree of polymerization. The polymer molecules in a single preparation do not have identical
molecular weight but have instead a relatively narrow or broad distribution,
depending upon the synthetic method. Two most common average molecu-
lar weights used are the number average, $\overline{M}_n$, and the weight average, $\overline{M}_w$.
They are defined by

$$\overline{M}_n = \sum_i N_i M_i \Big/ \sum_i N_i = \sum_i W_i \Big/ \sum_i \left(\frac{W_i}{M_i}\right) \qquad (1.3)$$

and

$$\overline{M}_w = \sum_i N_i M_i^2 \Big/ \sum_i N_i M_i = \sum_i W_i M_i \Big/ \sum_i W_i, \qquad (1.4)$$

where N_i is the number of polymer molecules of species *i* of molar mass M_i,
and $W_i = N_i M_i/N_A$ with N_A being the Avogadro number.

A polymer molecule does not have the simple structure implied by its ideal formula. Depending on the monomer and the method of polymerization, a polymer molecule may contain various kinds of imperfections, such as head-to-head versus head-to-tail placements, branchings, and geometric isomeric and stereoisomeric units. These irregularities also occur statistically. A polymer possessing backbone flexibility and/or structural irregularities usually forms an amorphous solid. It becomes a glassy material below the glass transition temperature. Thus the physical and mechanical properties are strongly affected by the chain irregularity and mobility but are relatively independent of molecular weight above a certain value, say, 10,000.

A polymer with nearly perfect structure, in which regular packing arrangement is possible, can be crystallized. However, perfect crystalline polymers are not encountered in practice, but instead a given sample will contain varying proportions of ordered and disordered regions. These semicrystalline polymers usually exhibit melting transitions. The melting transition occurs at higher temperature and over a narrower temperature range the more orderly the structure. A number of distinct morphological features have been identified for various semicrystalline polymers. The X-ray pattern usually displays discrete maxima, which arise from the scattering by small regions of three-dimensional order called crystallites. The sizes of these crystallites are usually very small, rarely exceeding 100 nm. A number of polymers including polyesters, polyamides, polyolefins, and cellulose acetate form well-defined single crystals by very slow cooling of dilute polymer solutions. Electron microscopy reveals them to be made up of thin lamellae, often lozenge shaped and about 10–20 nm thick. The long polymer chain folds back and forth with the direction of the chain axis across the thickness of the platelet. There are three models for the chain folding: adjacent regular reentry, adjacent but irregular reentry, and random reentry.

The crystallites themselves are often arranged regularly into spherulites displaying Maltese cross optical extinction patterns. The size of the spherulite can attain quite large dimensions, in some instances a few millimeters in diameter. The spherulites are built up from bundles of fibrous subunits that grow outward from a central nucleus. The space between fibrils or crystallites is filled with amorphous polymers.

Many polymers can be spun into fibers, which upon drawing became highly oriented, possessing great tensile strength in the draw direction. They can also be blown or drawn in two directions to give biaxially oriented films or extruded or molded into objects of variegated shape and size. The processing conditions are largely determined by the melt rheological properties of a polymer.

Synthetic polymers, being organic molecules, are subjected to thermal, oxidative, photochemical, and other environmentally induced degradations. Therefore, most commercial polymers contain one or more additives in order to enhance their stability for intended applications.

Although most synthetic efforts on organic polymers are directed toward improvement in physical, mechanical, thermal, and processing properties, there have been attempts to synthesize polymers possessing metallike conductivity. The basic idea was to synthesize macromolecules having an extended π-electron system in the backbone or having ordered π-electron pendant groups or both. It was thought that in the former case the π electrons would be delocalized and could act as charge carriers. In the latter case, overlap of π orbitals was proposed to provide a pathway for electronic conduction. These expectations are not justifiable on close scrutiny. Polymers synthesized based on these ideas have been found to be either insulators or semiconductors. Thus perfectly conjugated, macroscopic single crystals of poly(diacetylenes) have been prepared by solid-state polymerization of diacetylene monomers (Wegner, 1969). These materials have $\sigma_{RT} \sim 10^{-10}$ $(\Omega \text{ cm})^{-1}$. EPR studies by Bloor *et al.* (1974) showed that the poly(diacetylenes) contain no measurable amount of free carriers. Holob *et al.* (1972) postulated that carrier generation in conjugated polyenes requires rotation of the backbone carbon atoms to sever the double bonds; this mechanism is not possible in poly(diacetylenes) where the chain is locked in the crystal lattice. It will be shown that the formation of topological defects occurs with great ease in polyacetylene. Poly(phenylacetylene) has a disappointingly low conductivity of $\sim 10^{-15}$ $(\Omega \text{ cm})^{-1}$ (Hankin and North, 1967). The bulky substituent was thought to introduce conformational defects destroying the planarity of the π system. Conjugated ladder polymers, such as polyacene quinines (Hartman and Pohl, 1968), display semiconducting properties, but these materials are highly intractable.

It is now known that an extended π-electron system is a necessary but not a sufficient condition for conductivity. Even highly crystalline polyacetylene is only a semiconductor. There must be introduced, such as by doping, a large number of free carriers in the polymer before it can conduct electricity. Therefore, it was a very significant discovery in 1977 when Shirakawa *et al.* demonstrated that polyacetylene upon doping exhibits metallic conductivity. This was followed by reports of several other organic polymers that share this characteristic. It can be said that a new class of materials has been discovered. They are being referred to as "organic metals," "synthetic metals," or "low-dimensional conductors."

There are many questions that must be answered in order to establish the relations between the molecular and supermolecular structures of organic polymers and their electrical transport properties. Among the known organic metallic systems, polyacetylene is most suited for detailed investigation. It is the simplest conjugated macromolecule and therefore amenable to theoretical treatment. The polymer is highly crystalline and its structure can be determined by diffraction techniques. Doped polyacetylenes have the highest conductivities among all the conducting polymers synthesized to date. Therefore, the number of publications on polyacetylene far exceeds those on all other conducting polymers combined. One hopes that with sufficient basic understanding of the structure and properties of polyacetylene guidelines for the synthesis of new candidate polymers may be formulated.

1.2   ENERGY BANDS

Electrons in isolated atoms are described by their atomic orbital wave functions. Their energies are determined solely by the interactions with the nucleus and other electrons of the atom. However, when the atoms are placed in ordered arrays, the electrons can be thought of as belonging to the whole crystal. Even though the electrons of each orbital type have different energies, the difference between them becomes so small that the orbitals merge into bands. There are several equivalent ways of giving mathematical substance to the theory of energy bands.

The fact that a crystalline material has a periodic arrangement of the atoms on a crystal lattice gives rise to a periodic potential within which the electrons exist. This periodic potential leads to a series of allowed energy bands separated by energy gaps in which electron states are not allowed, i.e., forbidden energy bands. This description is possible without the need to consider any details of the system. The point of transition between allowed and forbidden bands occurs when $k = m\pi/a$, where k is the wave vector, m is an integer, and a is the lattice parameter, which is usually the equilibrium separation between the atoms. This condition is the same as that set forth by the Bragg reflection rule. Thus electron waves with $k = m\pi/a$ are not able to propagate in the crystal, and hence correspond physically to forbidden energies for the electrons.

Energy bands can also be viewed as the effect on the energy levels of atoms caused by interaction between atoms as they are brought together to form crystals. This is illustrated for the case of sodium (Fig. 1.1). At infinite separation (Fig. 1.1a) all the atomic orbitals have discrete energies. As the atoms are brought together, their orbitals begin to overlap, and energy bands result (Fig. 1.1b). When the interatomic distance becomes comparable to or less than the spatial extension of the electronic wave function associated

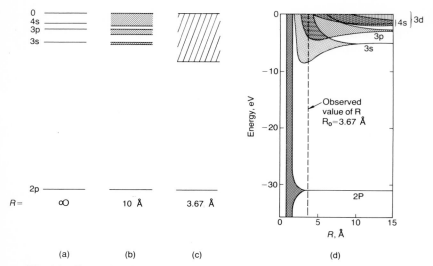

Fig. 1.1 Energy bands developing in metallic sodium: (a) large atomic separation, (b) atoms separated by 10 Å, (c) actual interatomic separation in metallic sodium, (d) energy bands developing as a continuous function of interatomic distance. [After Slater (1934).]

with a particular atom, the electrons do not belong to any particular atom but to the entire crystal (Fig. 1.1c). On the other hand, according to the Pauli exclusion principle, the electrons of a given atomic orbital cannot have identical energy in the crystal. The result is that they form a band of energy levels with very small energy differences between individual levels. In the case of sodium atoms in a crystal at an equilibrium interatomic separation of 3.67 Å, all the 3s, 3p, and 4s orbitals up to continuum merge into one band. The development of energy bands as a continuous function of interatomic distance is illustrated in Fig. 1.1d (Slater, 1934).

Yet another approach to energy bands is to consider the consequence of a small periodic potential on an electron gas. The energy of a free electron given by $E = \hbar^2 k^2 / 2m_e$ is not quantized because k can assume any value. However, the imposition of a periodic potential restricts those electron waves with $k = m\pi/a$ to standing waves rather than traveling waves. Energy gaps are opened up whenever $k = m\pi/a$ in the electron wave, whereas previously it had a continuous energy $E(k)$.

There are two common ways to present the energy band structures, the extended representation and the reduced representation. They are illustrated for the simple case of a free-electron model perturbed by a small periodic potential. Figure 1.2a shows a plot of the energies of electrons $E(k)$ as a function of k explicitly for all values of k. The gaps are introduced by the periodic potential at each Bragg reflection condition. Any given segment of

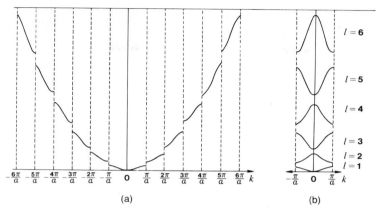

Fig. 1.2 Band structure for a free-electron model in the presence of a periodic potential causing Bragg reflections for $k = m\pi/a$: (a) extended representation, (b) reduced representation.

this broken curve can be considered still to possess free-electron-like behavior except for values of k near the gap. This is the extended representation.

It is possible to convert the extended representation of energy bands to a more compact reduced representation shown in Fig. 1.2b because of the periodic nature of the crystalline structure. The solution for a lattice wave is not altered by displacing k by $m2\pi/a$. Therefore, the segments in Fig. 1.2a can be shifted by integral multiples of $2\pi/a$ so that all values of $E(k)$ are brought into the range $-\pi/a \leqslant k \leqslant +\pi/a$. Each segment is designated with an index l. For $l = 1$, the curves for the two representations are the same. The curves in the reduced zone, with $l = 2$ and 3, are obtained by a $2\pi/a$ translation of the two corresponding segments in Fig. 1.2a, and $4\pi/a$ translations yielded the curves for $l = 4$ and 5, etc. In this one-dimensional reduced representation, the zone of $l = 1$ is called the first Brillouin zone; the second Brillouin zone is for $l = 2$; etc. Brillouin zones for a real three-dimensional lattice can be quite complicated; they reflect the actual crystal structure of the substance.

1.3 METALS, SEMICONDUCTORS, AND INSULATORS

Pristine *cis*-polyacetylene can almost be considered an insulator, whereas *trans*-polyacetylene has transport properties of a semiconductor. Heavily doped polyacetylenes have metallic conductivities. The purpose of this section is to discuss the transport properties of the conventional metals, the crystalline inorganic semiconductors, and the typical insulators, and how they are described by the band theory. The reader may find it useful to return to this section when the properties of polyacetylenes are discussed, and theoretical models for polyacetylenes are presented.

1.3.1   Metals

According to the band theory the electrical properties of a material are the direct consequence of zone filling and zone spacing. Let us consider the model of free electrons of mass m_e in a three-dimensional box, which is capable of describing many of the properties of metals. The electrons are confined in a cubic box of dimensions L in each of x, y, and z coordinates; the potential inside the box is zero and is infinite outside. The wave equation is

$$\frac{1}{X}\frac{d^2X}{dx^2} + \frac{1}{Y}\frac{d^2Y}{dy^2} + \frac{1}{Z}\frac{d^2Z}{dz^2} + \frac{2m_e}{\hbar^2} E = 0. \tag{1.5}$$

Its solution has eigenvalues $E_{n_x n_y n_z}$ corresponding to allowed energies for the electrons

$$E_{n_x n_y n_z} = \frac{\hbar^2 \pi^2}{2m_e L^2} (n_x^2 + n_y^2 + n_z^2), \tag{1.6}$$

where n_x, n_y, and n_z are the quantum numbers. Each state specified by this set of quantum numbers can accommodate two electrons.

We now define the density of states $N(E)$ such that $N(E)dE$ is the number of orbital states with energies lying between E and $E + dE$. If we also define $\mathcal{N}(E)$ to be the number of orbital states with energy less than E, then

$$\mathcal{N}(E) = \int_0^E N(E)\, dE \tag{1.7}$$

and

$$N(E) = \frac{d\mathcal{N}(E)}{dE}. \tag{1.8}$$

The value of $\mathcal{N}(E)$ can be obtained by simple geometrical consideration by rewriting Eq. (1.6) as

$$n_x^2 + n_y^2 + n_z^2 = \frac{2m_e L^2}{\hbar^2 \pi^2} E, \tag{1.9}$$

which in n space is the equation of a sphere with radius $(2m_eL^2/\hbar^2\pi^2)^{1/2}$. Because the quantum numbers are positive integers, the number of states $\mathcal{N}(E)$ is given by the volume of the positive octant of this sphere, or

$$\mathcal{N}(E) = \frac{1}{8}\frac{4\pi}{3}\left(\frac{2m_eL^2}{\hbar^2\pi^2}E\right)^{3/2}. \tag{1.10}$$

Taking the derivative with respect to energy, we obtain

$$N(E) = \frac{1}{2\pi^2}\left(\frac{2m_e}{\hbar^2}\right)^{3/2}E^{1/2} \tag{1.11}$$

for the density of allowed energy states per unit volume, including spin.

To obtain the probability that a state is occupied, we introduce the term of Fermi energy, E_F. If we fill the allowed energy states of the system at 0 K with electrons, E_F is the energy for the last added electron. In other words, E_F marks the occupied states from the higher-lying unoccupied states. The probability that a given state E is occupied at temperature T is given by the Fermi distribution function $f(E)$, which is

$$f(E) = \frac{1}{\exp[(E - E_F)/k_BT] + 1}. \tag{1.12}$$

The density of occupied state $N_0(E)$ is simply the product of $N(E)$ and $f(E)$, or

$$N_0(E) = \frac{1}{2\pi^2}\left(\frac{2m_e}{\hbar^2}\right)^{3/2}\frac{E^{1/2}}{\exp[(E - E_F)/k_BT] + 1}. \tag{1.13}$$

The characteristic of the Fermi distribution function is that $f(E) = 1$ for $E \ll E_F$; it has a value of $\frac{1}{2}$ for $E = E_F$ and is zero for $E \gg E_F$. Therefore,

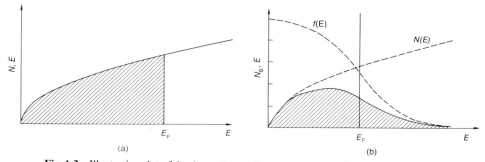

Fig. 1.3 Illustrative plot of the dependence of total density of states per unit volume $N(E)$, Fermi distribution function $f(E)$, and density of occupied states per unit volume $N_0(E)$ with energy: (a) at T = 0 K (b) at $T > 0$ K.

$f(E)$ is weakly temperature dependent. Figure 1.3 illustrates the variation of $f(E)$, $N(E)$, and $N_0(E)$. At absolute zero all allowed energy states are filled up to the Fermi energy (Fig. 1.3a), and states above E_F become occupied as temperature increases (Fig. 1.3b).

The basic difference between metals and other solids, i.e., semiconductors and insulators, is that the highest allowed energy band occupied by electrons is only partially occupied in metals (Fig. 1.4a), whereas they are filled in semiconductors and insulators (Figs. 1.4c and d).

In an electric field the electrons in metals are able to drift in the direction of the field with a mobility μ, which is the velocity per unit electric field. The electrical conductivity σ is given by

$$\sigma = \mathbf{n}q\mu, \tag{1.14}$$

where $\mathbf{n}$ is the density of free carriers contributing to the conductivity and q is the charge. In a metal $\mathbf{n} = N_0(E)$ and corresponds to a large number of the order 10^{22} cm^{-3}. The carrier mobility is also high [$\sim 10^2 - 10^3$ cm^2 (V sec)$^{-1}$]. They combine to give large values of conductivity of $10^5 - 10^6$ (Ω cm)$^{-1}$ as shown in Fig. 1.5.

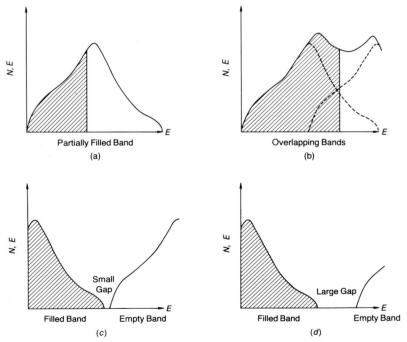

Fig. 1.4   Description of materials according to band spacing and band filling: (a) metals, (b) semimetals, (c) semiconductors, (d) insulators.

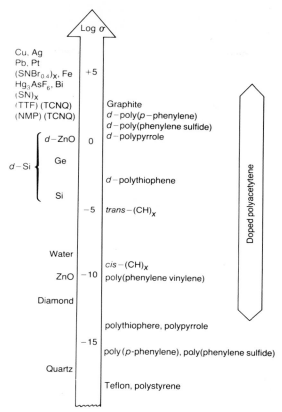

Fig. 1.5 Comparison of electrical conductivities of metals, semiconductors, and insulators; *d* indicates doped material.

The progress of the electrons in a metal is continually disrupted, on the average, after traveling a distance corresponding to the mean free path or a time corresponding to the scattering relaxation time, τ_s. The mobility is related to τ_s by

$$\mu = (q/m_e^*)\tau_s, \tag{1.15}$$

where m_e^* is the effective electron mass. The dominant scattering mechanism is by interaction with lattice atoms as they vibrate with thermal energy. In the range where acoustic lattice scattering is elastic,

$$\tau_s \propto \mu \propto T^{-1}, \tag{1.16}$$

which is responsible for the increase of σ for metals with decrease of temperature.

The purpose of the foregoing discussion is to show how a conventional metal behaves. Its conductivity increases with decreasing temperature. It will be shown that even when polyacetylene is doped to the metallic regime, the conductivity decreases with decrease of temperature. Therefore, the material cannot be considered as a true metal in the usual sense.

1.3.2 Intrinsic Semiconductors

Semiconductors have filled valence bands and empty conduction bands (Fig. 1.4c). The gap energy E_G is small as compared to the larger gap in insulators (Fig. 1.4d). Electrons in the valence band can be thermally excited into the conduction band. A corresponding hole is created in the valence band. Both the free electrons and free holes are carriers. In the intrinsic semiconductor, the density of the electron (hole), $\mathbf{n_e}$ ($\mathbf{n_h}$), are equal. The intrinsic carrier density, $\mathbf{n_i}$, is given by

$$\mathbf{n_i} = \mathbf{n_e}\mathbf{n_h} = N_c N_v \exp(-E_G/k_B T), \tag{1.17}$$

where $N_c(N_v)$ is the effective density of states in the conduction (valence) band defined by

$$N_c = 2(2\pi m_e^* k_B T/h^2)^{3/2},$$
$$N_v = 2(2\pi m_h^* k_B T/h^2)^{3/2}, \tag{1.18}$$

where m_e^* (m_h^*) is the effective mass for the electron (hole). The requirement that $\mathbf{n_e} \equiv \mathbf{n_h}$ gives

$$\mathbf{n_i} = N_c(m_h^*/m_e^*)^{3/4} \exp(-E_G/2k_B T), \tag{1.19}$$

and the Fermi energy is

$$E_F = E_G/2 + (k_B T/2) \ln(N_c/N_v) = E_G/2 + (3k_B T/4) \ln(m_e^*/m_h^*). \tag{1.20}$$

The conductivity of an intrinsic semiconductor is thus strongly affected by temperature according to

$$\sigma_i = N_c q(m_h^*/m_e^*)^{3/4}(\mu_e + \mu_h) \exp(-E_G/2k_B T), \tag{1.21}$$

where μ_e (μ_h) is the mobility of the electron (hole). Equation (1.21) shows exponential temperature dependence for the conductivity. Even though the carrier mobility in such material is affected by the same scattering process as in a metal and contributes toward temperature dependence of conductivity according to Eqs. (1.15) and (1.16), the $\mathbf{n_i}$ dependence on T of Eq. (1.19) dominates the overall dependence of conductivity on T [Eq. (1.21)] over a wide temperature range. Intrinsic semiconductors have low room temperature conductivities (Fig. 1.5) because there are few carriers.

To obtain the thermopower we define conductivity as

$$\sigma = -\int \sigma_E \frac{\partial f}{\partial E} \, dE. \tag{1.22}$$

The current dj due to electrons with energy between E and $E + dE$ in a field F is

$$dj = -\sigma_E \frac{\partial f}{\partial E} F \, dE. \tag{1.23}$$

The free energy carried by this current is

$$\frac{-(E - E_F) \, dj}{e} = \frac{1}{e} \frac{\partial f}{\partial E} \sigma_E (E - E_F) F \, dE. \tag{1.24}$$

The total electronic heat transport is equal to Π_j, with Π being the Peltier coefficient. Thermopower $S = \Pi/T$ can be obtained from

$$S\sigma = \frac{k_B}{e} \int \sigma_E \frac{E - E_F}{k_B T} \frac{\partial f}{\partial E} \, dE. \tag{1.25}$$

For an intrinsic semiconductor

$$S = \frac{k}{|e|} \left(\frac{E_c - E_F}{k_B T} + \frac{5}{2} + r \right), \tag{1.26}$$

where E_c is the energy separating the localized and nonlocalized states, and

$$r = d(\ln \tau_s)/d(\ln E). \tag{1.27}$$

Pristine polyacetylene has large band gaps; E_F is about 0.7 eV for *trans*-polyacetylene and about 0.9 eV for *cis*-polyacetylene. Therefore, conductivity by thermal excitation of carriers is unimportant. Also the thermopower of undoped polyacetylene is temperature independent in contrast to Eq. (1.26) for intrinsic semiconductors. Whereas the band edges are sharp in crystalline inorganic semiconductors, they are not in polyacetylenes.

1.3.3 Extrinsic Semiconductors

The number of carriers in a semiconductor can be increased by the introduction of impurities (dopants). The resulting material is an extrinsic semiconductor whose carrier density is determined more by the concentration and property of the dopant and less affected by temperature. For example, silicon is an intrinsic semiconductor whose conductivity can be increased upon doping. Substitution of small amounts of gallium (a group-III element) for silicon (a group-IV element) in a silicon framework provides acceptor sites that allow the formation of charge deficiencies or positive holes in the valence band of silicon. Because holes with their positive charges are responsible for conduction, these materials are referred to as *p*-type semiconductors. On the other hand, partial replacement of silicon by arsenic (a group-V element) provides donor sites and thus a small excess of electrons in the empty conduction band of silicon. Such a material is referred to as an *n*-type semiconductor.

Three important points concerning extrinsic semiconductors should be noted. First, the donor and acceptor states are usually very close to the conduction and valence bands, respectively, allowing electron promotion to occur nearly completely at room temperature. Second, these materials have only one type of majority carrier, i.e., electrons or holes, in contrast to intrinsic semiconductors, which possess equal numbers of both. Finally, the impurity states are not localized.

The energy level diagram of extrinsic semiconductors is given in Fig. 1.6. The ionization of dopants can be represented by

$$D^\circ \rightarrow D^+ + e^- \tag{1.28}$$

$$A^\circ \rightarrow A^- + h^+ \tag{1.29}$$

with ionization energy E_D (E_A) for the donor (acceptor). The Fermi energy for the case of *n*-type semiconductors is given by

$$E_F = (E_D/2) + (k_B T/2) \ln(gN_C/N_D), \tag{1.30}$$

where g is the degeneracy factor of the specific level.

The negative carrier density in a donor-only case is

$$\mathbf{n_e} = N_{D^+} = N_D - N_{D^\circ}. \tag{1.31}$$

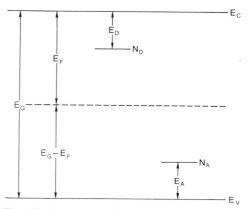

Fig. 1.6   Energy level diagram for semiconductors.

Similarly for acceptor-only semiconductors, the positive carrier density is

$$\mathbf{n_h} = N_{A-} = N_A - N_{Ao}. \tag{1.32}$$

The donor or acceptor is referred to as "impurity."

Charge transport in an extrinsic semiconductor is by means of impurity conduction and behaves like

$$\sigma = \sigma_3 \exp(-\epsilon_3/k_B T), \tag{1.33}$$

where σ_3 depends on the concentration of impurity. The conductivity consists of four parts. The first is the probability (p) per unit time that an electron jumps from one impurity site to another situated at a distance R away and with energy higher than E_D for the donor

$$p = p_o \exp(-2\alpha R - E_D/k_B T), \tag{1.34}$$

where α^{-1} is the radius of the impurity center, R is typically of the order 300 Å, and p_o may be of the order of 10^{12} sec^{-1}. The hopping energy ϵ_3 is

$$\epsilon_3 = \frac{e^2}{\kappa}\left(\frac{1}{R_D} - \frac{1.35}{R_A}\right), \tag{1.35}$$

where κ is the background dielectric constant, R_D (R_A) is the average distances between donors (acceptors). The conductivity is then given by

$$\sigma = e^2 N(E_F) p R_D^2. \tag{1.36}$$

The ac conductivity (σ_{ac}) involves interaction with phonons and is related to σ_{dc} by

$$\sigma_{ac} = \sigma_{dc} + A\omega^s, \tag{1.37}$$

where A is a parameter little dependent on T and the exponent s is close to 0.8. The semiconductor to metal transition occurs when $\epsilon_3 \to 0$ as the dopant concentration becomes very high. In these systems the dc conductivity is strongly dependent on impurity concentration (C_{im}) because $R = (4\pi C_{im}/3)^{-1/3}$.

Doped polyacetylenes do not have the above dependence on temperature and dopant concentration for conductivity and should not be treated simply as extrinsic inorganic semiconductors. The conductivities of some examples of intrinsic and extrinsic semiconductors are given in Fig. 1.5.

1.3.4 Insulators

In insulators the electrons are strongly localized between the atoms forming chemical bonds. There are no significant orbital overlaps between adjacent atoms that are not bonded to each other. The valence band is completely filled. Furthermore, a large energy separates it from the conduction band, which is not accessible by thermal excitation (Fig. 1.4d). Quartz is an insulator because there are no strong interactions between Si and O atoms not bonded to one another. All organic compounds and polymers having only σ bonds are good insulators (cf. Fig. 1.5). Even if charged carriers are created by excitation, for instance by electromagnetic radiation, they cannot move freely because there are no energy bands. The molecules retain their identity; as a result, the carrier must "hop" from molecule to molecule by thermally activated processes. The carrier mobility is extremely low inasmuch as the carriers reside for long periods of time on the individual molecules. These restrictions may be somewhat eased for aromatic organic substances. For example, there is sufficient intermolecular π-orbital overlap in an anthracene crystal for the formation of energy bands. Carriers produced by photoexcitation have quite high mobility [$\mu \sim 1 \text{ cm}^2 \text{ (V sec)}^{-1}$], and single crystals of anthracene exhibit photoconductivity.

It is worthwhile to note that the band gap of an insulator can be substantially reduced or even eliminated by compressing the material, which may reduce internuclear distances and increase orbital overlap. For example, diamond, which is normally an insulator, displays metallic properties at a pressure of 600,000 atm.

1.4 SEMIMETALS: DIMENSIONALITY

In a limited number of materials in which all valence bands are filled through formation of covalent bands, these bands overlap with the conduction bands to generate two incompletely filled bands, as shown schematically in Fig. 1.4b. This allows metallike conductivity and the material is sometimes referred to as a semimetal.

The most common example of a semimetal is graphite (Vogel, 1979). In graphite, the sp^2 hybridized orbitals constituting σ bonds form a planar hexagonal network. The remaining p_z electrons, one per carbon atom, form the π bonds and also help to hold adjacent planes loosely together. This weak interplanar interaction permits adjacent planes to slide with respect to each other, producing the lubricating properties of graphite. The conductivity of graphite is $\sim 10^4 \text{ (}\Omega \text{ cm)}^{-1}$ at room temperature and increases with decreasing temperature, as does a metal. The electrical conductivity is anisotropic, being greater in the aromatic plane than perpendicular to it, as a direct consequence of good in-plane π overlap. The anisotropy $\sigma_{\parallel}/\sigma_{\perp}$, is $\sim 10^4$ at room temperature.

This example of anisotropy serves to introduce the concept of dimensionality in discussing electrically conducting materials that are not classical metals. Common metals have uniform conductivity in all directions and are referred to as three-dimensional metals. On the other hand, the preceding description of graphite is that for a two-dimensional metal. As dimensionality decreases, electrical conductivity becomes progressively more sensitive to defects and disorder within the material, especially if they occur along the conduction pathway.

The conductivity of graphite can be further increased upon doping with electron donors (alkali metals) or acceptors (nonmetallic molecules such as Br_2, HNO_3, or metallic molecules such as $FeCl_3$, AsF_5), with the graphite lattice acting as a giant polyanion or polycation, respectively. Extremely high electrical conductivity is observed for selected graphite intercalation compounds. The conductivity for the graphite–SbF_5 complex (Vogel, 1977) is $\sim 10^6$ $(\Omega \text{ cm})^{-1}$ with $\sigma_{\parallel}/\sigma_{\perp} \sim 10^6$. Doping with some alkali metals was found to raise the superconducting critical temperature.

Another interesting semimetal is $[SN]_x$, which was first synthesized by F. P. Burt of University College London in 1910. Its structure is

$$\diagdown S \diagup^{\displaystyle N} \diagdown S \diagup^{\displaystyle N} \diagdown S \diagup^{\displaystyle N} \diagdown S \diagup^{\displaystyle N} \diagdown \qquad\qquad (1.38)$$

$[SN]_x$ has a room-temperature conductivity of ~ 2000 $(\Omega \text{ cm})^{-1}$ and becomes superconducting below 0.3 K (Gooding, 1976). This property can be understood from its crystal structure. The polymer chains are parallel to one another with good overlap between the π orbitals, creating a pathway for conduction throughout the length of the polymer chain. Conductivity perpendicular to the chains is also quite high ($\sigma_{\parallel}/\sigma_{\perp} \sim 15$ at room temperature), indicating that there is significant overlapping of orbitals between chains. The temperature dependence of conductivity for $[SN]_x$ resembles that of metals.

1.5 CHARGE-TRANSFER SALTS

A class of organic salts is known to possess unusually high conductivity. They have the general fundamental structure of parallel stacks of molecules in which orbital overlap is strong along the direction of the stacks. Because the greatest degrees of intermolecular overlap between orbitals of organic molecules are those of π-electron clouds, the constituents of the salts are flat, aromatic molecules of low ionization potentials as the donor and of high electron affinity as the acceptor, such as tetrathiafulvalene (TTF) and tetracyanoquinodimethane (TCNQ), respectively (Garito and Heeger, 1974).

$$\text{(1.39)}$$

TTF   TCNQ

Each constituent separately is devoid of carriers. But when they combine to form a salt by means of charge transfer, partially filled bands and high conductivity result. The complex has a herringbone structure with alternating stacks of the two planar donor (TTF) and acceptor (TCNQ) molecules. Within each stack the molecules are parallel to one another, but they are inclined to the axis of the stack. Orbitals that extend above and below the plane of each molecule overlap, giving rise to an electronic conduction band along the stack. The charge transfer is, on the average, 0.59 of an electron per TTF $\cdot$ TCNQ. The room temperature conductivity is ~ 650 $(\Omega \text{ cm})^{-1}$ parallel to the stack direction, which increases with decrease of temperature, reaching a sharp maximum at 50–60 K, approaching that of copper. The anisotropy is $\sigma_{\parallel}/\sigma_{\perp} \sim 500$. The negative sign of the thermopower coefficient indicates that the acceptor stacks dominate the conduction mechanism. The sharp decrease of conductivity below 58 K corresponds to a metal–insulator transition. This unusual temperature dependence cannot be easily explained by a simple band model. Charge density waves have been implicated as a possible rationale.

In addition, aromatic molecules such as perylene form donor–acceptor complexes with iodine (Perlstein, 1977). This substance attains a conductivity of $\sim 10^{-2} (\Omega \text{ cm})^{-1}$ in the stack direction, with $\sigma_{\parallel}/\sigma_{\perp} \sim 10^3$. Radical ion salts can have very high conductivities. For example, Bechgaard $et~al.$ (1980) found salts of di(tetramethyltetraselenafulvalene) radical anion with PF_6^-, AsF_6^-, etc., to have $\sigma_{RT} \sim 800$ $(\Omega \text{ cm})^{-1}$ and to become superconducting below 0.9 K at a pressure of 12 kbar (Jerome $et~al.$, 1980). This represents the first example of superconductivity in a synthetic organic material.

1.6   ORGANIC CONDUCTIVE POLYMERS

Following the fundamental observations of Shirakawa $et~al.$ (1977) and Chiang $et~al.$ (1977) that doping can increase the conductivity of polyacetylene by 13 orders of magnitude, a number of polymers have subsequently been shown to possess this characteristic also. Some examples are given in Table 1.1. There is no doubt that new ones will be added to the current list in the near future. All the polymers in the table have very low conductivities in the undoped states. There are one or more serious difficulties associated with every one of these polymers: (a) the molecular weight and morphology are unknown; (b) the structure of the polymer is uncertain; (c) the molecular weight and morphology of the polymer may be significantly altered after doping; (d) doping can be extremely nonuniform; and (e) the doped conductive polymer is very unstable when exposed to air and moisture. We shall not deal fully with these problems for all the polymers in Table 1.1. Instead, the nature of the problem will be illustrated by two examples.

Poly(phenylene vinylene) (Wnek *et al.*, 1979b; Gourley *et al.*, 1983) and poly(ferrocenylene vinylene phenylene vinylene) (Gooding *et al.*, 1983) were synthesized by Wittig reactions and are, in fact, oligomers. Their molecular weights can be estimated from end-group analysis. Only AsF_5 was found to be an effective dopant for poly(phenylene vinylene). Doping is complicated by probable chain extension because AsF_5 can catalyze Friedel–Crafts reactions. Exposure of a pressed pellet of poly(phenylene vinylene) to AsF_5 vapor caused an immediate change of its color from yellow to black. However, the interior of the pellet retained its original color. Apparently, the chemical transformation of the surface of the specimen creates an impermeable barrier to the dopant. The doped polymer is very sensitive to air and moisture.

On the other hand, poly(ferrocenylene vinylene phenylene vinylene) showed only a slight increase in conductivity upon reaction with AsF_5. Iodine was found to be a dopant, but the interaction is weak. The dopant can be readily removed by evacuation. However, the iodine-doped polymer is stable toward air and moisture.

1.7 POLYACETYLENE

Acetylene was first polymerized by Natta and co-workers (1958) using a $Ti(OBu)_4$–$AlEt_3$ catalyst. The polymer obtained was a gray infusible powdery material, which is not soluble in any common solvents. It was not amenable to most characterization methods and did not possess the interesting electrical properties that were anticipated. Several other catalyst systems were shown to polymerize acetylene, yielding similar products.

Using the same catalyst employed by Natta *et al.*, but different experimental conditions, Shirakawa and co-workers (Shirakawa and Ikeda, 1971; Shirakawa *et al.*, 1973; Ito *et al.*, 1974) succeeded in preparing free-standing polyacetylene films of metallic luster. Very thin films of polymer obtained at low temperature are red in color and have the cis structure. High-tempera-

TABLE 1.1

Some Conducting Polymers

Structural unit	Dopant	σ_{RT} $(\Omega\ cm)^{-1}$	References
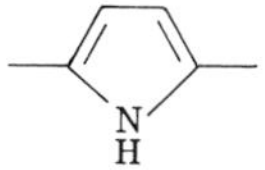	BF_4^-	100	Kanazawa et al. (1979)
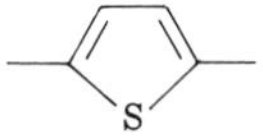	I_2	3.4×10^{-4}	Yamamoto et al. (1980)
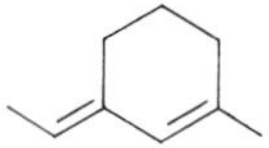	I_2	0.1	Gibson et al. (1980)
	AsF_5 K	145 7	Shacklette et al. (1979)
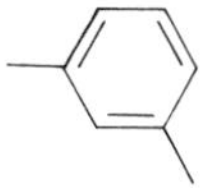	AsF_5	10^{-3}	Baughman et al. (1980)

20

	Dopant	Conductivity	Reference
—CH=CH— (p-phenylene vinylene structure)	AsF_5	3	Wnek *et al.* (1979b, 1981) Gourley *et al.* (1983)
—S— (p-phenylene sulfide structure)	AsF_5	1	Rabolt *et al.* (1980)
—O— (p-phenylene oxide structure)	AsF_5	10^{-3}	Chance *et al.* (1980)
Ferrocene —CH=CH— structure	I_2	10^{-4}	Gooding *et al.* (1983)

21

ture polymerization leads to blue-colored trans polymer films. Films of various thickness can be prepared by adjusting the polymerization conditions.

Polyacetylene film has a density of 0.4 g cm^{-3}. Electron microscopy showed it to comprise fibrils of $\sim$ 20 nm in diameter and X-ray diffraction showed it to be highly crystalline. The pristine polyacetylenes are direct-gap, wide bandwidth semiconductors. The conductivities are about 10^{-10} and 10^{-6} (Ω cm)$^{-1}$ for the cis and trans polymers, respectively. From the large positive value for the thermopower coefficient, the carriers are deduced to be positively charged.

Polyacetylene offers an excellent model to illustrate the formation of energy bands. Figure 1.7 shows schematically the increase in the number of π MOs and the decrease in their energy difference as the length of the [CH]$_x$ chain increases. For very high molecular weight, i.e., x is a large number, and if the backbone has uniform bond order, the polymer should be an intrinsic metal inasmuch as every orbital is occupied by one electron and the highest energy band will be only half filled. In the case of *trans*-polyacetylene the doubly degenerate ground state is unstable and undergoes Peierls distortion, which is equivalent to the Jahn–Teller effect familiar to chemists. The result is an alternation of bond lengths creating a band gap of $\sim$ 1.4 eV. In the case of *cis*-polyacetylene, the ground-state structure is nondegenerate, and the valence band is naturally filled.

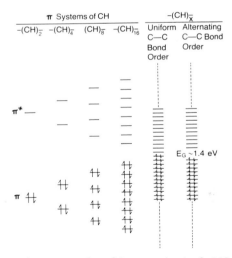

Fig. 1.7   Diagrammatic representation of the energy levels of π MOs with increasing size of the molecule for [CH]$_x$.

There are very basic differences between the cis and trans forms of polyacetylene. It is through careful investigations of these differences that novel physics emerges. The trans polymer is endowed with highly mobile spins, which will be referred to as neutral solitons; it is photoconducting and does not show band edge luminescence. On the other hand, *cis*-polyacetylene shows nearly the opposite properties; the unpaired spins are delocalized but not diffusive. It is nonphotoconducting and emits band-edge luminescence characteristic of electron–hole recombination.

The large increase in electrical properties upon doping suggests that polyacetylene cannot be treated as a simple system. It is useful to consider that there exist several regimes distinguishing the responses in conductivity increases with increasing doping. It stands to reason that the mechanism of conduction is different for the different regimes. To understand the mechanisms of conduction will be one of the main goals of experimental and theoretical investigations for some time to come.

Polyacetylene can be synthesized, leading to molecular weights ranging from about 500 to 220,000. Studies on molecular weight property relationships would be most instructive. Under different polymerization conditions, using $Ti(OBu)_4 - AlEt_3$ catalyst or initiating with other types of catalysts, the polymer obtained has diverse morphologies and structures that seem to influence strongly the electronic properties of the materials.

Chapter 2

Polymerization

2.1 INTRODUCTION

Acetylene polymerizations has been initiated by radiation, by anionic, cationic, and radical initiators, and by a variety of transition and rare-earth metal catalysts. There are many side reactions, depending on the method of initiation. One prominent side reaction is cyclization to benzene; other processes lead to low molecular weight oligomers, branched polymers, and cross-linked polymers. Though it is easy to quantify the cyclization yield, the extent of other reactions is difficult to measure because the products are insoluble.

The preferred catalyst system is $Ti(OBu)_4 - AlEt_3$ because under certain conditions it produces almost exclusively linear, high molecular weight, and crystalline polymers in the form of mechanically strong free-standing films, which upon doping become highly conducting. The other catalysts produce polyacetylenes that are deficient in one or more of these characteristics, as will be discussed in Section 2.8. Even with this particular Ziegler–Natta catalyst, variations in the catalyst concentration, Al/Ti ratio, aging of catalyst, monomer pressure, and temperature of polymerization give polyacetylenes differing in some properties. Furthermore, polyacetylene readily undergoes physico–chemical changes during storage and handling. Oftentimes discrepancies in reported properties and structures of polyacetylene can be traced to sample differences. For this reason a detailed description of a procedure for the preparation of high-quality polyacetylene film and its purification, storage, and handling is given in Section 2.2. Methods for the preparation of polyacetylene of various densities and other useful forms are also described. Purity criteria for polyacetylene are recommended in Section 2.2.9.

Some investigators have used the Luttinger catalyst to polymerize acetylene and claimed it to be superior to the Ziegler–Natta catalyst. Recently, Haberkorn *et al.* (1982) reported quite the opposite (Section 2.8). The procedure for the Luttinger catalysis is described in Section 2.3. Methods for the homo- and copolymerization of methylacetylene are given in Section 2.4.

Ziegler–Natta catalysts are extensively used in the polymerization of ethylene and α-olefins. Studies made in the past 30 years had shown the catalyst system to be of extreme complexity (Chien, 1959, 1963a,b, 1975; Chien and Hsieh, 1976; Chien *et al.*, 1982d, 1983a,b; Chien and Wu, 1982a,b). This is also the case for acetylene polymerization. By way of illustration, the same concentration of $Ti(OBu)_4-4AlEt_3$ was used to polymerize acetylene at 197 and 293 K (Table 2.1) under quiescent conditions. The polymer obtained at the lower temperature has a cis content of 74% and a $\sigma_{RT} = 105-250$ $(\Omega\ cm)^{-1}$ upon doping. The room-temperature polymerized product is only 28% cis and has a doped conductivity of $50-80$ $(\Omega\ cm)^{-1}$. The two polymers also differ significantly in oxygen content. Polymerizations carried out by bubbling acetylene into the catalyst solution and under quiescent conditions, methods B and A in Table 2.1, respectively, gave products that differ in isomeric composition, crystallinity, oxygen content, and σ_{RT} after doping.

Because polyacetylene is insoluble in any solvent, conventional methods for molecular weight determination are inapplicable. The determination of $\overline{M}_n$ by radioquenching is given in Section 2.6; the kinetics of acetylene polymerization are discussed in Section 2.7. This chapter concludes with a brief survey of other catalysts for acetylene polymerization and different polymers obtained therefrom.

2.2 POLYMERIZATION OF ACETYLENE CATALYZED BY Ti(OBu)₄−AlEt₃

Acetylene was first polymerized by Natta *et al.* (1958) to a linear conjugated macromolecule using the $Ti(OBu)_4-AlEt_3$ catalyst. At a low concentration of a few millimolar and in a stirred reactor, i.e., conditions standard for olefin polymerization, they obtained an insoluble, infusible,

TABLE 2.1

Comparison of Properties of Polyacetylenes Prepared by Different Methods[a]

	Polymerization conditions					Properties			
Catalyst	$[M]^b$ (mM)	Temperature (K)	Solvent	Method[c]	cis (%)	Oxygen (%)	X-ray peak width[a]	Morphology	$\sigma_{RT}{}^e$ $(\Omega\ cm)^{-1}$
Ti(OBu)$_4$–4AlEt$_3$	12.5	197	Toluene	A	74	1.5	1.2	Fibrillar	105–250
Ti(OBu)$_4$–4AlEt$_3$	12.5	197	Toluene	B	98	1.8	1.0	Fibrillar	500–750
Ti(OBu)$_4$–4AlEt$_3$	12.5	197	Toluene	A	0[f]	1.2	1.8	Fibrillar	50–80
Ti(OBu)$_4$–4AlEt$_3$	12.5	293	Toluene	A	28	0.5	1.1	Fibrillar	50–80
Ti(OPr)$_4$–4AlEt$_3$	12.5	197	Heptane	A	29	4.2	1.8	Fibrillar	50–80
WCl$_6$/0.5H$_2$O	4.0	280	Toluene	B	24	1.3	5.0	Globular with short fibrils	50–80
MoCl$_5$/0.5H$_2$0	6.0	197	Toluene	B	24	8.0	2.0[g]	Globular	70–100
TiCl$_4$/1.2AlEt$_3$	7.3	197	Toluene	B	3	1.7	5.0[g]	Globular with short fibrils	6 × 10^{-3}
TiCl$_4$/1.2AlEt$_3$	7.3	293	Toluene	B	7	1.7	3.5[g]	Globular	5 × 10^{-4}
TiCl$_4$/1.2AlEt$_3$	7.3	293	Heptane	B	0	5.2	3.5[g]	Globular	5 × 10^{-4}

[a]After Deits *et al.* (1981).

[b]Concentration of the transition metal.

[c]Methods of polymerization: (A) a mixture of catalyst and solvent were exposed to C$_2$H$_2$ (initial pressure ~ 500 torr) for 72 hr according to the method of Wnek *et al.* (1981). (B) C$_2$H$_2$ was bubbled for 2.5 hr through a mixture of catalyst and solvent.

[d]Relative half-peak width at $2\theta = 23$ to 25°.

[e]Measured at maximum iodine uptake ($y = 0.15$–0.25).

[f]Sample annealed at 471 K for 1.5 hr.

[g]A second WAXS broad peak at $2\theta = 16°$.

gray powder. The polymer was thought to be partially cross-linked, cyclized, and oxidized. These postulates have now been shown to be incorrect. We have now shown (Chapter 4) that the morphology and other properties of polyacetylene obtained by Natta's procedure are not significantly different from those of the free-standing films produced by improved techniques to be detailed below.

Using the same $Ti(OBu)_4 - AlEt_3$ catalyst but changing concentration, solvent, and reaction temperature, Hatano *et al.* (1961) were able to prepare polyacetylene with varying degrees of crystallinity. They reported a Bragg reflection of 23–25° for the polymer, which corresponds to a *d*-spacing of 3.5–3.8 Å (Hatano, 1962). Less than 1% of the reaction leads to benzene (Ikeda, 1967); therefore this catalyst produces higher yields of linear polyacetylene than other systems investigated.

The breakthrough, at least in so far as current research on the electronic properties of polyacetylene are concerned, occurred when Ito *et al.* (1974) succeeded in preparing high-quality, free-standing film. These workers employed a very high concentration of $Ti(OBu)_4 - AlEt_3$ and allowed acetylene to polymerize on the wall of the reaction vessel. The product obtained at 195 K was ∼ 88% *cis*-polyacetylene; the polymer has increasingly higher trans content as the polymerization temperature is raised. At 423 K the polymer obtained has a predominantly trans structure. This and other procedures for the syntheses of pure polyacetylenes in various forms are detailed below.

2.2.1 Equipment

Ziegler–Natta catalysts are poisoned by a variety of substances that can react with the titanium σ-carbon bonds; examples are air, water, electron donors, and protic and polar compounds. Even though a substoichiometric amount of an electron donor is sometimes used to enhance the stereospecificity of polymerization of α-olefins and a trace of oxygen can be beneficial if the active species is a Ti^{4+} complex, these marginal effects have not been studied for acetylene polymerization. Therefore, all possible catalyst poisons must be removed. However, these precautions are not necessary for some of the Ni-, Co-, W-, and Mo-based catalysts.

Whenever rubber tubing is required because of its flexibility, those made of butyl rubber (used for the inner tubes of tires) are preferred. In particular, tygon tubings made of plasticized poly(vinyl chloride) should be avoided because of its high oxygen permeability and the volatile plasticizer it contains, both of which are deleterious to the catalyst and to the polyacetylene as well. For handling of reactive dopants (e.g., halogens, AsF_5, $HClO_4$, etc.) greaseless Teflon stopcocks should be employed, and Viton O-ring seals are preferred over standard-tapered joints.

A Schlenk tube is the most valuable piece of apparatus employed in the synthesis, transfer, and storage of air-sensitive materials. Although a variety of shapes and sizes were used, the essential common features are (1) a side arm with a high-vacuum stopcock for introduction of inert gas, (2) an O-ring joint or Teflon-sleeved male ground joint to which it can be attached, and (3) a complementary joint with a high-vacuum stopcock for evacuation.

A good drybox ($<$ 1 ppm O_2) is essential for this type of work. Under certain situations we used crown-topped bottles, covered with a self-sealing butyl rubber liner, and a metal cap crimped over it. The metal cap has two holes for the insertion of a syringe needle in order to evacuate or flush the bottle or for the introduction of catalyst, monomer, and solvent. For additional protection against leakage, a rubber serum cap is wrapped over the crimped metal cap. A constant stream of purified argon flows in the space between the two caps providing a positive barrier against air contamination.

2.2.2   Inert Gas and Solvent Purification

Argon was the inert gas of choice. It was purified by passage through a 50×4.5-cm heated column (383–393 K) of "BTS"-supported copper catalyst (BASF Corp.) for removal of oxygen, followed by two 60×3 cm columns containing molecular sieves and KOH pellets, respectively, and small amounts of Drierite as an indicator for moisture removal.

Aliphatic hydrocarbon solvents, i.e., pentane or heptane, were freed of olefinic impurities by stirring with concentrated H_2SO_4 overnight. The brown H_2SO_4 layer was then removed and fresh acid added. The process was repeated until the H_2SO_4 layer remained colorless after overnight stirring. The solvent was then washed once with distilled H_2O, thrice with 10% NaOH, and then with more distilled H_2O until the aqueous layer was neutral to litmus paper. The solvent was dried over anhydrous $MgSO_4$ overnight, filtered, transferred into a still containing CaH_2 (Fig. 2.1) under a stream of argon, and brought to reflux. A slow stream of argon was passed through the still at least overnight to remove oxygen. After this time, the stopcock to the argon line was closed, although the solvent was always kept under continuous reflux. The still is totally grease free; stopcocks are fitted with Teflon plugs, and Teflon sleeves provide seals at the ground glass joints of the three-neck flask.

Aromatic solvents, such as toluene, were also purified by this procedure,

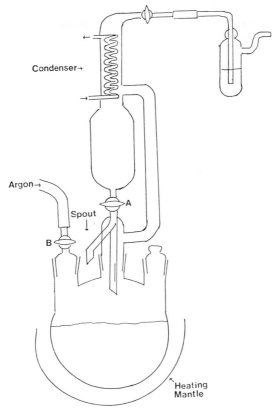

Fig. 2.1 Solvent still.

except that each treatment with H_2SO_4 should not exceed 12 hr. This precaution is necessary to avoid sulfonation.

2.2.3 Acetylene

To prevent explosion, commercial acetylene is always stabilized with acetone and phosphine. Acetone and phosphine are removed by a purification train (Fig. 2.2) consisting of two bubblers of concentrated H_2SO_4 and a U-tube containing P_2O_5 to remove moisture. The inlet of the train was connected to a tank of prepurified nitrogen and a cylinder of acetylene, and its outlet was connected to the vacuum line.

A 2-liter bulb equipped with a cold finger and Teflon stopcock serves as a

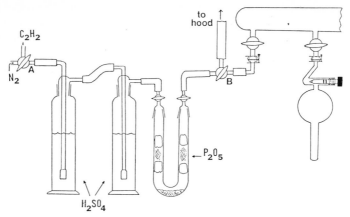

Fig. 2.2 Acetylene purification system.

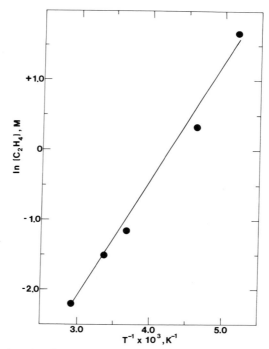

Fig. 2.3 Arrhenius plot of concentration of acetylene in toluene at 760 torr versus T^{-1}.

storage vessel for purified acetylene and is attached to the submanifold of the vacuum line. The pressure of acetylene in the storage bulb should not exceed 1 atm because an explosion may spontaneously occur at pressures near 2 atm. Immediately prior to polymerization, the acetylene in the bulb was condensed into the cold finger at 77 K and pumped to remove traces of N_2 or air, which may have entered the system during the interim.

The solubility of acetylene in toluene as a function of temperature was determined at one atmosphere of monomer pressure by a standard *P-T-V* method. Figure 2.3 gave the Arrhenius plot of concentration of acetylene in toluene as a function of T^{-1}. From the slope the enthalpy of mixing was found to be -0.82 kcal mole^{-1}.

2.2.4 Catalyst Preparation

Titanium tetrabutoxide was purified by reduced-pressure distillation under a slow stream of argon. A low-boiling, colorless liquid (presumably *n*-butanol) was collected and discarded. The titanium tetrabutoxide was distilled as a light yellow, viscous liquid, collected in the Schlenk tube, and stored under argon. The tube was wrapped with black electrical tape to protect its contents from light. Triethylaluminum (Ethyl Corp.) was used as received.

The same vessel was used for catalyst preparation and polymerization (Fig. 2.4). Its attachment to the vacuum line was by a 12−18 cm length of flexible stainless steel tubing having a ball and socket joint on either end to couple with the ball joint of the reactor to a similar one on the vacuum line. After being thus connected and evacuated to a high vacuum overnight, the reactor was cooled to 77 K, and 20 ml of toluene containing $\sim 0.5\%$ AlEt₃ was distilled into it. The AlEt₃ serves to scavenge any impurities. The apparatus was disconnected from the vacuum line after closing stopcock C, and the frozen toluene was allowed to warm to room temperature. Next, purified argon was introduced at stopcock A (after purging), and stopcocks B and C were opened. The top of the reactor was removed at O-ring F. Using a syringe, 1.7 ml of titanium tetrabutoxide was added to the toluene in the reactor. The bottom of the reactor was then cooled to 195 K (dry ice−acetone bath), and 2.7 ml AlEt₃ was added dropwise by a syringe over a period of about 1 min. During the addition, the color of the solution changed from light yellow to deep orange-red. The tip of the syringe needle was used occasionally to stir the solution as the AlEt₃ was added to assure homogeneous reduction of the titanium tetrabutoxide. The reactor was capped at O-ring F and, while stopcock C remained open, the argon flow was turned

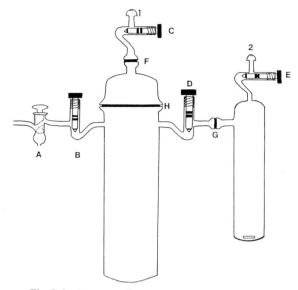

Fig. 2.4 Apparatus for the synthesis of (CH)$_x$ films.

down to a low rate, after which stopcocks C, B, and A were immediately closed.

The reactor was removed from the 195-K bath and the catalyst solution was allowed to warm to room temperature and aged for 30 min. The color of the solution changed from deep red-orange to dark brown during this process. After the aging period, the reactor was immersed up to the stopcock "arms" in a dry ice–acetone bath, and connected to the vacuum line. It is now ready for polymerization.

2.2.5 Preparation of Free-Standing *cis*-Polyacetylene Film

The reactor containing the aged catalyst solution was evacuated with gentle shaking for 30–90 min at 195 K. The catalyst solution became progressively more viscous. The reactor was briefly removed from the dry ice–acetone bath and shaken to wet as much of the reactor walls with the catalyst solution as possible. It was then immediately placed back into the 195-K bath, and stopcock C was opened, allowing acetylene to enter. All surfaces wet with the catalyst solution and which were below the 195-K bath turned bright red due to the formation of *cis*-polyacetylene, which developed a coppery luster within a few minutes as the thickness of the polymer film increased. A more silvery film (higher in trans isomer content) was formed on surfaces above the level of the bath. The film thickness was very dependent upon the interval of time between wetting the reactor walls with catalyst and the introduction of the acetylene. For thin ($\sim 30 - 50$ μm) films suitable for infrared spectroscopy, at least 30 sec should elapse before introducing the acetylene. Typical reaction times were 10 – 30 min, with uptake of acetylene in the range of 100 – 200 torr. After the desired time period, stopcock C was closed and the remaining acetylene in the submanifold was condensed back into the bulb at 77 K.

The apparatus (at 195 K) was disconnected from the manifold, and argon was introduced, after which the top was removed at O-ring F. The argon flow removed the remaining acetylene in the reactor. A 50-ml syringe with a long needle was used to punch several holes in the polyacetylene film, which had grown on the surface of the catalyst solution at the bottom of the reactor. The catalyst solution under this film was then withdrawn by the syringe and discarded. The top was replaced, and all stopcocks were closed.

The apparatus (with the reactor at 195 K) was attached to the vacuum line and evacuated (stopcocks C and D open) to remove all argon and residual acetylene. Next, 30 – 50 ml of pentane (kept in vacuo over CaH₂ in a solvent storage tube) was distilled into the reactor. A small frit filter equipped with ball and socket joints was placed between the pentane tube and manifold to prevent bumping of the CaH₂ into the manifold. The pentane was stirred with a magnetic bar to facilitate transfer. After the distillation, stopcock C was closed and the apparatus was disconnected from the manifold.

The reactor was then removed from the 195-K bath and the washing arm was placed in the bath for several seconds. With stopcock D open, the apparatus was turned upside down several times, allowing the brown pentane – catalyst solution to flow into the washing arm. Stopcock D was then closed and the reactor was placed back in the cold bath. The solution in the side arm was then stirred with a magnetic bar, and stopcock D was opened slowly, allowing fresh pentane to distill back into the reactor. The distillation rate was controlled by the stirring rate and the size of the opening at stopcock D. This washing procedure was repeated until the pentane in the reactor was colorless ($\sim 8 - 12$ cycles). If the glass surfaces were clean, much of the film separated from the reactor walls during washing. After every five or so cycles, the reactor was connected to the vacuum line and pumped briefly. It should be noted that the washing arm was equipped with a stopcock (E). In the event of a leak in this portion of the apparatus, which would most likely be at O-ring G, the washing arm could be disconnected and the O-ring replaced, followed by connection of a clean washing arm. The arm could then be pumped out via stopcock E, followed by distillation of fresh pentane from the storage tube. However, this process was seldom necessary.

2.2.6 Transfer and Storage of *cis*-Polyacetylene Film

The reactor was removed from the 195-K bath and allowed to warm to room temperature. The washing arm was disconnected and the catalyst solution was decomposed by pouring it into a large quantity of methanol. All moisture was wiped from the outside of the reactor, and it was then placed in a glove bag or drybox along with scissors, a long spatula and forceps, and a long Schlenk tube (with ₮29/42 ground glass joints), which had been previously evacuated.

In the glove bag or drybox, the large O-ring (H) of the reactor was disconnected and a long spatula was gently played along the reactor wall to free the polymer film. Forceps were used to pull the film (in one piece) from the reactor. The side of the film that grew on the surface of the reactor wall displayed a coppery luster, whereas the backside was dull maroon in color. The bottom of the film (shiny green) that grew on the surface of the catalyst reservoir was cut away with scissors, along with any silvery film that was formed above the level of the dry ice–acetone bath. The remaining cis-polymer film was cut into small pieces ($\sim 3 \times 1$ cm) and placed in the long Schlenk tube. At this time, samples were cut for microanalysis, doping, etc. It was during these operations that certain amounts of isomerization (~ 10–12%) took place. Therefore, the time that the polymer was manipulated at ambient temperature should be kept to a minimum. The specimens were transferred into a Schlenk tube, which was capped in the drybox and then stored in a Dewar at 195 K to preserve the cis-isomer content of the film. Subsequent removal of film for any reason was performed in the glove bag or drybox. The film should be used within two weeks of preparation. We found that *cis*-polyacetylene stored at 195 K for a long time loses stretchability, and there were changes in the elemental analysis and reductions in conductivity upon saturation doping.

2.2.7 Preparation of Gel and Low-Density Foam of Polyacetylene

The parent catalyst was preapred in a Schlenk tube by adding 1.7 ml of $Ti(OBu)_4$ and 2.7 ml of $AlEt_3$ to 20 ml of dry toluene, followed by aging at room temperature for ~ 30 min. This solution was diluted by transferring

1 ml of the catalyst to another Schlenk tube (the reaction vessel) containing ~ 40 ml of dry toluene. The remaining parent catalyst was stored in a refrigerator at ~ 278 K for future reactions. The Schlenk tube containing the diluted catalyst was cooled to 195 K, and the solution was degassed on a vacuum line with occasional shaking. The quiescent catalyst solution, still at 195 K, was then exposed to acetylene (2-liter bulb, ~ 650 – 700 torr initial pressure). A reddish film (gel) appeared immediately on the surface of the catalyst solution and grew to a depth of 1 – 2 cm after about 18 hr. Typically, 250 – 300 torr of acetylene was consumed during the reaction. The upper surface and sides of the gel in contact with the glass reactor possessed a coppery luster, although the bulk of the material was reddish-maroon.

Using Schlenk-tube techniques, the gel (kept at 195 K) was covered with ~ 30 ml of dry toluene (or pentane), and then a syringe needle was used to free a small part of the gel from the glass reactor wall. The toluene – catalyst solution in the bottom of the reactor was removed through this opening with a syringe and was discarded. Care was taken not to allow the toluene level in the reactor to drop below the surface of the gel inasmuch as partial drying can collapse the gel. This process was repeated until the toluene was colorless, and then the gel was completely freed from the reactor wall, using a syringe needle. The toluene-soaked gel was allowed to stand overnight at 195 K under argon, after which time the toluene became yellow in color, indicating that additional catalyst was extracted. The washing procedure was repeated until the toluene remained colorless after standing overnight.

The gel was removed in one piece, using forceps, and then cut into sections with a razor blade. The material was quite flexible. Pieces of the gel that were not to be used immediately were placed back into the toluene to preserve the solvent-swollen state. Films were easily prepared by squeezing a piece of the toluene-soaked gel between glass slides coated with Teflon tape. The surfaces of the films displayed a dull golden or coppery luster. Solvent removal upon squeezing was completely irreversible, i.e., the material could not be reswollen into gel by toluene. The thicknesses of the films were dependent upon the amount of manual pressure used to squeeze the gel.

Low-density, foamlike material was prepared from the original toluene-soaked gel by first removing as much toluene as possible with a syringe while still keeping the gel "wet", followed by additon of ~ 30 ml of dry benzene, using Schlenk-tube techniques. This process was repeated about eight times so as to replace the toluene in the gel with benzene. The gel was kept at room temperature during these washings. It was then washed twice with fresh benzene by stirring gently with a magnetic bar overnight at room temperature. The Schlenk tube was then placed in an ice bath to freeze the benzene (mp 278.5 K) completely and then connected to a vacuum line to remove the frozen benzene by sublimation.

The freeze-dried, foamlike polyacetylene possessed essentially the same dimensions as the original gel "plug." The surface displayed a dull, silvery-gray luster, whereas the interior possessed a maroon coloration. The material was extremely light and spongy and possessed typical densities in the range of 0.02–0.04 g ml^{-1}. Samples of intermediate densities can be obtained by pressing the toluene-soaked gels to smaller thickness, solvent-exchanging with benzene, and freeze-drying as described above.

2.2.8   Ultrathin In-Situ Film

Acetylene polymerization is highly exothermic. The nascent morphology of polyacetylene may be modified by local heating during the preparation of thick film (Section 4.3). Removal of catalyst residues by washing becomes increasingly inefficient as the sample thickness increases. Sample handling at room temperature and brief exposure to air may affect the properties of the polymer to be investigated. Furthermore, for direct-transmission electron microscopy (EM), ultrathin specimens are needed. An in situ polymerization technique was developed in our laboratories, which produces films <100 nm thick at the same time eliminating the other problems mentioned above.

EM grids (3-mm disks of 300-mesh gold screen) were soldered with indium onto a fine gold wire and hung above the catalyst solution in the reactor, which had been previously flame dried and evacuated overnight. The catalyst solution [7 mM Ti(OBu)$_4$, 28 mM AlEt$_3$] was introduced onto the grid as a thin liquid film by shaking. Acetylene gas (total pressure ranging from 10 to 760 torr of acetylene) was admitted for less than 1 min and pumped off immediately. The thickness of the polymer film may be controlled by varying the catalyst concentration, acetylene pressure, and the time and temperature of the reaction. The ultrathin specimen thus obtained was washed to remove catalyst residues by repeated pentane distillation. The sample was used directly for transmission EM and electron diffraction studies. For scanning EM and scanning transmission EM, the polymerized film was gold coated either by sputtering or evaporation. The thickness of the gold coating was carefully controlled and kept in the 3–7 nm range.

2.2.9   Purity of Polyacetylene

The quality of polyacetylene affects its properties both before and after doping. The following standards are recommended:

(a)   Elemental analysis should have C + H > 99% and Ti < 0.5%

(b)   Conductivity of the undoped state at room temperature should lie between 10^{-10} and 10^{-11} $(\Omega\ cm)^{-1}$ for *cis*-polyacetylene and have values between 10^{-5} and 10^{-6} $(\Omega\ cm)^{-1}$ for the trans polymer.

(c)   The cis polymer should have a cis content of $\sim 90\%$ as measured by the cis band at 740 cm^{-1} and the trans band at 1015 cm^{-1}. The cis content is given by

$$\text{Percent cis} = \frac{1.30 A_{\text{cis}}}{1.30 A_{\text{cis}} + A_{\text{trans}}} \times 100, \tag{2.1}$$

where A is the absorbance of the afore mentioned infrared bands.

(d)   There should be no IR absorptions in the region of aliphatic C–H stretching vibrations at 2957, 2924, and 2553 cm^{-1}, C–H deformation vibrations at about 1553, 1463, and 1377 cm^{-1}, or of the carbonyl groups at 1736 and 1718 cm^{-1}.

(e)   There should be no signal above noise in magic-angle spinning cross polarization ^{13}C FTNMR attributable to sp^3 type carbon atoms.

(f)   The wide-angle X-ray scattering should correspond to > 70% crystallinity.

(g)   The room temperature conductivity after doping should exceed 200 $(\Omega\ cm)^{-1}$.

2.2.10   Polymerization of Acetylene onto Substrates

Polyacetylene composite can be prepared by the use of another polymer as the matrix. Galvin and Wnek (1982) soaked low-density polyethylene films (0.3 mm thick) for 24 hr in dry toluene to remove additives. It was then immersed in a solution of the Ziegler–Natta catalyst in a Schlenk tube under Ar and heated to 343 K under a stream of Ar for ~ 1.5 hr to impregnate the substrate with the catalyst. Surface catalyst residues were then washed away. Acetylene was admitted and polymerized between 195 and 383 K. The product is low-density polyethylene containing 2–10% polyacetylene. The method is obviously applicable to many other polymeric substrates.

A convenient method for the preparation of Schottky diodes or p–n heterojunctions is to polymerize acetylene directly onto the surface of a single crystal semiconductor (Section 12.2) or a metal substrate. The thin polyacetylene film can be subsequently doped if it is to serve as the metallic contact.

Direct synthesis of oriented polyacetylene had been demonstrated.

Woerner *et al.* (1982) degassed biphenyl, melted it under vacuum, and introduced the Ti(OBu)$_4$–AlEt$_3$ catalyst at room temperature under a stream of dry argon. The toluene of the catalyst solution was removed by evacuation. Melting of the biphenyl affected dissolution of the catalyst in the substrate to obtain a thick, polycrystalline layer of the catalyst in solid biphenyl. Admission of the monomer at 195 K produced a polyacetylene film that is shiny on both sides, indicating a more compact material than those produced on the reactor wall (Section 2.2.5). Patterns from the biphenyl substrate appear replicated in the polymer film with regions varying in size up to 0.5 × 0.1 mm. The polyacetylene film is birefringent when viewed between crossed polarizers as the specimen is rotated. Therefore, the film is apparently highly oriented.

2.3 POLYMERIZATION OF ACETYLENE BY THE LUTTINGER CATALYST

The Luttinger catalyst was prepared by dissolving 10 mg of NaBH$_4$ in 55 ml of Ar-saturated absolute ethanol and diluted with 25 ml of ether. An aliquot of this solution was introduced into the reactor and brought to the polymerization temperature (195 or 243 K). Aliquots of 1 wt % solution Co(NO$_3$)$_2$ in Ar-saturated absolute ethanol were added. The catalyst mixture was degassed quickly and acetylene immediately added. The Co/NaBH$_4$ molar ratios of the catalyst were 0.7 at 243 K and 0.07 at 195 K.

During the polymerization the catalyst solution color changes from pink to magenta, blue, blue-black, and finally black. But polymerization can be interrupted at any of these stages by removal of the monomer. The rate of polymerization obviously depends on monomer pressure, temperature, and catalyst concentration.

In situ ultrathin polyacetylene film was obtained with the Luttinger catalyst as described above for the Ziegler–Natta catalyst, except that the specimen was washed repeatedly with absolute ethanol instead of pentane.

2.4 METHYLACETYLENE HOMO- AND COPOLYMERS

Methylacetylene is the simplest hydrocarbon derivative of acetylene, yet the two *sp*-hybridized carbon atoms are different by virtue of the methyl group on one of them. Therefore, poly(methylacetylene) should have a strongly alternating backbone, and comparison of its properties with those of polyacetylene would be of interest. Furthermore, poly(methylacetylene) is soluble in common organic solvents and its chemical reactions can be more readily studied than polyacetylene. Finally, copolymers of the two monomers of different composition can reveal the effect of interruption by strongly bond-alternating methylacetylene units on the properties of acetylene units.

2.4.1 Homopolymerization of Methylacetylene

The $Ti(OBu)_4 - AlEt_3$ catalyst solution was prepared at 298 K in a Schlenk tube containing a stirring bar. A mercury bubbler was connected to the top of the tube with butyl rubber tubing. After 30 min aging, the catalyst solution was cooled to 195 K for $\sim 30-60$ min of degassing, and then brought up to 273 K. Upon admission of methylacetylene, the orange-brown catalyst solution assumed a dark brownish-red color and become viscous as the polymerization proceeded. Typical reaction times were $60-90$ min (Chien *et al.,* 1981).

The polymerization was stopped by dropwise addition of 6% HCl in methanol, which serves both to precipitate the polymer and to decompose the catalyst. After the polymer had settled, as much as possible of the supernatant was removed with a syringe and discarded. Enough toluene was added to dissolve the product. The solution was filtered with a Schlenk apparatus to remove small amounts of black gummy materials. The deep orange-colored filtrate was mixed with 6% HCl in methanol. The precipitated polymer was washed several times with Ar-saturated anhydrous methanol and dried by pumping overnight. The yields of flaky orange-colored poly(methylacetylene) ranged from 0.3 to 0.6 g from a catalyst solution containing 1.7 ml of $Ti(OBu)_4$.

2.4.2 Copolymerization of Acetylene and Methylacetylene

A 2-liter monomer storage bulb was evacuated and then filled with methylacetylene to a desired pressure. Acetylene was allowed into the system to bring the total gas pressure to ~ 720 torr.

Copolymerization conditions were selected for the following reasons. The boiling point of methylacetylene is ~ 250 K, thus the polymerization temperature used was 263 K in order to avoid condensation of this mono-

TABLE 2.2

Comonomer Feed and Copolymer Composition

Sample[a]	C_2H_2 in feed (mole %)	Chemical analysis of polymers[a]					C_2H_2 in polymer (mole %)
		$C(\%)^b$	$H(\%)^b$	Total C,H(%)	H/C		
$[CH]_x$	100	84.16	7.61	91.77	1.08 (1.00 theor.)		(100)
AMA-61	85	86.31	8.15	94.46	1.13		70
AMA-31	75	87.31	8.61	95.92	1.18		55
AMA-11	50	86.36	8.94	95.30	1.24		33
AMA-13	25	84.53	9.30	93.83	1.31		15
$(C_3H_4)_x$	0	88.81	10.95	99.76	1.48 (1.33 theor.)		(0)

[a] Code = Acetylene – MethylAcetylene – mole ratio C_2H_2/C_3H_4 in feed.
[b] Galbraith Laboratories, Inc., Knoxville, Tenn.

mer. Second, the catalyst solution used in the preparation of free-standing $[CH]_x$ film was observed to "thin out" on the reactor wall very rapidly. A more concentrated catalyst solution, made of 2.5 ml of $Ti(OBu)_4$, 4.0 ml of $AlEt_3$, and 10 ml of toluene, was found to be sufficiently viscous at 263 K to produce copolymer films. Under these conditions the comonomer mixture was polymerized. The resulting film was subjected to 15–20 toluene wash cycles over a period of two days. Table 2.2 gives a summary of the polymerization conditions and an analysis of the copolymers. Figure 2.5 shows the variation of copolymer composition with the comonomer feed.

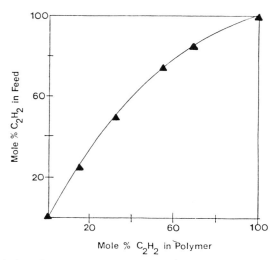

Fig. 2.5 Variation of acetylene content in feed with acetylene in copolymers of acetylene and methylacetylene.

2.5 POLYMERIZATION MECHANISM

Among addition polymerizations, it may be said that the mechanistic complexity is the greatest for the Ziegler–Natta catalysts. In fact, after more than three decades of intensive studies, the detailed mechanism of Ziegler–Natta polymerization cannot be said to be known with certainty. There are still differing views concerning whether one or two metals are required for the active site and what factors control its stereospecificity. However, it is generally accepted that the propagating chain is coordinated to the transition metal and that the monomer is inserted between the metal and the chain during polymerization. In this section, we shall attempt to discuss the polymerization mechanism. It should be recognized that much of what follows is drawn from our accumulated knowledge of olefin polymerizations by Ziegler–Natta catalysts.

2.5.1 Active Catalytic Species

The $Ti(OBu)_4$–$AlEt_3$ combination constitutes a very versatile catalyst. It has been used to polymerize a number of monomers such as ethylene (Bawn and Symcox, 1959), styrene (Takeda *et al.*, 1968), dienes (Dawes and Winkler, 1964; Natta *et al.*, 1964), and even the polar methyl methacrylate (Koide *et al.*, 1967). A mixture of $Ti(OBu)_4$ and $AlEt_3$ turns brown in color rapidly with evolution of ethylene and ethane (Natta *et al.*, 1964; Bawn and Symcox, 1959; Dawes and Winkler, 1964; Dzhabiev *et al.*, 1971). Infrared spectroscopy has been used by Takeda *et al.* (1968) to follow the reactions between $Ti(OBu)_4$ and $AlEt_3$. There are decreases in intensity of the 1120 cm^{-1} band assigned to the Ti—O—C stretching vibration with increase of the Al/Ti molar ratio; this is accompanied by an increase of the 1045 cm^{-1} intensity due to the Al—O—C stretching vibration.

The reaction between $Ti(OBu)_4$ and $AlEt_3$ resulted in the reduction of titanium to its lower oxidation states. The active species are probably some Ti^{3+} complexes. In styrene polymerization, Takeda *et al.* (1968) showed that there was a fairly good correlation between the polymerization yield and the intensity of the Ti^{3+} EPR signal.

Chien *et al.* (1980a) recorded the EPR spectra of the catalyst mixture directly in the microwave cavity. Figure 2.6a, depicting the spectrum recorded at 195 K for a mixture of 7 mM $Ti(OBu)_4$ and 28 mM $AlEt_3$ aged for 30 min at 298 K, shows the complexity of the system. Four EPR resonances with g-values of 1.981 (**1**), 1.976 (**2**), 1.965 (**3**), and 1.945 (**4**), and respective

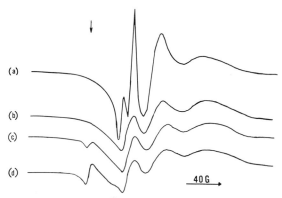

Fig. 2.6 (a) EPR spectrum of Ti(OBu)$_4$–4AlEt$_3$, [Ti]$_0$ = 7mM at 195 K. (b) Spectrum after admission of acetylene and polymerized for 20 min. (c) Bringing (b) to 298 K under anaerobic conditions or introduction of a few millimeters of O$_2$ at 195 K for 1 min. (d) Spectrum of sample at 298 K and more oxygen added. Marker is for DPPH.

line widths of 5.0 ± 0.5, 5.0 ± 0.5, 20 ± 2, and ∼ 57 ± 5 G were observed. The linewidth of resonance **4** was only an estimated value because it was never completely resolved. Even though the five catalyst samples were identically prepared, there were significant variations in the relative intensities of the four signals (Table 2.3). It has been well documented that the EPR signals of this catalyst system are very sensitive to Al/Ti ratios, temperature, and aging time. Up to eleven different Ti^{3+} resonances have been described (Takeda *et al.*, 1968; Hirai *et al.*, 1972).

Signal **1**, upon warming to room temperature, showed 21 hyperfine lines (Fig. 2.6a) with hyperfine splitting of 2.4 G; species **2** has no resolved hyperfine structures. Species **3** showed 11 well-solved hyperfine lines with a coupling constant of 2.6 G (Fig. 2.6b); resonance **4** has a very broad linewidth and is devoid of hyperfine structure. The EPR spectra can be analyzed by

$$H = g\beta H_0 \mathbf{S} + A\mathbf{S} \cdot \mathbf{I},   \qquad (2.2)$$

for a species with a nuclear spin **I** for the paramagnetic ion. It is

$$H = g\beta H_0 \mathbf{S} + A(\mathbf{S}_1 \cdot \mathbf{I}_1 + \mathbf{S}_2 \cdot \mathbf{I}_2) + J\mathbf{S}_1 \cdot \mathbf{S}_2   \qquad (2.3)$$

for two paramagnetic ions where β is the Bohr magneton, H_0 is the applied field, **S** and **I** are the electron and nuclear spin angular momentum, respectively, subscripts 1 and 2 indicate the magnetic species, A is the Fermi contact hyperfine splitting constant, and J is the fine-structure splitting constant.

Let us first discuss the structures of the catalytic species with hyperfine

TABLE 2.3

EPR Data for In-Situ Acetylene Polymerization by $Ti(OBu)_4 - 4AlEt_3$ at 195 K

	Relative EPR intensities peak to peak[a]			
	Species **1**	Species **2**	Species **3**	Species **4**
g-Value	1.981	1.967	1.965	1.945
Sample A				
initial	1	0.29	0.23	0.52
$+C_2H_2$ (60 min)	0.65	0.06	0.06	0.52
Sample B				
initial	1	0.02	1.43	0.12
$+C_2H_2$ (60 min)	0.35	0.02	0.15	0.49
Sample C				
initial	1	3.89	2.0	0.2
$+C_2H_2$ (25 min)	0.62	1.38	0.89	0.24
Sample D				
initial	1	1.05	0	1.87
$+C_2H_2$ (20 min)	0.89	shoulder	shoulder	1.87
Sample F				
initial	1	0	5.0	0
$+C_2H_2$ (10 min)	0.71	shoulder	3.0	0

[a] The signal for species **1** is taken to be unity for each preparation.

splittings. Species **3** has 11 hyperfine lines (Fig. 2.7b) and is attributed to a single paramagnetic ion coupled to two ^{27}Al nuclei ($I = \frac{5}{2}$). Species **2** and **4** have no hyperfine splittings and they have very broad linewidths, the resonances are probably broadened due to short spin–lattice relax-

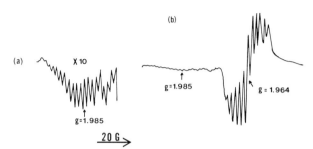

Fig. 2.7 (a) EPR spectrum of $Ti(OBu)_4 - AlEt_3$ catalyst solution with added C_2H_2 warmed to room temperature showing 21 hyperfine lines of species **1** obtained at high gain. (b) EPR spectrum of the same sample showing 11 hyperfine lines of species **3** obtained at high gain.

ation times, dipolar interactions, or both. Species **1** has 21 hyperfine lines
(Fig. 2.7a). Two interpretations are possible: the first and the preferred one
is that there are two Ti^{3+} and two Al ions in the complex with rapid
electron exchange between the paramagnetic ions, i.e., $J \gg A$. In this case
there should be $4I + 1 = 4(5) + 1 = 21$ lines with $1:2:3: \cdots 2I:$
$2I - 1: \cdots :3:2:1$ intensity ratios, in agreement with Fig. 2.7a (Chien
and Westhead, 1971). However, there is a second possibility where the
catalytic species has one Ti^{3+} coupled to two ^{27}Al nuclei and an additional
nucleus with $I = \frac{1}{2}$, such as a hydride ion or the α proton of the propagating
chain. Two observations mitigate against this possible interpretation. First,
the 21-line spectrum was observed by merely warming up the catalyst
solution in the absence of monomer, and second, the spectrum was not
altered in the presence of perdeuterated acetylene prepared from calcium
carbide and D_2O. Whereas β-hydrogen elimination to give titanium hydride
species is facile with $(i\text{-Bu})_3Al$, the same reaction was not observed for Et_3Al
under mild conditions. It seems that the first interpretation is the more likely
one.

EPR spectral interpretation is further assisted by considering the g value.
Under rapidly tumbling conditions for a d^1 system in a distorted octahedral
ligand field, the g value is given by

$$g = 2.0036 - \frac{4\zeta}{\Delta E(d_{x^2-y^2} - d_{xy})}, \tag{2.4}$$

where ζ is the spin–orbit coupling coefficient. Therefore, the smaller the
separation between the ground state and the excited state, the smaller the g
value. This condition corresponds to small values of 10 Dq, which is the
crystal field splitting. By inference the paramagnetic species with a small g
value is coordinately more saturated, which incidentally should also have a
short spin–lattice relaxation time and a broad linewidth.

It is with the above considerations that Chien *et al.* (1980a) assigned the
following structures to species **1** and **3**:

$$\tag{2.5}$$

where X denotes the ligand, which is either OBu or Et. EPR spectra are
incapable of differentiating them. However, since OBu is a better bridging
ligand than Et, it is more likely to be the bridging group. The remaining
ligands comprise both OBu and Et, depending on the stoichiometric ratio of
Al/Ti.

The assignments of species **2** and **4** are much more speculative. We favor a more asymmetric Ti^{3+} for species **2**

$$
\begin{array}{ccccc}
X & & X & & X \\
\diagdown & \diagup & & \diagdown & \diagup \\
& Ti^{3+} & & Al & \\
\diagup & & \diagdown & \diagup & \diagdown \\
X & & X & & X \\
\end{array}
\qquad (2.6)
$$

2

because of its small **g** value. An additional aluminum alkyl may be complexed to **2**. Species **4** has the broadest linewidth and may have a structure such as

$$
\begin{array}{ccccc}
X & & X & & X \\
\diagdown & \diagup & & \diagdown & \diagup \\
& Ti^{3+} & & Ti^{3+} & \\
\diagup & & \diagdown & & \diagdown \\
X & & X & & X \\
\end{array}
\qquad (2.7)
$$

4

Probably species **1**, **2**, and **3** are all catalytically active. Their EPR signals all decrease in intensity (Table 2.4) to varying degrees during the course of a polymerization (Fig. 2.6b).

Based on the above experimental evidence, one can propose the following mechanism for the formation of the catalytic species **1** to **3**:

$$
\begin{aligned}
Ti(OBu)_4 + AlEt_3 &\rightarrow Ti(OBu)_3Et + AlEt_2(OBu) \\
2Ti(OBu)_3Et &\rightarrow 2Ti(OBu)_3 + C_2H_4 + C_2H_6 \\
Ti(OBu)_3 + AlEt_3 &\rightarrow Ti(OBu)_2Et + AlEt_2(OBu) \\
Ti(OBu)_2Et + AlEt_3 &\rightarrow \mathbf{1+2} \\
Ti(OBu)_2Et + 2AlEt_3 &\rightarrow \mathbf{3}
\end{aligned}
\qquad (2.8)
$$

With this particular Ziegler–Natta catalyst the optimum Al/Ti ratio is about four for acetylene polymerization. This favors **1** or **2** as the active catalytic species. It will be shown below that only a few percent of the titanium complexes of all kinds take part in acetylene polymerization, that **2** is probably the most active species, and that **3** is the least active species.

At Al/Ti ratios much less then four, there is incomplete reduction of titanium and little or no polymerization of acetylene. At much larger ratios, Ti^{3+} is reduced to lower oxidation states

$$
2Ti(OBu)_2Et \rightarrow Ti(OBu)_2 + C_2H_4 + C_2H_6 \qquad (2.9)
$$

Ziegler–Natta catalysts with Ti oxidation states $\leqslant +2$ are found to be very poor initiators for ethylene and propylene polymerizations. It will be shown in Section 2.7 that polyacetylene yields decrease with increasing Al/Ti when the ratio is less than four. One can conclude that Ti oxidation states $\leqslant +2$ are also poor initiators for, or inactive in, acetylene polymerization.

2.5.2   Propagation Step

Like the ethylene and propylene polymerizations initiated by Ziegler—
Natta catalysts (Chien, 1959, 1963a,b), polyacetylene was also produced by
the cis insertion of the monomer into the Ti—C bond of the complex
catalyst. Experimental support has been provided by D'yachkovskii et al.
(1964) who showed by NMR that in the $(\pi\text{-}C_5H_5)_2TiCl_2 - Me_2AlCl$-catalyzed
polymerization of phenylacetylene the monomers were inserted in the
Ti—Me bond. Furthermore, based on the structure of the polyenes produced (Ikeda, 1967; Berlin and Cherkashin, 1971; Simionescu *et al.,* 1974) it
was concluded that the monomer, after being activated on π complexation,
undergoes cis opening of the triple bond, which has a Möbius transition state
leading to addition to the Ti—C bond, as shown in Eq. (2.10), where lines
represent unspecified ligands and the square a vacant coordination position.

$$(2.10)$$

This mechanism leads to a *cis*-transoid structure for polyacetylene obtained
at low temperature. Polymerizations at elevated temperatures produce
products with higher trans contents as given in Table 2.4.

TABLE 2.4

Effect of Temperature of Polymerization on
the Isomeric Content of Polyacetylene[a]

Temperature (K)	cis (%)
195	98.1
255	95.4
273	78.6
291	59.3
323	32.4
373	7.5
423	0

[a]After Ito *et al.* (1974).

There are two possible explanations for the results in Table 2.4. One is that the trans units are formed by trans opening of the triple bond during monomer insertion. The second is that they result from isomerization of the *cis*-transoid units. The former is unlikely because the concerted trans insertion, having a Hückel structure for the transition state, is symmetry forbidden and sterically unfavorable. On the other hand, *cis*-polyacetylene is readily isomerized to the trans polymer, as will be extensively discussed in Chapter 5. Therefore, the results shown in Table 2.4 are best attributed to propagation by cis insertion followed by thermally activated isomerization to various trans contents, depending upon the temperature of polymerization before the chain segment is crystallized.

The propagation scheme (2.10) is shown for the growth of a single polyacetylene chain. Actually, the catalytic species consist of clusters of Ti and Al atoms bridged by ethyl and butoxyl groups, such as the structure proposed for **1**. Titanium alkoxides are known to exist as trimers and tetramers. Complexation to even higher oligomers may occur at low polymerization temperatures. Furthermore, upon deposition on the wall of the reaction vessel, the oligomeric catalyst complexes are likely to aggregate into larger clusters. Consequently, a number of polyacetylene chains are initiated in close vicinity. Thus they can form nuclei for crystallization of polyacetylene. Polymerization and crystallization probably occur nearly simultaneously to produce microfibrils about 2–3 nm in diameter, as described in Chapter 4. The same mechanism had been postulated in the Ziegler–Natta-catalyzed polymerizations of ethylene and propylene. Even though the $Ti(OBu)_4 – AlEt_3$ system has been said to be homogeneous, the polymerization of acetylene to form a free-standing film on the reactor wall should be considered as a pseudoheterogeneous process.

2.5.3   Chain Transfer

Several chain transfer reactions are known for Ziegler – Natta catalysis in olefin polymerizations. We list them here for acetylene polymerization.
   Transfer to titanium butoxide

$$Ti^{3+}(C_2H_2)_nEt + Ti(OBu)_4 \longrightarrow Ti^{3+}(OBu) + Ti(OBu)_3(C_2H_2)_nEt \qquad (2.11)$$

$$Ti^{3+}(OBu) + AlEt_3 \longrightarrow Ti^{3+}Et + Et_2Al(OBu) \qquad (2.12)$$

Transfer to aluminum alkyls

$$Ti^{3+}(C_2H_2)_nEt + AlEt_3 \longrightarrow Ti^{3+}Et + Et_2Al(C_2H_2)_nEt \qquad (2.13)$$

Transfer with monomer

$$Ti^{3+}(C_2H_2)_nEt + C_2H_2 \longrightarrow Ti^{+3}CH{=}CH_2 + CH{\equiv}C(C_2H_2)_{n-1}Et \qquad (2.14)$$

Transfer by β-hydrogen elimination

$$Ti^{3+}(C_2H_2)_nEt \longrightarrow Ti^{3+}H + CH{\equiv}C(C_2H_2)_{n-1}Et \qquad (2.15)$$

The reactions depicted in Eqs. (2.14) and (2.15) require significant activation energies and are probably unimportant for acetylene polymerization at low temperatures. The analogous processes occur only at elevated temperatures (above 375 K) in ethylene and propylene polymerizations. The only significant chain-transfer reactions in olefin polymerization at moderate temperatures are those with aluminum alkyls [Eq. (2.13)] (Chien, 1959, 1963a,b). It will be shown in Sections 2.6 and 2.7 that such is probably also the case for the present system.

2.5.4   Chain Termination

Whether chain termination occurs in a Ziegler – Natta polymerization is apparently determined by two principal considerations: the presence of solubilized propagation species and the presence of growing chains in the immediate vicinity of each other on a heterogeneous catalyst. In the polymerization of ethylene by a soluble $(\pi\text{-}C_5H_5)_2TiCl_2 - AlMe_2Cl$ catalyst (Chien, 1959) there is bimolecular termination. On the other hand there is no apparent chain termination in the propylene polymerization catalyzed by $\alpha\text{-}TiCl_3$ (Chien, 1963a), probably because the active centers are far separated. But in the high-activity, $MgCl_2$-supported catalysts (Chien, 1980), there is bimolecular termination probably because either the propagating species became detached from the support or termination occurs between adjacent propagating centers.

We have found that bimolecular termination prevails in the $Ti(OBu)_4$ – $AlEt_3$-catalyzed acetylene polymerization (Section 2.7). The possible processes are:

Reductive terminations

$$2Ti^{3+}P \rightarrow 2Ti^{2+} + P_{-H} + P_{+H} \tag{2.16}$$

$$Ti^{3+}P + Ti^{3+}Et \rightarrow 2Ti^{2+} + P_{+H} + C_2H_4(P_{-H} + C_2H_6) \tag{2.17}$$

Reduction by alkylation

$$Ti^{3+}P + AlEt_3 \rightarrow Ti^{3+}P(Et) \rightarrow Ti^{1+} + PH + C_2H_4 \tag{2.18}$$

Oxidative coupling

$$2Ti^{3+}P + C_2H_2 \rightarrow PTi^{4+}C_2H_2Ti^{4+}P. \tag{2.19}$$

The reactions of Eqs. (2.16), (2.17), and (2.19) are favored by the presence of soluble $Ti^{3+}P$ or $Ti^{3+}Et$ species, where P is a short polyacetylene chain, and by the availability of aluminum alkyls for the reaction of Eq. (2.18).

Transfer reactions tend to lower the molecular weight of the polyacetylene, whereas termination reactions limit the molecular weight as well as the yield of polymers. In order to assure maximum uniformity of products, it is customary to carry out Ziegler–Natta polymerization of olefins with thorough agitation. However, the procedure described above for the preparation of free-standing polyacetylene film precludes any agitation. As a result, the polymerization is nonuniform and there are three reaction zones: the reactor wall, the surface of the catalyst solution, and the bulk of the catalyst solution. The concentrations of soluble propagating species and of ethylaluminum compounds differ in these zones and increase in the order given. Consequently, chain-transfer and termination processes are most important in the catalyst medium and least on the reactor wall. The polyacetylenes produced in these three zones differ greatly in their molecular weights, as will be shown in Section 2.6.

2.5.5   Catalyst Residues

It is common knowledge that in the absence of the transfer processes of Eqs. (2.14) and (2.15) and the termination reactions of Eqs. (2.16)–(2.18), each polymer chain in Ziegler–Natta polymerization has either an Al or a Ti atom for its terminus. It is a customary practice to sever these linkages by alcoholysis or even more efficiently by treatment with acidic alcohol. However, since a protonic acid can dope polyacetylene, its usage in polymer purification is generally avoided. Even purification with alcoholysis is not usually employed. Repeated and thorough washing of polyacetylene with pentane can, at best, remove pentane-soluble catalyst residues. The process is incapable of serving the Al–polymer and Ti–polymer bonds.

We have found that the average of many pentane-washed polyacetylene samples contained (0.2 $\pm$ 0.015%) of Ti, which corresponds to (42 $\pm$ 3) $\times$ 10^{-6} mole of Ti per g of polymer. If we take the $\overline{M}_n$ of polyacetylene film to be 11,000 (Section 2.6), there is 0.46 $\pm$ 0.03 Ti atom per polyacetylene chain. On the average, half of the polyacetylene chains have a Ti terminus, in agreement with the proposed mechanism. The Ti content can be greatly reduced by alcoholysis. Washing with 10% HCl in methanol was found to lower the Ti content to 0.011%; treatment with 30% HCl in methanol gave the still lower Ti content of 0.008%. The fact that there is Ti bound to polyacetylene will loom large when the question of intrinsic impurity in polyacetylene is considered.

2.6 MOLECULAR WEIGHT DETERMINATION

Polyacetylene is insoluble in any solvent. Consequently, the molecular weight of polyacetylene cannot be determined by the usual methods of osmometry, light scattering, ultracentrifugation, viscometry, or gel permeation chromatography. Attempts have been made to convert polyacetylene to a soluble derivative in order to determine its molecular weight indirectly through that of the derivative. But these methods have severe drawbacks. Radioquenching had been used to determine $\overline{M}_n$ directly, and we believe to be far more reliable. These methods are compared in this section.

2.6.1 Indirect Determination by Derivatization

An obvious method of determining the molecular weight of polyacetylene indirectly is to hydrogenate the polymer completely and to determine the molecular weight of the resulting polyethylene. But it is not easy to hydrogenate polyacetylene. We have tried a variety of homogeneous and heterogeneous catalysts but failed to achieve a significant degree of hydrogenation. Shirakawa *et al.* (1980a) demonstrated that sodium-doped polyacetylene can be hydrogenated at elevated temperature and pressure. This reaction, which is not without complications, will be discussed in Section 7.2.3. At 470 K, the temperature of hydrogenation, the polyacetylene film retained the initial shape. This indicates partial crosslinking of the poly-acetylene and/or the hydrogenated polymer. About 60% of the product is soluble in refluxing tetralin. Infrared examination showed the polymer to be polyethylenelike but to contain detectable amounts of $C{=}C$ bond vibra-tions. The GPC-determined $\overline{M}_n$ value was 6200 with $\overline{M}_w/\overline{M}_n = 3.44$, but there was a high molecular weight component in the GPC curve (Fig. 7.17), indicating bimodal distribution.

This indirect hydrogenation method for the determination of polyacety-lene molecular weight has several drawbacks. Only 60% of the hydrogena-tion product was tetralin soluble. This makes it uncertain whether the molecular weight obtained from the polyethylene corresponds to the aver-age molecular weight of the parent polyacetylene or is the low molecular weight portion of the polyacetylene that becomes solubilized upon hydro-genation. Judging from $\overline{M}_n$ values determined for the parent polyacetylene by radioquenching (Section 2.6.2), it appears that the value of $\overline{M}_n$ obtained from the hydrogenation product is the lower limit. In fact, since cross-link-ing probably occurred during hydrogenation, as evidenced by the fact that the polyacetylene film retains its original slope above the melting tempera-ture of polyethylene, the reaction would have a fractionation effect. Finally, under the severe hydrogenation condition, the possibility that chain scission is also taking place cannot be entirely discounted.

Natta *et al.* (1958) first chlorinated polyacetylene at 273–278 K to a colorless and insoluble product. Enklemann *et al.* (1981b) reported that when polyacetylene was prepared with the Luttinger catalyst at $T \leqslant 243$ K and chlorinated immediately after the monomer was removed, a completely soluble, chlorinated polymer was obtained. The product was found to have an $\overline{M}_n$ value of 5900. However, the polymer was insolubilized upon storage at 243 K; after about 15 and 50 hr of storage, 20 and 30% of the chlorination products were insoluble, respectively. The molecular weights of the soluble fraction increased to 9100 after 2 hr and 12,800 after 16 hr, with propor-tionate broadening of molecular weight distribution. Wegner (1981) inter-preted the preceding results as due to spontaneous cross-linking of polyacet-ylene. However, it will be shown in various parts of this book that poly(methylacetylene) and also probably polyacetylene do not cross-link readily and spontaneously. Haberkorn *et al.* (1982) showed that Luttinger-catalyzed polyacetylene contains sp^3 carbons, which are absent in polymers obtained with the Ziegler–Natta catlyst. The insolubilization reported may be due to reactions caused by catalyst residues, oxygen, or other agents. The work should be repeated for verification.

2.6.2   Direct Determination by Radioquenching

The basis of $\overline{M}_n$ determination by radioisotope techniques had been established previously for Ziegler–Natta-catalyst initiated olefin polymerizations. For instance, in the ethylene polymerization by $(\pi\text{-}C_5H_5)_2TiCl_2 - AlMe_2^*Cl$, a ^{14}C-labeled methyltitanium species,

$$(\pi\text{-}C_5H_5)_2Ti \overset{\overset{\displaystyle Me^*}{|}}{\underset{\displaystyle Cl}{\diagdown}} \overset{\displaystyle Cl}{\underset{\displaystyle }{\diagup\diagdown}} AlMe_2^* \, ,$$

was formed, and polymerization proceeds via insertion of ethylene into the Ti—Me* bond. Quenching of polymerization with $^{131}I_2$ labeled the polyethylene chain with ^{131}I and/or ^{14}C (Chien, 1959). This elaborate double radiolabeling was employed to differentiate and quantify various chain-transfer and termination processes. The values of $\overline{M}_n$ thus obtained were in agreement with values obtained by fractionation. Similarly, ^{14}C-labeled $Al(C_2^*H_5)_2Cl$ was synthesized and used to produce the active site $TiCl_2C_2^*H_5(s)$ where s represents the surface of $TiCl_3$ crystallites. The values of $\overline{M}_n$ for polypropylene obtained by radioassay and by fractionation were in excellent agreement (Chien, 1963b). However, the iodine radioquenching method is inapplicable for the determination of $\overline{M}_n$ of polyacetylene because of the doping reaction. The use of radioactive aluminum alkyls is prohibitive for acetylene polymerization because of the large quantity required. Consequently, radioquenching using tritiated methanol is the best available method for the direct determination of $\overline{M}_n$ of polyacetylene.

In this determination of polyacetylene molecular weight, an amount of tritiated methanol (CH_3OH^*) in excess of $4[Ti] + 3[Al]$ was introduced at the end of a polymerization to react with all polymer–metal bonds:

$$TiP + CH_3OT \rightarrow TiOCH_3 + PT \tag{2.20}$$

$$TiP + CH_3OH \rightarrow TiOCH_3 + PH \tag{2.21}$$

$$AlP + CH_3OT \rightarrow AlOCH_3 + PT \tag{2.22}$$

$$AlP + CH_3OH \rightarrow AlOCH_3 + PH \tag{2.23}$$

By radioassay of the polyacetylene, knowing the specific activity of the tritiated methanol, and with the determination of the kinetic isotope effect, values of $\overline{M}_n$ can be obtained directly. The kinetic isotope effect is needed to correct the faster rates of the reactions of Eqs. (2.21) and (2.23) as compared to those of Eqs. (2.20) and (2.22).

One complication of acetylene polymerization is that polymers differing in gross morphology were formed in different reaction zones. In the case of acetylene polymerization to obtain free-standing polyacetylene film, polymers were formed on the reactor wall in the form of film (**F**), at the surface of the catalyst solution in the form of gel (**G**), and within the bulk of the reaction medium in the form of powder (**P**) (Shirakawa and Ikeda, 1979; Wnek *et al.*, 1979a). The relative amounts of these polymers produced depend strongly on the catalyst concentration. Between a wide concentration range for Ti(OBu)$_4$ of 5×10^{-3} to $1.4\,M$ at Al/Ti $= 4$, free standing films of polyacetylene can be prepared. At $5 \times 10^{-1}\,M >$ [Ti] $> 5 \times 10^{-4}\,M$, polyacetylene gel was obtained (Section 2.2.7). Between 5×10^{-3} and $5 \times 10^{-1}\,M$ [Ti], both film and gel were formed. At [Ti] $< 10^{-3}\,M$, only powdery polyacetylene was produced; between 5×10^{-4} and $10^{-3}\,M$ of [Ti], acetylene polymerizes to both gel and powder states. Though the gross appearances of these materials are very different, the basic morphology is fibrillar in all cases except that in some of the powdery polyacetylene the fibrils were almost unrecognizable. A distinct advantage of $\overline{M}_n$ determination by radioquenching is that small specimens from different reaction zones can be sampled. Radioassay gave $\overline{M}_n$ for the polymer formed in a particular reaction zone.

2.6.2.1   *Quenching Rate*

For the radioquenching method to be valid, all the metal–polymer bonds must react completely with methanol. The rates of reaction between tritiated methanol and metal–polymer bonds have been determined at

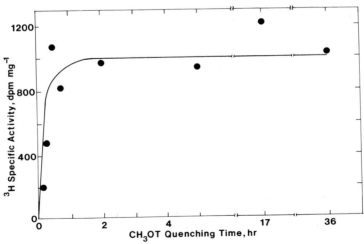

Fig. 2.8   Specific activity of tritium in polyacetylene as a function of time of reaction with tritiated methanol.

room temperature. A number of acetylene polymerizations were carried out under identical conditions: 2.0×10^{-5} mole of Ti(OBu)$_4$, 10^{-4} mole of AlEt$_3$, 20 ml of toluene, 530 torr of acetylene, 298 K, for 2 hr. An excess of tritiated methanol (5×10^{-3} mole) was added to each polymerization and the polymer worked up after various reaction times. Figure 2.8 showed the variation of specific activity with reaction time with CH$_3$OH*. A constant specific activity was attained for reactions of a half hour or longer. Therefore, the reactions shown in Eqs. (2.20)–(2.23) are complete in about 30 min.

2.6.2.2 Kinetic Isotope Effect

The reaction rates between CH$_3$OT and metal–polymer bonds [Eqs. (2.20) and (2.22)] are expected to be slower than the corresponding reactions of CH$_3$OH [Eqs. (2.21) and (2.23)] due to the well-known kinetic isotope effect. Therefore, when a polymerization mixture is quenched with an

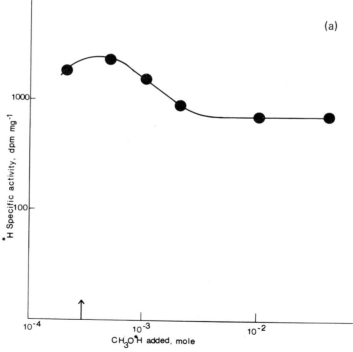

Fig. 2.9 Variation of specific activity of polyacetylene with amount of tritiated methanol. Polymerization conditions: (a) Ti(OBu)$_4$ = 2×10^{-5} mole; $T = 263$ K; C$_2$H$_2$ pressure, 460 torr; time, 2 hr; (●) powder product; (b) Ti(OBu)$_4$ = 2×10^{-3} mole; $T = 195$ K; C$_2$H$_2$ pressure, 460 torr; time, 2 hr; (■) film product; (▲) gel product.

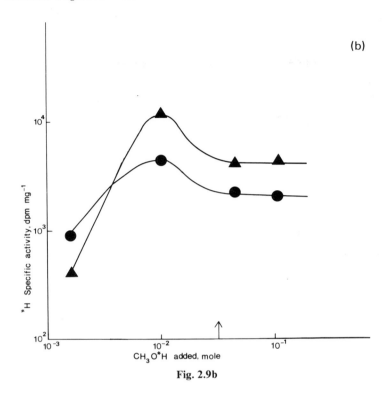

Fig. 2.9b

excess of tritiated methanol, the reactions between metal polymer bonds and CH_3OH are favored over those with CH_3OT. The kinetic isotope effect, k_H/k_T, can be determined by choosing reaction conditions under which both CH_3OH and CH_3OT can react with equal probability. The specific activity in polyacetylene obtained in this manner divided by the specific activity obtained by using an excess of tritiated methanol gives the ratio k_H/k_T. Two methods have been used to determine this effect. In the first method, a polymerization was stopped by removing the monomer and titrating with 17.5-μmole aliquots of CH_3OH^*. Each aliquot was allowed to react for 1 hr before the addition of the next aliquot; the experiment took 36 hr. A total of 1.4 mmoles of tritiated methanol was added; this is a 3.5-fold excess of the equivalent total number of potential metal–ligand bonds. The tritium specific activity of this polyacetylene is about 1.7 times larger than that obtained for an identical polymerization but quenched by a single addition of an excess of CH_3OH^*. Therefore, $k_H/k_T \sim 1.7$.

In the second method, a number of identical polymerizations were stopped by monomer removal, and various amounts of CH_3OH^* were added to each reaction. Typical results are shown in Fig. 2.9. At 263 K and a low catalyst concentration, the polyacetylene has a powdery morphology. The maximum kinetic isotope effect was obtained at an amount of methanol nearly stoichiometric to the total moles of possible metal–ligand bonds. The equivalent point is shown by the arrow in the figure; the value of k_H/k_T is 3.4. At 195 K and high catalyst concentration, both film and gel forms of polyacetylene were obtained. The kinetic isotope effect is 2.7 for the gel and 2.3 for the film products. The maximum specific activities were obtained at about one-third of the stoichiometric amounts. Taking the average of all the aforementioned k_H/k_T values, we use a value of 2.8 for the kinetic isotope effect in the determination of $\overline{M}_n$ for polyacetylene, with a standard deviation of $\pm 14\%$.

2.6.2.3 Polyacetylene Formed on the Reactor Wall

Because acetylene polymerized on the reactor wall was initiated by the coating of a viscous catalyst solution, the latter cannot be of uniform thickness, and thus the amount of catalyst on the wall must also differ. A study had been made to find out whether the polyacetylenes formed on different parts of the reactor wall are the same.

Acetylene was polymerized under the normal conditions: $[Ti] = 0.20\ M$, $Al/Ti = 4$, $P_{C_2H_2} = 630{-}560$ torr, temperature $= 195$ K for 30 min, and quenching by CH_3OH^*. A 1-cm-wide sample of polyacetylene was cut from the free-standing film immediately above the surface of the catalyst solution. Then 0.5-, 0.5-, and 1-cm-wide sections of the film were cut at heights of 3, 5.5, and 8 cm, respectively, from the surface of the catalyst solution. Each sample was radioassayed in quadruplicate. Virtually identical $\overline{M}_n$ values were found for the three lower specimens; the average value was 10,500 $\pm$ 150 with a very small standard deviation. The topmost specimen had larger values: $\overline{M}_n = 11,700 \pm 3200$. The polyacetylene formed at the surface of the catalyst solution has much lower molecular weight: $\overline{M}_n = 5300 \pm 1200$. More will be said about the molecular weight of polyacetylenes found at the various reaction zones.

2.6.2.4 Effect of Catalyst Concentration

Chien et al. (1982f, 1983d) had determined the $\overline{M}_n$ of polyacetylenes **F**, **G**, and **P** produced at various catalyst concentrations. The results are summarized in Table 2.5 and Fig. 2.10. The highest molecular-weight polyacetylene was obtained at the lowest catalyst concentration under which conditions only powdery polymers were isolated. As the catalyst concentration was increased, polyacetylene having the other morphologies emerged. The polyacetylene gel (**G**) fraction has $\overline{M}_n$, which is only slightly dependent on [Ti]. This suggests that there is no significant chain transfer or

TABLE 2.5

Effect of Catalyst Concentration on Acetylene Polymerization[a]

[Ti] (mM)	Powder			Gel			Film
	Yield (mg)	[MPB] (mM)	$\overline{M}_n$	Yield (mg)	[MPB] (mM)	$\overline{M}_n$	$\overline{M}_n$
1	60.6 ± 14	0.15	20,100	—	—	—	—
5.1	—	—	—	462 ± 15	3.0	7,600	—
5.1	978 ± 71[b]	—	3,800	—	—	8,400	—
41.0	129 ± 10	24	550	340 ± 59	6.0	5,700	—
204	—	—	—	298 ± 16[c]	—	5,800	11,800

[a] Al/Ti = 4, T = 195 K, $P_{C_2H_2}$ = 456 ± 40 torr, time = 2 hr, values for yields were average of four runs.
[b] Total yield of gel and powder.
[c] Total yield of gel and film.

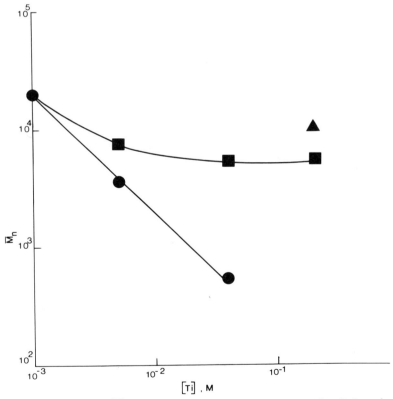

Fig. 2.10 Variation of $\overline{M}_n$ of polyacetylene with catalyst concentration. Polymerization conditions: Al/Ti = 4; T = 195 K; C_2H_2 pressure, 460 torr; time, 2 hr; (●) powder; (■) gel; (▲) film.

termination. A probable reason is that the gel is dense and acts as a barrier against diffusion, and thus the catalyst species cannot diffuse into the gel layer. Consequently, the frequency of chain transfer or termination is greatly reduced and the $\overline{M}_n$ is insensitive to catalyst concentration for $(CH)_x(G)$.

In contrast, the powder polymer produced beneath the gel layer shows strong $\overline{M}_n$ dependence on [Ti]. Of course, there is an abundance of catalyst in the solution to cause chain transfer and termination. In addition, the diffusion of monomer from the gas phase into the catalyst solution is limited by the gel layer. The solution phase is thus monomer starved. At [Ti] = 41 mM, the $(CH)_x(P)$ formed in this phase has $\overline{M}_n$ of only 550, or a degree of polymerization of 20. It remains to be shown whether such very low molecular weight polyacetylene will crystallize in lamellae form.

At the high catalyst concentration of [Ti] = 0.2 M, free-standing film of polyacetylene was produced. It has an average $\overline{M}_n$ of 11,000. For the same reason given for the gel product, the $\overline{M}_n$ for the free-standing film of polyacetylene would be insensitive to the catalyst concentration, and it was shown above (Section 2.6.2.3) that the value of $\overline{M}_n$ is uniform for nearly the entire film.

2.6.2.5 Effect of Polymerization Temperature

It has been well established that the molecular weight of polyolefins produced in a Ziegler–Natta polymerization decreases with increasing

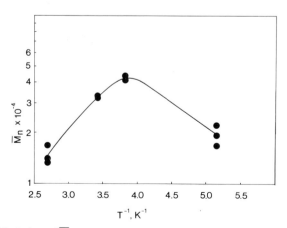

Fig. 2.11 Variation of $\overline{M}_n$ with temperature of polymerization. Conditions: Ti(OBu)$_4$ = 2 × 10^{-5} mole, C$_2$H$_2$ pressure 460 torr; time, 2 hr.

TABLE 2.6

Effect of Temperature on Acetylene Polymerization[a]

Temperature (K)	Polyacetylene yield (mg)	[MPB] (mM)	$\overline{M}_n$
343	115.2	0.35	16,300
343	78.0	0.30	13,100
343	112.4	0.40	14,000
293	228.1	0.36	37,700
293	242.6	0.38	31,600
263	203.5	0.24	43,100
263	198.0	0.24	42,000
195	81.5	0.23	17,500
195	56.9	0.12	23,000
195	52.9	0.14	19,100

[a] $[Ti]_0 = 1$ mM, Al/Ti = 4, time = 2 hr, $P_{C_2H_2} = 456$ torr, $[C_2H_2] = 0.06\ M$ (343 K), 0.13 M (293 K), 0.28 M (263 K), 2.1 M (195 K).

temperature. This is because the combined activation energies for initiation and termination exceeded the activation energy for propagation by a significant amount. However, studies of this effect had been limited to room temperatures and above. In the case of acetylene polymerization, the temperature used ranges from 195 to as high as 423 K. Chien *et al.* (1982f) had determined the molecular weights of polyacetylene obtained from 195 to 348 K. The results are shown in Fig. 2.11 and summarized in Table 2.6.

There is a threefold increase of $\overline{M}_n$ for about a 70° decrease of temperature above 263 K. In this temperature region, polyethylenes obtained by means of $(\pi\text{-}C_5H_5)_2TiCl_2 - 2AlMe_2Cl$ showed a 2.6-fold increase of $\overline{M}_n$ for a 30° decrease of temperature (Chien, 1959). Similar dependence was found in the $\alpha\text{-}TiCl_3 - AlEt_2Cl$-initiated polymerization of propylene (Chien, 1963b). The effect is attributable to a combination of the temperature dependence of initiation, propagation, transfer, and termination processes. The decrease of $\overline{M}_n$ with decreasing temperatures below 263 K is attributable to the decline of rate of propagation as will be shown in Section 2.7.

2.6.2.6   *Effect of Al/Ti Ratio*

The Al/Ti ratio can affect acetylene polymerization in two ways (Table 2.7). The first is the activation of Ti(OBu)$_4$ by reducing it to a trivalent

TABLE 2.7

Effect of Al/Ti Ratios on Acetylene Polymerization[a]

Al/Ti	Polyacetylene yield (mg)	[MPB] (mM)	$\overline{M}_n$
1	0	—	—
2	0	—	—
4	300	0.4	40,000
7	314	2.5	6,200
10	171	2.8	3,050

[a] $[Ti]_0 = 10^{-3}\ M$, $P_{C_2H_2} = 1$ atm, $T = 298$ K, time = 120 min, catalyst aging time, 1 hr.

complex. Thus no acetylene polymerization occurred for Al/Ti $\leqslant$ 2. At higher Al/Ti ratios, chain transfer becomes more frequent thus lowering the polymer molecular weight. At Al/Ti = 7 there was a more than sixfold decrease in $\overline{M}_n$ as compared to Al/Ti = 4. With a tenfold excess of AlEt$_3$ over Ti(OBu)$_4$, the molecular weight was lowered to half of those obtained at Al/Ti = 7.

2.6.2.7 Effect of Catalyst Aging

It is common in the preparation of a Ziegler–Natta catalyst for a specific polymerization purpose, to include some grinding, heat treatment, or aging process. The effect of aging on the present Ti(OBu)$_4$–AlEt$_3$ catalyst had been examined in our laboratory. The results are given in Table 2.8. Unaged

TABLE 2.8

Effect of Catalyst on Acetylene Polymerization[a]

Aging time (hr)	Polyacetylene yield (mg)	[MPB], (mM)	$\overline{M}_n$
0	380	1.0	18,900
1	300	0.4	40,000
192	180	0.33	27,200
1200	74	0.069	53,100
6216	71	0.1	34,300

[a] $[Ti]_0 = 10^{-3}\ M$, $P_{C_2H_2} = 760$ torr, $T = 298$ K, time = 2 hr. Results are average of four runs.

catalyst produced low molecular weight polyacetylenes. Catalyst aged 1 hr or longer gave polymers with comparable $\overline{M}_n$.

2.6.2.8 *Effect of Monomer Concentration*

The effect of monomer concentration was studied for acetylene polymerization at 195 K. The reaction medium was stirred magnetically in order to minimize physical nonuniformity. In runs 1–6 of Table 2.9, the reproducibility of duplicate polymerizations was shown to be quite good and typical for Ziegler–Natta polymerizations. At the low catalyst concentration of $[Ti]_0 = 1$ mM and $P_{C_2H_2} \leq 380$ torr, only powdery polyacetylene was formed. A twofold increase of $P_{C_2H_2}$ increase $\overline{M}_n$ by 1.8-fold. At 1 atm of monomer, polyacetylene gel was produced, and agitation did not break up the gel. Therefore, polymerization occurs mainly at the surface of the catalyst solution. There was a large increase in $\overline{M}_n$ in runs 5 and 6, more than might be expected based on simple proportionality-to-monomer concentration. This is believed to be a physical effect rather than a change of mechanism. For example, if there were chain transfer to monomer, runs 5 and 6 should produce polymers with lower molecular weights.

In runs 8–10 of Table 2.9, the toluene was saturated with acetylene before the catalyst was injected. Compared to the results of runs 5–7, the yield was about twice as large and $\overline{M}_n$ also was higher when the solvent was

TABLE 2.9

Effect of Acetylene Pressure on Its Polymerization[a]

No	$P_{C_2H_2}$ (torr)	Time (min)	Polyacetylene yield (mg)	[MPB] (mM)	$\overline{M}_n$
1	190	120	26	0.15	8,500
2	190	120	31	0.13	11,600
3	380	120	48	0.13	18,700
4	380	120	39	0.10	19,200
5	760	120	118	0.06	100,000
6	760	120	128	0.16	40,000
7	760	120	101	0.09	85,000
8	760[b]	20	144	0.07	105,000
9	760[b]	75	201	0.08	121,600
10	760[b]	240	235	0.10	120,400

[a] $[Ti]_0 = 1$ mM, Al/Ti $= 4$, $T = 195$ K acetylene admitted after catalyst has been first added.

[b] Same as in *a* except acetylene was equilibrated with toluene and then catalyst injected.

saturated with monomer. Consequently, the subsequent kinetic experiments were performed in this manner.

2.6.2.9   Advantages and Limitations

The radiotagging technique for molecular weight determination has both advantages and limitations. On the plus side, the labeling is done before the polymer is subjected to conditions that may lead to chemical and/or physical transformations. Therefore, any post-polymerization events such as crystallization, crosslinking, or autoxidation cannot alter the specific activity of the labeled polymer and thus the initial molecular weight. Individual molecular weights can be obtained from various regions of polymerization, i.e., film, gel, and powder. The method is not complicated by possible fractionation effects.

There are, however, several limitations in the radioquenching technique. Foremost is the fact that the method gives only $\overline{M}_n$ but no information about the width of molecular weight distribution. The determination of the kinetic isotopic effect assumes it to be the same for the Ti—P and Al—P bonds. This assumption was made in all previous uses of CH_3OH^* in Ziegler–Natta polymerizations. Another possible limitation of the radioquenching determination of $\overline{M}_n$, which can be of a serious nature, is the fact that those polyacetylene chains that are not bound to a metal, either Ti or Al, are not tagged. These molecules can be formed by the reactions of Eqs. (2.14)–(2.18). However, it will be shown in Section 2.7 that they are not important in these acetylene polymerizations.

2.7   KINETICS

2.7.1   Active Center Counting

The number of active centers in olefin polymerization by the Ziegler–Natta catalyst can be determined by using radioactive C^*O. The method is based on the reaction

$$TiP + C^*O \rightarrow Ti\overset{\overset{\displaystyle O}{\|}}{C}^*P \tag{2.24}$$

followed by methanol workup

$$Ti\overset{\overset{\displaystyle O}{\|}}{C}^*P + CH_3OH \rightarrow TiOCH_3 + P\overset{\overset{\displaystyle O}{\|}}{C}^*H \tag{2.25}$$

and radioassay.

The technique is not as straightforward as it appears due to the following complications. The first is the reinsertion of monomer:

$$
\begin{matrix} O & & O \\ \parallel & & \parallel \\ \text{TiC*P} + n\text{C}_2\text{H}_2 & \rightarrow & \text{Ti(C}_2\text{H}_2)_n\text{C*P} \end{matrix} \qquad (2.26)
$$

The occurrence of this reaction is demonstrated by steady increase of specific activity with time of reaction with C*O in the presence of monomer attributable to

$$
\begin{matrix} O & & O & O \\ \parallel & & \parallel & \parallel \\ \text{Ti(C}_2\text{H}_2)_n\text{C*P} + \text{C*O} & \rightarrow & \text{TiC*(C}_2\text{H}_2)_n\text{C*P} \end{matrix} \qquad (2.27)
$$

resulting in multiple incorporation of C*O into a polyacetylene molecule. If the unchanged acetylene is removed by evacuation prior to the introduction of C*O, the specific activity increases for a time and then remains constant.

The rate of reaction of C*O with a titanium polymer bond in olefin polymerizations had been claimed by some to be extremely rapid and by others to be very slow. We have carried out identical 2-hr acetylene polymerizations, removed the excess acetylene, added C*O, and allowed the reaction mixture to stand at 298 K for various lengths of time before workup and radioassay. The results are given in Fig. 2.12. The ^{14}C specific activity in the polymer increased with increase of reaction time, reaching a constant value after about 4–5 hr. Therefore, the reaction of Eq. (2.24) is relatively slow. In the work to be described below, C*O was allowed to react with the polymerization mixture in the absence of monomer for 6 hr before workup.

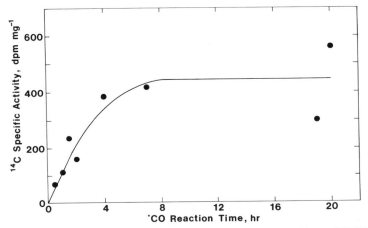

Fig. 2.12   Rate of incorporation of C*O. Polymerization conditions: $[\text{Ti(OBu)}_4]_0 =$ 1 mM, Al/Ti = 4, T = 298 K; $P_{\text{C}_2\text{H}_2}$ = 530 torr, time = 2 hr. At the end of polymerization C$_2$H$_2$ was evacuated, and 1.2 ml of C*O at STP was injected.

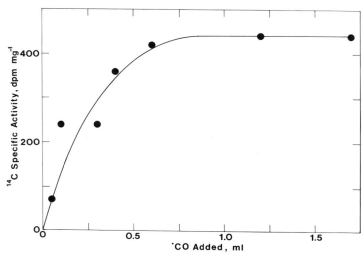

Fig. 2.13 Variation of ^{14}C specific activity in polyacetylene with the quantity of C*O added. Conditions are the same as in Fig. 2.12; the contact time for C*O is 6 hr.

Sufficient amount of C*O must be added to react with all Ti—σC bonds. This was determined empirically. In a 20-ml reaction of acetylene polymerization initiated by 1 mM Ti(OBu)$_4$ and 4 mM AlEt$_3$ at 298 K and 1 atm of acetylene, the monomer was evacuated after 2 hr, various amounts of C*O were added, and the mixture was worked up after 6 hr of contact time. Figure 2.13 gave the results, which showed that the maximum constant ^{14}C specific activity was reached with 0.5–0.6 ml of C*O at STP. This corresponds to 22–27 μmole of C*O. There was present 20μmole of Ti(OBu)$_4$. If each Ti atom was alkylated by AlEt$_3$, then the result implies that C*O reacts only with Ti—Et or Ti—P bonds but not with Ti—OBu bonds nor with organoaluminum compounds. The last point is of utmost importance. If C*O also reacts with Al—P bonds, then the technique will be no different from the radiotagging with CH$_3$OH* except that there would be no appreciable kinetic isotope effect for the C*O reaction. The foregoing results showed that C*O counts the number of active centers and CH$_3$OH* counts the total metal–polymer bonds [MPB].

The specific activity of C*O was determined by quantitative oxidation over heated copper oxide catalyst to C*O$_2$. This is achieved by passing 5 ml of C*O with purified argon carrier gas flowing at 10 ml min^{-1} over the catalyst at 923 K and allowing the effluent to bubble into 20 ml of 1 N NaOH. Calibration with nonradioactive CO showed the conversion efficiency to be 100 ± 2%. The radioactivity of the Na$_2$C*O$_3$ solution was determined by liquid scintillation counting against a standard ^{14}C sample. The specific activity of the C*O used in this work was 13.3 mCi mole^{-1}.

2.7.2 Factors Influencing Acetylene Polymerization Kinetics

There was no polymerization of acetylene when catalysts with Al/Ti ratios of one and two were used. At Al/Ti = 4, we obtained the highest yield of polyacetylene having the highest molecular weight. These results are in accord with Eq. (2.8). At this cocatalyst ratio the active species **1** or **2** is formed. At Al/Ti ratios ⩽ 2, there were insufficient amounts of AlEt$_3$ to reduce Ti(OBu)$_4$ to the proper reduced oxidation and coordination states. At a ratio of 7:1, the polymerization yield was about the same as with 4:1, but there was a sharp drop in $\overline{M}_n$ values (Table 2.7). There was a decrease of polymer yield as well as further lowering of $\overline{M}_n$ when an Al/Ti ratio of ten was used, explainable by overreduction of Ti by the excess AlEt$_3$:

$$EtTi(OBu)_2 + AlEt_3 \rightarrow Et_2Ti(OBu) + AlEt_2(OBu)$$
$$Et_2Ti(OBu) \rightarrow Ti(OBu) + C_2H_4 + C_2H_6 \tag{2.28}$$

In these reactions, it is understood that the titanium atoms are complexed to aluminum alkyls and/or other titanium compounds via electron-deficient bridges.

The [MPB] was 0.4 mM at the end of 2 hr of polymerization with an Al/Ti ratio for four. This corresponds to 40% of the [Ti(OBu)$_4$]$_0$. The [MPB] increased sixfold at Al/Ti = 7, with comparable decrease in $\overline{M}_n$. This is strong evidence of chain transfer with aluminum alkyls [Eq. (2.13)]. In addition, chain transfer can take place with AlEt$_2$(OBu) such as

$$TiP + AlEt_2(OBu) \rightarrow TiEt + PAlEt(OBu) \tag{2.29}$$

and even with AlEt(OBu)$_2$, albeit the efficiency is expected to decrease with the fewer number of ethyl groups on aluminum. If at Al/Ti = 4 there still remain two moles of alkyl aluminum compounds, then at Al/Ti = 7 there are five moles available for chain transfer. The foregoing results suggest that

$$\frac{d[AlP]}{dt} = k_{tr}[C][AlR]^2, \tag{2.30}$$

where [C] is the active-center concentration. Catalyst aging has a strong influence on the polymerization of acetylene. Table 2.8 shows decreases of polymer yield, [MPB], and $\overline{M}_n$ for catalysts that had been aged for too long a time.

The effect of temperature on the polymer yield is given in Table 2.7 for

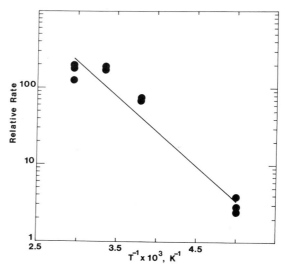

Fig. 2.14 Arrhenius plot for the relative rate of polymerization: $[Ti]_0 = 1$ mM, Al/Ti = 4, Al/Ti = 4, $P_{C_2H_2} = 456$ torr, time = 2 hr.

reactions carried out all with $[Ti]_0 = 1$ mM, Al/Ti = 4, for 2 hr. The yield of polyacetylene increases with increased temperature, then decreases at the highest temperature (Table 2.6). However, there is a significant temperature dependence for the solubility of acetylene in toluene (Fig. 2.3). The relative rate of polymerization is then obtained as polymer yield/$[C_2H_2]$, and Fig. 2.14 is an Arrhenius plot of the relative rate. The overall activation energy for polymerization of acetylene is 4.2 kcal mole^{-1}.

The total [MPB] is the sum of [C] and [AlP]. It showed small increases with increased temperature (Fig. 2.15). The apparent activation energy is only 0.85 kcal mole^{-1}.

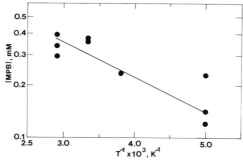

Fig. 2.15 Arrhenius plot for [MPB]. Polymerization conditions as in Fig. 2.14.

2.7.3 Kinetics of Polymerization

Over 60 polymerizations were carried out at 298 and 195 K under otherwise identical conditions of $[Ti(OBu)_4]_0 = 1$ mM, Al/Ti = 4, catalyst aging time = 30 min, and $P_{C_2H_2} = 760$ torr. The polymerizations were stopped after a predetermined time and radioquenched with either CH_3OH^* or C^*O.

The yields of polymer as a function of time of polymerization are shown in Figs. 2.16a and 2.17a. Most of the polyacetylene was produced in the initial period. At 298 K, 60% of the polymer was obtained during the first 20 min as compared to the total yield after 240 min. The yield at 25 min of polymerization at 195 K was about 75% of that after 240 min. In other words, the rate of polymerization was very rapid initially and became very slow afterwards (Figs. 2.18a and 2.19a). At both temperatures, R_p decreased about 30-fold during the first 30 min of polymerization. On the other hand, the active site concentration, [C], as determined by C^*O quenching, remained virtually constant during the first period and began to decrease only after $\sim$ 50 min of polymerization (Figs. 2.20a and 2.21a). After 240 min [C] decreased to one-half to one-third of the maximum initial value. These basic characteristics must first be discussed before the detailed kinetic data can be analyzed. There are three possible causes for the observed behavior.

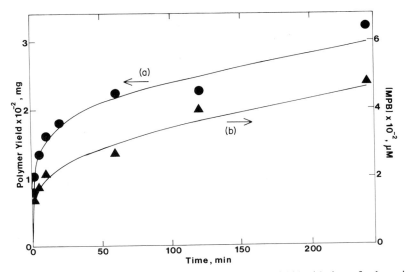

Fig. 2.16 Variation of (a) polymer yield (●) and (b) [MPB] (▲) with time of polymerization: $[Ti]_0 = 1$ mM, Al/Ti = 4, catalyst aging time = 30 min, $T = 298$ K, $P_{C_2H_2} = 760$ torr.

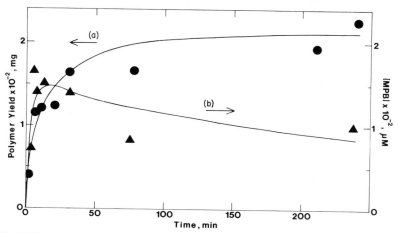

Fig. 2.17 Variation of (a) polymer yield (●) and (b) [MPB] (▲) with time of polymerization: $[Ti]_0 = 1$ mM, Al/Ti = 4, catalyst aging time = 30 min, $T = 195$ K, $P_{C_2H_2} = 760$ torr.

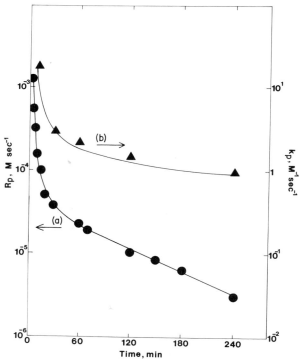

Fig. 2.18 Variation of (a) R_p (●) and (b) k_p (▲) with time of polymerization; conditions are the same as in Fig. 2.16.

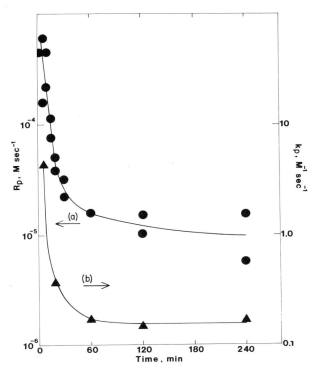

Fig. 2.19 Variation of (a) R_p (●) and (b) k_p (▲) with time of polymerization; conditions are the same as in Fig. 2.17.

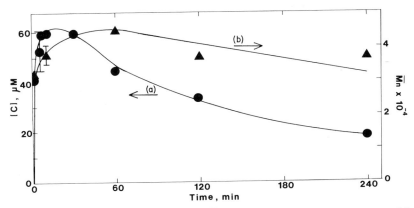

Fig. 2.20 Variation of (a) [C] (●) and (b) $\overline{M}_n$ (▲) with time of polymerization; conditions are the same as in Fig. 2.16.

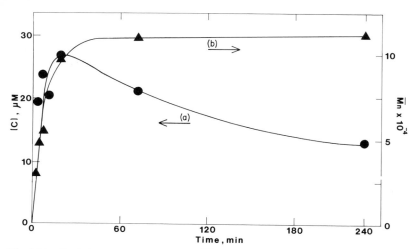

Fig. 2.21 Variation of (b) [C] (●) and (b) $\overline{M}_n$ (▲) with time of polymerization; conditions are the same as in Fig. 2.17.

The first explanation is that the polymer produced forms an increasingly thick barrier to the diffusion of monomer at the active site. Schmeal and Street (1971) and Singh and Merrill (1971) proposed diffusion limitation models to account for the polydispersity of polymers obtained with catalysts such as the Ziegler–Natta type. We have critically considered this problem and defined criteria for diffusion limitation in coordination polymerization (Chien, 1979a). The findings were that with the possible exception of the very active catalysts, most olefin polymerizations are not diffusion limited. Recently, we addressed the question with direct experimental tests. The new MgCl₂-supported Ziegler–Natta catalyst has an activity several thousand times that of the classical TiCl₃ catalyst and shows very rapid rate decay. We have compared side-by-side polymerizations at 323 K, producing insoluble polypropylene and soluble poly(decene-1). The latter polymer has a melting point of only 313 K. The polymerizations have nearly the same rates and virtually the same decay of R_p. Consequently, diffusion limitation is highly unlikely for the present polymerization of acetylene, which is very much slower.

A second rationalization for the results in Figs. 2.8–2.11 is that there is one kind of active center, which terminates very rapidly. This is not viable because when R_p decreases most rapidly [C] remains unchanged.

Consequently, we are led to the third possible explanation, which postulates the presence of two types of active sites: a very active kind C_1 and a much less active type C_2. Hereafter, subscript 1 (2) will be used to denote the values for active sites C_1 (C_2). There are two subcases to this postulate: (a) C_1 and C_2 are independent of one another and (b) C_1 is the precursor to C_2. The fact that [C] remained relatively constant during the period of rapid rate decline favors case b. It will be shown below that the total kinetic analysis is consistent with this basic assumption.

2.7.3.1   Low Activity C_2 Sites

We first discuss the second slow stage of polymerization occurring between $t = 60$ min and the end of polymerization. It is assumed that because after 30 min of polymerization the steep decline of R_p had markedly leveled off, all of the more active C_1 sites had decayed and only C_2 remained.

The rate of termination for C_2 can be obtained directly from the C*O quenching results; the decrease in ^{14}C activity in polyacetylene may be attributed to the termination of C_2 sites. Figures 2.22 and 2.23 showed linear plots of second-order disappearance of C_2 at 298 and 195 K, respectively. Referring back to Fig. 2.10a and 2.11a, one sees that polymerization commenced immediately with maximum rate upon the injection of aged catalyst. Furthermore, Table 2.2 showed that the catalytic activity is not lost

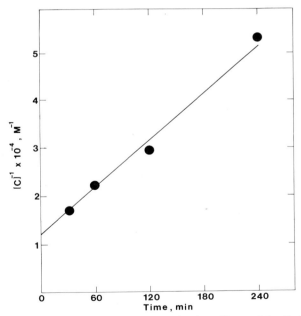

Fig. 2.22   Second-order kinetics plot for the termination of low activity C_2 sites at 298 K.

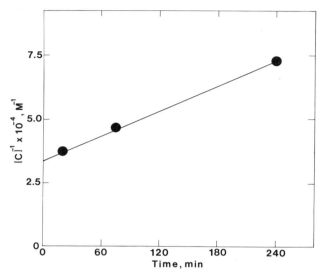

Fig. 2.23   Second-order kinetics plot for the termination of low activity C_2 sites at 195 K.

appreciably unless aged for very long times. These observations imply that the active sites are stable in the absence of monomer and suggest the involvement of the monomer in the termination process. The rate of termination may be written

$$\frac{-d[C_2]}{dt} = k_{t_2}[C_2]^2[M].$$ (2.31)

From the slopes of Fig. 2.22 and 2.23 and the monomer concentrations, we find values for k_{t_2} to be 0.7 M^{-2} sec^{-1} and 16.8 M^{-2} sec^{-1} at 195 and 298 K, respectively. The activation energy ΔE_{t_2} for the termination of C_2 sites is about 3.6 kcal mole^{-1}.

The rate of propagation for a Ziegler–Natta polymerization is given by

$$R_p = k_p[C][M].$$ (2.32)

We calculated and show the Fig. 2.11b that k_{p_2} at 195 K has a relatively constant value of 0.16 M^{-1} sec^{-1} between $t = 60$ and 240 min; the value is ~ 1.1 M^{-1} sec^{-1} between $t = 120$ and 240 min at 298 K. Thus the activation energy is only about 2 kcal mole^{-1}.

The [MPB] increases gradually at 298 K. The rate of increase toward the end of polymerization is about 6.5×10^{-9} M sec^{-1} (Fig. 2.16b). This suggests transfer of the polyacetylene chain from Ti to Al. The number average molecular weight is given by

$$\overline{M}_n = \left(\frac{R_p}{R_t + R_{tr}}\right)26. \tag{2.33}$$

where R_t and R_{tr} are the ratios of termination and transfer, respectively.

The rate of polymerization is 8.5×10^{-6} M sec^{-1} at $t = 150$ min and 3.3×10^{-6} M sec^{-1} at $t = 240$ min; [C_2] is 31 and 19 μM at $t = 150$ and $t = 240$ min, respectively. Using $k_{t_2} = 31$ M^{-1} sec^{-1}, we obtained $\overline{M}_n$ values of 2.5×10^4 and 1.1×10^4 at $t = 150$ and 240 min, respectively. The agreement with experimental values of $\overline{M}_n$ (Fig. 2.20b) is reasonable, though the calculated values are somewhat lower. This suggests the possibility that the chain transfer process may be reversible, i.e.,

$$TiP + AlEt \rightleftharpoons TiEt + AlP \tag{2.34}$$

The results at 195 K indicate that the process is favored at lower temperatures. Figure 2.9b showed that there was actually a slow decline of [MPB] with time while $\overline{M}_n$ remained constant (Fig. 2.13b).

2.7.3.2 High Activity C_1 Sites

Figures 2.24 and 2.25 are expanded scale plots of the initial portion of Figs. 2.18 and 2.19 in order to show the results better during the early stage of polymerizations and to include all those data points that could not be accommodated in Figs. 2.18a and 2.19a.

Estimates for k_{p_1} can be obtained by assuming that the maximum initial [C] corresponds to [C_1]. This is the consequence of our postulate that C_1 is the precursor of C_2. With this assumption and according to Eq. (2.20) we find the values for k_{p_1} to be 96 M^{-1} sec^{-1} and 4.1 M^{-1} sec^{-1} at 298 and 195 K, respectively. An approximate value for ΔE_{p_1} is 3.5 kcal mole^{-1}. Therefore, C_1 is about 65 times more active than C_2 at 298 K; the activity differential is eightfold at 195 K. The more active catalytic site requires about three times higher energy of activation for propagation than the less active catalytic sites.

The rate of termination of C_1 can be obtained indirectly from the rate of decrease of R_p because $R_p \propto C_1$ during this initial period of polymerization. Figures 2.26 and 2.27 show plots of R_p^{-1} versus time, indicating that C_1 also disappears according to second-order kinetics. The slope is $k_{p_1} k_{t_1}[M]^2$ if we assume that the monomer is involved in the termination of C_1, as for C_2. Substituting values of k_{p_1} and [M] into Eq. (2.27), we find values of k_{t_1} to be 1.6×10^4 and 0.96×10^2 M^{-2} sec^{-1} at 298 and 195 K, respectively. The activation energy ΔE_{t_1} is about 5.7 kcal mole^{-1}.

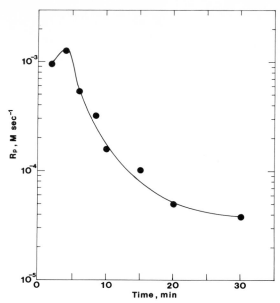

Fig. 2.24 Initial variation of R_p with time for Fig. 2.18a.

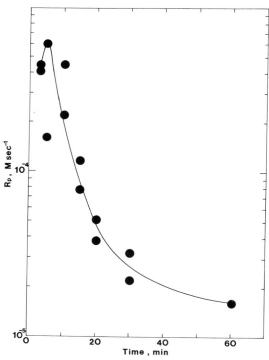

Fig. 2.25 Initial variation of R_p with time for Fig. 2.19a.

74

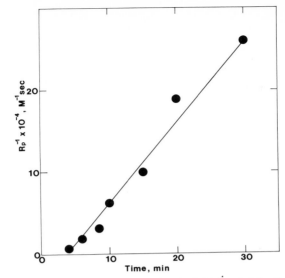

Fig. 2.26 Plot of R_p^{-1} versus time during the initial period for Fig. 2.24.

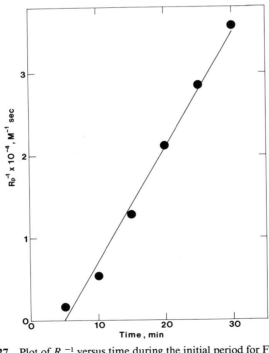

Fig. 2.27 Plot of R_p^{-1} versus time during the initial period for Fig. 2.25.

The relative concentration $[C_1]$ at the beginning of a polymerization and $[C_2]$ just after most of the former have decayed, i.e., k_{p_2} becomes more or less constant, can be obtained from

$$[C_1]_0/[C_2]_0 \sim (R_{p_1})_0 k_{p_2}/(R_{p_2})_0 k_{p_1}. \tag{2.35}$$

The ratio is 1.2 at 298 K and 1.4 at 195 K. Therefore, the two active sites are comparable in numbers in support of the postulate that C_2 is derived from C_1.

The maximum values of $[C]$ are only ~ 61 and 27 μM at 298 and 195 K, respectively. This indicates that only 2.7–6.1% of the $Ti(OBu)_4$ is trans-formed to active sites for acetylene polymerization. This is, however, rather efficient utilization of Ti. In many Ziegler–Natta polymerizations less than 0.1% of the Ti participates in the initiation of polymerization.

The results of Figs. 2.16b and 2.17b showed that during the initial stage of polymerization at 298 K, $[MPB] \sim 3\,[C]$. In other words, on the average the active sites had undergone three chain transfers. During a similar period of polymerization at 195 K the chain-transfer efficiency is about seven.

The kinetics suggest that conversion of the more active species to the less active species may be written

$$\tag{2.36}$$

That is, C_1 and C_2 correspond to species **2** and **1**, respectively, postulated in Section 2.5.1.

2.8 COMPARISON OF VARIOUS CATALYST SYSTEMS

Simple cationic initiators can polymerize acetylene. Soga *et al.* (1982) obtained polyacetylene with the strong Lewis acid AsF_5 as the initiator at 198 to 75 K. The mechanism is probably

$$CH{\equiv}CH + AsF_5 \rightarrow HAsF_5^+CH{=}C^- \xrightarrow{n-C_2H_2} (C_2H_2)_nCH{=}C^- HAsF_5^+ \tag{2.37}$$

In this polymerization the product is simultaneously doped so that typically the polymer produced has a composition of $[CH(AsF_5)_{0.011}]_x$ with a $\sigma_{RT} = 4.8 \times 10^{-2}\ (\Omega\ cm)^{-1}$.

A number of transition metal acetylacetonates have been compared for acetylene polymerization (Kambara *et al.*, 1962). With $AlEt_3$ as the activator in all cases, titanium and vanadium complexes were found to be more active than the others: Ti, V $\gg$ Cr > Fe > Co. The main competing reaction to chain growth is the trimerization of acetylene to benzene. For example, the catalyst system $Ti(acac)_3 - AlEt_2Cl$ produces mostly benzene with traces of ethylbenzene and polyacetylene (Shirakawa and Ikeda, 1974). In comparison, the yield of linear polyacetylene is >99% with the $Ti(OBu)_4 - AlEt_3$ catalyst (Ikeda, 1967).

Other titanium-based catalysts have been examined for acetylene polymerization. $TiCl_4$ and $TiCl_3$ with aluminum alkyls polymerize acetylene with varying yields of benzene (Ikeda and Tamaki, 1966) and high polymers. The morphology of the polymer is largely globular (Table 2.1), and the conductivities of iodine-doped polymers are only 5×10^{-4} to 6×10^{-3} $(\Omega \text{ cm})^{-1}$. Furthermore, Deits *et al.* (1981) found that the $TiCl_4 - 1.2AlEt_3$ catalysts yielded polyacetylene with 0 to 7% cis structure at both 197 and 293 K. It is possible that this catalyst is too active and that local heating caused almost complete isomerization, melting of fibrils into globular morphology, and other side reactions causing low-doped conductivities (Table 2.1).

A particularly interesting catalyst is μ-($\eta^1 : \eta^5$-cyclopentadienyl)-1-tris(η-cyclopentadienyl)dititanium (Pez, 1976). It is prepared by the reduction of $(\pi\text{-}C_5H_5)_2TiCl_2$ with potassium naphthalide at low temperatures. The reaction times are 3 days at 193 K, followed by 2 days at 228 K and, finally, 16 hr at 250 K. The product is gray–black and pyrophoric. Crystalline derivatives containing two coordinated tetrahydrofurans were grown for X-ray structure study (Pez, 1976). A 3.3×10^{-4}-M hexane solution of this catalyst polymerizes acetylene to a gel product (Hsu *et al.*, 1978). The polymer has a predominantly trans structure obtained at room temperature; a cis-rich polyacetylene was produced at 193 K. The properties of undoped and doped polyacetylenes obtained with this catalyst are very similar to those obtained with the $Ti(OBu)_4 - 4AlEt_3$ system.

Tetrabenzyltitanium dissolved in tolune polymerizes acetylene very slowly at room temperature (Aldissi *et al.*, 1982b). The polymer has fibrillar morphology with an average diameter of about 200 nm.

Rare-earth coordination compounds had been used to polymerize acetylene (Cao and Qian, 1982). At ambient temperatures various rare-earth metal naphthanates – $AlEt_3$ catalysts gave polymers with 71 – 80% cis content. The corresponding isopropoxide complexes gave products with somewhat lower cis content.

Different catalysts based on molybdenum and tungsten have been used to polymerize acetylene. Equimolar mixtures of tungsten hexachloride,

WCl_6, or molybdenum pentachloride, $MoCl_5$, with tetraphenyltin, $SnPh_4$, aged at 303 K for 15 min, polymerize acetylene slowly at room temperature. The rate of polymerization with $WCl_6 - SnPh_4$ is faster than that of $MoCl_5 - SnPh_4$ but still very much slower than that of the $Ti(OBu)_4 - AlEt_3$ catalyst. The polyacetylenes have the trans structure with fibrillar morphology; the fibril diameter is 30 nm for the molybdenum catalyst and is exceedingly large at 12,000 nm with $WCl_6 - SnPh_4$. The polymers have σ_{RT} of $10 - 20$ $(\Omega \text{ cm})^{-1}$ when doped with I_2, SbF_5, or CF_3SO_3H.

$MoCl_5$, WCl_3, and $MoCl_4$ will polymerize acetylene in the absence of $SnPh_4$ (Voronkov *et al.*, 1980). The polymer obtained displays a strong carbonyl band at 1705 cm^{-1}, corresponding to one oxygen atom per six CH units. It has a very low decomposition temperature of $413 - 423$ K. By comparison, the free-standing polyacetylene film does not decompose significantly until heated above 600 K (Section 7.1.1). In the presence of trialkysilanes these catalysts cause hydropolymerization of acetylene to give small amounts of polyethylene according to ^{1}H NMR.

Water can be a cocatalyst for WCl_6 or $MoCl_5$ (Table 2.1). They gave trans-rich polyacetylenes with globular morphology and $\sigma_{RT} = 50 - 100$ $(\Omega \text{ cm})^{-1}$ for iodine-doped materials.

The mechanism of acetylene polymerization by group-VB metals may be different from that proposed for Ti above and Ni below. Woon and Farona (1974) had studied the polymerization of phenylacetylene with $ArM(CO)_3$ where Ar is arene and M = Cr, Mo, or W. At early stages of a slow polymerization with Ar = mesitylene and M = Cr or W, a ladder compound and poly(phenylacetylene) were both formed. The former is a light-yellow, low-melting solid of molecular weight 2030. It has an aliphatic C–H stretching band at about 2900 cm^{-1}; other IR bands, such as 620 cm^{-1} (δ, ring), and 960 and 1010 cm^{-1} (v, C—C), indicate the presence of fused four-membered rings. ^{1}H NMR spectra consisted of a broad multiplet ($\delta = 7.0 - 7.8$) for phenyl protons, a singlet ($\delta = 1.27$) for aliphatic protons. The two have an intensity ratio of 5 : 1. This evidence prompted the authors to propose a mechanism of metathesis catalysis.

Merriwether (1961) had investigated the polymerizations of a large number of mono- and disubstituted acetylenes by nickel – carbonyl – phosphine complexes. He proposed cis insertion of the triple bond as in Eq. (2.10). Daniels (1964) found $NiX_2 \cdot 2R_3P$ to be active in acetylene polymerization, where X = Br or I, and R = Ph or *n*-Bu. The polymerization can be carried out in ethanol, tetrahydrofuran, benzene, or acetonitrile. The yield of polymer and its crystallinity are both higher in ethanol than in THF. The highest yield obtained was 12.7 g polymer/g catalyst at 290 K. Neither NiX_2 nor R_3P alone will polymerize acetylene; $CoX_2 \cdot 2Ph_3P$ and $PdCl_2 \cdot 2(n\text{-}Bu_3P)$ are totally inactive. The mechanism of polymerization was proposed as

$$C_2H_2 + R_3P \longrightarrow \underset{\underset{Br}{|}}{\overset{\overset{R_3P}{|}}{NiBr}} \xrightarrow{-R_3P} R_3P \longrightarrow NiBr \xrightarrow{HC\equiv CH} R_3P \longrightarrow \underset{\underset{Br}{|}}{\overset{\overset{|}{|}}{NiCH}=CHBr} \tag{2.38}$$

$$\xrightarrow{C_2H_2} R_3P \longrightarrow \underset{\underset{Br}{|}}{Ni(CH=CH)_2Br}, \text{ etc.}$$

Luttinger (1962) found that the tris(2-cyanoethyl)phosphine complex of nickel chloride together with $NaBH_4$ polymerizes acetylene in either ethanol, acetonitrile, or water. The high molecular weight polymer has the trans structure. Also formed were brown-colored amorphous products that may be related to the random acetylene polymer cuprene but believed to be largely linear. The polymer was found to contain large amounts of Ni and B in a 2:1 molar ratio, which can be removed by washing with hydrochloric acid.

A cobalt-based catalyst was discovered by Luttinger (1960). It is an ethanol solution of $Co(NO_3)_2$ (1 wt %) and of $NaBH_4$ (0.04 wt %) mixed with ether (Section 2.3). Polymerization of acetylene can be carried out over a wide temperature range. The polymer also contains catalyst residue, which can be removed by methanolic HCl. The morphology of this polymer will be discussed in detail in Section 4.4.

Haberkorn *et al.* (1982) reported a comparison of the structures of polyacetylene obtained with different catalysts. They used high concentrations of $Ti(OBu)_4 - 4AlEt_3$ to obtain $ZN-(CH)_x-(F)$ and a dilute solution of this catalyst to produce $ZN-(CH)_x-(P)$, both at 195 K, where ZN denotes Ziegler–Natta catalyst. Luttinger catalyst was used to give $L-(CH)_x-(P)$ (L for Luttinger catalyst), the $TiCl_4-AlEt_3$ catalyst was used by Kambara *et al.* (1962) to form $K-(CH)_x-(P)$ (K for Kambara catalyst), and cuprene was synthesized by Reppe's method (1948). The products were compared for cis–trans content by infrared spectroscopy (IR), fraction of sp^3 carbon atoms ($f\text{-}sp^3$) by solid ^{13}C NMR, percent crystallinity (W_c) by X-ray diffraction (WAXS), and σ_{RT} by doping to saturation with iodine. The results are summarized in Table 2.10.

The IR spectra of $ZN-(CH)_x$ are the same as reported by other investigators. (Section 6.1.2). But in $L-(CH)_x$ and $K-(CH)_x$ there were strong absorptions between 2900 and 3000 cm^{-1} due to C–H stretching vibrations of CH_2 and CH_3 groups, and the absorption bands of CH_3 groups ($\delta\text{-}CH_3$) at about 1375 and 1450 cm^{-1} were clearly present.

The ^{13}C NMR spectrum of $ZN-(CH)_x$ has only one peak, 40 Hz in width, found at 126.87 ppm for cis and 136.49 ppm for trans units. In the

TABLE 2.10

Structure, Crystallinity, and Conductivity of Polyacetylene Obtained with Various Catalysts[a]

Polymer	IR, (% cis)	^{13}C NMR $f\text{-}sp^3$ (%)	W_c, (%)	$\sigma_{RT}[CHI_y]_x$, $[(\Omega \text{ cm})^{-1}]$
ZN-$(CH)_x$-(F)	95	<2	77	14.5
ZN-$(CH)_x$-(P)	95	<2	76	5.5
L-$(CH)_x$-(P)	50	10-12	65	0.5
Cuprene-(P)	5	25	49	1.3×10^{-4}
K-$(CH)_x$-(P)	20	40	23	2.5×10^{-5}

[a] After Haberkorn *et al.* (1982).

case of L-$(CH)_x$ both resonances were observed, and the trans signal was asymmetric and was shifted to 135.61 ppm. The reason for this shift was proposed to be either due to short or isolated trans sequences or to the existence of trans-transoid forms. In addition, there were observed resonances at 33.4 and 16.4 ppm, attributable to sp^3 carbon atoms. The spectrum of K-$(CH)_x$ resembles that of L-$(CH)_x$. The trans resonance was shifted to 135.8–134.8 ppm. In the aliphatic carbon region there was a very broad and intense signal at about 46 ppm, and superimposed on it were three narrow peaks at 14, 33, and 66 ppm, assigned respectively to $CH_3(CH_2)_x$, CH_3, and/or $CH_2C{=}C$, and CHOH, CCH_2OH, or CH_2OR groups.

There appeared to be a strong dependence of W_c on $f\text{-}sp^3$, as shown in Fig. 2.28, which suggests that if $f\text{-}sp^3 \geqslant 0.5$, polyacetylene would be completely amorphous. The conductivity of iodine-doped polymer is sensitive to both the degree of crystallinity and saturated carbons. Figure 2.29 shows

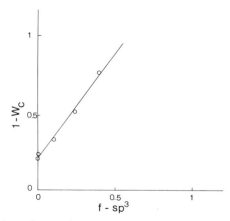

Fig. 2.28 Variation of amorphous content $(1 - W_c)$ with $f\text{-}sp^3$ content for different polyacetylene samples. [After Haberkorn *et al.* (1982).]

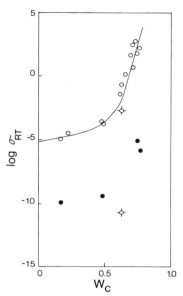

Fig. 2.29 Relationship between conductivity and degree of crystallinity for: (●) undoped *trans*-polyacetylene; (○) saturation doped $(CH_y)_x$. [After Haberkorn *et al.* (1982).]

conductivity to increase with W_c only if the crystallinity is $\geqslant 50\%$. One discrepancy noted with $L–(CH)_x$ is that, whereas both IR and NMR gave a cis to trans ratio of about 1 : 1, the WAXS diagram showed only cis reflections and a crystallinity of 65%.

The very high values of *f-sp³* for some of the polyacetylenes are very surprising. Aside from $L–(CH)_x$, which is initiated by a catalyst containing $NaBH_4$, the catalysts for $K–(CH)_x$ and cuprene do not contain sources of hydrogen. One wonders whether the solid ^{13}C NMR technique has been properly calibrated for quantitative measurements. For example, the nuclear Overhauser enhancement factors for *sp²* and *sp³* carbon atoms may be different and should be separately determined. Also the Hartmann–Hahn matching conditions for cross polarization may not be simultaneously satisfied for the different types of carbon atoms. If the *sp³* contents given in Table 2.10 are correct and they are randomly distributed, then $L–(CH)_x$ should be lighter in color than the black polymer usually obtained. As far as $K–(CH)_x$ is concerned, the polymer should be white because it contains only 60% unsaturated carbon atoms according to the ^{13}C NMR results. More measurements should be made on these polymers—such as EPR, magnetic susceptibility, pyrolysis GC–MS, XPS, and other techniques.

The very high values for *f-sp³* can have three origins. First, the $\overline{M}_n$ of the polymer is low. The Et end group may cause a significant *f-sp³*. Second, there may be extensive branching resulting from the following processes:

$$Ti(CH=CH)_aEt \longrightarrow TiH + CH\equiv C(CH=CH)_{a-1}Et$$
$$TiH + AlEt \longrightarrow TiEt + AlH \cdot$$

$$(2.39)$$

$$TiEt + CH\equiv C(CH=CH)_{a-1}Et \longrightarrow TiCH=\overset{\overset{\displaystyle Et}{|}}{C}(CH=CH)_{a-1}Et$$

Short chain branches can be formed by intramolecular rearrangement:

$$(2.40)$$

The presence of low molecular weight polyacetylene could lead to

$$TiEt + b(CH=CH) \longrightarrow Ti(CH=Ch)_bEt$$
$$Ti(CH=CH)_bEt + CH\equiv C(CH=\text{⇌} \text{|||}_{a-1}Et \longrightarrow TiCHC(CH=CH)_bEt$$

$$(2.41)$$

where a and b are small integers. Finally, crosslinking reactions can also contribute toward sp^3 carbon atoms; they are discussed in Section 7.6.

2.9 GRAFT AND BLOCK COPOLYMERS

Graft copolymers have been shown to exhibit properties that can be quite unlike those of homopolymer blends, such as elevated critical point and microphase domain separations. Bates and Baker (1983a) have polymerized acetylene with the $Ti(OBu)_4 - AlEt_3$ catalyst in a toluene solution of carrier polymers. Two carrier polymers were employed: polyisoprene (80% cis 1-4, 15% trans 1-4, 15% 3-4) and polystyrene. Both polymers are nearly monodisperse with molecular weight of 2×10^5 and were modified through oxidation to form electrophilic sites. The polymerizations were performed at 195 and 298 K in the presence of polyisoprene and polystyrene, respectively, to give intensely maroon and blue solutions.

The graft reaction has been attributed to the nucleophilic attack of the Ti terminus of the growing polyacetylene chain on electrophilic sites on the carrier polymer. The graft copolymer was separated from residual homopolymers, which should comprise both ungrafted carrier polymer and free polyacetylene. The former is present because an excess of carrier polymer was used during acetylene polymerization, which also suggests that the graft copolymers contain no more than one polyacetylene branch per carrier polymer chain. Based on the infrared absorptivities of the graft copolymer and the known molecular weight of the carrier polymer, the molecular weights of the polyacetylene block are estimated to be between 3×10^4 and 6×10^4. However, this estimate is based on one graft per carrier polymer molecule. The actual value may be lower if there is more than one graft chain on some of the carrier polymer molecules or the removal of homopolyacetylene is incomplete. The molecular weight may be higher than the estimate if there is less than one graft chain per carrier polymer molecule or incomplete removal of the ungrafted carrier polymer. A transmission electron micrograph showed polyacetylene microdomains of 10–15 nm diameter for OsO_4 stained poly(styrene-*g*-acetylene).

The visible absorption peaks in both the solution and film of the graft copolymers are shifted to higher energy as compared to the usual *trans*-polyacetylenes by about 0.25 eV. This shift has been attributed to some bond rotation in the solution and the presence of amorphous polyacetylene blocks trapped in the microdomains of the graft copolymer films.

Diblock copolymers of styrene and acetylene have been synthesized (Bates and Baker, 1983b). The polystyrene block is probably prepared by living anionic polymerization to a molecular weight of 1×10^5. It has a single terminal methyl ketone functionality presumably introduced by terminating the polymerization with a reagent such as acetyl chloride. The polystyrene was used as the carrier polymer for acetylene polymerization at 195 K initiated by $Ti(OBu)_4 - AlEt_3$. Terminal graft is assumed to occur as described above to give poly(styrene-di-*b*-acetylene). The length of the polyacetylene block was estimated to be about 6×10^4. However, the value is subject to qualifications similar to those in the case of graft copolymers.

The synthesis of soluble graft copolymers of polyacetylene suggests that the entire growing polyacetylene chain may exist for a short time in the dissolved state prior to crystallization. On the other hand, since the diblock copolymer of styrene and acetylene have crystal structure entirely different from that of the normal *cis*-polyacetylene (Section 3.9), the growing polyacetylene chains apparently crystallize before the proposed equilibrium conformation of helical *cis*-cisoid is attained.

Chapter 3

Structures

3.1 INTRODUCTION

The structures of small molecules are well defined by the atomic composition and the bond lengths and angles between the atoms. The structures of a polymer are more complicated and in some aspects not precisely defined. Acetylene can be polymerized to cis or trans structures. It was pointed out in Section 2.5.2 that the propagation proceeds via cis insertion of the monomer to give a cis-transoid structure for the polymer. However, the other isomeric structures have energies very close to those of the cis-transoid type (Section 3.2). This is one of the reasons for the ease of isomerization of *cis*-polyacetylene to the trans polymer (Chapter 5).

Polyacetylene has been said to be highly crystalline. This implies a regular backbone structure (Section 3.3). However, all crystalline organic polymers contain some amorphous phase. In Section 3.4 we describe the wide angle X-ray diffraction results analyzed according to the method of Ruland. The free-standing cis films are about 76–84% crystalline; the crystallinity decreases slightly to 71–79% upon isomerization. Polyacetylene gel has a much lower degree of crystallinity and undergoes significant further loss of order upon cis–trans isomerization (Section 3.4).

The crystal structures of polyacetylene are, of course, of paramount interest. Very few X-ray reflections can be observed from unoriented film. With the help of packing analysis, Baughman *et al.* (1978) proposed a crystal structure for *cis*-polyacetylene. We were able to carry out electron-diffraction measurements on aligned fibrils and assign the chain-axis direction in the unit cell. However, the limited number of reflections leaves uncertain the precise crystal structures for polyacetylenes.

Finally, estimates for bond alternation are presented, and the differences obtained with X-ray and electron diffraction techniques are noted.

3.2 GEOMETRIC ISOMERISM

Four energetically nonequivalent backbone geometries can be envisioned for polyacetylene, three of which are shown below:

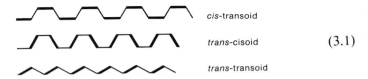

$$cis\text{-transoid}$$
$$trans\text{-cisoid}$$ $$(3.1)$$
$$trans\text{-transoid}$$

The fourth, cis-cisoid, has been thought to be sterically too unfavorable to warrant the calculation of its energy. However, diffraction data on lamallar crystals of poly(styrene-di *b*-acetylene) are consistent with a helical cis-cisoid structure for the polyacetylene chains in the diblock copolymers. In the above structures, the heavy lines represent short bonds (slightly longer than the C=C bond in ethylene) and the light lines denote long bonds (somewhat shorter than the C—C bond in ethane). This representation will be used throughout this book.

The term *cis*-polyacetylene usually refers to the polymer prepared and stored at < 195 K, with minimum handling in a drybox, and used within a week of preparation. However, even brief handling at room temperature during sample preparation can result in a certain degree of isomerization. Therefore, such materials do not have 100% cis-transoid units and are merely cis rich. A sample is said to be *cis*-polyacetylene, herein, if its cis content is ⩾ 88%.

The cis–trans content in polyacetylene can be varied by the conditions of polymerization. For instance, high Al/Ti ratios for the catalyst and low temperatures favor the formation of *cis*-polyacetylene; the converse leads to high trans-content materials (Shirakawa and Ikeda, 1971). Essentially pure trans polymer can be obtained by thermal, chemical, or electrochemical isomerization of the cis polymer, which is discussed in Chapter 5.

It is important to note that the cis-transoid and trans-cisoid types are distinct structures; they are not related by any symmetry operations. On the other hand, the two structures shown below for trans-transoid **A** and **B** are related by a screw-axis symmetry and are identical in energy:

$$
\text{[waveform diagram A]} \quad \text{A}
$$
$$
\text{[waveform diagram B]} \quad \text{B}
$$
(3.2)

One may consider the trans-transoid to have a doubly degenerate ground state if the end groups are ignored. In reality, a polyacetylene chain most probably has an ethyl group at one end and a Ti or Al atom at the other. The neglect of end groups is generally harmless if the degree of polymerization is large, except when we consider the question of the intrinsic carriers. Cis and trans polymers usually differ in physical and mechanical properties. For example, *cis*-poly(butadiene-1,4) is amorphous and an elastomer, whereas the trans polymer is crystalline. Nevertheless, the differences in various electronic and spectroscopic properties of *cis*- and *trans*-polyacetylene are truly remarkable and fascinating.

The ease of isomerization suggests small energy differences between the three isomeric structures in Eq. (3.1) (Brédas, 1979). Karpfen and Höller (1981) carried out ab initio calculations on several cis and trans isomers of polyacetylene, using a minimal STO-3G basis set. Bond distances and bond angles were optimized to obtain the equilibrium structures. The results are summarized in Table 3.1. There are very small differences between the bond lengths of the various isomers; the deviation of the

$$
\text{C}\overset{\text{C}}{\diagup\diagdown}\text{C}
$$

bond angles from the idealized 120° value are more pronounced. The degree of bond alternation by this calculation is estimated to be 0.16 Å, which is much larger than the experimental results (Section 3.7). This is a known shortcoming of the STO-3G basis set.

The relative energies of the isomers of polyacetylene, referred to zero for the bond-alternating trans-transoid form, are given in Table 3.2 (Karpfen

TABLE 3.1

Equilibrium Structure of Polyacetylene Isomers[a]

	Bond distance (Å)			Bond angle (deg)	
Structure	C—C short	C—C long	C—H	C—C—C	C—C—H
Trans-transoid	1.327	1.477	1.085	124.2	119.5
Cis-transoid	1.329	1.480	1.083	126.8	117.3
Trans-cisoid	1.325	1.489	1.084	126.3	114.2

[a] Karpfen and Höller (1981); Karpfen and Petkov (1979a,b).

TABLE 3.2

Relative Energies of Polyacetylene Isomers

Structure	Skeleton	Energy (kcal mole^{-1})
Trans-transoid	Alternant bonds	0
Trans	Equidistant bonds	7.3
Cis-transoid	Alternant bonds	1.9
Trans-cisoid	Alternant bonds	2.1
Cis	Equidistant bonds	7.2

and Höller, 1981). Also included are the uniform bond-length structures. The most stable structure is the bond alternant trans-transoid, which is about 2 kcal mole^{-1} more stable than the cis isomers. The cis-transoid is slightly lower in energy than the trans-cisoid; the difference is not thought to be significant. However, Yamabe *et al.* (1979) estimated energies of cis-transoid and trans-cisoid to be 1.9 and 4.3 kcal mole^{-1}, respectively, which makes the former quite a bit more stable than the latter.

For the trans isomer the uniform bond-length structure is less stable than the trans-transoid by about 7 kcal mole^{-1}. It is well known that Hartree–Fock instabilities occur for the former structure. The symmetry-broken bond order and length-alternation structure is necessarily lower in energy.

The cis-transoid and trans-cisoid structures constitute local mimima on the energy hypersurface. The uniform bond-length cis structure lies on the top of the barrier of this unsymmetrical double minimum potential and must be relatively higher in energy. The extent of this destabilization is comparable to that for the trans case.

3.3   CHAIN REGULARITY

The polyacetylene as polymerized at low temperatures has nearly a perfect cis-transoid structure. If this were not so and a chain is interrupted by the other isomeric units, there would be unpaired spins formed at the point of structural alteration, i.e.,

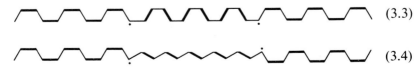

$$(3.3)$$

$$(3.4)$$

Chien *et al.* (1980a) showed that *cis*-polyacetylene prepared by direct polymerization in an EPR sample tube at 195 K is free of a $g = 2$ signal. (Fig. 2.6b). Because EPR is a very sensitive technique, a nearly perfect *cis*-transoid structure may be assumed for this substance. However, by simply warming up the specimen or by exposing it to air, a $g = 2$ signal with ~ 6 *G* linewidth immediately appeared (Figs. 2.6c and d). Isomerization must have occurred even under these conditions; this is expected in view of the very small energy differences between the different isomeric structures (Table 3.2).

One often finds in articles on polyacetylene statements that the polymer is cross-linked to explain changes in some physical properties. There is no concrete evidence substantiating the presence of significant numbers of cross-links in polyacetylene. On the contrary, there is much evidence against it. However, cross-linking does occur during pyrolysis at high temperatures. The subject of cross-linking will be fully discussed in Sections 7.1 and 7.6. Another type of chain irregularity is chain branches. Those catalysts that appear to produce branched polyacetylenes and the probable mechanisms are discussed in Section 2.8.

The high crystallinity of polyacetylene obtained with the $Ti(OBu)_4$– AEt_3 catalysts indicates that this polymer contains very few branches or crosslinks, if any. Experimentally, it would be difficult to detect the chain irregularities if they are present in small numbers. Finally, it will be shown that even frequent interruption of the weakly alternating conjugated backbone by strongly alternating units does not seem to affect significantly the transport properties of polyacetylene, as in the copolymers of acetylene and methylacetylene (Section 11.3.10).

3.4   DEGREE OF CRYSTALLINITY

An organic polymer can not be completely crystalline. Even single crystals of polyethylene with lamellar morphology contain amorphous materials present as chain folds and cilia. Therefore, a crystalline polymer is always a semicrystalline substance comprising crystalline, paracrystalline, and amorphous regions. There are several ways to determine the degree of crystallinity. One of these is the detailed analysis of the X-ray diffraction data, using the method of Ruland (1961). The observed total scattering intensity from the crystalline and the amorphous regions, $I(s)$, is plotted in the form of $s^2 I(s)$ versus s. The apparent crystalline scattering intensity, $I_c(s)$, is obtained by subtracting from $I(s)$ the amorphous scattering $I_a(s)$. We define

$$X' = \int_{s_1}^{s_2} s^2 I_c(s)\ ds \bigg/ \int_{s_1}^{s_2} s^2 I(s)\ ds \tag{3.5}$$

and

$$K = \int_{s_1}^{s_2} s^2 \langle F^2 \rangle\ ds \bigg/ \int_{s_1}^{s_2} s^2 \langle F^2 \rangle\ \exp[-ks^2]\ ds, \tag{3.6}$$

where $\langle F^2 \rangle$ is the monomer unit structure factor, which is calculated from atomic structure factors, and k is the disorder parameter, which is a constant characteristic of a given polymer sample. The degree of crystallinity X_c is given by

$$X_c = KX'. \tag{3.7}$$

The value of this analysis is that it determines not only the degree of crystallinity but also how perfectly ordered the crystalline regions are. A small value of k corresponds to highly ordered crystalline regions; a large value of k indicates disordered crystalline material.

A single film of polyacetylene is not sufficiently thick for accurate measurement of scattering intensities. Thus several films from the same preparation had to be stacked together for wide angle X-ray diffraction

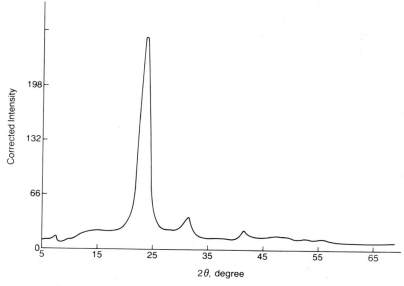

Fig. 3.1 Wide-angle X-ray diffraction patterns for *cis*-polyacetylene film.

experiments. The diffraction patterns for the *cis*- and *trans*-polyacetylenes are shown in Figs. 3.1 and 3.2, respectively. The corresponding Ruland plots are given in Figs. 3.3 and 3.4. Table 3.3 gives the results of Ruland analysis; the correct degree of crystallinity is the one that is independent of the interval of *s* for integration in Eqs. (3.5) and (3.6). In the case of *cis*-polyacetylene, the standard deviation in X_c is 2.2% for both $k = 5$ and 6. The polymer is deduced to be about 76–84% crystalline.

The results for *trans*-polyacetylene showed the materials to be 71–79% crystalline with a disorder factor of 5–6. Akaishi *et al.* (1980) had performed similar analyses for *trans*-polyacetylene and reported values of 81% crystallinity and $k = 5$. The trans polymer of Fig. 3.2 was isomerized from cis by heating in vacuo at 473 K for 2 hr. There is reason to believe now that this isomerization condition may not be an optimal one. It has been shown by electron diffraction (Sections 3.6 and 3.7) that improved isomerization conditions produce *trans*-polyacetylene, which is more highly crystalline than the starting cis polymer. However, this result has yet to be confirmed by X-ray diffraction.

Degree of crystallinity was also determined for polyacetylene gel, which was obtained by the procedure given in section 2.2.7. The X-ray diffraction patterns and Ruland plots for the *cis*- and *trans*-polyacetylene gels are given in Figs. 3.5–3.8 and the results of the Ruland analysis are summarized in Table 3.4. The polyacetylene gels have fibrillar morphology, as in the

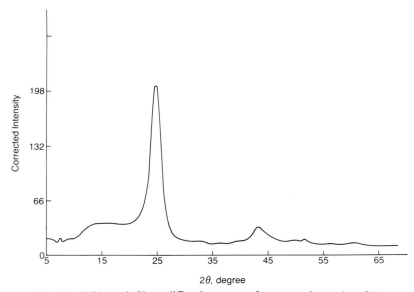

Fig. 3.2   Wide-angle X-ray diffraction pattern for *trans*-polyacetylene film.

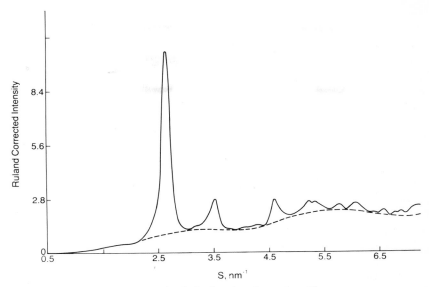

Fig. 3.3   Ruland plot for *cis*-polyacetylene film.

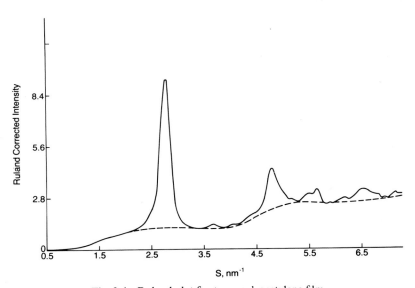

Fig. 3.4   Ruland plot for *trans*-polyacetylene film.

TABLE 3.3

Ruland Analysis of X-Ray Diffraction Data for *cis*- and *trans*-Polyacetylene Films

| | | | | X_c | | | | | | | | | | | | |
| | | | | $k = 1$ | | $k = 2$ | | $k = 3$ | | $k = 4$ | | $k = 5$ | | $k = 6$ | |
Polymer	s interval	$I(s)$	$I_c(s)$	K	X_c	K	X_c	K	X_c	K	X_c	K	X_c	K	X_c
cis-(CH)$_x$	0.1–0.3	0.465	0.282	1.05	0.64	1.05	0.64	1.21	0.74	1.21	0.74	1.30	0.79	1.38	0.84
	0.1–0.4	0.690	0.347	1.08	0.54	1.12	0.56	1.32	0.66	1.38	0.69	1.51	0.76	1.65	0.82
	0.1–0.5	0.990	0.397	1.12	0.47	1.23	0.52	1.46	0.61	1.60	0.67	1.80	0.76	2.00	0.84
	0.1–0.6	1.290	0.442	1.18	0.40	1.35	0.46	1.65	0.56	1.87	0.64	2.15	0.73	2.44	0.83
	0.1–0.7	1.610	0.489	1.24	0.37	1.48	0.44	1.87	0.56	2.19	0.66	2.56	0.77	2.95	0.88
												Av 0.76 ± 0.022^a		Av 0.84 ± 0.022^b	
trans-(CH)$_x$	0.1–0.3	0.525	0.303	1.05	0.61	1.05	0.61	1.21	0.70	1.21	0.70	1.30	0.75	1.38	0.80
	0.1–0.4	0.700	0.318	1.08	0.49	1.12	0.50	1.32	0.59	1.38	0.62	1.51	0.68	1.65	0.74
	0.1–0.5	1.054	0.420	1.12	0.45	1.23	0.49	1.46	0.58	1.60	0.64	1.80	0.72	2.00	0.80
	0.1–0.6	1.444	0.469	1.18	0.38	1.35	0.43	1.65	0.53	1.87	0.60	2.15	0.69	2.44	0.78
	0.1–0.7	1.852	0.513	1.24	0.35	1.48	0.41	1.87	0.52	2.19	0.61	2.56	0.72	2.95	0.83
												Av 0.71 ± 0.028^a		Av 0.79 ± 0.03^b	

[a] Average X_c for $k = 5$.
[b] Average X_c for $k = 6$.

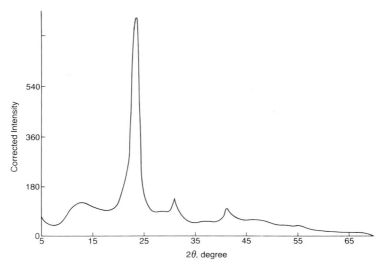

Fig. 3.5 Wide angle X-ray diffraction pattern for *cis*-polyacetylene gel.

free-standing film. However, the fibrillar diameters are much greater than they are in the film. The X-ray diffraction results show the cis polymer to be less crystalline in the gel state than it is for the film. Even more striking is the reduction of X_c from 0.76 to 0.48 upon isomerization at 423 K. These

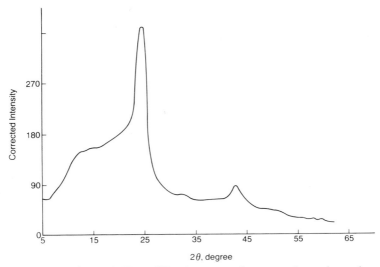

Fig. 3.6 Wide angle X-ray diffraction pattern for *trans*-polyacetylene gel.

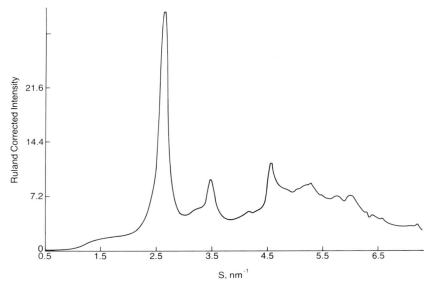

Fig. 3.7   Ruland plot for *cis*-polyacetylene gel.

differences are not unexpected. The gels were obtained in the reaction region where chain-transfer and termination reactions are more probable than the polymerization on the reactor wall. The fibrils of the gel appeared to be swollen by the solvent and may contain more amorphous materials. In contrast, the polymerization occurring on the wall of the reaction vessel is

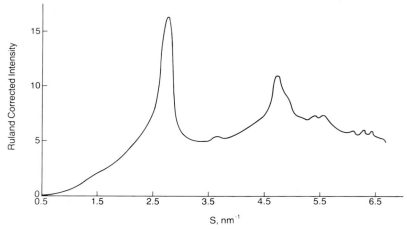

Fig. 3.8   Ruland plots for *trans*-polyacetylene gel.

TABLE 3.4

Ruland Analysis of X-Ray Diffraction for *cis*- and *trans*-Polyacetylene Gel

Polymer	s interval	$I(s)$	$I_c(s)$	k = 1		k = 2		k = 3		k = 4		k = 5		k = 6	
				K	X_c	K	X_c	K	X_c	K	X_c	K	X_c	K	X_c
cis-(CH)$_x$	0.1–0.3	0.625	0.363	1.05	0.61	1.05	0.61	1.21	0.70	1.21	0.70	1.30	0.75	1.38	0.80
	0.1–0.4	0.930	0.420	1.08	0.49	1.12	0.51	1.32	0.60	1.38	0.62	1.51	0.68	1.65	0.74
	0.1–0.5	1.33	0.486	1.12	0.41	1.23	0.45	1.46	0.53	1.60	0.58	1.80	0.66	2.00	0.73
	0.1–0.6	1.76	0.523	1.18	0.37	1.35	0.40	1.65	0.49	1.82	0.54	2.15	0.64	2.14	0.73
	0.1–0.7	2.01	0.544	1.24	0.34	1.48	0.40	1.87	0.51	2.19	0.59	2.56	0.69	2.95	0.80
												Av 0.68 ± 0.04[a]		0.76 ± 0.04[b]	
trans-(CH)$_x$	0.1–0.3	0.85	0.330	1.05	0.41	1.05	0.41	0.21	0.47	1.21	0.47	1.30	0.50	1.38	0.54
	0.1–0.4	1.27	0.348	1.08	0.30	0.12	0.31	1.32	0.36	0.38	0.38	1.51	0.41	1.65	0.45
	0.1–0.5	1.90	0.437	1.12	0.26	0.23	0.28	1.46	0.34	0.37	1.60	1.80	0.41	2.00	0.46
	0.1–0.6	2.44	0.461	1.18	0.22	0.35	0.26	1.65	0.31	0.34	1.82	2.15	0.41	2.44	0.46
												Av 0.43 ± 0.05[a]		0.48 ± 0.06[b]	

[a] Average X_c for k = 5.
[b] Average X_c for k = 6.

heterogeneous, the catalyst is probably deposited as clusters so that more orderly crystallization can take place simultaneously with chain growth.

3.5   CRYSTAL STRUCTURE OF *cis*-POLYACETYLENE

It must be pointed out at the outset that different diffraction investigations had led to dissimilar crystal structures for polyacetylenes. In principle, the precise crystal structure can be determined by diffraction techniques if there are a sufficient number of reflections. In the case of polyacetylenes,

TABLE 3.5

Indices for Reflections for Pristine *cis*-Polyacetylene

E.D.[a] Aligned fibrils		X-ray[b] Randomly oriented film		E.D.[c] Polymer suspension	
d(Å)	(*hkl*)	*d*(Å)	(*hlk*)[d]	*d*(Å)	(*hkl*)
				4.47	(001)
3.84	(200)	3.80	(200)	3.85	(200)
	(110)		(101)	3.78	(110)
3.06	(011)	3.13	(011)		
2.92	(201)				
2.89	(210)	2.87	(201)	2.89	(210)
		2.41	(211)		
	(020)				
2.23	(120)			2.21	(310)
	(310)				
2.19	(002)	2.19	(002)	2.16	(020)
		2.11	(102)	2.08	(120)
1.98	(220)	2.00	(311)		
	(400)				
		1.90	(202)	1.89	(220)
1.70	(320)	1.74	(401)	1.77	(410)
	(410)				
		1.65	(411)	1.66	(320)
1.42	(130)(230)				
	(420)(510)				
1.28	(330)(520)			1.26	(520)
1.12	(004)				

[a] Chien *et al.* (1982a,e).

[b] Baughman *et al.* (1978).

[c] Lieser *et al.* (1980a).

[d] (*hlk*) instead of (*hkl*) because in Baughman's model the **b**-axis was assumed to be the chain axis.

fewer than ten reflections can be observed. This is further complicated by the fact that the specimens are randomly oriented, resulting in some instances in erroneous assignment of the chain-axis direction. Finally, the crystallites are formed during the course of chain propagation. Therefore, the atomic arrangement in the crystallites may not be the thermodynamically most favorable one. This consideration renders packing arrangement analysis an interesting exercise but not necessarily relevant to the crystal structure of as-polymerized polyacetylene.

There have been three studies on the crystal structures of *cis*-polyacetylene, one by X-ray diffraction (Baughman *et al.*, 1978), and two by electron diffraction (Chien *et al.*, 1982a, e; Lieser *et al.*, 1980a). Table 3.5 summarizes the indexing of reflections and Table 3.6 the unit cell parameters arrived at in the three studies.

Let us first present our electron diffraction results because they were obtained on oriented specimens. It is usually not possible to obtain single crystals of most organic polymers. If the polymer sample contains randomly oriented crystallites, only Debye rings are obtained in diffraction experiments. Very little structural information can be derived from this. Therefore, it is customary to draw the polymer into fibers from which fiber diffraction patterns can be obtained. In fact, the crystal structures of most stereoregular poly(α-olefins) and that of DNA were determined from their fiber diffraction patterns. In the course of our study of the nascent morphology of polyacetylene with minimal exposure to environment and specimen

TABLE 3.6

Unit Cell Parameters for *cis*-Polyacetylene

	E.D.[a] aligned fibrils	X-ray[b] randomly oriented film	E.D.[c] polymer suspension
Lattice type	Orthorhombic	Orthorhombic	Orthorhombic
a(Å)	7.68	7.61	7.74
b(Å)	4.46	4.47	4.32
c(Å)	4.38[d]	4.39	4.47
ρ(g cm^{-3})	1.15	1.16	1.16
C—C—C	125°	127°	
setting angle ϕ	32°[e]	59°	

[a] Chien *et al.* (1982a,e).

[b] Baughman *et al.* (1978) took the **b** to be along the molecular axis.

[c] Lieser *et al.* (1980a).

[d] Molecular chain axis and fiber axis.

[e] Value for the cis-transoid structure.

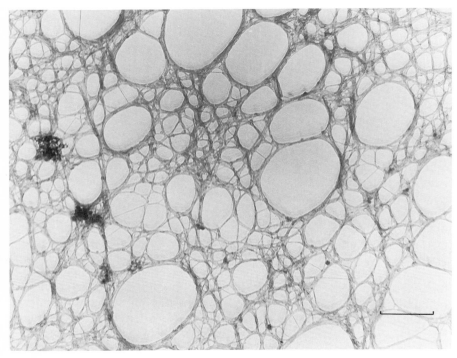

Fig. 3.9   Electron micrograph of nascent *cis*-polyacetylene. Marker is 500 nm.

handling (Karasz *et al.*, 1979), it was found that very thin films of polyacetylene can be directly polymerized onto EM grids. Probably as the result of
repeated washing and solvent evacuation, regions with aligned bundles of
fibrils were formed (Fig. 3.9). An electron beam focused on this selected
region gave the fiber electron diffraction pattern shown in Fig. 3.10 (Chien *et
al.* 1982a, e). There are also mats of polyacetylene fibrils lying randomly on
top of each other in these specimens, which give rise to Debye rings.

Figure 3.11a is a schematic representation of the fiber electron diffraction pattern. The data require that the polymer chain have a 2_1 screw axis in
the primitive cell, the reciprocal lattice of which is given in Fig. 3.11c. All the
observed reflections fall on the reciprocal lattice sites. The unit cell of
cis-polyacetylene is orthorhombic with *Pnam* space group (Table 3.7). The
cell parameters are given in column 2 of Table 3.6. The pattern showed
unequivocally that **c** is the molecular axis, which is along the fiber axis. With
one CH per asymmetric unit, there are eight CH units per unit cell,
corresponding to a calculated density of 1.15.

Fig. 3.10   Electron diffraction patterns of aligned fibrils of nascent *cis*-polyacetylene.

   The observed intensities of the equatorial reflections are shown in Fig. 3.12; the first one, (110) + (200), is much more intense than all the others. The theoretical equatorial intensities were calculated for both the cis-transoid and trans-cisoid structures, using the following parameters: C=C, 1.35 Å; C—C, 1.46 Å; C—H, 1.09 Å; and repeat distance, 4.38 Å (Fig. 3.13). The structural factors of (*hk*0) reflections were calculated as a function of the setting angle ϕ. In cases of overlapping reflections, their calcu-

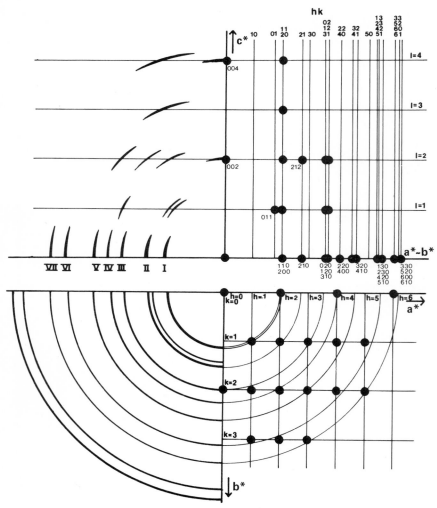

Fig. 3.11 Counter-clockwise starting at the upper left quadrant: (a) schematic representation of the electron diffraction of oriented *cis*-polacetylene fibrils; (b) schematic representation of the Debye rings of randomly packed regions; (c) lattice for the **a*b*** plane, (●) observed reflections; (d) projection of the reciprocal lattice of the first five layer **a*b*** planes.

lated intensities are summed. The setting angle is the one that gives the best reliability factor, R (Fig. 3.14):

$$R = \sum_{r=2}^{7} (|F_{obsd}| - |F_{calcd}|) \bigg/ \sum_{r=2}^{7} |F_{obsd}| \tag{3.8}$$

TABLE 3.7

Reflections for *cis*-Polyacetylene

	Reflections		
	Present	Absent	
(*hkl*)	(111)(112)(113) (114)(121)(122) (212)(311) . . . all		*P*
(0*kl*)	(011)(022)? $k + l$ = even	(012)(021)? $k + l$ = odd	*n* (⊥**a**)
(*h*0*l*)	(201)(202)(203) h = even	(101)(102) (301)(302) h = odd	*a* (⊥**b**)
(*hk*0)	(110)(120)(210) (220)(310)(420) (510) . . . all		*m* (⊥**c**)
(*h*00)	(200)(400)	(100)(300)	
(0*k*0)	(020)(040)	(010)(030)	
(00*l*)	(002)(004)	(001)(003)	

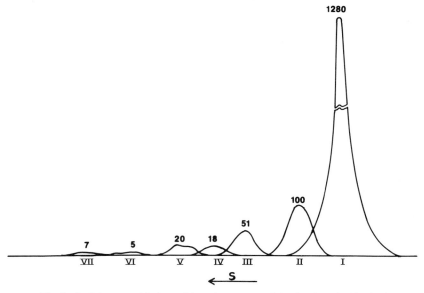

Fig. 3.12   Photographic intensities of the equatorial reflections in Fig. 3.10.

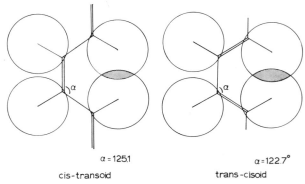

α = 125.1   α = 122.7°

cis-transoid   trans-cisoid

Fig. 3.13   Atomic parameter used to calculate reflection intensities.

where r is the reflection number, F_{obsd} is proportional to $(I_{obsd} \sin 2\theta)^{1/2}$, I is the intensity, and the sum is over equatorial reflections 2 to 7. The first equatorial reflection is omitted because its very strong intensity tends to dominate the calculated intensities leading to a shallow minimum in the R versus ϕ plot (Fig. 3.14). The agreement between observed and calculated intensities is shown in Table 3.8.

Figure 3.15 shows the molecular conformation of *cis*-polyacetylene in

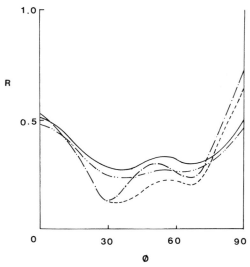

Fig. 3.14   Variation of the reliability factor with setting angle ϕ: (——) cis-transoid for reflections I to VII; (— · — · —) cis-transoid reflections II to VII; (— · · — · · — · ·) trans-cisoid for reflections I to VII; (-----) trans-cisoid for reflections II to VII.

TABLE 3.8

Observed and Calculated Electron Diffraction Intensities for Pristine *cis*-Polyacetylene

Spacing		Reflection	Observed	Calculated intensity	
In Å	Direction[a]	(h k l)	intensity	Cis-transoid $\phi = 32°$[b]	Trans-cisoid $\phi = 35°$[b]
3.84	e	(110)(200)	963	727	829
3.06	off	(011)			
2.92	off	(201)			
2.89	e	(210)	100	100	100
2.33	e	{(020)(120) (310)	66	59	68
2.19	m	(002)			
1.98	e	(220)(400)	26	20	22
1.70	e	(320)(410)	34	3	4
1.42	e	{(130)(230) (420)(510)	10	8	8
1.28	e	{(330)(520) (600)(610)	16	22	23
1.10	m	(004)			

[a] e = equatorial, m = meridional, off = off-meridional.
[b] ϕ = setting angle.

the crystal. The black regions in the figure correspond to CH separation of only 2.7 Å, which is less than their van der Waals radii of 1.80 Å + 1.77 Å = 2.97 Å. Therefore, the cis polymer appears to have a rather unstable crystal structure produced by the crystallization–polymerization process. This instability may contribute to the facile cis–trans isomerization, which begins at temperatures lower than 250 K.

Baughman *et al.* (1978) observed ten reflections by X-ray diffraction

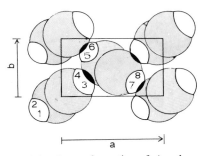

Fig. 3.15 Molecular conformation of *cis*-polyacetylene.

(Table 3.5) on polycrystalline, unoriented polyacetylene films. Therefore, it was not possible to determine the chain axes directly. The observed diffraction spacings and packing calculations are consistent with the orthorhombic unit cell *Pnam* (Table 3.6) with dimensions identical to those from electron diffraction. However, they assigned **b** as the chain axis in disagreement with the directly determined chain axis **c** by electron diffraction. Therefore, the Miller indices given by the authors are really (*hlk*) reflections. Their intensity analysis gives $\phi = 59°$. This setting angle is that with the **a** axis in the **ac** plane, whereas the one obtained by electron diffraction is the angle of the C—C bond projection with the **a** axis in the **ab** plane.

   In the other electron diffraction study, the specimen is said to consist of irregularly shaped lamellalike particles, with dimensions typically a few hundred angstroms in diameter and thicknesses of 50–100 Å, which aggregate in a pseudofibrillar arrangement (Lieser *et al.*, 1980a). Only equatorial reflections were observed at normal electron-beam intensities. In rare cases, where a set of subunits is tilted from the substrate such that the electron beam impinges on the particle edge-on, were intense (001) reflections observed. Because this reflection should be extinct for the *Pnam* space group, the data require *cis*-polyacetylene to have some other space group. The authors proposed a folded-chain crystal structure in which the **c** axis is perpendicular to the direction linking the particles. These results are complicated by the fact that the observed particles may be artifacts, as will be detailed in Chapter 4. One important result that is not subject to error is that in polyacetylene fibrils the **c** axis is the molecular chain axis as well as the fiber axis.

3.6   CRYSTAL STRUCTURE OF
   trans-POLYACETYLENE

   There have been two reported electron diffraction studies and one X-ray diffraction study on the crystal structure of *trans*-polyacetylene. The one by Shimamura *et al.* (1981) was made on ultrathin film directly polymerized and subsequently isomerized on EM grids. Figure 3.16 shows an electron micrograph of a *trans*-polyacetylene film. The selected region electron diffraction pattern of aligned fibrils is given in Fig. 3.17a and the Debye ring for randomly packed fibrils is shown in Fig. 3.17b.

   Along the meridian of the fiber diffraction pattern the (002) reflections is intense, there is very weak intensity along the (001) layer line. The latter will be used to estimate bond alternation in this polymer (Section 3.7). The

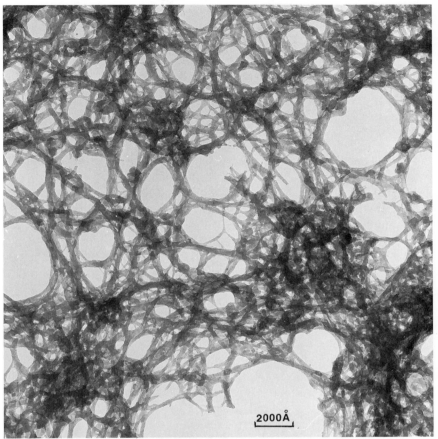

Fig. 3.16 Electron micrograph of an ultrathin film of *trans*-polyacetylene.

reciprocal lattice is approximated to be "orthorhombic." However, the presence of (001) intensity would require certain modifications not permitted by the present data. The unit-cell parameters are given in Table 3.9. As in the cis polymer, the **c** axis is the molecular axis of *trans*-polyacetylene, which is along the fiber axis.

Six meridional reflections were observed. These and other reflections fall on the approximated orthorhombic reciprocal lattice sites (Fig. 3.18c). Using the same procedures described in Section 3.5, the (*hk*0) reflection intensity (Fig. 3.19) was calculated as a function of the setting angle and compared with the calculated intensities in Table 3.10. The best setting angle was found to be ∼ 24°, according to Eq. (3.14) (Fig. 3.20). The

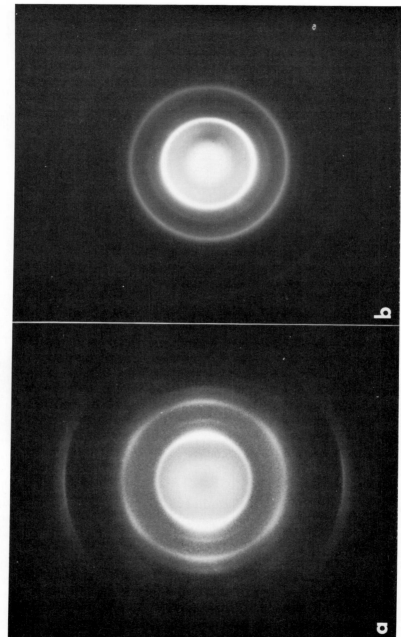

Fig. 3.17 Electron diffraction patterns of pristine *trans*-polyacetylene film (a) for aligned fibrils; (b) for randomly packed region.

TABLE 3.9

Unit Cell Parameters for *trans*-Polyacetylene

Lattice type	E.D.[a] aligned fibrils "Orthorhombic"[d]	E.D.[b] Model I unoriented Orthorhombic	E.D.[b] Model II Monoclinic $\gamma = 98°$	X-ray[c] Monoclinic $\beta = 91-93°$
a(Å)	7.32	5.62	3.73	4.24
b(Å)	4.24	4.92	3.73	7.32
c(Å)	2.46	2.59	2.44	2.46
ρ(g cm^{-3})	1.13	1.20	1.27	
C—C—C angle	122°	135°	120°	
Setting angle ϕ	24–28°			55°

[a] Shimamura *et al.* (1981).
[b] Lieser *et al.* (1980b).
[c] Fincher *et al.* (1982).
[d] Approximate lattice type.

molecular conformation of *trans*-polyacetylene in the crystal is given in Fig. 3.21. Figure 3.22 shows the packing arrangement for the chains. The adjacent monomer units along the **b** axis are precisely parallel to each other and there is no atomic separation smaller than the sum of van der Waals radii as there was in the *cis*-polyacetylene structure. The crystal structure suggests the possibility of maximum π-orbital overlap between polyacetylene chains and perhaps a minimum of interchain resistance for electrical conductivity.

Fincher *et al.* (1982) analyzed the X-ray diffraction data of *trans*-polyacetylene according to a monoclinic unit cell, and the lattice parameters are given in Column 5 of Table 3.9. The space group was assumed to be $P2_1/n$. The **a** and **b** axes of the monoclinic unit cell are interchanged for the approximate "orthorhombic" unit cell.

Lieser *et al.* (1980b) reported electron diffraction unit cell structures for two modifications of *trans*-polyacetylene; one orthorhombic (model I) for polymers obtained under shear flow conditions (Section 4.5) and model II for materials polymerized with more common procedures. No intensities or (*hkl*) assignments were given. Although model I is based on fiber diffraction patterns of oriented specimens, there are two obvious difficulties. First, in each quadrant there is a pair of off-meridional reflections that lie in the vicinity of, but do not belong to, the layer line. Therefore, the sample probably contains more than one structural modification. Second, model I predicts an unusual dihedral angle for the backbone of 135°, which is quite

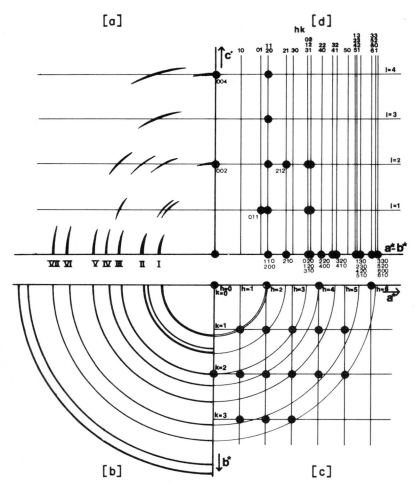

Fig. 3.18   (a) Schematic representation of the electron diffraction of oriented *trans*-poly-acetylene fibrils; (b) schematic representation of the electron diffraction of randomly packed region; (c) lattice for the **a*b*** plane (●) observed reflections (○) predicted but not observed reflections; (d) projection of the reciprocal lattice of the first five layers of the **a*b*** plane.

untenable. Model II is based on Debye-ring reflections only and the analysis is consequently unreliable. The density calculated for model II is much greater than the experimental floatation value. We have tried to locate our observed reflections in the proposed reciprocal lattice sites of either model I or II of Lieser *et al.* (1980b). Figure 3.23 shows that some of the reflections do not fall on the lattice sites of either model. One cannot be certain from the

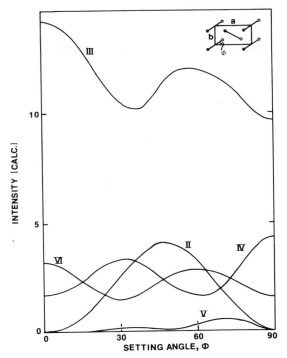

Fig. 3.19   Variation of calculated diffraction intensities for *trans*-polyacetylene with the setting angle, the (*hkl*) values for II to VI are given in Table 3.10.

publications which catalyst, i.e., Ziegler–Natta or Luttinger, was used to initiate the polymerizations.

It is evident that there are discrepancies for the crystal structures of polyacetylene. The differences may be real. Many polymers exist in more than one crystalline form, e.g., polyethylene, poly(vinylidene fluoride), polypropylene, and poly(butylene terephthalate). It is possible that poly-actylene exists in polymorphic forms. On the other hand, there are some problems special to polyacetylene. The cis polymer crystallizes during polymerization. Therefore, its crystal structure may be strongly influenced by the kinetics of polymerization and crystallization. In the case of *trans*-polyacetylene, proper isomerization procedure is of the utmost importance because the polymer may assume more than one crystal modification during transformation, a process that is likely to be kinetically controlled. The changes in *d*-spacings of polyacetylene with time of heating are discussed in Section 3.8.

TABLE 3.10

Observed and Calculated Intensities of Electron Diffractions for *trans*-Polyacetylene

Spacing		Reflection (hkl)	Observed intensity	Calculated intensity $\phi = 24°$[b]
In Å	Direction[a]			
3.68	e	(110)(200)	100	100
2.75	e	(210)	2.2	1.8
2.12	e	{(020)(120) (310)	10.3	11.5
2.08	off	{(011)(111) (201)		
1.83	e	(220)(400)	3.1	3.1
1.64	e	(320)(410)	0	0.1
1.56	off	(121)(311)		
1.36	e	{(130)(230) (420)(510)	0.9	2.1
1.23	m	(002)		
1.16	off	{(102)(112) (202)		

[a] e = equatorial, m = meridional, off = off-meridional.
[b] ϕ = setting angle.

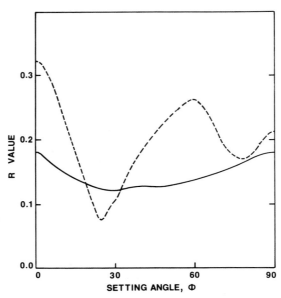

Fig. 3.20 Variation of the reliability factor R with ϕ; solid line includes all six equatorial reflections of *trans*-polyacetylene; broken line is the plot including equatorial reflections II to VI.

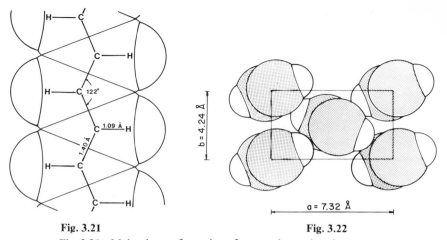

Fig. 3.21

Fig. 3.22

Fig. 3.21   Molecular conformation of *trans*-polyacetylene in crystal.

Fig. 3.22   Packing arrangement for *trans*-polyacetylene projected onto **ab** plane.

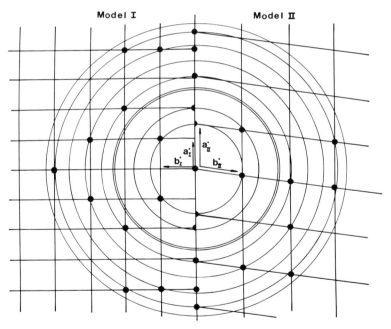

Fig. 3.23   The observed reflections of Shimamura *et al.* (1981) and the reciprocal lattice sites of the two structural models proposed by Lieser *et al.* (1980b) for *trans*-polyacetylene.

3.7   BOND-LENGTH ALTERNATION

As pointed out in Section 3.2 and emphasized in Chapter 9 and else-where, trans-polyacetylene cannot possibly have uniform bond lengths or it would be a quasi-one-dimensional metal, and its transport properties would be vastly different from what they are. Physicists refer to polyacetylene as Peierls-unstable: it undergoes distortion of index 2 to a bond-alternating incommensurate state. However, the bond-length alternation is expected to be small and consequently difficult to determine with any precision. Never-theless, there have been two studies that afforded estimates of its magnitude. Basically there are three possible structures for *trans*-polyacetylene, as shown in Fig. 3.24, all with two chains per unit cell: (**A**) uniform bond length, (**B**) in-phase alternation between adjacent chains, and (**C**) out-of-phase bond alternation.

In the electron diffraction study (Chien *et al.*, 1982h), we note that structures **A** and **C** require the total extinction of the (001) reflection. No (001) intensity was observed for *trans*-polyacetylene isomerized at 473 K for 2 hr (Figs. 3.17a and 3.18a). However, as we optimize the isomerization condition, faint (001) intensity begins to appear. The (001) intensity is most pronounced when the ultrathin film is heated at 443 K for 20 min (Fig. 3.25). Its intensity weakens and becomes absent when the polymer is isomerized at either higher or lower temperatures. Questions of multiple scattering and other artifacts of electron diffraction experiments were con-sidered. Electron diffraction measurements, using identical conditions, were made on *cis*-polyacetylene and polyethylene. The (001) reflections should be strictly extinct in the two polymers, and none were detected. This result is significant for two reasons: (1) the (001) intensity in *trans*-polyacetylene

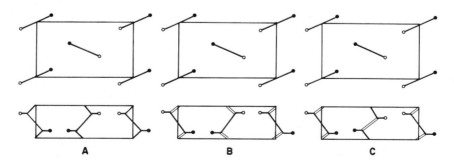

Fig. 3.24   Possible structures for *trans*-polyacetylene: (**A**) uniform bond length, (**B**) in-phase bond alternation between adjacent chains, (**C**) out-of-phase bond alternation between adjacent chains.

cannot be attributed to experimental artifacts; and (2) its unit cell structure deviates from orthorhombic and has a space group of lower symmetry.

Let us proceed with the analysis of the electron diffraction data of *trans*-polyacetylene. The observed (001) reflection, though weak in intensity, favors structure **B** of Fig. 3.24. The in-phase bond alternation on structure has an $I_{(001)}/I_{(002)}$ reflection intensity ratio of

$$\frac{I_{(001)}}{I_{(002)}} = \left[\frac{f_{(001)}}{f_{(002)}}\right]^2 \left[\sin^2\left(\frac{2\pi\Delta}{1.14c}\right) \Big/ \cos^2\left(\frac{4\pi\Delta}{1.14c}\right)\right] \qquad (3.9)$$

where Δ is the deviation from uniform bond length. Figure 3.26 shows the $I_{(001)}/I_{(002)}$ as a function of Δ. The meridional intensity distribution is given in Fig. 3.27. The first one is the sum of (001), (011), (111), and (201) reflections. The (011), (111), and (201) reflections contribute toward meridional intensity because of imperfect crystallite orientation. Their contribution can be subtracted because the (001) reflection occurs at a smaller scattering angle. The profile of the (011), (111), and (201) reflections can be

Fig. 3.25   Electron diffraction of optimally isomerized *trans*-polyacetylene shown intensity at (001) reflection.

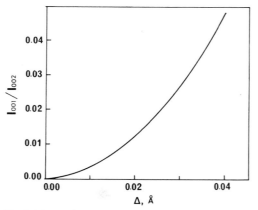

Fig. 3.26   Variation of $I_{(001)}/I_{(002)}$ as a function of Δ.

measured along off-meridional directions, and subtraction gives the profile of the (001) reflection, shown as the shaded region in Fig. 3.27. The second peak comprises the (002), (112), and (202) reflections. By a similar procedure the profile of the (002) reflection is obtained (Fig. 3.27).

Corrections for the Lorentz factor and for photodensitometry were first made. The background intensity distribution was assumed to be Gaussian, $f = A \exp(-Bs^2)$, with its center located at the incident-beam position of the

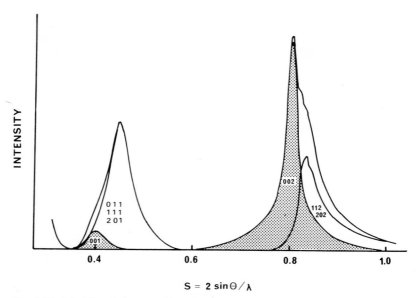

Fig. 3.27   Meridional electron-diffraction intensity distributions for optimally isomerized *trans*-polyacetylene.

diffraction pattern. Intensity at $s = 0.6$ and $s = 1.4$ were used to determine the constants A and B. At the beginning of the calculations, it was assumed that $\Delta = 0$. From the repeat distance along the chain, θ was estimated to be 122°. Using this value, i.e., assuming the denominators in Eq. (3.9) to be 1.14c, differences in bond lengths were estimated based on the $I_{(001)}/I_{(002)}$ relation. This gives a new value θ', and the process is iterated until 1.14c is constant within experimental accuracy. The final ratio of $I_{(001)}/I_{(002)}$ was found to be 0.017, which gives Δ to be $\sim 0.024 \pm 0.002$ Å from Fig. 3.26. Consequently, the bond lengths in *trans*-polyacetylene are 1.38 and 1.43 Å, compared to regular $C{=}C$ and $C{-}C$ bond lengths of 1.35 and 1.54 Å, respectively. This value of Δ is much smaller than the value of $\Delta = 0.10$ calculated by Karpfen and Höller (1981), using a minimal STO-3G basis set (Section 3.2).

Fincher *et al.* (1982) used a triple-axis X-ray spectrometer to estimate the symmetry-breaking parameter. Polyacetylene films were stretch aligned and isomerized at 473 K for 2 hr; ten such strips were clamped together for a total thickness of ~ 1 mm, and placed in an evacuated Be can. The scattering geometry has the average polyacetylene fibril axis lying in the scattering plane (the **H** direction) for ($hk0$) reflections with a rectangular **H K** coordinate system. The observed reflections are shown in Fig. 3.28; the inset shows the peak at 5.1 Å^{-1}, along **H**, corresponding to the carbon–carbon spacing along the backbone. The arrangement of the chains when projected onto a plane normal to the chain direction was obtained from a scan normal to **H** (Fig. 3.29). The fit to the data was good. However, the number of reflections, their indexing, and their symmetry were not given. The results were interpreted to give a *Pgg* symmetry and a setting angle of 55°. However, there were only four peaks in the equator (0, ζ) and their assignment to ($hk0$) indices were unclear. This results in some doubts in the subsequent estimates of the setting angle and unit-cell vectors without showing the variation of the reliability factor with the setting angle. Moreover, the intensity distributions around (2.55, 1.72) are due to several reflections. The conclusions may be strongly influenced by the particular (hkl) reflections contributing to them. The specimen has orientation distribution with full width at a maximum of $\sim 35°$. This means that reflections about (2.55, 1.72) have the shape of an arc. It seems evident that an orientation effect should be included to get the best fit shown in Fig. 3.29. The authors' statement that at $Q = 1.72$ Å^{-1}, the full width at half maximum is independent of the magnitude of momentum transfer is curious because at $Q = 1.72$ Å^{-1} the peak comprises at least two reflections. The authors did not analyze whether the intensity comes from (110) + (220) or (200) + (400) reflections.

Two space groups were considered by Fincher *et al.* (1982), $P2_1/c$ and $P2_1/n$. The former was rejected because the (001) reflection was beyond detection. Best fit was obtained by varying β, the monoclinic angle, and u, the amplitude of longitudinal distortion. We have reanalyzed these data and

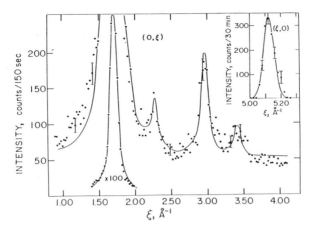

Fig. 3.28 X-ray diffraction data from scan in the plane normal to the oriented direction of the *trans*-polyacetylene film. Inset shows data from scan along oriented direction at the (002) reflection. [After Fincher *et al.* (1982).]

found no distinct minimum. Fincher *et al.* (1982) concluded that $\beta = 91.4°$, $\Delta = 0.026 \pm 0.01$ Å, and the polymer has the out-of-phase bond alternation structure **C**.

The failure to detect (001) intensity by X ray may be due to lack of sensitivity. If one assumes a monoclinic angle of 90°, calculation of the structure of this unit cell gives

$$\frac{I_{001}}{I_{002}} = \left(f_{001} \sin 2\pi \, \frac{0.03 \cos 30°}{2.40} \right) \bigg/ \left(f_{002} \cos 4\pi \, \frac{0.03 \cos 30°}{2.46} \right) \simeq 0.01$$

$$(3.10)$$

where 0.03 corresponds to the suggested value for u_0. The reported data

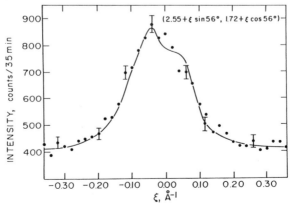

Fig. 3.29 Data obtained by $\theta - 2\theta$ scan through (2.56, 1.72) point in scattering plane. Solid lines represent the fit to the data. [After Fincher *et al.* (1982).]

(Fincher *et al.*, 1982) around (ζ, 0) corresponding to (002) is very weak (320 counts/30 min) so that it is very difficult to detect (001) intensity around ($\frac{1}{2}\zeta$, 0) because natural fluctuation of X-ray photons would overwhelm the very weak (001) signal (S/N $\simeq$ 0). This precludes the detection of any weak (001) intensity for space group $P2_1/c$. Also, the X-ray diffraction pattern of a semicrystalline polymer must be corrected for the degree of crystallinity. Some of the intensity distributions would be very strong and can influence the background.

A reasonable explanation of the electron and X-ray diffraction results is that *cis*-polyacetylene is first isomerized to the in-phase bond alternation structure **B**. Further prolonged heating at higher temperature changes the structure to **C** with the disappearance of (001) intensity in both diffraction measurements.

In conclusion, bond alternation has been demonstrated for *trans*-polyacetylene. The actual structure may depend upon how the specimen is prepared. According to the theory (Su *et al.*, 1979), the band gap is related to the bond alternation by the electron-lattice coupling constant α by

$$E_G = 8\alpha\Delta/\cos 30°. \qquad (3.11)$$

Using $1.6 \text{ eV} \leqslant E_G \leqslant 1.8$ eV for the band gap, the values of Δ found above by either diffraction methods give $\alpha \simeq 6.7 - 8.1$ eV Å^{-1}.

3.8 THERMAL ISOMERIZATION AND CRYSTAL STRUCTURE CHANGE

Comparison of the crystallographic spacings for *cis*-polyacetylene (Table 3.5) and *trans*-polyacetylene (Table 3.10) show the changes upon isomerization. Robin *et al.* (1983) followed the changes in spacings as a function of isomerization at 373 K. The ratio of cis-to-trans isomer was determined by infrared and NMR analysis. The diffraction data were obtained with the X-ray synchrotron radiation at LURE, Orsay. Electron diffraction results (Tables 3.5 and 3.10) showed the (110) (200) reflections at 3.84 Å in *cis*-polyacetylene to change to 3.68 Å in *trans*-polyacetylene; the corresponding changes are 2.89 to 2.75 Å for the (210) reflection and 2.20 to 2.12 Å for the (310) (020) (120) reflections. Robin *et al.* (1983) found that these changes in *d* spacings are linear with respect to the increase in the *trans* content. The authors used the *hlk* indexing of Baughman *et al.* (1978), so the above-mentioned (110) reflections etc. were transposed to (101) and so forth. Only one set of Debye–Scherrer patterns were observed for all degrees of isomerization. The result demonstrates that the isomerization proceeds homogeneously through the polyacetylene crystal. It is not consistent with the heterogeneous growth of isolated *trans* polyacetylene domains.

From the angular width at half height of the Debye–Sherrer ring of the (110) (200) reflection, the intrinsic broadening due to the finite size of the crystalline areas that contribute to Bragg reflections was estimated after a Gaussian correction to give the coherence length using the Scherrer formula. The transverse coherence length was said to be 150 ± 30 Å, independent of isomerization. The result indicates that there is no degradation of crystalline size and crystallinity, and is consistent with a simple model of linear shift of transverse lattice parameter with the trans content.

Additional observations on the effect of heating on the crystal structures of polyacetylene are described in Section 4.5.

3.9 CRYSTAL STRUCTURE OF DIBLOCK COPOLYMERS OF STYRENE AND ACETYLENE

A diblock copolymer containing 35–40% by weight of polyacetylene undergoes phase separation from solution to form lamellar single crystals (Bates and Baker, 1983b). The formation of crystals about 0.5 μm in size is consistent with the upper critical solution temperature phase behavior exhibited by block copolymers characterized by van der Waals interactions. A description of the diffraction results can be given by a hexagonal lattice with $a = 5.12$ Å and $c = 4.84$ Å. The placement of *cis*-transoid chains in this unit cell would result in an unreasonably low crystal density of 0.79 g cm^{-3}. However, the actual density was not determined. Bates and Baker (1983b) suggested instead that the chains have a distorted *cis*-cisoid 2*3/1 helical conformation. When packed in the hexagonal unit cell the crystal density was calculated to be 1.18 g cm^{-3}. Based on estimated molecular weight of about 6×10^4 and the 100-Å thickness of the crystal, it was proposed that the polyacetylene crystallizes with chain folding on the average of 35 folds per chain.

Both the orthorhombic unit cell structure of the usual *cis*-polyacetylene and the hexagonal unit cell of the block copolymer have nearly the same density. However, the latter contains 36% more (00*l*) surface area per unit cell than the former. Chain folding would increase the effective (00*l*) crystal surface area per molecule so that decreasing the fold length results in an increase in the interfacial surface area, which is needed to accommodate the polyacetylene block. That the block copolymer is soluble permits the polyacetylene to crystallize in a distorted *cis*-cisoid helix conformation. In the case of the Shirakawa synthesis polymerization is followed almost immediately by crystallization in the planar *cis*-transoid conformation dictated by the propagation mechanism (Section 2.5).

Chapter 4

Morphology

4.1 INTRODUCTION

Because polyacetylene is formed by simultaneous polymerization – crystallization process, the morphology of the polymer should depend on how it was obtained. For instance, different catalysts may produce polymers with different morphologies (Table 2.1 and Section 2.8). Factors that can influence the morphology are the catalyst concentration, the temperature of polymerization, and the location in the reactor where the polymers are formed (Section 2.6.2). The basic fibril morphology is discussed in Section 4.2 and the possible effect of heat of polymerization described in Section 4.3.

In the laboratories of Wegner, polyacetylenes with pseudolamellar morphology were consistently obtained. A folded-chain lamellar structure was proposed. On the other hand, polyacetylene has been reported to have the fibrillar morphology by many investigators. The question of fibrillar versus lamellar morphologies will be discussed in Section 4.4.

When polymerization is conducted in the presence of a shear flow field, polymers with interesting morphologies were obtained, as described in Section 4.5. The results of alignment of polyacetylene are presented in Section 4.6.

4.2   FIBRILS AND MICROFIBRILS

When Ito *et al.* first prepared freestanding polyacetylene films in 1974, they showed that the material is composed of fibrils of about 20 nm in diameter. Figure 4.1 shows the scanning electron micrographs (SEM) of such a film. The highly reflecting side of the film facing the reactor wall has flattened fibrils, whereas the dull side of the film facing away from the wall is composed of a loose web of polyacetylene fibrils. This is consistent with the initiation of polymerization by catalyst deposited on the reactor wall from where the polymer fibrils grew. By gentle rubbing, the dull side can be made highly reflecting as the fibrils are compacted.

In order to eliminate alteration of polymer morphology due to any aspects involved in specimen handling, we have polymerized acetylene directly onto EM grids to capture the nascent morphology of polyacetylene (Section 2.2.8). Two typical TEM micrographs of such films are shown in Figs. 3.9 and 3.16. In Fig. 3.16 the trans polymer film is about 100 nm thick; most of the polymer appears to consist of flat ribbons of average 20 nm in width. There are regions where parallel bundles of fibrils aggregate; fiber electron diffraction patterns described in Section 3.6 were obtained by directing the electron beam at these bundles. The *cis*-polyacetylene film shown in Fig. 3.9 is even thinner. In such photomicrographs very small-diameter (2–3 nm) microfibrils are present in abundance. From the unit-cell structure for *cis*-polyacetylene (Section 3.5) one deduces that the 2-nm microfibrils contain ~ 13 polyacetylene chains and ~ 60 polyacetylene chains for the 3-nm microfibrils. These microfibrils are undoubtedly the ultimate morphological entity. Very low catalyst concentration was used to polymerize acetylene in the preparation of these specimens. Therefore, it was relatively easy to wash away the catalyst residues with toluene. Even then one can clearly see some remains of the catalyst on the left side of Fig. 3.9 and smaller deposits of foreign substances here and there. Thermal isomerization of these thin polyacetylene films cause virtually no discernible change in the fibril morphology.

The microfibrils in Fig. 3.9 were straight and taut. They are probably separated from the fibrils during washing and evacuation cycles. The microfibrils aggregate into fibrils by apparently several ways. Close examination of

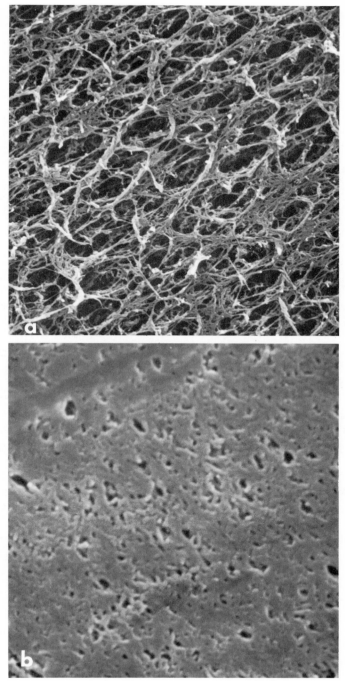

Fig. 4.1 Scanning electron micrographs of free-standing polyacetylene film: (*a*) dull face away from the reactor surface; (*b*) shiny face toward the reactor surface. Magnification 50,000×.

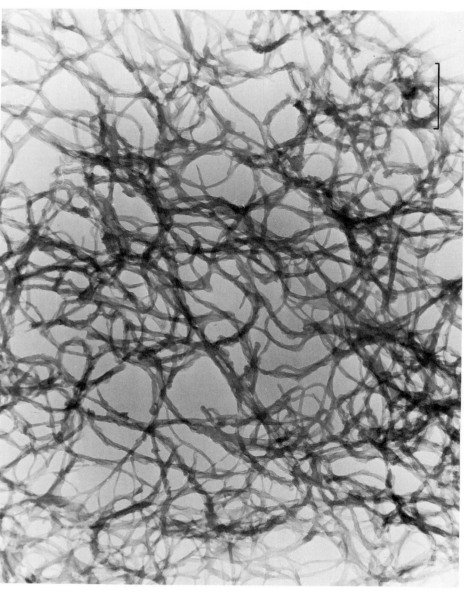

Fig. 4.2 Electron micrograph of nascent polyacetylene polymerized directly onto the EM grid with the Luttinger catalyst. Marker is 500 nm.

Fig. 3.9 showed some fibrils of 15–20 nm to be uniform in photodensity across and along the fibril. Larger-diameter fibrils gave the appearance of containing twisted microfibrils. There is, however, no regular pitch to it. This twisted feature is more pronounced for larger aggregates of fibrils. But in general the fiber axes are partially aligned inasmuch as they gave fiber electron-diffraction patterns, as discussed in Sections 3.5 and 3.6.

Figure 3.16 shows that one bundle of fibrils is connected to two or more others, which suggests that the microfibrils can grow and become incorporated into different fibrils. Therefore, the interactions between the microfibrils in a fiber are weak. In fact, at high magnification of EM, the broken end of a fibril has the appearance of a tree with leafless branches at the top or even branches farther down from the end. Finally, the aggregates of bundles of fibrils are definitely not uniform in density, as seen by the differences in photodensity across an aggregate.

The fibrillar morphology is not unique to polyacetylene prepared with the $Ti(OBu)_4$–$AlEt_3$ catalyst. It is shared at least by the polyacetylene polymerized directly onto the EM grid with the Luttinger catalyst. Figure 4.2 is the electron micrograph of such an ultrathin film, which resembles Fig. 3.16 closely with some subtle differences. While the very small-diameter entities are fibrillar, the larger aggregates appear to be more flat ribbons. It seems that the microfibrils in the ribbons produced by this catalyst are less well aligned than those in the bundles of fibrils formed by the Ziegler–Natta catalyst. This may be rationalized by the active species being bridged complexes of titanium butoxides and alkyl aluminum compounds in the Ziegler–Natta catalyst. Such bridged complexes can promote parallel growth of polyacetylene chains and subsequent crystallization. It is unlikely that catalytic species derived from $Co(NO_3)_2$ and $NaBH_4$ in ethanol would be highly associated. The electron diffraction patterns of the polyacetylene obtained with the Luttinger catalyst (Fig. 4.3) are distinctly more diffuse than those of polyacetylene produced with the Ziegler–Natta catalyst (Fig. 3.10). Haberkorn *et al.* (1982) showed the former to be of low crystallinity and to contain an appreciable fraction of sp^3 carbon atoms.

Ziegler–Natta catalysis of α-olefin polymerization has been known to lead directly to crystalline polymers that precipitate from the diluent. The degree of crystallinity of the nascent polyolefins is higher than that of the same polymer after any kind of processing. Fibrillar morphology was observed for polyethylene (20–30 nm) obtained with a $TiCl_4$–$AlEt_3$ catalyst (Ingram and Schindler, 1968) and also for polypropylene (30–50 nm) with the same catalyst (Guttman and Guillet, 1970). It was proposed by Wristlers (1973) that for these heterogeneous catalysts the polyolefin chains formed on adjacent sites in the α-$TiCl_3$ surface act as nucleation centers,

Fig. 4.3   Selected region electron diffraction pattern of partially aligned acetylene ribbons of Fig. 4.2.

producing the microfibrils. The present $Ti(OBu)_4 - AlEt_3$ catalyst has been said to be either a homogeneous or a heterogeneous system, depending upon the experimental conditions, i.e., Ti/Al ratio, catalyst concentration, the catalyst aging procedure, and the temperature of polymerization (Koide *et al.*, 1967; Chien *et al.*, 1980a; Hirai *et al.*, 1972; Hiraki *et al.*, 1972). If the catalyst is heterogeneous, crystallization of chains initiated by nearby active sites of the catalyst particle is possible. Alternatively, and for the homogeneous case, microfibrils were probably formed with a number of cis-transoidal segments of several polyacetylene chains acting as nucleation centers. Further development of the morphology occurs with the aggregation of the microfibrils into larger-diameter fibrils.

4.3   GLOBULAR MORPHOLOGY AND EFFECT OF HEAT OF POLYMERIZATION

It has been shown by many laboratories that the side of free-standing film facing the reactor wall is shiny and has a metallic luster. Scanning EM

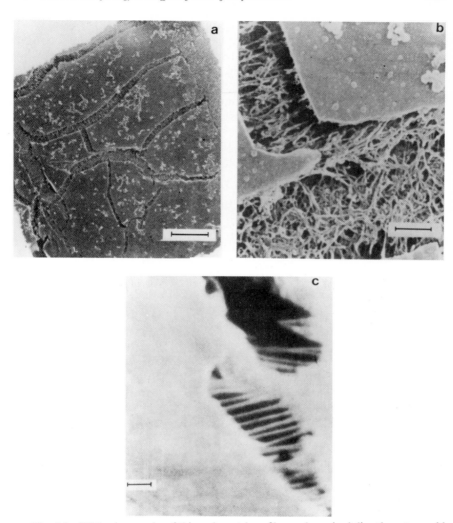

Fig. 4.4   SEM micrographs of thin polyacetylene films polymerized directly onto a gold grid at 760 torr of acetylene for ∼ 1 min., dried and sputter coated with 3 nm of gold: (a) ×900, scale bar = 10 μm; (b) ×9000, scale bar = 1 μm; (c) ×60,000, scale bar = 0.1 μm.

showed that the fibrils were flattened (Fig. 4.1a). It seems plausible that this was caused by poor dissipation of the heat of polymerization. Support for this possibility came from direct polymerization onto EM grids under certain conditions (Karasz *et al.,* 1979).

Direct polymerizations of acetylene on EM grid at high monomer pressure (760 torr) yields an extremely smooth film (Fig. 4.4a). This may be attributed to fusion of the fibrils by liberated heat of polymerization because there is no diluent to act as a heat sink. This occurs, however, only at the surface. Figure 4.4b showed that where the smooth film is fractured the usual fibrils underneath were revealed. In fact, this fracture can result in highly oriented fibrils of 100 nm diameter in some regions (Fig. 4.4c).

Sometimes acetylene polymerizes into globular morphology as shown,

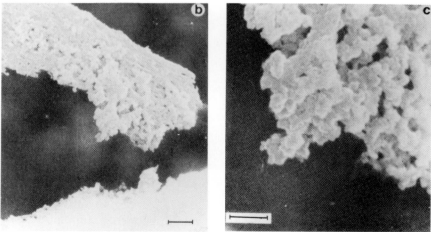

Fig. 4.5 SEM micrograph of another thin polyacetylene film polymerized as in Fig. 4.39: (a) ×1800, scale bar = 5 μm; (b) ×4800, scale bar = 1.25 μm; (c) ×18,000, scale bar = 0.5 μm.

with increasing magnification, in Fig. 4.5. Examination of Figs. 4.5a and b shows that the globular growth appears to have been separated from a portion of the polyacetylene film as a result of melting contraction. The texture is shown better in Fig. 4.5c at higher magnification.

These results showed that smooth-film surface and globular morphology can be obtained in acetylene polymerization. They are probably the result of rapid polymerization and poor heat dissipation. It is interesting to note that once polyacetylene has crystallized, it is not fusible. This would require cooperative crystal melting. On the other hand, during polymerization individual polymer chains or segments of microfibrils can apparently be melted or solubilized.

4.4 LAMELLAR MORPHOLOGY

In sharp contrast to the morphologies described above, Lieser *et al.* (1980a) reported that polyacetylene obtained with either the Ziegler–Natta or the Luttinger catalysts consists of irregularly shaped lamellarlike particles of typically a few hundred Angstroms in diameter and a thickness of 50–100 Å as the smallest morphological entity. These particles were said to aggregate to form loose networks or sponges in a pseudofibrillar arrangement. Enkelmann *et al.* (1979, 1981a) and Lieser *et al.* (1981) developed the theme further. They argued that polyacetylene is morphologically far more complex than the basic fibrils observed by many workers. Questions were raised regarding the structural purity of the polymer. The authors felt that the catalysts used in the polymerization are also likely to catalyze cyclization or isomerization and other thermally or photochemically induced intermolecular and intramolecular reactions. Lieser *et al.* (1981) and Wegner (1981) proposed very detailed models of folded-chain lamellar morphology for polyacetylene. Because of the obvious significance of this possible morphology to the fundamental understanding of polymer crystallization and because the observations are contrary to all other reports, independent confirmation of these experimental findings was sought by us.

We consider first polyacetylene obtained with the Ziegler–Natta catalyst. Lieser *et al.* (1980a) polymerized acetylene with a minimum amount of the $Ti(OBu)_4$–$AlEt_3$ catalyst but did not mention the actual concentration used. A suspension of deep-red particles of polymer was produced in a few minutes at 203 K; the monomer pressure was not given. There was no description of methods used to remove the catalyst. The specimen was scooped up with a copper grid and shadowed with platinum. Chien *et al.* (1982g) carried out similar polymerizations with varying catalyst concen-

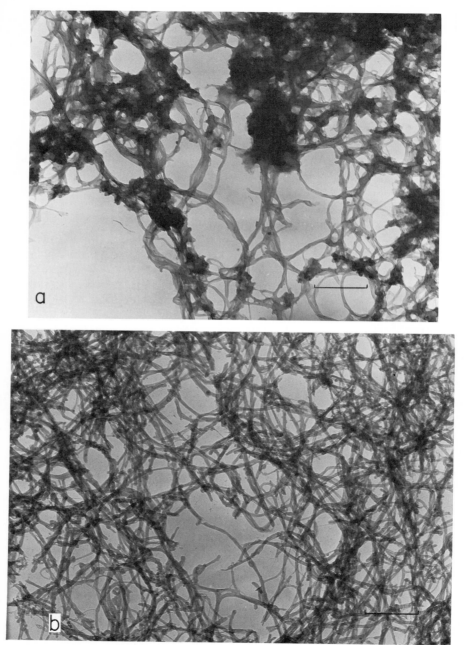

Fig. 4.6 Electron micrograph of polyacetylene polymerized in the diluent with the Ziegler–Natta catalyst, [Ti] = 1.02 mM: (a) polymer washed with toluene; (b) washed with methanolic HCl. Scale bar is 500 nm.

trations, and the specimen was purified by different procedures. The results are presented below.

Acetylene was polymerized in a stirred reactor at 195 K and initiated with the Ziegler–Natta catalyst, 1.02 mM in Ti(OBu)$_4$ and Al/Ti = 4. The polymer was washed thoroughly with toluene, and specimens were collected with an EM grid. The TEM of the sample (Fig. 4.6a) shows the presence of fibrils, ribbons, and dark spongy masses. The spongy substances are found mostly where a multitude of fibrils and ribbons of polyacetylene aggregate, as in the region at the upper portion of the electron micrograph. No pseudolamellae particles are seen in this preparation. An aliquot of this toluene-washed polymerization mixture was subsequently washed with methanolic HCl. A specimen was collected on a carbon coated grid; it was found to have very clean fibrillar morphology (Fig. 4.6b). There is no spongy material in this specimen; apparently the spongy substance is soluble in methanolic HCl. Comparison of Figs. 4.6a and b suggests that the ribbons seen in the former EM may be collections of fibrils held together by small amounts of catalyst residue or low molecular weight polymers. As the "glues" were solubilized by methanolic HCl, the ribbon became separated into the constituent fibrils.

At a relatively high catalyst concentration of [Ti] = 40.8 mM, the polymer formed even after very thorough toluene washing was contaminated by large quantities of spongy mass. Figure 4.7a is an electron micrograph of such a specimen. In addition, there are aggregates of irregularly shaped lamellar particles at the extreme left and right of the figure. Also present are fibrils that appear to be striated. In fact, this micrograph resembles closely those published by Wegner and co-workers: Fig. 1 of Lieser *et al.* (1980a), Fig. 3 of Enkelmann *et al.* (1979), Fig. 2 of Lieser *et al.* (1981), Fig. 2 of Enkelmann *et al.* (1981b), and Fig. 4 of Wegner (1981). By washing this polymer sample with methanol alone, the resulting polyacetylene contains a much smaller amount of the spongy mass, as shown in Fig. 4.7b. Most of the lamellalike particles remained. When this material was further washed with methanolic HCl, the micrograph shown in Fig. 4.7c was obtained. The density of the fibrils is high, but its fibrillar nature cannot be disputed. Running though the middle of the figure are dark regions that probably represent a remnant of the contaminants. The catalyst concentration used in this experiment is still lower than that commonly used to prepare free-standing polyacetylene films, which is 0.2 M. At this point it is appropriate to mention that this catalyst concentration is two to four orders of magnitude greater than that used to polymerize olefins by the traditional Ziegler–Natta catalyst (δ-TiCl$_3$ · 0.33 AlCl$_3$) and the high activity catalyst (MgCl$_2$ supported catalysts), respectively. It is common practice to remove catalyst residues from the polymer by methanolic HCl washing. In some

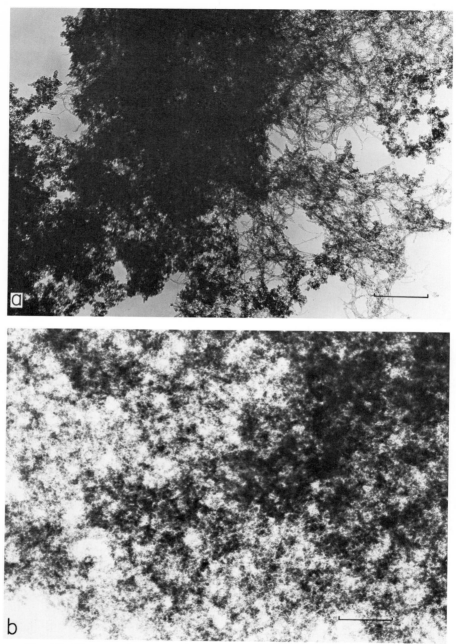

Fig. 4.7 Electron micrograph of polyacetylene polymerized in the diluent with the Ziegler–Natta catalyst, [Ti] = 40.8 n*M*: (a) polymer washed with toluene; (b) washed with methanol; (c) washed with methanolic HCl. Scale bar is 500 nm.

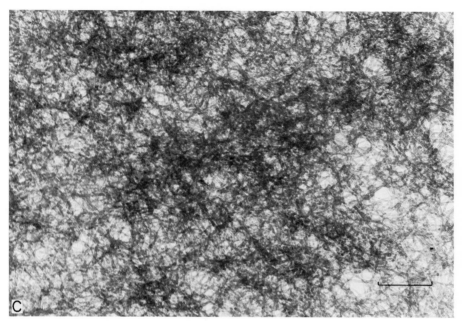

Fig. 4.7 *(Continued)*

industrial manufacturing processes the polyolefins are extracted with steam and detergent to reduce the catalyst residue further.

At an intermediate catalyst concentration of 5.1 mM Ti, the toluene-washed sample gives the electron micrograph shown in Fig. 4.8. This particular sample was subjected to ultrasonic vibration during washing. Therefore, the amount of spongy mass is not as much as one might expect, and the fibrils and ribbons appear to be somewhat entangled. Treatment with methanolic HCl converted the polymer morphology to the fibrillar type.

We next studied the Luttinger-catalyzed polymerization. Acetylene was polymerized with the Luttinger catalyst, as described by Lieser *et al.* (1980a). The product, after washing with ethanol, gave the electron micrograph shown in Fig. 4.9a, which exhibits mostly dense spongy mass and maybe some pseudolamallae particles similar to the ones reported by Wegner and co-workers. Further washing with methanolic HCl greatly reduced these foreign substances (Fig. 4.9b); fibrillar morphology was seen to emerge, but there were no well-developed fibrils. Apparently, the residue of the Luttinger catalyst is more difficult to remove than that of the Ziegler–Natta catalyst. When, on the other hand, a tenfold lower concentrated catalyst was used,

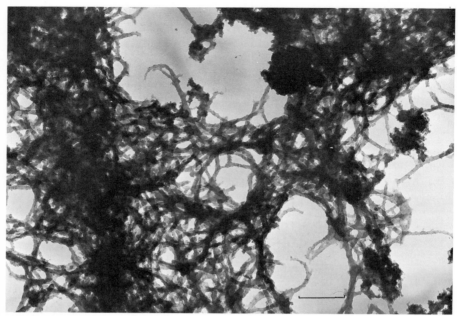

Fig. 4.8 Electron micrograph of polyacetylene polymerized in the diluent with the Ziegler–Natta catalyst, [Ti] = 5.1 mM, washed with toluene. Scale bar is 500 nm.

the polyacetylene obtained and washed only with ethanol contained a mixture of spongy mass and polymer fibrils (Fig. 4.10a). Further treatment with methanolic HCl yielded clean fiber morphology with only a little contaminant remaining (Fig. 4.10b). Finally, at every dilute concentration of Luttinger catalyst the polyacetylene after methanolic HCl washing showed only fibrils of the polymer (Fig. 4.11b).

The foregoing results, together with the morphology observed on ultrathin films polymerized directly on the grid, using either the Ziegler–Natta or the Luttinger catalyst, showed that the basic morphological entities of polyacetylene are microfibrils and fibrils. On the other hand, there may be lamellar crystallites in the preparations of Wegner and co-workers. At high catalysts concentrations, very low molecular weight polyacetylenes were produced in the reaction medium (Section 2.6), which may be sufficiently soluble to form lamellar crystals as in poly(styrene-di*b*-acetylene).

Meyer (1981b) observed that polyacetylene subjected to low-voltage electron-beam damage or OsO$_4$ staining displayed fibrils with regularly spaced striations. This was interpreted to be due to lamellae packed like stacks of salami slices. Chien *et al.* (1982g) observed similar aberrations of

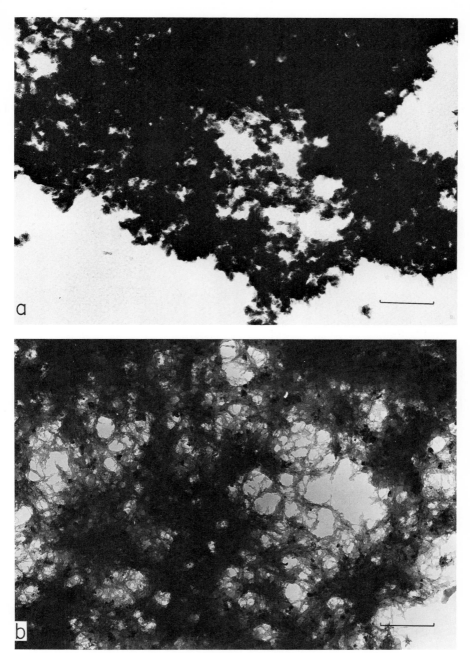

Fig. 4.9 Electron micrograph of polyacetylene polymerized in the diluent with the Luttinger catalyst, [Co] = 36 mM: (a) polymer washed with ethanol; (b) washed with methanolic HCl. Scale bar is 500 nm.

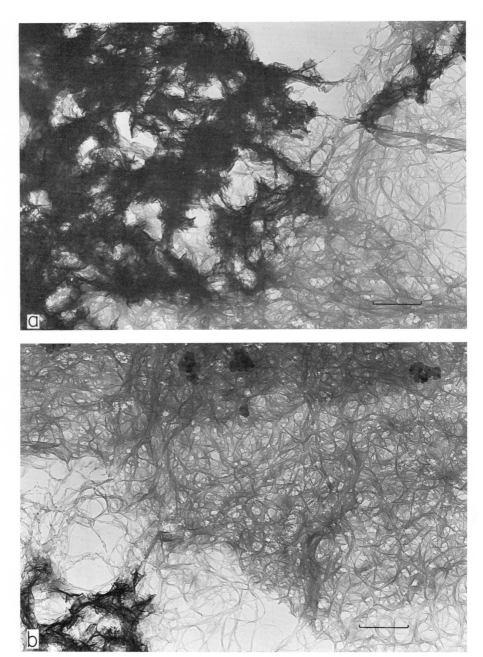

Fig. 4.10 Electron micrograph of polyacetylene polymerized in the diluent with the Luttinger catalyst, [Co] = 2.6 mM: (a) polymer washed with ethanol; (b) washed with methanolic HCl. Scale bar is 500 nm.

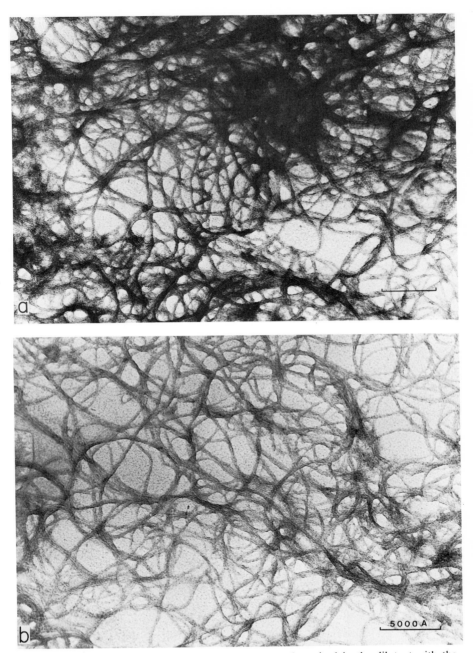

Fig. 4.11 Electron micrograph of polyacetylene polymerized in the dilutent with the Luttinger catalyst, [Co] = 0.3 mM: (a) polymer washed with ethanol; (b) washed with methanolic HCl. Scale bar is 500 nm.

polyacetylene fibrils in samples that had been kept in cold storage for extended periods of time and in some samples washed with methanolic HCl (Fig. 4.12). These striations may be caused oxidation- or solvent-induced stress cracking. Such cracks can appear as regularly spaced voids along the original fibril. The regularity of the cracks may be explained by the release of stress in their immediate vicinity. Freshly prepared polyacetylene specimens never exhibit such striation. The same interpretation may be applied to the observations made by Meyer (1981b). OsO_4 is a very strong oxidizing agent and may well cause stress cracking of the polyacetylene fibrils. Electron damages could also create the observed striation.

Even very brief exposure to oxygen can cause discernible modification of polyacetylene morphology. The fibrils in the ultrathin film of pristine polyacetylene, such as that shown in Fig. 4.13a, are smooth along the edges and have well-defined textures. Exposure of this specimen film to air at room temperature for 30 min, and observed at 123 K, gives the electron micrograph shown in Fig. 4.13b. The fibrils developed rough edges and have

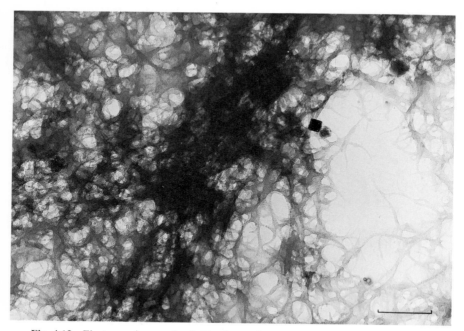

Fig. 4.12   Electron micrograph of old polyacetylene specimen showing striations, which are probably due to stress cracking. Scale bar is 500 nm.

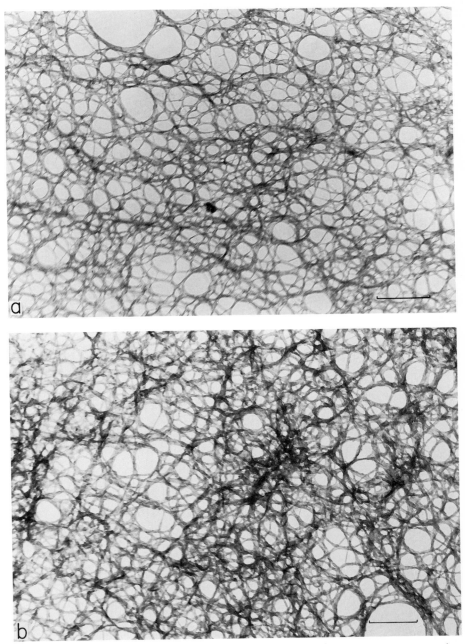

Fig. 4.13 Electron micrograph of an ultrathin *trans*-polyacetylene film, (a) as prepared, (b) after exposure of 30 min to air at room temperature.

nonuniform photographic darkness. It appears that the fibrils became wrinkled or twisted. This morphological change caused by oxygen is not reversible.

In conclusion, polyacetylene presents special problems to investigators. It is chemically very reactive, so that handling and exposure to the environment can result in artifactual morphology. The high catalyst concentration used to polymerize acetylene can also give rise to features not belonging to the polymer but to catalyst debris if they are not removed from the specimen.

4.5   WHISKERLIKE RIBBON CRYSTAL MORPHOLOGIES

Polyacetylene with macroscopic ribbon crystal morphology was produced when the polymerization was carried out under a hydrodynamic flow field (Chien *et al.* (1983c). In these experiments acetylene was polymerized with the Ziegler – Natta catalyst, 1 mM in Ti(OBu)$_4$ at room temperature. The catalyst solution was stirred at $\sim$ 800 rpm. Macroscopic ribbons of polymers started to grow from the ends of a magnetic stirring bar (Fig. 4.14a). As polymerization progresses, some of the ribbons broke off from the stirring bar (Fig. 4.14b). Individual macroscopic "whiskers" are over a centimeter long and at least 0.1 mm wide. Figure 4.15 shows that the whisker consists of a central ribbon with dendritic growth radiating from it and that it is highly birefringent.

The dendrites were examined under high magnification by TEM and were found to be flat ribbons comprised of the basic 20-nm fibrils. There are, however, several variations of this basic morphology.

(a)   *Loose collection of fibrils.*   Figure 4.16 shows those flat ribbons that are collections of loose fibrils. Whereas some of the fibrils are running more or less parallel to the long direction of the ribbon, others are found across this direction, leaving one ribbon to enter into another. That the basic morphological entity is the microfibril cannot be doubted as they can be seen clearly in the space between the two ribbons in this figure.

(b)   *Dense collection of aligned fibrils.*   Some of the ribbons are densely packed fibrils that have the fibril axes parallel to the long direction of the ribbon. They are very wide—the one shown in Fig. 4.17 has a width in excess of 10 μm. There are only a few fibrils meandering across the ribbon.

Fig. 4.14 Polymerization of acetylene under a hydrodynamic flow field: (a) polymers grown from ends of a magnetic stirring bar, 1.35 × magnification; (b) pieces broken off from the stirrer.

The crystallinity and alignment of such ribbons were demonstrated by the electron-diffraction pattern obtained at 123 K (Fig. 4.18). The majority of reflections in this pattern can be indexed with the *cis*-polyacetylene unit cell (cf. Section 3.5). In addition, there is the (002) reflection originating from the trans polymer. This is to be expected because the polymerization temperature was 298 K.

Isomerization of these ribbon crystals for 3 min at 448 K gave the electron diffraction pattern of Fig. 4.17b. Most of the reflections can now be indexed with the *trans*-polyacetylene unit cell (cf. Section 3.6) including faint (001) intensity. There remained faint intensity from the (002) reflection of the *cis*-polyacetylene. This is attributable to the very short time of isomerization used at the optimum temperature.

A very important observation is made by comparing Figs. 4.18a and b. The cis diffraction pattern has discrete arcs in the equator but diffuse and weak arcs in the layer lines, suggesting that the specimen possesses relatively good lateral order but poor order along the fiber direction. In comparison, the trans diffraction pattern has sharper layer lines, indicating an improvement in the longitudinal order. This increase in order was further confirmed

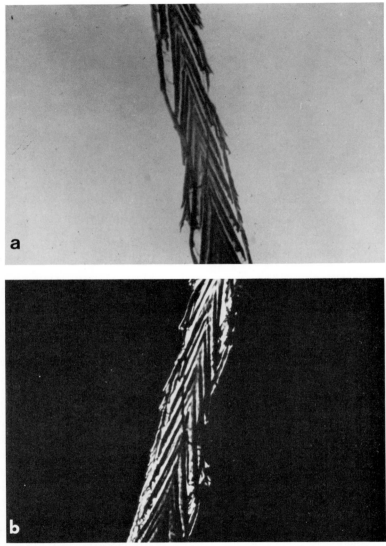

Fig. 4.15 Polyacetylene "whisker": (a) polarizer perpendicular to the long direction, (b) polarizer and analyzer crossed.

by dark-field images taken with the (200) and (110) piled reflections of the cis ribbon crystals (Fig. 4.19a) and of the trans ribbon crystals (Fig. 4.19b). The abundance of white spots in the latter corresponds to crystalline

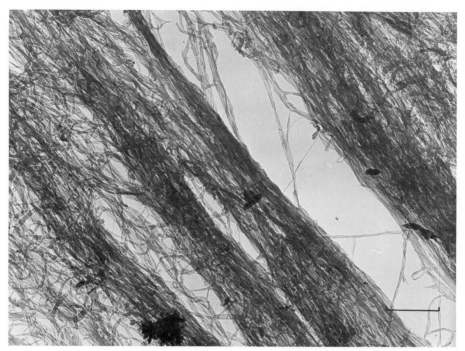

Fig. 4.16 Dendritic ribbons from the "whiskers," which is a loose collection of fibrils. Scale bar is 500 nm.

domains of about 70–100 nm in length. The dark-field image for the cis material is by comparison much less ordered in this regard. Obviously, there is annealing of the ribbon crystallites during this thermal treatment. This observation is contrary to the belief that polyacetylene undergoes spontaneous crosslinking even at low temperature. If this were true, then this annealing process would be unlikely to occur.

(c) *Intertwined ribbons.* Even though the ribbons radiate out from the central ribbon, they can become intertwined and fold over one another, as shown in Fig. 4.20. The ribbons do not all have uniform widths. Furthermore, other EM micrographs showed that the ribbons can be twisted as they are formed.

(d) *Woven fibrils.* Some of the ribbons appear to be woven from fibrils (Fig. 4.21).

Formation of such macroscopic whiskerlike ribbon crystals is uncommon in a polymerization process. Undoubtedly, the hydrodynamic shear

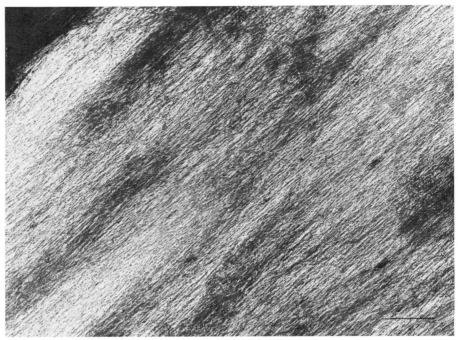

Fig. 4.17   Electron micrographs of dendritic ribbons made of densely and parallel packed fibrils. Scale bar is 500 nm.

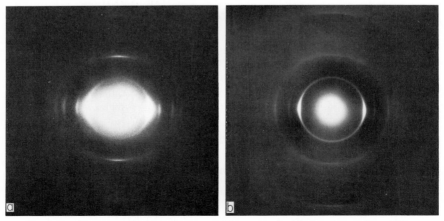

Fig. 4.18   Electron-diffraction pattern of ribbon crystals: (a) *cis*-polyacetylene; (b) isomerized for 3 min at 448 K.

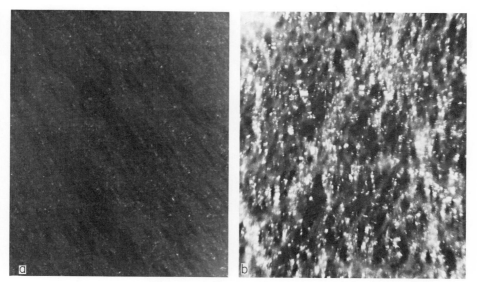

Fig. 4.19   Dark field images taken with the (200) and (110) piled reflections of ribbon crystals: (a) *cis*-polyacetylene; (b) *trans*-polyacetylene.

flow field is promoting this crystal growth. Similar effects have been observed in the growth of polymer crystals from solution. Pennings *et al.* (1973) observed that longitudinal growth of macrofibrillar polyethylene crystals was effectuated in a 0.4% solution of the polymer in *p*-xylene contained between two coaxial cylinders at 377 K, one of which rotates to create a velocity gradient as low as 0.7 sec^{-1}. The product can be wound up at a speed equal to the longitudinal growth rate. Continuous macrofibrils of polyethylene several hundred meters in length and diameter in the micron range could be obtained with this technique. Zwijnenburg and Pennings (1976) analyzed the phenomenon and showed that the relative velocity of the flowing polymer solution with respect to the growing fibers is equal to the sum of the velocity of the solution and the takeup speed minus the crystal growth rate.

Because of the similarity between the observations of Pennings and co-workers on polyethylene crystallization and ours on acetylene polymerization, it may be assumed that the same mechanism applies. In the latter case, the catalyst is adsorbed on the stirring bar and polymerization is initiated. The polymer chains grow from the stirring bar out into the diluent. As monomers are inserted, the shear field orients the newly formed segments. Each chain provides nucleation sites for adjacent chains to crystallize together to form the whiskers. As the shear field is diverging from the tip of

Fig. 4.20   Intertwined ribbons of polyacetylene. Magnification 12,000×.

the stirring bar, dendritic growth occurs to produce the ribbon crystals radiating from the central ribbon. According to the calculation of Zwijnenburg and Pennings (1976), the process requires only a very small elongational stress. In our experiment, the minimum and maximum radial veloc-

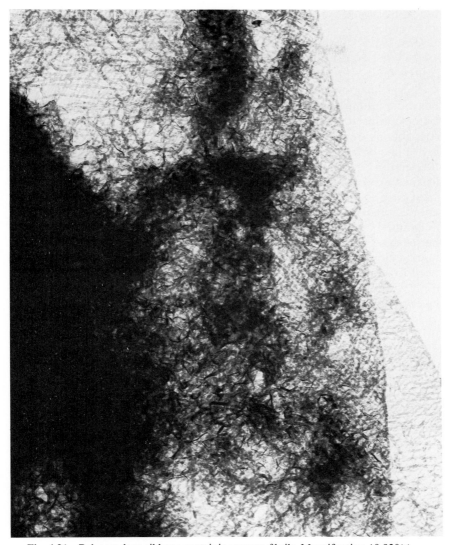

Fig. 4.21 Polyacetylene ribbons containing woven fibrils. Magnification 19,920×.

ity of the stirring bar ranges from 10^7 to 10^{10} Å sec^{-1}. This corresponds to a maximum shear rate of 10^2 sec^{-1} at the stirring-bar tips. The result of Mackley (1975) indicates very large hydrodynamic velocity graidents along the side surface and tips of a prolate ellipsoid aligned in such a flow field. The dominant force for orientation is not the transverse velocity gradient but is rather the elongational velocity gradient, which is about 10^6 sec^{-1} for a polyacetylene fibril of 20-nm diameter at a distance of 100 nm from the surface. This is of the same order as observed by Zwijnenburg and Pennings (1976) for longitudinal crystallization of polyethylene in Poisenille and Couette flow.

Meyer has polymerized acetylene under a more controlled shear field that ours, produced by a concentric Couette-type apparatus with a Teflon drum rotating at 3000–4000 rpm. In one paper (Meyer, 1981a) polyacetylene ribbons were obtained but that they have the trans structure, giving Debye-like diffraction patterns. Upon tilting the specimen with respect to the electron beam, Meyer observed *d*-spacings twice as large as the repeat distance expected for *trans*-polyacetylene. The author concluded that the chain direction is not along the fiber axis. In another paper, Meyer (1981b) reported obtaining a polyacetylene that was described as strings that have no preferential orientation of the polymer chain axis parallel to the long axis of the strings. The material is only weakly dichroic. A crystal structure "trans modification II" was proposed (Lieser *et al.*, 1980b). By increasing the rotor speed to 4000 rpm, Meyer (1981b) obtained highly oriented polyacetylene fibrils with their axes parallel to the chain axes. Lieser *et al.* (1980b) found the crystal structure to be trans modification I, which has a very unusual

$$\text{c}^{\diagup\text{c}^{\diagdown}}\text{c}$$

bond angle of 135°. Finally, Meyer (1981b) used a stator with concentric grooves on its inner surface, attempting graphoepitaxial polymerization. In this case, aligned fibrils were obtained with trans modification II structure.

Therefore, even though Meyer used Couette apparatus for acetylene polymerization under controlled shear field, the polymer obtained did not have as high a crystallinity as those obtained by us and certainly not the macroscopic whiskerlike polyacetylene ribbons.

4.6   MECHANICALLY ALIGNED POLYACETYLENES

4.6.1   Stretch Alignment

Shirakawa and Ikeda (1979) reported the stretch alignment of polyacetylene film. The maximum draw ratio was $l/l_o = 3.0$ for the cis-rich film. The prestretched film largely maintained its initial high cis content with a few percent of cis–trans isomerization occurring during the stretching. Electron microscopy showed the fibrils to become aligned along the stretching direction. The extensibility of polyacetylene is sensitive to the quality of the material. It must be freshly prepared with a high initial cis content, i.e., > 88%, and with a minimal exposure to air. Otherwise the film will break at low draw ratios. Figure 4.22 showed that elongation to the breaking point decreases rapidly with decrease of cis content, and it was not possible to stretch the polymer of high trans content without fracturing. There is a much smaller decrease in tensile strength with cis content. However, the tensile strength of a material such as the polyacetylene film does not have the usual significance because of its fibrillar morphology and large void volume.

After the threefold stretching, the cis film can be further elongated by maintaining a stress of 0.53 kg mm^{-2} and heating the polymer in vacuo at a rate of 2–3 K min^{-1}. Most of the additional elongation occurred around the temperature range of 403 to 433 K. This coincides with the temperature range for the fast rate of cis–trans isomerization. The final l/l_o obtained was about four.

The stretching process decreases both the pore size and integrated pore

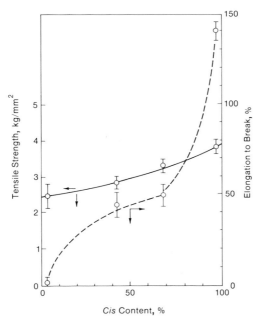

Fig. 4.22   Mechanical properties of polyacetylene film as a function of cis content. (After Shirakawa and Ikeda, 1979).

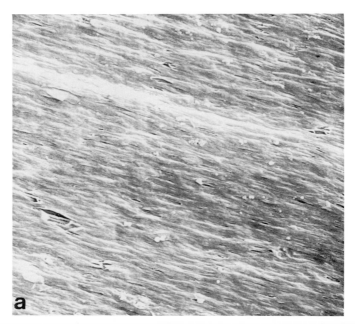

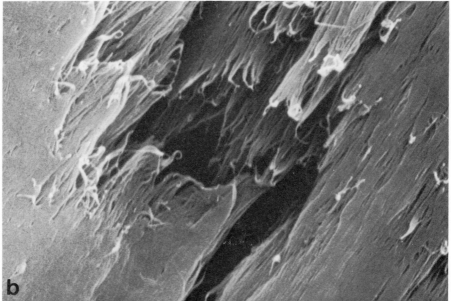

Fig. 4.23   Coextruded polyacetylene film: (a) smooth surface, magnification 5000×; (b) fractured area, magnification 8000×.

volume of the specimen. The integrated pore volume decreased from 1.5 cm^3 g^{-1} for the unstretched film to about 0.3 cm^3 g^{-1} for $l/l_o = 2.34$. This indicates that fibrils are more closely packed upon stretching. At $l/l_o = 2.34$ the fibril occupies 71% of the total volume of the film as compared to only one-third of the volume before stretching.

4.6.2 Coextrusion

Polyacetylene has been coextruded in our laboratory. In this technique, a piece of freshly prepared, high-purity polyacetylene film was placed between the two halves of a cylindrical polyethylene billet. The assembly was introduced under inert atmosphere into an Instron rheometer. Drawing conditions were varied; temperature, 323–403 K; speed, 0.1–1 cm min^{-1}; draw ratio, 2.9–4.8; at maximum pressures of 150–1400 kg cm^{-2}. The best coextrusion conditions were found to be 383 K, 0.1 cm min^{-1}, 3.4 draw ratio, and 350–800 kg cm^{-2} pressure. The specimen has $l/l_o \sim 2.7$. Figure 4.23a shows the smooth surface of the drawn material with parallel striations suggesting orientation along the draw direction. Under less than optimum conditions the extrudate had surface cracks and microfractures. Figure 4.23b is a SEM view into such a fracture region showing clearly that the fibrils beneath the surface of the extrudate were highly aligned by this process.

4.7 SURFACE AREAS

We have measured the surface area of typical free standing polyacetylene films. The results are summarized in Table 4.1.

TABLE 4.1

Surface Areas of Polyacetylene Film from
B.E.T. Measurements

(CH)$_x$ Sample	Surface area (m^2/g)
1	55.6
2	70.6
3	72.5
	Average = 66.0

The theoretical surface area of polyacetylene film, assuming the fibril diameter (d) to be 200 Å, was determined as follows. First, we take the length (l) of the fibrils to be, say, 2000 Å. This choice is completely arbitrary and any other value will lead to the same final result. The surface area S_A of one such fibril is given by

$$S_A = \pi dl$$

Hence, $S_A = \pi(2 \times 10^{-8} \text{ m})(2 \times 10^{-7} \text{ m}) = 1.25 \times 10^{-14} \text{ m}^2$. The volume of the fibril is simply $\pi r^2 l$ or $\pi(1 \times 10^{-8} \text{ m})^2 (2 \times 10^{-7} \text{ m}) = 6.3 \times 10^{-23} \text{ m}^3$. The amount of polyacetylene in such a fibril, taking the density (flotation) to be 1.2 g cm^{-3}, is

$$\text{Amount (CH)}_x = 1.2 \times 10^6 \times 6.3 \times 10^{-23}$$
$$= 7.6 \times 10^{-17} \text{ g/fibril}$$

The surface area in m^2 g^{-1} is

$$\frac{1.25 \times 10^{-14} \text{ m}^2}{7.6 \times 10^{-17} \text{ g}} = 165 \text{ m}^2 \text{ g}^{-1}$$

However, the density of the polymer is only 0.4 g cm^{-3}, indicating that the fibrils fill only about one-third of the total volume of the sample. The surface area then is

$$165 \text{ m}^2 \text{ g}^{-1} \times \frac{0.4 \text{ g cm}^{-3}}{1.2 \text{ g cm}^{-3}} \simeq 55 \text{ m}^2 \text{ g}^{-1}$$

This value is in very good agreement with the average value of 66 m^2 g^{-1} obtained experimentally.

Chapter 5

Isomerization,
Neutral Defects,
and Solitons

5.1 INTRODUCTION

Cis–trans isomerization of substituted olefinic compounds occurs with two types of unimolecular kinetics. The first type is characterized by a low-frequency factor and low activation energy

$$k \simeq 10^4 \exp\left(\frac{-25,000}{RT}\right) \quad \text{sec}^{-1} \tag{5.1}$$

Examples are maleic acid and its esters and butene-2. The mechanism is thought to involve the triplet state as an intermediate for rotation about the double bond. The low-frequency factor is attributable to the "forbidden" nature of the transition because a change in spin multiplicity is involved. The second type requires high activation energy but with a normal frequency factor

$$k \simeq 10^{11} \exp\left(\frac{-45,000}{RT}\right) \quad \text{sec}^{-1} \tag{5.2}$$

observed for stilbene and its derivatives and cinnamate esters. The activated state differs from the initial state by the torsional vibration only. In addition, isomerization can be catalyzed by molecules such as iodine and paramagnetic compounds.

Thermal cis–trans isomerization of polyacetylene occurs at rates much faster than those shown in Eqs. (5.1) or (5.2). Since the method is used widely to convert the cis polymer to the trans isomer, the process is discussed in detail in Section 5.2, and optimum conditions are recommended.

Doping results in the isomerization of polyacetylene. This can be achieved by using chemical doping, electrochemical doping, or a combination of the two, as discussed in Section 5.3. It should be pointed out that several direct and indirect methods have been used to follow the progress and extent of cis–trans isomerization. The most commonly used method is the relative intensity of the infrared bands at 740 cm^{-1} and 1015 cm^{-1} for the cis and trans C–H vibrations, respectively (Section 2.2.9). It will be shown below that isomerization of polyacetylene results in the formation of neutral defects characterized by a $g = 2$ EPR signal. However, the intensity of the resonance is not directly proportional to the isomeric content of the polymer. In ^{13}C NMR the cis and trans structures have resonances at 126 and 136 ppm from TMS, respectively. Therefore, this can be used to monitor the degree of isomerization, but the sensitivity of the technique is not great. A very sensitive indirect measure of isomerization is by electrical resistivity because the two isomers differ by as much as five to six orders of magnitude in this property. However, the transport property is really governed by the concentration of carriers and its mobility, both of which may depend on the isomer content. Consequently, they may be related by some high-order dependency. No one has yet made a quantitative comparison between the various methods.

The neutral defects in cis and trans polymers differ greatly in their diffusivity. *cis*-Polyacetylene has immobile defects. The defects in *trans*-polyacetylene are highly diffusive and have been referred to as neutral solitons. Their characterizations by EPR are the subjects of Section 5.4. The questions about how mobile the solitons are, both along the polymer chain and transverse to it, have been examined by EPR saturation, EPR linewidth, dynamic nuclear polarization, nuclear-spin relaxation, EPR-spin echo, and other methods. The diffusion coefficients estimated thusly are compared in Section 5.5 on spin dynamics.

Finally, we propose in Section 5.6 some mechanisms of soliton formation and of cis–trans isomerization, emphasizing the chemical aspects of the processes.

5.2   THERMAL ISOMERIZATION

5.2.1   Energetics

The common procedure for the conversion of *cis*-polyacetylene to the trans polymer is by heating in vacuo. Ito *et al.* (1975) were the first to study the process with IR spectroscopy. Their results showed that the process is quite complicated. The characteristic 740 cm^{-1} band for the cis C–H out-of-plane deformation decreased with heating. At any given temperature the rate of decrease is initially rapid but comes nearly to a stop after a certain level of trans content is reached (Fig. 5.1). Heating to higher temperatures increases the extent of isomerization further; but the process is irreversible. The rate curve did not follow any simple kinetic order; the authors used the initial rates to obtain activation energies. When the experiments were carried out using samples that have an initial cis content of greater than 88%, the Arrhenius plot of the initial rate of isomerization versus T^{-1} gave an activation energy of 17 kcal mole^{-1}. As the cis content of the polymer decreases, the activation energy for isomerization increases rapidly. This increase becomes more gradual between 40 and 60% trans content, but at higher trans content the activation energy climbs steeply again to as high as 38 kcal mole^{-1} (Fig. 5.2). The results suggest that isomerization of short segments is relatively easy but it is more difficult for long segments due to steric hindrance. Between 40 and 60% trans the crystallites probably change over from the cis to the trans structure. Finally, rotation of the cis unit

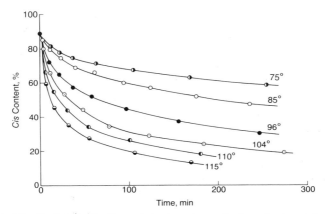

Fig. 5.1   Kinetic plots for the thermal cis–trans isomerization of polyacetylene at various indicated temperatures in degrees Celsius. [After Ito *et al.* (1975).]

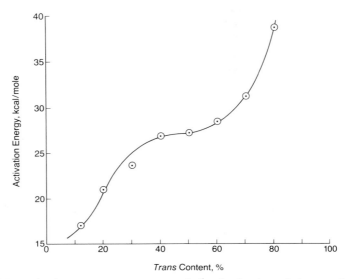

Fig. 5.2   Activation energy–trans content relation for the thermal cis–trans isomerization. [After Ito *et al.* (1975).]

between two long trans segments requires high activation energy. There is an analogy for this effect in low molecular weight polyenes. Thus the activation energy is 17 kcal mole^{-1} for thermal cis–trans isomerization of 1,6-diphenylhexatriene-1-*cis*-3-*trans*-5-*trans*; it is 24 kcal mole^{-1} for 1,6-diphenylhexatriene-1-*trans*-3-*cis*-5-*trans*. The changes in activation energy for isomerization of polyacetylene is accompanied by density changes. The density at first decreases with increasing trans content, reaching a minimum value at $\sim$ 50% trans, followed by an increase in density with further isomerization.

Complete isomerization is difficult to achieve thermally. A weak 747 cm^{-1} band persisted even after prolonged heating at 473 K. This absorption may be assigned to cis C–H out-of-plane deformation in a single or two consecutive cis double bonds flanked on both sides by longer trans sequences. The maximum trans content attainable by thermal isomerization appears to be 98%. In contrast, polymerization of acetylene above 423 K produces 100% trans polymer, as judged by the absence of the 747 cm^{-1} band. This is understandable, based on the mechanism of polymerization (Section 2.5). At 423 K, cis insertion of the monomer creates a new cis-transoid unit at the end of a growing polyacetylene chain, which almost immediately undergoes bond rotation to the trans-transoid unit before the segment containing it is crystallized. Two factors may be contributing toward the ease of isomerization during chain propagation. There are no lattice constraints against the bond rotation. Furthermore, the newly inserted monomer unit is bound to a Ti^{3+} ion. It is a well-known fact that paramagnetic species can catalyze cis–trans isomerization.

The conversion of a cis crystal structure to the trans is a phase transition. The transition temperature was found at 418 K (Fig. 5.3, curve 2) by differential scanning calorimetry. There is no corresponding exotherm for the trans polymer (curve 3). The higher temperature exotherm and endotherm will be discussed in Section 7.1.

5.2.2 Rate

The rate of thermal isomerization has been measured by Montaner *et al.* (1981a) using IR. The integrated absorption of the *i*th band is obtained by

$$A_i = l^{-1} \int \log T(v) \, dv \tag{5.3}$$

where *l* is the sample thickness and $T(v)$ is the transmittance at wave number *v*. As polyacetylene is heated, A_i changes with time. The absorbances of the trans bands increase while those of the cis bands decrease with time of

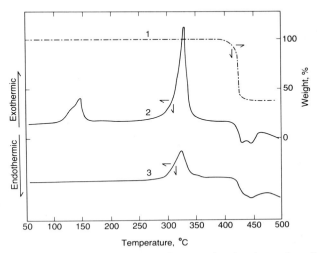

Fig. 5.3 DSC and TGA thermograms: (1) TGA curve for *cis*-polyacetylene; (2) DSC curve for *cis*-polyacetylene; (3) DSC curve for *trans*-polyacetylene. [After Ito *et al.* (1975).]

heating. The time constant τ_i for the ith band is obtained from

$$A_i(t) = A_i(0) \exp(-t/\tau_i), \tag{5.4}$$

and

$$A_i(t) = A_i(0) + [A_i(\infty) - A_i(0)][1 - \exp(-t/\tau_i)], \tag{5.5}$$

for the cis and trans bands, respectively.

The time constants decrease with increase of temperature (Table 5.1) but are not the same for the various bands. In the case of the cis bands, this has been explained by assuming that the 740 cm^{-1} band depends mainly on nearest neighbor interaction, whereas the 446 cm^{-1} band is associated with angular deformation of the cis carbon chain, involving longer range interactions. Thus during isomerization the fraction of long cis sequences decreases more rapidly than the fraction of isolated cis units. The trans vibrations, on the other hand, are very readily saturated. The time constants for the formation of trans units obtained by Montaner *et al.* (1981a) cannot be relied upon heavily. The absorption of the 1015 cm^{-1} band of the specimen employed was so intense that there was little transmitted radiation. The accuracy in the measured intensity of the 1292 cm^{-1} band was probably quite low because of severe overlapping with adjacent bands. It will be worthwhile to repeat these experiments using thinner specimens.

High-resolution ^{13}C NMR had been used by Bernier *et al.* (1981a) to follow the rate of isomerization. At 453 K the cis peak at 126 ppm disappeared very rapidly, accompanid by a corresponding increase of the trans resonance at 136 ppm from TMS. In one min of heat treatment, the polymer was isomerized to 90% trans. The trans resonance intensity did not increase any further after 10 min at this temperature, indicating complete isomerization. Therefore, cis–trans isomerization is extremely rapid at 453 K. There is no resonance in the region between 40 and 100 ppm, where various kinds of sp^3 carbon atoms should have their chemical shifts, either

TABLE 5.1

Time Constants for Different IR Bands as a
Function of Isomerization Temperature[a]

Isomeric unit	ν_i (cm^{-1})	τ_i, min at		
		388 K	415 K	433 K
Cis	446	154	19	2.7
	740	454	46	6
Trans	1015	15	—	—
	1292	50	—	—

[a] After Montaner *et al.* (1981a).

in the initial *cis*-polyacetylene or after heating at 453 K for 60 min. There-fore, there is little, if any, cross-linking during this experiment.

Lefrant and Aldissi (1983) followed the thermal isomerization by Raman scattering. At 388 K the cis bands decreased in intensity monotoni-cally and disappeared after 400 min. The trans bands at 1100 and 1480 cm^{-1} were broad with rather flat tops as they first emerged. With time of heating the intensities of these bands increased, shifted toward lower frequency, and assumed a sharp maximum, but retained a high-frequency shoulder. At 413 K the cis bands disappeared in less than 7.5 min; none was observed at 463 K, indicating instantaneous cis–trans isomerization. At this temperature the trans bands become broader and shifted toward higher frequencies with time, which was interpreted as a decrease in conjugation length.

5.2.3   Optimum Conditions

A very sensitive method for monitoring changes in the structure of polyacetylene during thermal isomerization is by measuring its resistivity.

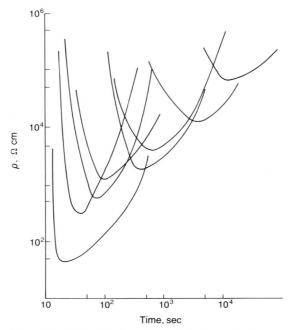

Fig. 5.4   Variation of resistivity of polyacetylene during isomerization process with time at different temperatures between 423 and 533 K. Temperature in K for curves from right to left: 533, 528, 523, 510, 483, 468, 438, 423. [After Rolland *et al.* (1980a).]

Rolland *et al.* (1980a) showed that at a given temperature of isomerization the variation of resistivity with time of heating curve is concave in shape (Fig. 5.4). At low temperatures it takes a long time (t_{min}) for a sample to reach the minimum resistivity before it increases. At high temperature the minimum was reached quickly. A semilog plot of the time to attain minimum resistivity at a given temperature versus T^{-1} gives an activation energy of 18 kcal mole^{-1} (Bernier *et al.*, 1981b). The optimum conditions appear to lie between 443 and 453 K for 5–20 min. This isomerization condition also produces *trans*-polyacetylene samples having the least diffuse electron-diffraction patterns, significant (001) intensity, and order along the fiber direction.

We can now compare the spectroscopic and resistivity results. At 388 K a few months is needed to isomerize *cis*-polyacetylene to the minimum resistivity, but the cis Raman bands disappeared within 400 min, and the IR time constants (Table 5.1) ranged from 15 to 450 min. At 423 K, t_{min} is about 3 hr, whereas the cis Raman bands disappeared in less than 7.5 min, and the IR time constants were also shorter than 6 min. The comparison indicates that the spectroscopic techniques are insufficiently sensitive to detect small amounts of cis units remaining, which may have important effect on the transport properties. Also, overisomerization cannot be seen until the polymer is severely degraded.

5.2.4 Excessive Heating

The effects of prolonged heat treatment in causing reduction in conductivity and modification of crystal structure have been described above. It also has a marked effect on increasing the EPR linewidth, ΔH_{pp}. Figure 5.5 showed the expected decrease in ΔH_{pp} for low-temperature heating as *cis*-polyacetylene is isomerized to trans. Beginning at 433 K the rapid initial decrease in ΔH_{pp} is followed by a slight increase with time of heating. At temperatures higher than 453 K, the initial line-narrowing phase occurred too rapidly to be followed, and the ΔH_{pp} increase due to excessive heating becomes greater with higher temperatures.

The narrow EPR linewidth of the unpaired spin in *trans*-polyacetylene is attributable to its high diffusivity (Section 5.5); conversely, the increase in linewidth is indicative of restrictions on its diffusion. The barriers to spin diffusion probably result from various reactions to be detailed in Chapter 7, which for the time being will be simply referred to as degradation.

The rate of linewidth increase at elevated temperatures cannot be de-

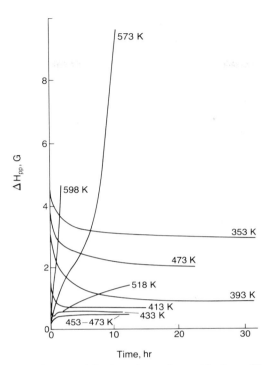

Fig. 5.5 Variation of ΔH_{pp} of S· in polyacetylene with time of heating at different temperatures. Initially the cis content is 95%. [After Bernier *et al.* (1981b).]

scribed by any simple kinetics. Bernier *et al.* (1981b) plotted the initial dH_{pp}/dt versus T^{-1} and found an activation energy of about 30 kcal mole^{-1}. This value represents the average energetic requirement for the thermal degradation of polyacetylene.

Another effect of excessive heating is to reduce the number of unpaired spins. Figure 5.6 shows that heating at 393 K increases [S·] as the polymer is isomerized. However, at temperatures $\geqslant 473$ K, [S·] first increases and then declines upon prolonged heating. This decrease in [S·] is greater, the higher the temperature.

In conclusion, excessive heating should be avoided in isomerizing *cis*-polyacetylene to *trans*-polyacetylene. Some of the properties are critically affected by it. There are also preliminary results (Rolland *et al.*, 1980a) showing that excessive heat treatment tended to reduce the conductivity of iodine-doped polymer as compared to those heated for a shorter period of time.

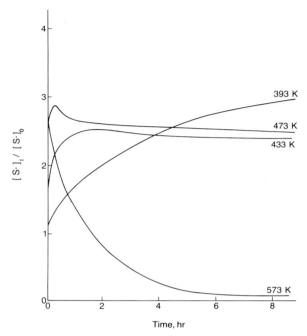

Fig. 5.6   Variation of [S·] with time at different temperatures. [After Bernier *et al.* (1981b).]

5.3   DOPING-INDUCED ISOMERIZATION

5.3.1   Chemical

It is apparently a general phenomenon that doping leads to cis–trans isomerization. For example, *cis*-polyacetylene is *n*-doped to the metallic state with sodium naphthalide solution and then compensated with 0.1 torr of AsF_5 vapor. When the visible–near IR spectrum of the specimen is followed during compensation, the undoping can be stopped at the right point before *p*-doping by AsF_5 occurs. After the salt produced is removed by washing, pure *trans*-polyacetylene can be obtained. The amount of trans units is negligible at low levels of doping and becomes appreciable when the polymer is doped to a few percent. Complete isomerization requires heavy doping of polyacetylene to the metallic state.

5.3.2   Catalyzed by Oxygen

Because it is virtually impossible to prevent even brief exposure of polyacetylene samples to oxygen, its effects on the structure properties of polyacetylene are always matters of great concern. In this section we present the results pertaining to catalysis of cis–trans isomerization by oxygen. When a defect-free polyacetylene sample prepared directly in an EPR tube was warmed up to 253 K, a $g = 2$ signal appeared (Section 2.5.1; Chien *et al.*, 1980a; Bernier *et al.*, 1980). When air was admitted to such a sample, the EPR signal became intensified (Fig. 2.6). The observation implies that oxygen can catalyze the formation of neutral defects and/or cis–trans isomerization.

The kinetics of oxygen-catalyzed isomerization was followed by Chien and Yang (1983a) with the characteristic IR bands. Free-standing *cis*-poly-acetylene film was placed in an IR cell filled with 1 atm of oxygen. The rates of isomerization from 313 to 343 K in oxygen are given in Fig. 5.7. Also given in the figure is the rate of isomerization under 1 atm of purified

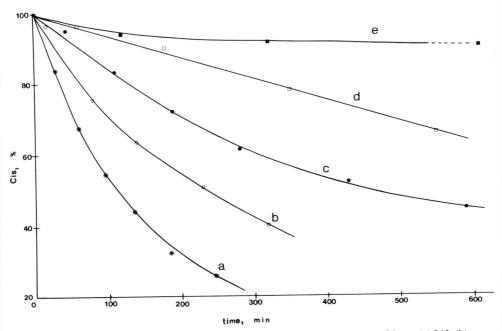

Fig. 5.7   Rate of cis–trans isomerization of polyacetylene under 1 atm of O_2: at (a) 343, (b) 333, (c) 323, and (d) 313 K. (e) Under 1 atm of N_2 at 333 K.

nitrogen at 333 K. By comparison, the process is much faster in the presence of oxygen. The rates of oxygen-catalyzed reactions obey first-order kinetics (Fig. 5.8). In contrast, the rate of isomerization in nitrogen is not strictly first order, but the rate at 333 K is so slow that the kinetics may be approximated as such. The first-order rate constant at 333 K is 5.0×10^{-5} sec^{-1} with oxygen, whereas it is 4.2×10^{-6} sec^{-1} without it. Therefore, oxygen increases the rate of isomerization at this temperature 12-fold. Figure 5.9 is an Arrhenius plot for the oxygen-induced cis–trans isomerization of polyacetylene, which gives 14.9 kcal mole^{-1} for its overall activation energy.

The oxygen-catalyzed isomerization of polyacetylene may be written as

$$cis\text{-}(CH)_x + O_2 \xrightleftharpoons{K} cis\text{-}(CH)_x \cdot O_2 \tag{5.6}$$

$$cis\text{-}(CH)_x \cdot O_2 \xrightarrow{k} trans\text{-}(CH)_x \cdot O_2 \tag{5.7}$$

The rate of isomerization is given by

$$\frac{-d[cis\text{-}(CH)_x]}{dt} = Kk[cis\text{-}(CH)_x][O_2] \tag{5.8}$$

The observed overall activation energy of 14.9 kcal mole^{-1} is the sum of the enthalpy change for the complexation of oxygen with polyacetylene [Eq. (5.6)] and the activation energy for isomerization [Eq. (5.7)]. Because the enthalpy change is not known, neither can the latter value be calculated.

According to Eq. (5.8), the rate of isomerization should be proportional to the oxygen concentration. This has been demonstrated. Figure 5.10 gives

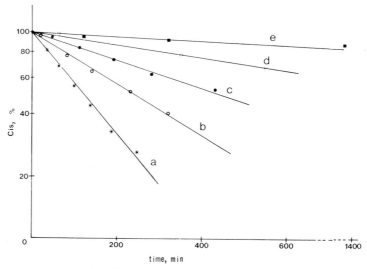

Fig. 5.8 First-order kinetics plots of data in Fig. 5.7.

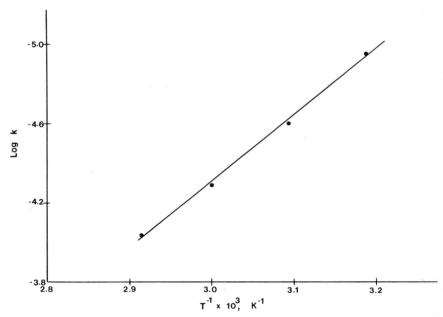

Fig. 5.9 Arrhenius plot for the oxygen-induced cis–trans isomerization of polyacetylene.

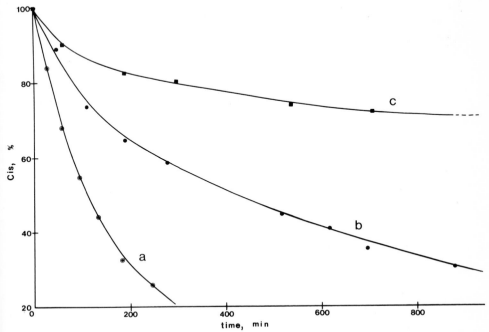

Fig. 5.10 Rate of cis–trans isomerization at 343 K: under (a) 1 atom of O_2, (b) 1 atm of air, and (c) 1 atm of N_2.

the data of isomerization rates at 343 K in oxygen, air, and nitrogen. The nitrogen results were used to correct for the thermally-induced isomerization. Figure 5.11 shows the first-order kinetics plot of the oxygen-catalyzed isomerization after correction for the thermal process. The ratio of the slopes of lines a and b in Fig. 5.11 corresponds closely to the oxygen partial pressures in the two reactions.

Let us compare the rates of isomerization with and without oxygen. The minimum activation energy for the thermal process is 18 kcal mole^{-1}. From the data of Montaner *et al.* (1981a) (Table 5.1), one calculates a rate constant value at 388 K of 3.7×10^{-5} sec^{-1} from the 740 cm^{-1} band. This band is chosen because the time constant obtained with it is more in line with the Raman results (Lefrant and Aldissi, 1983) than the 44.6 cm^{-1} band. The oxygen-catalyzed value at this temperature would be 1.4×10^{-3} sec^{-1}, which is about 40 times faster than the thermal process. Therefore, thermal isomerization can be strongly affected by oxygen contamination.

Lefrant *et al.* (1980) studied the oxygen-induced isomerization of poly-acetylene film with Raman spectroscopy. Figure 5.12a shows the spectrum of the freshly prepared cis polymer film, which remains unchanged after one

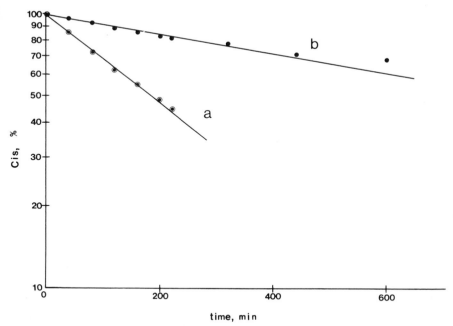

Fig. 5.11   First-order kinetics plots of data in Fig. 5.10 corrected for thermal isomeriza-tion: under (a) 1 atm of O$_2$ and (b) 1 atm of air.

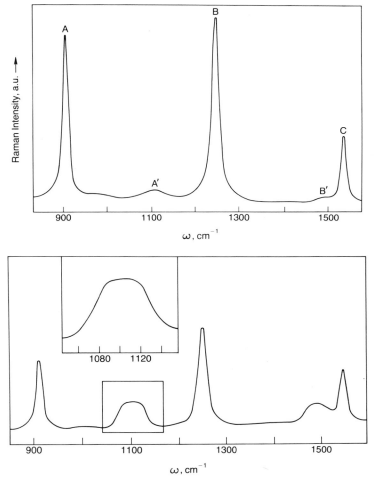

Fig. 5.12 Raman spectrum of cis-rich polyacetylene recorded at 78 K, excitation wavelength 600 nm: (a) after one month under vacuum at 298 K, (b) after one month in air at 298 K. [After Lefrant *et al.* (1980).]

month at 298 K in vacuo. The spectrum is characterized by the intense 908 cm^{-1} (A), 1247 cm^{-1} (B), and 1541 cm^{-1} (C) peaks for the cis polymer and the very weak 1080 cm^{-1} (A′) and 1450 cm^{-1} (B″) Raman bands for the trans units. After exposing the film to air at 298 K for one month, the cis bands had decreased in intensity, and the trans bands had intensified (Fig. 5.12b). The A′ band is very broad and has a flat top (inset of Fig. 5.12b). This has been interpreted as a distribution in the length of the trans sequence; the shorter the sequence, the higher the frequency. It seems that isomerization induced by oxygen leads to a large number of short trans segments according to this interpretation.

5.3.3 Electrochemical

After polyacetylene is either *n*- or *p*-doped electrochemically to the metallic state, it can be undoped by reversing the potential. The process converts *cis*-polyacetylene to *trans*-polyacetylene in a precisely controlled manner. This is possible because at a known potential applied to a poly-acetylene electrode, the polymer is doped to a corresponding level, as monitored by its visible–near-IR spectrum. Furthermore, from the spectral change one can determine when equilibrium or near equilibrium conditions are reached. The doped polymer can be undoped subsequently by changing the potential to that characteristic for the undoped trans polymer in a particular electrolyte.

Chung *et al.* (1982) used an electrochemical cell described in detail in Section 8.2.5. A fine glass frit separated the two electrodes. One of the electrodes was a conductive glass plate onto which a thin film of *cis*-poly-acetylene was polymerized. A lithium wire served as the other electrode. A "blank" spectrum of the cell plus the electrolyte was first recorded and stored in a computer. Its absorption was subtracted from the spectrum of the polymer, during doping and undoping. Electrochemical isomerization can be accomplished by several pathways.

5.3.3.1 Electrochemical n-Doping Followed by Electrochemical Undoping

To illustrate the process, *cis*-polyacetylene is *n*-doped with tetrabutyl-ammonium and then undoped to obtain pure *trans*-polyacetylene. The spectrum of the starting *cis*-polyacetylene is given in spectrum 1 of Fig. 5.13 using a 1.0 M solution of $(Bu_4N)(ClO_4)$ in THF as the electrolyte. *n*-Doping occurs spontaneously. A constant-voltage power supply was used simply to maintain the voltage applied to the polymer for a desired level of doping. Spectra 2–5 were obtained for ~ 1, ~ 3, ~ 5, and $\sim 8\%$ of doping at applied voltages of 1.5, 1.35, 1.3, and 1.25 V, respectively. The 2.1- and 2.3-eV visible absorption peaks characteristic of *cis*-polyacetylene contin-uously decrease while the midgap peak at 0.7–0.8 eV characteristic of

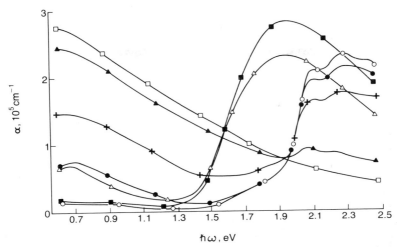

Fig. 5.13 Visible–near IR absorption spectral change during the electrochemical cis–trans isomerization (1) ○-○-○ ± $V_{oc} = 2.7$ V; *cis*-polyacetylene (2) ●-●-●, $V_{app} = 1.5$ V; (3) x-x-x, $V_{app} = 1.35$ V; (4) ▲-▲-▲, $V_{app} = 1.3$ V; (5) □ □, $V_{app} = 1.25$ V; (6) △-△-△-, $V_{app} = 2.0$ V; (7) ■ ■, $V_{app} = 2.8$ V. [After Chung *et al.* (1982).]

doped trans polymer increases with doping. At 1.24 V of applied electromotive force (EMF), the polymer exhibits spectrum 5, essentially identical to polyacetylene doped chemically to the metallic state such as $(CHNa_y)_x$. The process can be written

$$cis\text{-}(CH)_x + xye + xy(Bu_4N^+) \rightarrow trans\text{-}[(CH)^{-y}(Bu_4N^+)_y]_x \qquad (5.9)$$

To obtain pure *trans*-$(CH)_x$, one applies a potential up to 2.8 V. Spectrum 7 of Fig. 5.11 showed the material to be undoped *trans*-$(CH)_x$ free of midgap absorption. The undoping reaction is

$$trans\text{-}[(CH)^{-y}(Bu_4N^+)_y]_x \rightarrow trans\text{-}(CH)_x + xye + xy(Bu_4N^+) \qquad (5.10)$$

The doping–undoping isomerization need not be done in a single cycle. For instance, the polymer can be doped and undoped in steps, with increased extent of isomerization by each cycle. If the polymer is doped to y_1 concentration, then undoped to neutral polyacetylene, subsequent doping to y_2, with $y_2 < y_1$, results in no increase in isomerization, which shows that in this second doping cycle the existing trans units are preferentially doped over the cis units. However, if the redoping is to a higher level y_3, with $y_3 > y_1$, additional isomerization occurs as the remaining cis units are doped.

5.3.3.2   *Electrochemical* p-*Doping Followed by Electrochemical Undoping*

An isomerization procedure analogous to that desxcribed in Section 5.3.3.1 is to begin the process by *p*-doping. The *cis*-polyacetylene electrode was attached to the positive terminal of a constant-voltage power supply. A 1.0 *M* solution in $LiClO_4$ in propylene carbonate served as the electrolyte. With the applied EMF at 3.83 V, the polymer is transformed to the *p*-type metallic $[(CH)^{y+}(ClO_4^-)_y]_x$, which exhibits a spectrum very close to curve 5 of Fig. 5.13. The sample can be undoped at 2.0 V to produce pure undoped *trans*-polyacetylene. The reactions involved are

$$cis\text{-}(CH)_x - xye + xy(ClO_4^-) \rightarrow trans\text{-}[(CH)^{y+}(ClO_4^-)_y]_x \qquad (5.11)$$

$$trans\text{-}[(CH)^{y+}(ClO_4^-)_y]_x + xye \rightarrow trans\text{-}(CH)_x + xy(ClO_4^-) \qquad (5.12)$$

5.3.3.3   *Electrochemical Doping Followed by Chemical Undoping*

cis-Polyacetylene was *n*-doped to the metallic state. It was then exposed to very low pressure of AsF_5 while monitoring the IR spectrum all the time. Unreacted AsF_5 was pumped off as soon as the 1380 cm^{-1} peak characteristic for soliton disappears. The product is pure *trans*-polyacetylene. The processes can be represented by Eq. (5.9) followed by

$$2\ trans\text{-}[(CH)^{y-}(Bu_4N^+)_y]_x + xyAsF_5 \rightarrow$$
$$2\ trans\text{-}(CH)_x + 2xy(Bu_4N)F + xyAsF_3 \qquad (5.13)$$

$$xy(Bu_4N)F + xyAsF_5 \rightarrow xy(Bu_4N)(AsF_6) \qquad (5.14)$$

5.3.4   Summary

Even though thermal, chemical, and electrochemical isomerization of polyacetylene were presented above and elsewhere as if they were equivalent processes, our available knowledge about the chemistry and physics of the polymer suggests otherwise. Of the three methods, thermal isomerization is probably the most uniform. As the fibrils are heated uniformly there is no reason any portion of the polymer film would isomerize preferentially over another region. Electrochemical doping is known to commence at the surface of the polyacetylene film electrode and progresses inward. This process will be discussed in Section 8.2.5. Chemical doping is probably the least uniform of the three processes and is dependent on the particular dopant. Evidence will be found throughout this book that the most common dopant, AsF_5, is most likely to cause inhomogeneous doping and consequently nonuniform isomerization. The same consideration applies to the chemical compensation process. It is difficult to stop the latter precisely at the point of exact compensation, and the products of the doping and compensation reactions must be removed if pure undoped *trans*-polyacetylene is to be obtained.

The *trans*-polyacetylene obtained by the three methods probably also has different structures. The crystal structure was thermally transformed from cis to trans with a minimum of disturbance. In fact, there are indications (Section 4.5) that the longitudinal order is enhanced during isomerization under optimum conditions. The opposite is the case in chemical and electrochemical isomerizations. In these processes the dopant ions, depending on their sizes, enter between the polyacetylene chains and distort the interchain separation, as described in Section 9.2. When the polymer is undoped and the counter ions removed, it is doubtful that the lattice spacings would be the same as in thermally isomerized *trans*-polyacetylene. The dopant ions are probably randomly intercalated between the polymer chains, it is very likely that the *trans*-polyacetylenes derived from chemical and electrochemical doping have more disordered structures than thermally isomerized polymers. So far there have been no reports on the degree of crystallinity or the crystal structure of *trans*-polyacetylene prepared by the electrochemical or the chemical method.

The main drawback of thermal isomerization is chemical complications occurring at elevated temperatures. On the other hand, chemical reactions probably also take place in chemical doping. For instance, chain extension or crosslinking is possible with AsF_5 doping because it is an excellent Friedel–Crafts catalyst. Isomerization with oxygen can certainly lead to structural changes associated with autoxidation. It is the electrochemical isomerization that risks the least complication in this regard.

Finally, an intriguing question is how many cis units are isomerized when initiated by the isomerization of a single cis unit. Whereas no data pertaining to this point are available, common sense would suggest that this efficiency is greater for thermal isomerization than for the other processes. However, as isomerization approaches completion, the thermal process can become ineffectual and may require mechanisms represented by Eqs. (5.1) and (5.2). The chemical and electrochemical processes may not have this limitation.

5.4   ELECTRON PARAMAGNETIC RESONANCE AND NEUTRAL DEFECTS

5.4.1   Neutral Defects

When acetylene is polymerized directly in an EPR sample tube and never warmed up or exposed to oxygen, it is free of a $g = 2$ resonance (Fig. 2.6; Chien *et al.*, 1980a; Bernier *et al.* 1980). Therefore, *cis*-polyacetylene is devoid of any structural defects. Upon warming up to 253 K, a $g = 2$ signal appeared. Heating of a cis-rich sample resulted in a marked increase in the unpaired spin concentration [S·] as obtained by double integration of the EPR signal (Chien *et al.*, 1980b). These defects are located at the site where the π wave function changes phase and its amplitude vanishes, separating sequences of cis-transoid, trans-cisoid, and trans-transoid structures. Two of such kinds of defects are shown in Eqs. (3.3) and (3.4).

The neutral defect in *trans*-polyacetylene is referred to as a neutral soliton. It is understood that these defects are without charge, have spin one-half, and are delocalized and highly mobile. This paramagnetic species contributes toward Curie susceptibility. Figure 5.14 showed the variation of the integrated intensity of S· as a function of temperature (Schwoerer, 1980). The Curie law is obeyed from 300 to 1.5 K.

Chien *et al.* (1980b) studied the kinetics of formation of S·. The EPR intensity increased with time at a given temperature, reaching a constant value (Fig. 5.15). The production of neutral defects differs from the formation of trans structures in several respects, though the two processes are related. The formation of S· follows first-order kinetics (Fig. 5.16), whereas isomerization does not obey any simple kinetics. The rate of the former process at a given temperature is not dependent upon the initial cis content of the polymer. That is, the same rate constant was obtained at a particular temperature for cis-rich film or samples with significant trans content such as in an experiment in which the same specimen was used for the entire temperature range. In contrast, the rate and activation energy for thermal isomerization depend upon the trans content of the polymer.

The first-order rate constants gave an Arrhenius plot, shown in Fig. 5.17, with an activation energy of ~ 10 kcal mole^{-1}. This value is much smaller than the $17-18$ kcal mole^{-1} activation energy for the isomerization of cis-rich polyacetylene. There is no significant increase in activation energy for the production of S· with trans-rich polymer as observed for cis–trans isomerization. Therefore, the creation of neutral defects is necessary to initiate isomerization but is not directly associated with all the processes

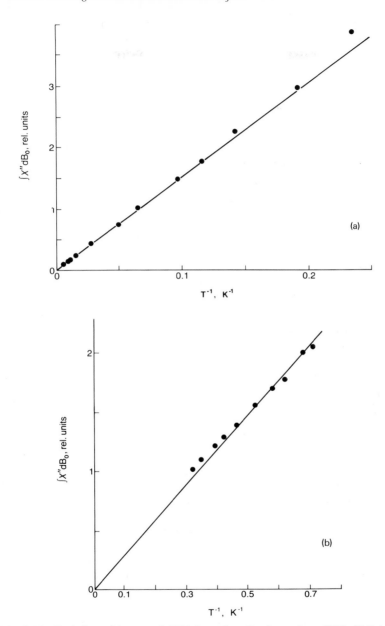

Fig. 5.14   Variation of integrated EPR intensity of polyacetylene [95% (C_2D_2) + 5% $(C_2H_2)]_x$ with temperature: (a) between 300 and 4.2 K; (b) between 3 and 1.5 K. [After Schwoerer *et al.* (1980).]

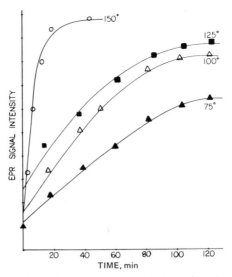

Fig. 5.15 Increase of free-spin concentration with time of heating at indicated temperature in degrees Celsius.

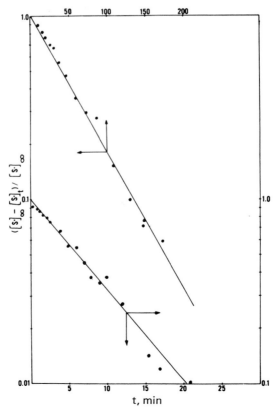

Fig. 5.16 First-order kinetics plots for the formation of unpaired spins in polyacetylene.

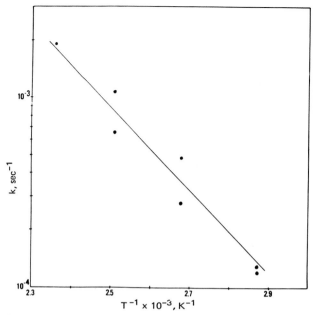

Fig. 5.17   Arrhenius plot for the rate constant of production of unpaired spins.

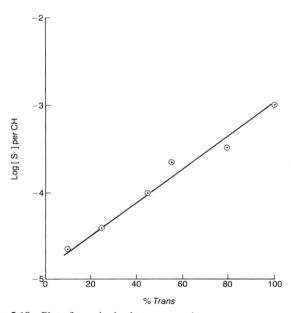

Fig. 5.18   Plot of unpaired spin concentration versus trans content.

173

involved in structural isomerization. Figure 5.18 shows the increase of [S ·] with the extent of isomerization to have a proportional relationship between % trans and log[S ·].

The low and constant activation energy required to form the neutral defect can be explained by the fact that the process involves merely interchange of adjacent short and long bonds, as illustrated below, starting with defect-free cis-transoid structure:

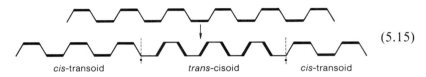

$$(5.15)$$

cis-transoid   *trans*-cisoid   *cis*-transoid

Similar processes can be visualized in the formation of neutral defects in a trans-transoid sequence.

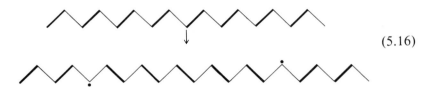

$$(5.16)$$

Compare the kinetics of oxygen-promoted cis – trans isomerization with that of the production of neutral defects; at 343 K the rate constants for the two processes are 1.05×10^{-4} sec^{-1} (Chien and Yang, 1983a) and 1.0×10^{-4} sec^{-1} (Chien *et al.*, 1980b), respectively. On the other hand, the latter has an activation energy of only 10 kcal mole^{-1}. Therefore, the rate of soliton formation is faster than oxygen-induced isomerization at temperatures lower than 343 K but slower by comparison at higher temperatures. The higher activation energy for the oxygen-induced isomerization may be that required to overcome steric hindrance, etc.

5.4.2   EPR Linewidth and Soliton Domain Width

When *cis*-polyacetylene is heated, the EPR linewidth decreases (Table 5.2). Therefore, the unpaired spins in cis and trans polymers have different linewidths. Weinberger *et al.* (1980) found that $cis\text{-}(CD)_x$ has a temperature-independent linewidth of ~ 4 G (Fig. 5.19), whereas $cis\text{-}(CH)_x$ has a linewidth of 8.5 G at 295 K and 11 G at 82 K. This shows that the linewidth is attributable to unresolved hyperfine splittings because the value of the

TABLE 5.2

Linewidth of S· as a Function of Temperature

Temperature (K)	298	323	348	373	398	423
ΔH_{pp} during heating (G)	10	7.4	4.8	3.5	1.3	0.7
ΔH_{pp} after cooling to 298 K (G)	10	7.6	5.0	4.5	2.0	1.2

hyperfine constant for an unpaired spin localized on a single CH unit is $a_o = 31$ G, whereas deuterium, which has a much smaller gyromagnetic ratio than a proton, would have a hyperfine constant of only 4.8 G for a spin localized on a single CD unit.

Because the observed linewidth of S· in *cis*-polyacetylene is only ~ 10 G, the spin must be delocalized over a number of carbon atoms. If this number is 2ξ, which is defined to be the domain width of a soliton, then the effective hyperfine constant is a fraction of a_o given by

$$a_{eff} = a_o|\psi(n)|^2, \tag{5.17}$$

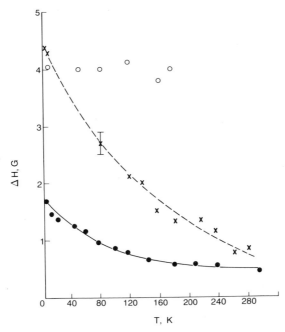

Fig. 5.19 First derivative peak-to-peak linewidths ΔH versus temperature: (x) *trans*-polyacetylene, (●) *trans*-$(CD)_x$, (○) *cis*-$(CD)_x$. [After Weinberger *et al.* (1980).]

where n denotes the nth monomer unit away from the domain center and $\psi(n)$ is the wave function for a nonbonded, single-particle state located at the center of the domain

$$\psi(n) = \xi^{-1/2} \operatorname{sech}(n/\xi) \cos(n\pi/2). \tag{5.18}$$

Weinberger *et al.* (1980) assumed a Lorentzian shape for each individual hyperfine line with a width of 0.45 G, which is the high-temperature asymptotic experimental value, and simulated the composite EPR spectrum. Figure 5.20 is a plot of ΔH_{pp} as a function of soliton half-width ξ. The simulated widths agree with experimental linewidths for a domain width of 12–16 lattice constants ($\xi = 6$–8). The linewidth of 90% ^{13}C-enriched *cis*-(^{13}CH)$_x$ is 14 G; the additional width is due to hyperfine interaction of the unpaired electron spin with the spin of the ^{13}C nucleus.

The results of the temperature dependence of EPR linewidth for protonated and deuterated *trans*-polyacetylene, shown in Fig. 5.19, contain a wealth of information. First, the linewidth of *trans*-(CH)$_x$ is always greater than that of *trans*-(CD)$_x$ showing nuclear hyperfine interaction to be the principal source of linewidth in the trans polymer as in the cis polymer. The linewidth decreases rapidly with increasing temperature, indicating motional narrowing. Thus the solitons in *trans*-polyacetylene are mobile. The

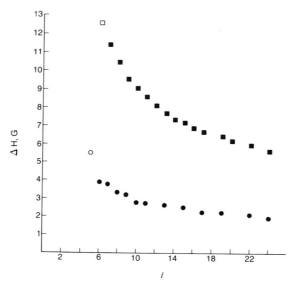

Fig. 5.20 ΔH of hyperfine simulations versus soliton half width l (ξ in text): (■) (CH)$_x$, $a_0^H = 31$ G; (●) (CD)$_x$, $a_0^D = 4.8$ G; (□,○) simulations that showed resolvable hyperfine structure. [After Weinberger *et al.* (1980).]

EPR linewidth of the neutral defects in *cis*-polyacetylene is temperature independent over the range observed; therefore the unpaired spins are immobile. The diffusion coefficient for solitons in *trans*-polyacetylene will be discussed in Section 5.5. Finally, at room temperature the linewidth of *trans*-(CH)$_x$ approaches that of *trans*-(CD)$_x$, suggesting that the soliton motion is comparable to hyperfine interaction. Tomkiewicz *et al.* (1979) had suggested that the narrow room temperature EPR resonance in *trans*-polyacetylene is due to exchange interaction. This is unlikely because exchange coupling is not expected to have the strong temperature dependence seen in Fig. 5.19. Also it is believed that two solitons in the same polyacetylene chain segment that experience exchange interaction would also tend to annihilate each other.

5.4.3   Spin–Lattice and Spin–Spin Relaxation Times

The saturation characteristics of S · and its spin–lattice relaxation time T_1 were obtained by recording EPR spectra as a function of microwave power (Chien *et al.*, 1982b). The microwave power in the TE$_{102}$ cavity H_1 was obtained by the method of a perturbing sphere. A metal sphere was placed in the EPR cavity and the shift in frequency Δv was measured as a function of klystron power W in watts. The H_1 in gauss was then calculated from

$$|H_1| = \frac{1}{2}\left[W\left(\frac{v^2 - v_0^2}{v_0^2}\right)\left(\frac{40}{\pi^2 \Delta v r^3}\right)\right]^{1/2}, \tag{5.19}$$

where $v_0 = 9.545$ GHz and $v = 9.554$ GHz are the initial and perturbed frequencies, and r is the radius of the sphere, which is 1.59 mm. We found the relation

$$|H_1|^2 = 0.49W. \tag{5.20}$$

From the plot of signal amplitude versus W, H_{1_m} is found as the maximum signal amplitude. Together with ΔH_{pp} below saturation, the spin–lattice and spin–spin relaxation times (T_1 and T_2, respectively) were calculated:

$$T_1 = 1.97 \times 10^{-7} \, \Delta H_{pp} \, [g \, (H_{1_m})^2]^{-1} \quad \text{sec}, \tag{5.21}$$

and

$$T_2 = 1.31 \times 10^{-7} \, [g \, \Delta H_{pp}]^{-1} \quad \text{sec}. \tag{5.22}$$

The saturation curves of *cis*-polyacetylene are given in Fig. 5.21. They showed that the resonance is inhomogeneously broadened (Chien *et al.*,

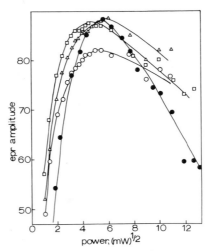

Fig. 5.21   Electron paramagnetic resonance saturation curves at room temperature: (○) *cis*-polyacetylene sample no. 1; (△) *cis*-polyacetylene sample no 2; (□) *cis*-polyacetylene sample no 3; (●) *cis*-(perdeuterated polyacetylene).

1982b). The relaxation times are $T_1 = (5.4 \pm 0.9) \times 10^{-5}$ sec, and $T_2 = (1.0 \pm 0.07) \times 10^{-8}$ sec. The saturation curve of *cis*-$(CD)_x$ is more homogeneously broadened than those of *cis*-$(CH)_x$. This is in agreement with the conclusion arrived at in Section 5.4.2 that the broadening of the EPR line is

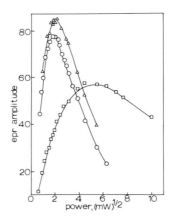

Fig. 5.22   EPR saturation curves at room temperature: (○) pristine *trans*-$(CH)_x$; (□) exposed to air for 15 min, (△) after overnight evacuation.

due to hyperfine interaction. *cis*-$(CD)_x$ has $T_1 \simeq 4 \times 10^{-6}$ sec and $T_2 \simeq 1.2 \times 10^{-8}$ sec.

The EPR saturation curve for *trans*-polyacetylene approaches the case of homogeneous broadening (Fig. 5.22). The signal Y_m is of the form

$$Y_m = z/(1 + z^2), \qquad (5.23)$$

where $z = \gamma H_1 (T_1 T_2)^{1/2}$. Figure 5.23 shows that linewidths increase with microwave power at both 77 and 298 K, as expected for homogeneous broadening. Measurements were made on *trans*-polyacetylene samples from many preparations. At ambient temperatures, the values of T_1 ranged from 1.9 to 6.6×10^{-5} sec with an average value of $(2.7 \pm 1.7) \times 10^{-5}$ sec and the values of T_2 ranged from 6 to 8.8×10^{-8} sec with an average value of $(7.8 \pm 1.0) \times 10^{-8}$ sec. Therefore, the variability in T_1 is much greater than it is for T_2 in *trans*-polyacetylene. At 77 K, T_1 was increased to 6.8×10^{-5} sec, whereas T_2 was reduced to 2.2×10^{-8} sec. Trans and cis polymers have nearly the same spin–lattice relaxation time, but the former has longer spin–spin relaxation time.

The effect of oxygen is also shown in Fig. 5.22. In this experiment, undoped *trans*-polyacetylene was exposed to air for 30 min at ambient temperature. The EPR of the neutral soliton becomes inhomogeneously broadened with values of $T_1 = 8.1 \times 10^{-6}$ sec and $T_2 = 6.6 \times 10^{-8}$ sec. Evacuation of the sample overnight at 10^{-6} torr restored the EPR saturation characteristics to that of the original material. Therefore, under these conditions, the effect of oxygen on the EPR of S $\cdot$ in polyacetylene is reversible.

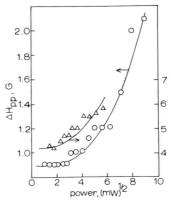

Fig. 5.23 Variation of EPR linewidth of *trans*-$(CO)_x$ with microwave power: ($\bigcirc$) at 298 K, ($\triangle$) at 77 K.

5.5   SPIN DYNAMICS

That the unpaired spins in *cis*-polyacetylene are immobile and those in *trans*-polyacetylene are highly diffusive is evident from their EPR linewidth, its temperature dependence, and saturation characteristics. The question is how mobile is the soliton domain in *trans*-polyacetylene? A number of techniques have been used to determine this mobility; the conclusions are not all in agreement. An associated phenomenon is the anisotropy of diffusion. Estimates differ on the relative in-chain and off-chain diffusivity.

Some of the discrepancies are probably sample related, particularly for *trans*-polyacetylene. For instance, whereas the linewidth of the EPR in *cis*-polyacetylene lies between 6 and 10 G for samples prepared by various laboratories, the range is 0.2–2.5 G for the trans polymer. This is not entirely attributable to possible variations in the thermal isomerization. For instance, different samples isomerized under identical conditions in our laboratories gave large variations in T_1 from 1.9 to 6.6 × 10^{-5} sec but a more reproducible T_2 of 6 to 8.8 × 10^{-8} sec. Thus sample variability affects the slower T_1 processes more than the faster T_2 processes.

5.5.1   EPR Saturation

The most direct way to measure spin dynamics is to introduce minute amounts of dopant, which can promote EPR relaxation. Iodine is such a dopant because of its large spin–orbit coupling constant. In these experiments (Chien *et al.*, 1982b) ppm amounts of $^{125}I_2$ were diffused very slowly into polyacetylene and the level of doping was quantitatively determined by radioassay. Figure 5.24 shows several typical EPR saturation curves of *trans*-polyacetylene doped with iodine. The change of T_1 with y is plotted in Fig. 5.25, and Fig. 5.26 gives the variation of T_2 with y. In these experiments, doping in the range of y from 3 × 10^{-6} to 3 × 10^{-3} used $^{125}I_2$, ordinary iodine was used for $y > 7 × 10^{-4}$. The two methods overlap between $y = 7 × 10^{-4}$ and 3 × 10^{-3}. If we assume that the very slow doping technique leads to uniform distribution of the dopant, then in order for the iodine dopant to affect the magnitude of T_1 (T_2), an S · must diffuse to close proximity of a dopant molecule in a time less than T_1 (T_2). The diffusion constant can be estimated with a formula for a random walk in one dimension:

$$\langle L^2 \rangle = 4Dt, \tag{5.24}$$

where L is the mean distance between dopant ions. We observed significant reduction in T_1 at levels of $y = 10^{-5}$; decrease in T_2 began only when y reached 10^{-3}.

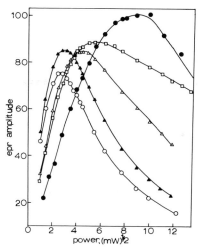

Fig. 5.24 Electron paramagnetic resonance saturation from curves at room temperature for *trans*-$(CHI_y)_x$: (○) $y = 3.8 \times 10^{-6}$; (▲) $y = 2.0 \times 10^{-5}$; (△) $y = 5.1 \times 10^{-4}$; (□) $y = 2.5 \times 10^{-3}$; (●) $y = 1.6 \times 10^{-2}$.

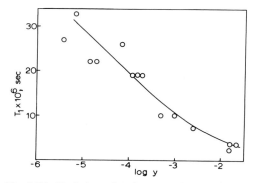

Fig. 5.25 Variation of T_1 for *trans*-$(CHI_y)_x$ with y.

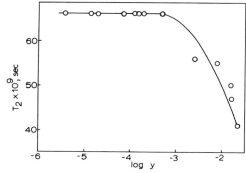

Fig. 5.26 Variation of T_2 for *trans*-$(CHI_y)_x$ with y.

Let us consider the magnitude of L by assuming the dopant to be I_3^- ions and that they are present more or less uniformly throughout the specimen. The number average molecular weight of the many polyacetylene samples prepared under standard condition (Section 2.2.5) have an average $\bar{M}_n$ value of 11,000 (Section 2.6.2). At $y(I_3^-) = 3 \times 10^{-6}$, there is only about one I_3^- ion per 393 molecules of polyacetylene. Therefore, only one neutral soliton out of a few hundred will encounter a dopant by intrachain diffusion only. If T_1 is to be affected by I_3^- ions, there must be interchain diffusion. In other words, the diffusion constants estimated from the data will be $D_\perp$. On the other hand, at $y(I_3^-) = 3 \times 10^{-4}$ when T_2 begins to decrease, there is about one I_3^- ion per four molecules of polyacetylene. Because the I_3^- ions are intercalated between polyacetylene chains (Section 9.2), it is highly probable that intrachain diffusion of the soliton would encounter a dopant and the diffusion constant estimated from the data would then be $D_\parallel$.

For $y(I_3^-) = 3 \times 10^{-6}$, $L \sim 4.2 \times 10^{-6}$ cm. Using $t = T_1 = 2.7 \times 10^{-5}$ sec for Eq. (5.24), we find $D_\perp = 1.6 \times 10^{-7}$ cm² sec⁻¹. The rate of diffusion is $\tilde{D}_\perp = 1.1 \times 10^9$ rad sec⁻¹. On the other hand, with $y(I_3^-) = 3 \times 10^{-4}$, $L \sim 1.3 \times 10^{-4}$ cm for a polyacetylene chain of 846 CH units. We find for $t = T_2 = 8 \times 10^{-8}$ sec, a value of $D_\parallel = 5.5 \times 10^{-2}$ cm² sec⁻¹ and $\tilde{D}_\parallel = 3.8 \times 10^{14}$ rad sec⁻¹. The velocity of the soliton is $v_s = D_\parallel/a = 4.6 \times 10^6$ cm sec⁻¹, where a is the lattice parameter 1.2 Å. The velocity is in excellent agreement with that estimated by consideration of the real time dynamics of the soliton

TABLE 5.3

Diffusion Coefficients of Solitons in *trans*-Polyacetylene

Technique	$\tilde{D}$ (rad sec⁻¹)			D (cm² sec⁻¹)	
	$\tilde{D}_\parallel$	$\tilde{D}_\perp$	$\tilde{D}_\parallel/\tilde{D}_\perp$	$D_\parallel$	$D_\perp$
Effect of dopant on EPR saturation	3.8×10^{14}	1.1×10^9	3×10^5	5.5×10^{-2}	1.6×10^{-7}
EPR line shape[a]	$\sim 10^{11}$	$\sim 10^8$	$\sim 2 \times 10^3$	2×10^{-5}	$\sim 10^{-8}$
Dynamic nuclear polarization[b,c]	$\sim 10^{11}$	—	—	—	—
Proton T_1:					
Nechtschein *et al.*[b]	6×10^{13}	$\leqslant 6 \times 10^7$	$\geqslant 10^6$	10^{-3}	—
Nechtschein *et al.*[c]	$(3-6) \times 10^{13}$	—	$\sim 10^4$	—	—
Alizon *et al.*[d]	$\sim 10^{11}$	$\sim 10^8$	$\sim 10^3$	—	—
Spin echo[e]	$\leqslant 10^{11}$	—	< 500	—	—

[a] Weinberger *et al.* (1980).
[b] Nechtschein *et al.* (1980).
[c] Nechtschein *et al.* (1983).
[d] Alizon *et al.* (1981).
[e] Shiren *et al.* (1983).

(Su and Schrieffer, 1980). The anisotropy of diffusion is $D_\parallel / D_\perp \sim 3 \times 10^5$. These results are summarized in Table 5.3.

The effect of iodine on the EPR saturation behavior of *cis*-polyacetylene is entirely different from that for *trans*-polyacetylene. Figure 5.27 shows some of the curves, and the relaxation times are summarized in Table 5.4. It is clear that doping of the *cis* polymer with iodine up to $y = 3.9 \times 10^{-4}$ does not significantly affect T_1 and up to $y = 2.2 \times 10^{-2}$ for T_2. One can conclude that the neutral defects in this polymer are immobile.

5.5.2 EPR Line Shape

Qualitative conclusions about spin mobility deduced from the temperature dependence of EPR linewidth have already been treated in Section 5.4.2. Additional information was obtained from line-shape analysis by Weinberger *et al.* (1980). The EPR line shape for *cis*-polyacetylene is nearly Gaussian, characteristic of an unnarrowed random distribution of hyperfine interactions, as shown in Fig. 5.28. In contrast, the *trans*-polyacetylene spectrum deviates from a pure Lorentzian function having substantial intensity in the low-field and high-field wings characteristic of a motionally narrowed resonance. The authors interpreted the deviation from Lorentz-

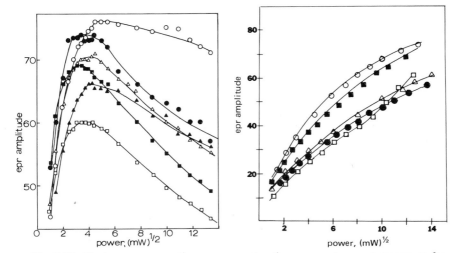

Fig. 5.27 Electron paramagnetic resonance saturation curves at room temperature for *cis*-$(CHI_y)_x$ for (a) (●) $y = 3.34 \times 10^{-5}$; (△) $y = 4.4. \times 10^{-5}$, (■) $y = 7.58 \times 10^{-5}$, (□) $y = 1.22 \times 10^{-4}$, (○,▲) $y = 3.9 \times 10^{-4}$; for (b) (△) $y = 1.4 \times 10^{-3}$, (●) $y = 4 \times 10^{-3}$, (■) $y = 8 \times 10^{-3}$, (○) $y = 1.3 \times 10^{-2}$, (□) $y = 2.2. \times 10^{-2}$.

TABLE 5.4

EPR Relaxation Data for cis-$(CH_y)_x$

y^a	$T_1 \times 10^5$ (sec)	$T_2 \times 10^8$ (sec)	ΔH_{pp} (G)
3.3×10^{-5}	100	1.1	6.0
4.4×10^{-5}	80	1.0	6.3
7.6×10^{-5}	122	1.0	6.2
1.2×10^{-4}	97	1.0	6.5
3.9×10^{-4}	53	1.0	6.5
3.9×10^{-4}	73	1.0	6.3
1.4×10^{-3}	—^{b}	1.0	6.5–7
4.0×10^{-3}	—	1.0	6.5–7
8.0×10^{-3}	—	1.0	6.5–7
1.3×10^{-2}	—	1.0	6.5–7
2.2×10^{-2}	—	1.0	6.5–7

^{a} The first five samples, $y = 3.3 \times 10^{-5}$–3.9×10^{-4}, were doped with $^{125}I_2$; the remaining specimens were doped with normal iodine.

^{b} No maximum in the saturation curve.

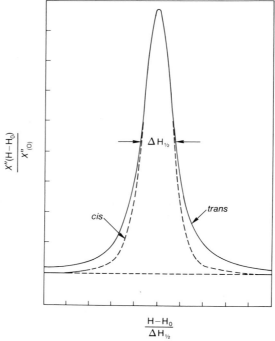

Fig. 5.28 Comparison of representative cis- and $trans$-polyacetylene absorption line shapes, normalized in amplitude and plotted versus half width. Both spectra obtained at 82 K. [After Weinberger *et al.* (1980).]

ian line shape as due to low dimensionality of the spin diffusion and the correlation time of the motion.

Two limiting cases were considered. If the diffusion is three dimensional, the line shape should be pure Lorentzian characterized by a correlation time τ_c. On the other hand, if the motion persists only in one dimension, the relaxation function for $t \gg \tau_c$ is

$$\rho(t) \propto \exp[-(\gamma t)^{3/2}], \tag{5.25}$$

where

$$\gamma \simeq \langle \Delta \omega^2 \rangle^{2/3} \tau_c^{1/3}. \tag{5.26}$$

The observed line shape is much closer to the three-dimensional case than it is to the one-dimensional limit as they are compared with the experimental results in Fig. 5.29. The intrachain diffusion constant was estimated from

$$D_\parallel = a^2 \tau_c^{-1}. \tag{5.27}$$

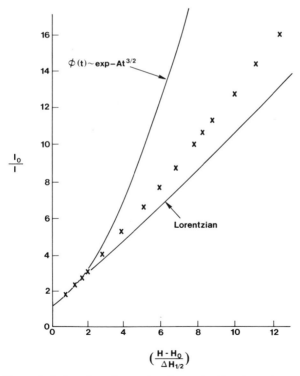

Fig. 5.29 Theoretical plots for a Lorentzian line and a line shape for one-dimensional diffusion. (x) Experimental data for *trans*-(CH)$_x$ at 82 K. [After Weinberger *et al.* (1980).]

The value of τ_c was obtained from the unnarrowed linewidth of *cis*-polyacetylene, which at room temperature corresponds to 10^{-11} sec; thus $D_\parallel = 2 \times 10^{-5}$ cm^2 sec^{-1}. Using the theory of Hennessy *et al.* (1973), Weinberger *et al.* (1980) find an anisotropy for spin motion of the order of 2×10^3 and an interchain diffusion constant of $\sim 10^{-8}$ cm^2 sec^{-1}. Therefore, the results of line-shape analysis differ from that obtained in Section 5.5.1 for the intrachain diffusion but agree within an order of magnitude on the interchain processes.

There are some flaws in the preceding analysis. The observed EPR linewidth is not an intrinsic property but depends strongly on paramagnetic contaminants such as oxygen and catalyst residues. Their presence can cause large variation in soliton EPR linewidth in *trans*-polyacetylene. Recently, linewidth values as small as 0.2 G have been reported. This small ΔH_{pp} would alter the estimates given by Weinberger *et al.* (1980) significantly. The rationale of using unnarrowed linewidth of *cis*-polyacetylene to estimate τ_c and then to calculate $D_\parallel$ for soliton in *trans*-polyacetylene is unclear.

A different interpretation was given by Holczer *et al.* (1981b), who showed that the EPR lineshape varies from sample to sample and cannot be described by a single function. They proposed a two-spin model. The initial postulate was that the majority of the spins are diffusive spins with linewidth $\Delta\omega_d/\gamma_e \sim 0.12-0.38$ G and that there are also localized spins with linewidth $\Delta\omega_s/\gamma_e$ of $5-10$ G (γ_e is the electron gyromagnetic ratio). Because only a single EPR line was observed, they argued that there must be exchange between the two spin species with an exchange frequency $\omega_{ex} > \Delta\omega_d \simeq 3 \times 10^{-6}$ rad sec^{-1}. This condition implies that the diffusive spin has a lifetime too short to make the linewidth equal to $\Delta\omega_d$. Subsequently, the model was refined to assume the mobile and trapped spins to be in thermal equilibrium.

According to the two-spin model (Holczer *et al.*, 1981a), the EPR spectrum was decomposed into fixed and mobile spin contributions by casting the memory function $K(\omega)$ as an additive of the mobile and fixed-spin memory functons

$$K(\omega) = n_d K_d(\omega) + (1 - n_d) K_s(\omega), \qquad (5.28)$$

where n_d is the number fraction of diffusive spins. This is illustrated by the following. The EPR signal was recorded with a 100-G scan, which corresponds to $140 \times \Delta H_{1/2}$ at 300 K and $25 \times \Delta H_{1/2}$ at 4.2 K. Letting $h = (\omega - \omega_e)\gamma_e^{-1}$ and $Y(h)$ be the EPR derivative signal, the function $F(h) = h^3 Y(h)$ is proportional to the real part of the complex memory function $K(\omega)$. Plots of $F(h)$ versus h at several temperatures are shown in Fig. 5.30. The curves for 4.2 and 20 K almost coincide, and it was assumed that all the spins are immobile at 4.2 K, i.e., $n_d(4.2 \text{ K}) = 0$, and $F(h, 4.2 \text{ K})$ contains only $K_s(\omega)$.

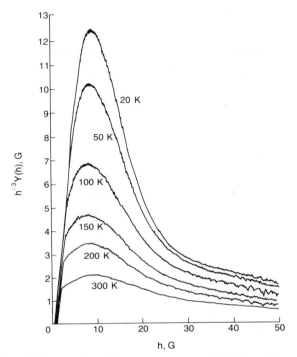

Fig. 5.30 $h^3 \times Y(h)$ versus h at different temperatures. $Y(h)$ is the derivative of the EPR line. [After Holczer *et al.* (1981a).]

This assumption is, however, not strictly valid according to the dynamic nuclear polarization results to be described in the next section. The number of mobile spins at any temperature $n_d(T)$ can be determined from the condition that for $h \gg \Delta h$ the function $F(h,T) - [1 - n_d(T)] \times F(h, 4.2 \text{ K})$ should be proportional to $h^{-1/2}$. The calculated values of n_d are shown in Fig. 5.31.

Alternatively, an independent way to determine $n_d(T)$ is from the temperature dependence of the EPR linewidth:

$$\Delta H = n_d \Delta H_d + (1 - n_d)\Delta H_s, \tag{5.29}$$

where ΔH is the observed line width and ΔH_d (ΔH_s) are the linewidths for diffusive (localized) spins. Assuming again $n_d = 0$ at 4.2 K and also $\Delta H_s = \Delta H(4.2 \text{ K})$ is temperature independent, then for $\Delta H_d \ll \Delta H_s$ one finds

$$n_d(T) = \frac{\Delta H(4.2 \text{ K}) - \Delta H(T)}{\Delta H(4.2 \text{ K})}. \tag{5.30}$$

The results of this calculation are also shown in Fig. 5.31.

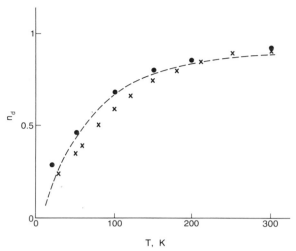

Fig. 5.31 Number fraction of diffusive spins, n_d: (●) from EPR line shape analysis, (x) from EPR linewidth, (–––) calculated for an activation energy of 3.4 meV. [After Holczer *et al.* (1981a).]

Nechtschein *et al.* (1983) attributed the trap site to oxygen impurities. They proposed that adsorbed oxygen acts as an electron acceptor to give an ion pair O_2^- – carbocation, and that attractive interaction between this localized charged soliton and a neutral soliton previously present in the chain represents a trap site. The main difficulty with this hypothesis is that O_2^- is a more stronger oxidant than O_2 by about 1.2 V. Therefore, a neutral soliton would be immediately oxidized by O_2^- to another carbocation on encounter.

An alternative possible trap site is the titanium ion. It has been shown in Section 2.5 that a significant fraction of the polyacetylene chains are bound to Ti^{3+} ions. It could cause dipolar broadening or attractive trapping of a neutral soliton on the chain. It was shown above that the addition of a minute amount of iodine to *trans*-polyacetylene reduces the scatter in T_1 from ± 63 to $\pm 20\%$. Presumably iodine acts to decouple the metal ion from the polyacetylene chain,

$$Ti^{3+}\text{-polyacetylene} + I_2 \rightarrow Ti^{4+}I^- + I\text{-polyacetylene} \qquad (5.31)$$

In conclusion, the deviation from Lorentzian EPR line shape had been analyzed as giving motional information of the spin by Weinberger *et al.* (1980) or attributed to an admixture of localized and mobile spins by Holczer *et al.* (1981a). Both analyses are complicated by possible impurity effects.

5.5.3   Dynamic Nuclear Polarization

Spin dynamics has been studied with dynamic nuclear polarization. In this experiment one observes the NMR signal (at the nuclear Larmor frequency) while pumping with microwave power near the electronic Larmor frequency. There are two limiting results, depending on whether the electron – nuclear dipolar coupling is static or dynamic. In the static case, $\tau_c \geq \omega_N^{-1}$, the forbidden transition at $\omega_e \pm \omega_N$ can be induced. One observes positive NMR enhancement when the system is pumped at $\omega_e - \omega_N$, i.e., at frequencies $< \omega_e$. Conversely, pumping at frequencies higher than ω_e results in NMR deenhancement at $\omega_e + \omega_N$. This is referred to as the "solid state effect." In the case of dynamic electron – nuclear dipolar coupling $\tau_c \leq \omega_e^{-1}$, there is a component of the frequency spectrum of the electron-spin motion at ω_e, and the NMR signal intensity is enhanced just by pumping the system at ω_e. This is the well-known electron Overhauser effect.

The limiting solid-state and electron Overhauser effects are ilustrated in Fig. 5.32. When dynamic nuclear polarization measurements were made on *trans*-polyacetylene by Nechtschein *et al.* (1980), only a pure electronic Overhauser effect was observed at 300 K. The results indicate that the spins are moving with a frequency of at least $\omega_e \simeq 5 \times 10^{10}$ rad sec^{-1}. At 5.5 K

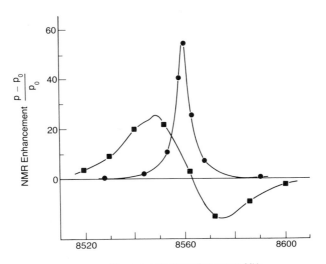

Microwave Pumping Frequency, MHz

Fig. 5.32   Enhancement of ^{1}H NMR amplitude in *trans*-polyacetylene as a function of the microwave pumping frequency ν_p near $\omega_e/2\pi = 8260$ MHz: (●) 300K, 5W; (■) 5.5K, 0.5W. [After Nechtschein *et al.* (1980).]

there are both positive and negative NMR enhancements, and the solid-state effect dominates. However, the positive enhancement at $\omega < \omega_e$ is larger than the negative enhancement at $\omega > \omega_e$. In other words, the electron Overhauser effect is still contributing at this low temperature. This was viewed as indicating that either a fraction of the solitons are still mobile and the remainder are trapped or that all the solitons are moving slowly and the frequency spectrum of this motion still has a small component at ω_e. Measurements were also made on partially isomerized polyacetylene samples with 50 and 65% trans contents. The results in Fig. 5.33 show that the specimens with higher cis content display more prominent solid-state effects, and the proportion of electron Overhauser to solid-state effects increases with the amount of trans structure. The observations suggest that during the isomerization some localized spins, which eventually become mobile, are formed.

Alizon *et al.* (1981) also performed dynamic nuclear polarization on *trans*-polyacetylene. They found it not to be a pure Overhauser effect, but the sample exhibits a weak solid-state effect at $\omega_e p 2\pi = 9.6 \times 10^9$ Hz. These DNP observations suggest that $\tau_c \sim \omega_e \sim 10^{-11}$ sec and $D_\parallel \sim 10^{11}$ rad sec^{-1}.

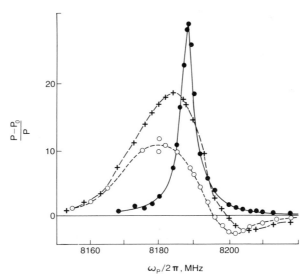

Fig. 5.33 Dynamic nuclear polarization results for polyacetylene of trans content: (○) 50%, (+) 65%, (●) 100%. [After Nechtschein *et al.* (1980).]

5.5.4 Nuclear Spin Relaxations

There have been several studies on proton–nuclear relaxation in *trans*-polyacetylene with the objective of shedding light on the electron-spin dynamics. The proton spin–lattice relaxation time can be expressed as

$$1/T_1 = k_B T\chi[\tfrac{3}{5}b^2 f(\omega_N) + (a_{iso}^2 + \tfrac{7}{5}b^2)f(\omega_e)], \tag{5.32}$$

where χ is the normalized spin susceptibility, $\chi = \chi_{molar}/N(g\mu_B)^2$, and a_{iso} (b) is the isotropic (dipolar) electron–proton hyperfine coupling. The $f(\omega_N)[f(\omega_e)]$ is the proton (electron) spectral density function, which is given by

$$f(\omega) = \frac{1}{k_B T\chi} \int_{-\infty}^{+\infty} \langle s^z(t)s^z(o)\rangle e^{i\omega t} \, dt. \tag{5.33}$$

The singular behavior of $f(\omega)$ comes from the slow decay at long times of the spin correlation function

$$\langle s^z(t)s^z(o)\rangle \sim (\tilde{D}t)^{-d/2}, \tag{5.34}$$

where d is the dimensionality.

The frequency spectrum is divergent at the nuclear Larmor frequency. Letting $\tilde{D}_1$ be the diffusion rate in the one-dimensional case and $\tilde{D}_2$ and $\tilde{D}_3$ be the other rates for higher dimensionalities, their $f(\omega)$s are given in Table 5.5. For the one-dimensional diffusion case and $\omega_e = 658\omega_N$, Eq. (5.32) can be cast into the form

$$T_1^{-1} = n_d(2\tilde{D}_\parallel\omega)^{-1/2}, \tag{5.35}$$

and a plot of T_1^{-1} versus $\omega^{-1/2}$ would be linear. At very low frequency, T_1^{-1} is independent of ω as hopping or interchain exchange causes the system to become three dimensional.

Nechtschein *et al.* (1980) measured T_1 at room temperature over the frequency range 10–340 MHz. The recovery of the nuclear magnetization

TABLE 5.5

Spin Correlation Functions

Dimensionality	1	2	3
Frequency range	$\tilde{D}_1 \gg \omega \gg \tilde{D}_2, \tilde{D}_3$	$\tilde{D}_1, \tilde{D}_2 \gg \omega \gg \tilde{D}_3$	$\tilde{D}_1, \tilde{D}_2, \tilde{D}_3 \gg \omega$
$f(\omega)$	$\dfrac{1}{(2\tilde{D}_1\omega)^{1/2}}$	$\dfrac{1}{2\pi(\tilde{D}_1\tilde{D}_2)^{1/2}} \ln\left(\dfrac{4\pi^2\tilde{D}_2}{\omega}\right)$	$\dfrac{1}{2\pi(\tilde{D}_1\tilde{D}_2)^{1/2}} \ln\left(\dfrac{4\pi^2\tilde{D}_2}{\tilde{D}_3}\right)$

was exponential over a decade of change in magnetization. The results plotted according to Eq. (5.35) are linear with intercept at the origin (Fig. 5.34). The straight line extends over the entire measurement range with no indication of a cutoff frequency for higher than one-dimensional diffusion. To obtain an estimate for diffusion rates from the slope of the plot, using Eqs. (5.32) and (5.35), the authors assumed that $b^2/a_{iso}^2 \simeq \frac{1}{4}$ to $\frac{1}{3}$ and took the electron–proton scalar coupling a_{iso} to have the same value as a free spin in a pure carbon $2p_z$ orbital, i.e., $|a_{CH}/\gamma_e| = 23.4$ G. Substitution of these values into Eq. (5.32) gave $\tilde{D}_\parallel \simeq 6 \times 10^{13}$ rad sec^{-1}. Because there was no three-dimensional behavior (i.e., cutoff frequency) at the lowest measuring frequency of 10 MHz, the authors took this as the upper imit for the interchain diffusion rate, i.e., $\tilde{D}_\perp \leqslant 6 \times 10^7$ rad sec^{-1}. The anisotropy for one-dimensional diffusion becomes $\tilde{D}_\parallel/\tilde{D}_\perp \geqslant 10^6$. These estimates were changed later (Nechtschein *et al.*, 1983) to $\tilde{D}_\parallel \sim (3-6) \times 10^{13}$ rad sec^{-1} and $\tilde{D}_\parallel/\tilde{D}_\perp \sim 10^4$, but no reasons were given for the new estimate of lower anisotropy.

In the study by Alizon *et al.* (1981) from 8 to 92 MHz, the temperature

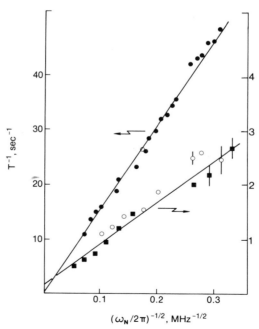

Fig. 5.34 Nuclear relaxation rate T_1^{-1} versus $(\omega_N/2\pi)^{-1/2}$ for undoped *trans*-polyacetylene; solid line is the theoretical prediction of Eq. (5.37). (●) Pristine *trans*-polyacetylene, (○,■) two samples of $[CH(AsF_5)_{0.1}]_x$. [After Nechtschein *et al.* (1980).]

was varied from 140 to 300 K. The results are shown in a similar T^{-1} versus $\omega^{1/2}$ plot (Fig. 5.35) for *trans*-polyacetylene. The results at 140 K obey the function dependence of Eq. (5.34) over the entire frequency range; in fact, the slope is identical to those of Fig. 5.34. However, the plots were not linear for temperatures $\geqslant 260$ K. There is a cutoff frequency at ~ 15 MHz at 300 K. Sample-related effects may be causing the different DNP results.

Alizon *et al.* (1981) did not use Eqs. (5.32) and (5.35) to estimate the diffusion rate. They just equated $\hat{D}_\parallel^{-1}$ with $\tau_\parallel$ and assumed $\tau_\parallel \simeq \omega_e^{-1} \simeq 10^{-11}$ sec to obtain $\tilde{D}_\parallel \sim 10^{11}$ rad sec^{-1}. At 140 K there was no deviation from one-dimensionality. A limit of less than 8 MHz was placed on the cutoff frequency. From this assumption, Alizon *et al.* (1981) obtained $D_\perp \simeq 2\pi\nu_c \simeq 10^8$ rad sec^{-1}. The anisotropy for diffusion is only about 10^3. However, a lower limit of ν_c would raise the estimate for anisotropy.

The temperature dependence of nuclear T_1 had been measured by Nechtschein *et al.* (1983) at 22.6, 84.5, and 170 MHz. The relaxation rate T_1^{-1} was found to decrease at first slowly with decreasing temperature, then more sharply at low temperatures. The dependence is greater for lower frequency. This result cannot be simply explained as a slowing down of spin

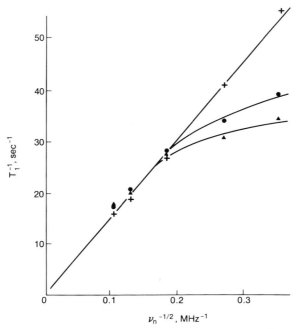

Fig. 5.35 Frequency dependence of T_1^{-1} for *trans*-polyacetylene at different temperatures: (+) 140, (●) 260, (▲) 300 K. [After Alizon *et al.* (1981).]

diffusion, i.e., decrease in $\tilde{D}$. If that were the case, one would expect an opposite effect, i.e., T_1 should decrease with decreasing temperature according to Eq. (5.35). On the other hand, lowering of temperature does increase EPR linewidth, indicating a slowing of soliton motion. The results can be rationalized by supposing that n_d, the number of diffusive spins, decreases with decreasing temperature. At low temperatures there are fewer mobile spins to induce nuclear relaxation and thus a decrease in T_1^{-1}. The increase in the number of localized spins would give rise to EPR line broadening. The above rationalization is viable provided that the temperature dependence of n_d is greater than that for $\tilde{D}^{1/2}$ and that even at very low temperatures the mobile and trapped spins exchange at a rate faster than the EPR linewidth at that temperature.

5.5.5   EPR Spin Echo

Whereas T_2 for the EPR spectrum of *trans*-polyacetylene is a monotonically increasing function of temperature, the spin phase-memory times need not behave similarly. Shiren *et al.* (1983) measured the spin echo at 10 GHz on *trans*-$(CD)_x$ samples. In a two-pulse sequence with pulse time 0, τ (the

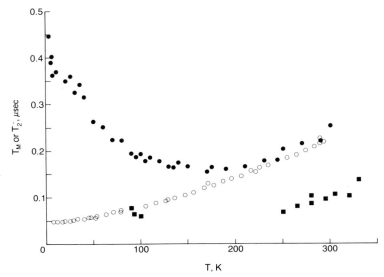

Fig. 5.36   Phase-memory time T_M (●) and T_2^* (○) versus temperature for *trans*-$(CD)_x$, and T_M (■) for *trans*-$(CH)_x$. [After Shiren *et al.* (1983).]

echo) occurs at 2τ. The phase-memory time T_M was determined by assuming that the echo amplitude $E(2\tau)$ decays exponentially with a time constant T_M. The value of T_M has a minimum value of $\sim 1.6 \times 10^{-7}$ sec at a temperature of 150 K (Fig. 5.36). This minimum separates the motional broadening regime below 150 K from the strongly narrowing regime above it. The data were analyzed according to the Gauss–Markov formalism, and the local field was modulated by hyperfine and electron–electron dipolar interactions. The correlation function, $\phi(t)$, was found to be

$$\phi(t) = \exp(-\tilde{D}_\perp t)[1 + \tfrac{1}{2}\tilde{D}_\parallel t]^{1/2}, \tag{5.36}$$

where $\tilde{D}_\perp$ is the Lorentzian interchain hopping rate and $\tilde{D}_\parallel$ is the one-dimensional intrachain soliton diffusion rate. The echo amplitude $E(2\tau)$ has the form,

$$E(2\tau) = E(0)\, \exp[-\delta^2\tau^2 F(\tilde{D}_\parallel\tau; \tilde{D}_\perp\tau)]. \tag{5.37}$$

By fitting the experimental data of $(CD)_x$ to Eq. (5.39), Shiren *et al.* (1983) estimated $\tilde{D}_\parallel < 10^{11}$ rad sec^{-1} and $\tilde{D}_\parallel/\tilde{D}_\perp < 500$. Similar measurements on $(CH)_x$ were not made over a wide enough temperature range to obtain a value for T_M.

5.5.6 Two-Spin Model

For the most recent version of the two-spin model Nechtschein *et al.* (1983) proposed that the solitons are either in a fixed state on a trap site or in a mobile state on regular sites. The relative soliton populations in regular and trapped states are n_d and $n_t = (1 - n_d)$, respectively. It was assumed that the traps have a broad distribution of energy from zero up to E_o with a constant value for the distribution function from zero to E_o. Then at thermodynamic equilibrium

$$n_t = \frac{p}{p + (1 - p)\exp(-E_o/k_B T)}, \tag{5.38}$$

where p is the relative concentration of the traps. Averaging over the trapping energy distribution, the authors obtained

$$n_t(T) = \frac{k_B T}{E_o} \log[1 + p(e^{E_o/k_B T} - 1)]. \tag{5.39}$$

Approximating that the same average applies to both the isotropic and dipolar contributions, Nechtschein *et al.* (1983) obtained for the EPR linewidth as a function of temperature,

$$\Delta H_{pp}(T) = \frac{5.6 \times 10^5}{\sqrt{\tilde{D}(T)}} + \frac{T}{E_o}\left[\left(3.3 - \frac{5.6 \times 10^5}{\sqrt{\tilde{D}(T)}}\right)\log(1 - p + pe^{E_o/k_BT})\right.$$

$$\left. + \left(1.2 - \frac{7.5 \times 10^5}{\sqrt{\tilde{D}(T)}}\right)\left(\frac{1 - e^{E_o/k_BT}}{1 - p + pe^{E_o/k_BT}}\right)\right]\left[\frac{p(1 - p)}{2}\right]. \quad (5.40)$$

To find the numerical coefficients in Eq. (5.40), the limiting low temperature linewidth is taken to be

$$\Delta\omega_t(0) = \Delta\omega_t^{dip}(0) + \Delta\omega_t^{hf}(0) \quad (5.41)$$

where the superscript dip (hf) refers to dipolar (hyperfine) contribution. The experimental values used were $\Delta\omega_t^{dip}(0)/\gamma = 1.2$ G and $\Delta\omega_t^{hf}(0)/\gamma = 2.1$ G. The linewidth from diffusive spins is

$$\Delta\omega_d(T) = \Delta\omega_d^{dip}(T) + \Delta\omega_d^{hf}(T) \quad (5.42)$$

with

$$\Delta\omega_d^{dip}(T) = \tfrac{1}{8}nR^2[2\tilde{D}_\parallel\omega_c]^{-1/2} \quad (5.43)$$

and

$$\Delta\omega_d^{hf}(T) = \tfrac{1}{8}(a_{iso}^2 + \tfrac{4}{5}b^2)[\tilde{D}_\parallel\omega_c]^{-1/2}. \quad (5.44)$$

The values of a_{iso} and b are those in Section 5.5.4, n is the concentration of free spins per CH unit, R is the effective dipolar coupling between one spin on a given chain and the spins in neighboring chains, which is estimated to the $R/\gamma = 2300$ G, and ω_c is taken to be 40 kHz. We note that this cutoff frequency seems to be sample dependent (Section 5.5.4).

Nechtschein *et al.* (1983) found good fit of experimental data with Eq. (5.42) by using $p = 0.05$ for the sample in vacuum, $p = 0.25$ for the sample in air, and $E_o/k_B = 650$ K. The fit is not too surprising, because there are three arbitrary adjustable parameters. Various laboratories reported different $\Delta H_{pp}(T)$ results. Nechtschein *et al.* (1983) had $\Delta H_{pp}(298$ K$) = 0.4$ G and $\Delta H_{pp}(4$ K$) = 3.2$ G. Weinberger *et al.* (1980) (cf. Fig. 5.19) found $\Delta H_{pp}(298$ K$) = 0.8$ G and $\Delta H_{pp}(4$ K$) = 4.3$ G; and Schwoerer *et al.* (1980) obtained $\Delta H_{pp}(300$ K$) = 4.1$ G and $\Delta H_{pp}(4.2$ K$) = 6.7$ G. Consequently, different values of p and E_o would be estimated from these results.

5.5.7   Soliton Domain Width in *trans*-Polyacetylene

From the EPR linewidth of spin in *cis*-polyacetylene it was deduced that the domain width is about $2\xi \simeq 12-16$ lattice parameters. However, the motional narrowing argument used in Eqs. (5.17) and (5.18) for spin delocalization are not applicable to the case of *trans*-polyacetylene, which has spin diffusing at a rate fast compared to the hyperfine interaction, and the spin delocalization term disappears from the linewidth expression. The domain width of a diffusing soliton can be estimated from a comparison of the EPR linewidth of fixed spins in $(CH)_x$ and $(CD)_x$ and from the cutoff frequency for proton relaxation time as influenced by the diffusive spins (Nechtschein *et al.*, 1983).

For the fixed spins the hyperfine contribution to the second moment of the EPR resonance is

$$(\overline{\Delta H^2})_A = \tfrac{1}{9}I(I+1)\,(A_{xx}^2 + A_{yy}^2 + A_{zz}^2)\sum_i \rho_i^2, \qquad (5.45)$$

where $\mathbf{A}$ is the hyperfine tensor and ρ_i is the spin density at the ith carbon atom. If we combine all other contributions to the second moment as $(\overline{\Delta H^2})_0$, which is the same for both $(CH)_x$ and $(CD)_x$ designated with superscripts H and D, respectively, then the observed second moments are

$$(\overline{\Delta H^2})^H = (\overline{\Delta H^2})_A^H + (\overline{\Delta H^2})_0 \qquad (5.46)$$

and

$$(\overline{\Delta H^2})^D = (\overline{\Delta H^2})_A^D + (\overline{\Delta H^2})_0. \qquad (5.47)$$

From which one obtains

$$\Delta(\overline{\Delta H^2}) = (\overline{\Delta H^2})_A^H - (\overline{\Delta H^2})_A^D$$
$$= \frac{1}{12}\left[1 - \frac{8}{3}\left(\frac{\gamma^D}{\gamma^H}\right)^2\right](A_{xx}^2 + A_{yy}^2 + A_{zz}^2)\sum_i \rho_i^2. \qquad (5.48)$$

The principal values of the hyperfine tensor are (McConnell *et al.*, 1960) $A_{xx}/\gamma_e = 1.07$, $A_{yy}/\gamma_e = 2.18$, and $A_{zz}/\gamma_e = 32.9$ G. Consequently,

$$\Delta(\overline{\Delta H^2}) = 140\sum \rho_i^2. \qquad (5.49)$$

The experimental second moment values are $(\overline{\Delta H^2})^H = 29 \pm 2$ G^2 and $(\overline{\Delta H^2})^D = 9 \pm 2$ G^2, thus $\Delta(\overline{\Delta H^2}) = 20 \pm 4$ G^2. If the spin–density distribution is assumed to be zero outside the domain wall and to be the same within it, then its width is given by $\sum_i \rho_i^2 = (2\xi)^{-1}$ or $2\xi = 7 \pm 1$, which is about half of the width estimated from the unresolved hyperfine EPR linewidth in *cis*-polyacetylene. If the spin wave function [Eq. (15.18)] is used instead of the box function, then $2\xi = 10 \pm 2$. Nechtschein *et al.* (1983) thought these were underestimates because they did not take into account both positive and negative spin densities. Alternate conjugated molecules usually have spin densities of opposite signs on adjacent carbon atoms.

In the case of diffusive spin, the effect of the hyperfine field on the proton relaxation time is not changed until the soliton has moved more than the distance of its domain width. The expression for $f(\omega)$ in one dimension in Table 5.5 is modified to be

$$f(\omega) = \frac{1}{(2\tilde{D}_{\|}\omega)^{1/2}} - \epsilon \frac{\xi}{\tilde{D}_{\|}}, \qquad (5.50)$$

where $\epsilon = 0.36$ for bond alternate solitons. Equation (5.50) should be linear with a nonzero intercept. Nechtschein *et al.* (1983) plotted T_1^{-1} data for different temperatures, and the intercept on the frequency axis gives the value for the high-frequency cut off as

$$\omega_{\max} = \tilde{D}_{\|}/2\epsilon^2\xi^2. \qquad (5.51)$$

Taking a value of 5.7×10^{10} rad sec^{-1} for $\tilde{D}_{\|}$ at 4.2 K, one finds $2\xi = 17$. Therefore, the defect domain widths are about the same in both cis and trans polymers.

5.5.8   Temperature Dependence

To obtain temperature dependence of spin diffusion rate, one combines Eqs. (5.34) and (5.37) to give

$$T_1^{-1}(T) = \frac{n_\mathrm{d}}{\sqrt{\tilde{D}_{\|}(T)F}} \frac{10^{-3}}{8\sqrt{\pi}} \left[\frac{3}{5}b^2 + \left(a_\mathrm{iso}^2 + \frac{7}{5}b^2\right)\frac{1}{\sqrt{660}}\right] \qquad (5.52)$$

where F is frequency in megahertz. At any temperature the value of $\tilde{D}_{\|}(T)$ relative to the value at 300 K is

$$\frac{\tilde{D}_{\|}(T)}{\tilde{D}_{\|}(300)} = \left(\frac{n_\mathrm{d}(T)}{n_\mathrm{d}(300)}\right)^2 \left(\frac{T_1^{-1}(300)}{T_1^{-1}(T)}\right)^2. \qquad (5.53)$$

At low temperature, n_d is not strongly dependent on p because

$$n_\mathrm{d}(T) \simeq \frac{k_\mathrm{B}T}{E_\mathrm{o}} \ln\left(\frac{1}{p}\right) \qquad (5.54)$$

For $T < 30$ K, $\tilde{D}_{\|}$ should be obtained from the slope of the T_1^{-1} versus $F^{-1/2}$ plot. On the other hand, when $T > 30$ K, measurement at one frequency is adequate. Figure 5.37 is a log–log plot of $\tilde{D}_{\|}$ versus temperature. The results showed that the diffusion rate is saturated near room temperature and decreases as temperature is decreased. Below 30 K, $\tilde{D}_{\|}$ varies with T^2. This dependence has been predicted by Wada and Schreiffer (1978) for the Brownian motion of a soliton domain wall assisted by second-order interaction with two phonons.

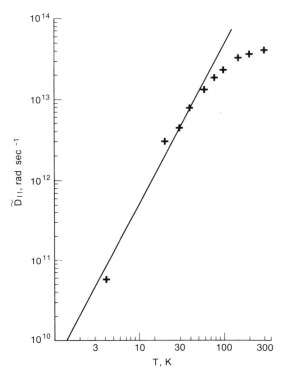

Fig. 5.37 Variation of spin diffusion rate with temperature. The solid line is the theoretical curve. [After Nechtschein *et al.* (1983).]

5.5.9 Summary

This section on spin dynamics shows clearly the complexity of polyacetylene, especially when the system is examined at the molecular level. The diffusion coefficients estimated by various techniques, summarized in Table 5.3, show large discrepancies. The contributions of trapped spins appeared to be very important, especially in the interpretations of the linewidth data.

The diffusion coefficients obtained from the effect of small amounts of dopant on T_1 and T_2 are probably least influenced by the impurities. This is because the dopant can remove the Ti^{3+} ions, as given in Eq. (5.31). Furthermore, iodine will displace any oxygen hopping site because iodine is a better acceptor dopant than oxygen. Consequently, the dominant interaction for the soliton would be with the intentionally introduced dopant ion. The use of T_1 to obtain values of $\tilde{D}_{\parallel}$ and of T_2 to estimate $\tilde{D}_{\perp}$ appears at first to be contrary to intuition. Careful consideration shows this to be reasonable. Small amounts of dopant affect the slower T_1 but not the faster T_2. Also, increase of dopant concentration caused a monotonic decrease in T_1 but has no effect whatsoever on T_2. This behavior would be as expected if there is interchain diffusion. Similar conclusions were arrived at in the intersoliton hopping model for conduction, as discussed in Section 11.1.6 (Kivelson, 1981a,b, 1982). The influence of dopant would influence spin relaxation due to in-chain diffusion if there is on the average one soliton for one to six polyacetylene molecules, depending on the number of chains with which the intercalated I_3^- is in contact. At $y(I_3^-) \simeq 3 \times 10^{-4}$ there is about one I_3^- for four polyacetylene chains and T_2 begins to decrease, from which a value of $D_\parallel$ was obtained. In all the other methods described in this section, separate estimates of $\tilde{D}_\parallel$ and $\tilde{D}_\perp$ were not possible.

The estimates of diffusion coefficients based on analyses of EPR line shape and linewidth seem to be full of uncertainties. Nechtschein *et al.* (1980) suggested that EPR line-shape analysis is incorrect if the contribution from fixed spins is not taken into account. On the other hand, Weinberger *et al.* (1980) pointed out that one should not use the proton hyperfine constant of a localized spin on a carbon atom as Nechtschein *et al.* (1980) did, but should take into consideration the delocalized nature of the neutral-defect wave function. The choice of a_{iso} is crucial because the NMR experiments actually measure the ratio $a_{iso}^2/(\tilde{D}_\parallel)^{1/2}$. Therefore, the value of $\tilde{D}_\parallel$, as obtained by the NMR experiments, will depend on the fourth power of a_{iso}. On this basis it was argued that the NMR and EPR line-shape analysis conclusion can be brought into agreement. But the large variation in observed EPR linewidths casts doubt on the relative merits of one analysis over the other.

Dynamic nuclear polarization seems to have an advantage over the EPR linewidth and line-shape measurements because the proton relaxation time should be influenced only by those mobile spins. Yet the results from different laboratories were not in entire accord.

Nevertheless, taking into consideration all of the results, the following conclusions may be made: the neutral defects in *cis*-polyacetylene are immobile, the neutral solitons in *trans*-polyacetylene have very high in-chain diffusion rates, the soliton velocity is close to the velocity of sound, and the diffusion is highly anisotropic.

5.6 MECHANISMS OF SOLITON FORMATION
AND OF CIS – TRANS ISOMERIZATION

Pristine *cis*-polyacetylene obtained by low-temperature polymerization may be assumed to have predominantly a cis-transoid structure and to be nearly free of structural defects. Neutral defects are formed initially by interchanges of adjacent long and short bonds, as shown in Eqs. (5.15) and (5.16) (Chien *et al.*, 1980b; Chien, 1979a,b, 1981, 1982). This process requires an activation energy of only 10 kcal mole^{-1} (Section 5.4.1). Su *et al.* (1979) estimated an energy of ~ 0.4 eV for the creation of neutral solitons in good agreement with experiment.

Equations (5.15) and (5.16) showed that, in fact, the neutral defects are produced thermally as soliton–antisoliton pairs $(S \cdot \bar{S} \cdot)$. Now we can distinguish the fate between a $S \cdot \bar{S} \cdot$ pair in *trans*- and *cis*-polyacetylene. In the former case, because of the mobility of solitons, there are several possible consequences. The $S \cdot \bar{S} \cdot$ may oscillate for a while, then be annihilated or may relax by multiphonon or nonradiative processes. A soliton can hop to an adjacent chain in $\sim 10^{-9}$ sec, as estimated from the value of $D_{\perp}$ (Table 5.3), thus separating it from the antisoliton. $S \cdot S \cdot$ pains may be converted to isolated solitons by autoxidation:

$$(5.55)$$

Finally, neutral solitons may be converted to charged solitons by O_2 and Ti^{3+} ion in conjunction:

$$(5.56)$$

One reasonable hypothesis is that a *trans*-polyacetylene chain probably contains either one $S \cdot$ or none. On the average, there may be present one-half of a neutral soliton per chain. Experimentally, *trans*-polyacetylene specimens isomerized by the optimum procedure have [S ·] values reported to lie between one S · per 1000–3000 CH units. Taking a value of 11,000 for $\overline{M}_n$ of free-standing polyacetylene films, we find that $\frac{1}{2}$S · per chain corresponds to one S · per 1700 CH units, in good agreement with observed soliton concentration. If there are barriers restricting soliton motion and therefore S · S̄ · annihilation, for instance, as the result of degradation or disruption of conjugation by twisting of the backbone, the soliton concentration may become substantially higher. In other words, it seems reasonable to expect that there should be about one-half of a neutral soliton for each effective conjugated trans sequence.

It is important to note that the hypothesis of a limit of one-half neutral soliton per *trans*-polyacetylene chain does not apply to spinless charged solitons. A chain can accommodate as many charged solitons of the same spin (S$^{\pm}$) as the procedure for doping can introduce and as permitted by Coulombic interactions.

Let us now consider the neutral defects in *cis*-polyacetylene. The defects do not diffuse apart because as a defect moves it converts more cis-transoid units to the trans-cisoid form, which is higher in energy. Therein lies the significant difference between the cis and trans polymers; the two π systems on either side of the soliton in the trans polymer are identical in energy by virtue of symmetry, but they are not in the case of *cis*-polyacetylene.

The principal event that occurs upon heating of *cis*-polyacetylene is isomerization of the trans-cisoid segments to the trans-transoid form [Eq. (5.15)]. Yamabe *et al.* (1981, 1982) had performed MINDO/3 calculations on this isomerization. They found that the energy barrier for the rotation to change the dihedral angle of one cis-transoid monomer unit by 180° [Eq. (5.57)] is ∼ 1 eV,

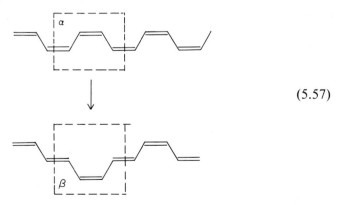

(5.57)

and that the structure β after the rotation is more stable by 0.5 eV than the initial structure α.

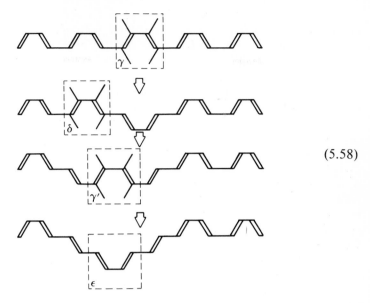

$$(5.58)$$

The conversion of trans-cisoid to trans-transoid occurs even more easily: The reason for this is that whereas structure β is derived from structure α by rotation about two carbon–carbon double bonds, the structure changes shown in Eq. (5.58) involve only rotation around two carbon–carbon single bonds. Calculation of the potential energy for rotations of γ and δ units showed the former process to be without a barrier and that the subsequent rotation of δ units proceeds along the very shallow potential of <0.1 eV (Fig. 5.38). It is anticipated that the rotation of the δ unit induces the further rotation of the γ' unit. The results suggest that once the cis-transoid units are isomerized to the trans-cisoid form, it is immediately converted to the trans-transoid structure.

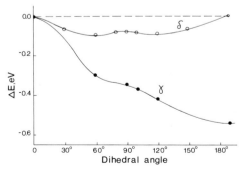

Fig. 5.38   Potential curve for the rotation of a four-carbon unit in a finite trans-cisoid unit to the trans-transoid structure. [After Yamabe *et al.* (1981).]

However, Yamabe *et al.* (1981, 1982) did not consider the role of the neutral defect. The neutral defect is flanked by two long bonds and should assist in the initiation of bond rotation. On the other hand, the calculation did not include the effect of steric constraints imposed by the crystal lattice for this cis–trans isomerization in the solid state.

The mechanism for chemical doping-induced isomerization can be depicted as

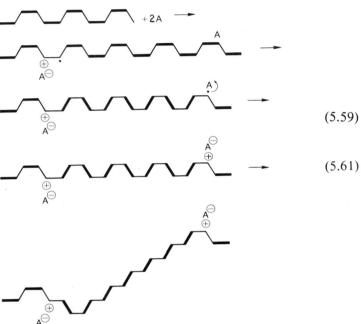

(5.59)

(5.61)

The sequence of events is entirely analogous to that described above for thermal isomerization. Finally, the electrochemical doping-induced isomerization differs from the chemical process only in that electrons were removed electrolytically, as shown below:

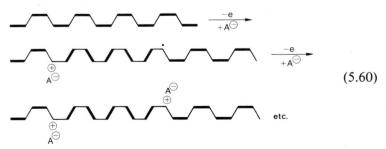

(5.60)

Chapter 6

Spectroscopic, Physical, and Mechanical Properties

This chapter presents some of the properties of undoped polyacetylene. The infrared and Raman spectra of the isomeric $(CH)_x$ and $(CD)_x$ were described in Section 6.1. In resonance Raman scattering experiments, increase of the laser frequency causes different dispersion effects in the spectra of cis and trans polymers, with *trans*-polyacetylene displaying anomalous resonance Raman profiles. In general, increase of excitation energy shifts the Raman profiles of the C=C and C—C stretching modes to higher frequencies, and the bands become asymmetric or will comprise two or more peaks. This has been interpreted as due to scattering from a distribution of conjugated sequences in the trans polymer, whereas the cis polymer was considered to be more uniform and longer in conjugation length. To fit the Raman dispersion profiles, distribution of conjugation lengths of 10–40 had been proposed for *trans*-polyacetylene. However, the effect is sample dependent. Polyacetylene isomerized by different procedures produces dissimilar Raman dispersion profiles. The shift of profile to short wavelengths is not monotonic with increase of trans content in partially isomerized polyacetylene specimens. The phenomenon has also been attributed to hot luminescence.

The results of optical absorption and reflection are described in Section 6.2. Kramers–Kronig analysis of the spectra yielded the optical dielectric constant, optical conductivity, and an estimate for the effective electron mass.

Very striking differences between *cis*- and *trans*-polyacetylene have been observed in their responses to photoexcitation. The cis polymer emits band-edge luminescence and is not photoconducting, whereas the trans polymer is photoconducting but does not luminesce. These behaviors were successfully explained (Section 6.3) by the confinement of electron–hole pairs in photoexcited *cis*-polyacetylene, which recombine radiatively. In the case of the trans polymer the electron–hole pair is transformed into a positive and negative soliton pair; their migration under an electric field results in photoconductivity. Eventually the charged solitons undergo neutralization, and the resulting neutral soliton–antisoliton pair relaxes to the ground state by multiphonon or nonradiative chemical processes.

The remainder of this chapter contains descriptions of the heat capacity, thermal conductivity, stress–strain, dielectric constants, and NMR results on polyacetylene. The second moments of ^1H NMR seem to be sample dependent. Furthermore, there is a report of an unusually sharp decrease in the linewidth at 155 K, a phenomenon somewhat affected by doping. More measurements at this temperature by other techniques are needed to verify whether there is some large amplitude motion and/or whether there are changes in long-range order in the vicinity of this transition temperature. Another NMR study failed to observe the transition.

6.1 VIBRATIONAL SPECTRA

6.1.1 Normal Vibrations

Based on the probable structures described in Chapter 3 for polyacetylene, normal vibrations have been analyzed by Shirakawa and Ikeda (1971) and Inagaki *et al.* (1975). The symmetry elements and the numbering of atoms are given in Fig. 6.1. The screw axis is denoted by C^s and the glide plane by σ_g. In the trans structure there is one acetylene per repeat unit; the factor group is isomorphous to the point group D_{2h} (character table is given in Table 6.1). The symmetry species for the vibrational modes are

$$\Gamma_{\text{trans}} = 4A_g + B_g + A_u + 2B_u \tag{6.1}$$

cis-Polyacetylene has two monomers per repeat unit with a factor group isomorphous to C_{2h}. The character table is given in Table 6.2; the symmetry species for the vibrational modes are

$$\Gamma_{\text{cis}} = 4A_g + B_{1g} + 4B_{2g} + 3B_{1u} + 1B_{2u} + 3B_{3u} \tag{6.2}$$

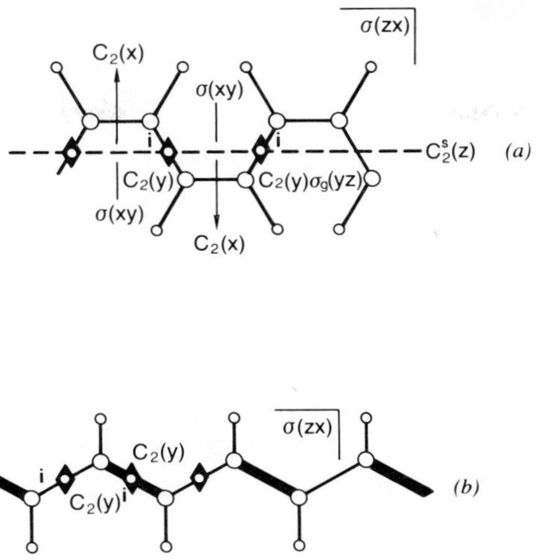

Fig. 6.1 Symmetry elements and numbering of atoms for: (a) *cis*-polyacetylene; (b) *trans*-polyacetylene. [After Shirakawa and Ikeda (1971).]

In Eqs. (6.1) and (6.2), the *g* modes are Raman active and the *u* modes are infrared active.

6.1.2 Infrared Spectra

The infrared spectra of $(C_2H_2)_x$, $(C_2D_2)_x$, $(C_2H_2\text{-}co\text{-}C_2D_2)_x$, and $(C_2H_2\text{-}co\text{-}C_2D_2\text{-}co\text{-}C_2HD)_x$ have been obtained by Shirakawa and Ikeda (1971). The spectra of the homopolymers are given in Figs. 6.2 and 6.3, and the band assignments are described in Table 6.3. Those bands whose intensities

TABLE 6.1

The Character Table for the Point Group C_{2h}, Symmetry Species, Selection Rules, and Number of Normal Modes for *trans*-Polyacetylene[a]

C_{2h}	E	$C_2(y)$	I	$\sigma(zx)$	Tr	Ro	Activity	N^b	n^b
A_g	1	1	1	1	—	—	R	4	4
B_g	1	−1	1	−1	—	R_z	R	2	1
A_u	1	1	−1	−1	T_y	—	IR	2	1
B_u	1	−1	−1	1	T_x, T_z	—	IR	4	2

[a] After Shirakawa and Ikeda (1971).
[b] N and n denote the number of normal modes and of internal vibrations, respectively.

TABLE 6.2

The Character Table for the Point Group D_{2h}, Symmetry Species, Selection Rules, and Number of Normal Modes for *cis*-Polyacetylene[a]

D_{2h}	E	$C_2^s(z)$	$C_2(y)$	$C_2(x)$	i	$\sigma(xy)$	$\sigma(zx)$	$\sigma_g(yz)$	Tr	Ro	Activity	N^b	n^b
A_g	1	1	1	1	1	1	1	1	—	—	R	4	4
B_{1g}	1	1	−1	−1	1	1	−1	−1	—	R_z	R	2	1
B_{2g}	1	−1	1	−1	1	−1	1	−1	—	—	R	4	4
B_{3g}	1	−1	−1	1	1	−1	−1	1	—	—	R	2	2
A_u	1	1	1	1	−1	−1	−1	−1	—	—	—	2	2
B_{1u}	1	1	−1	−1	−1	−1	1	1	T_z	—	IR	4	3
B_{2u}	1	−1	1	−1	−1	1	−1	1	T_y	—	IR	2	1
B_{3u}	1	−1	−1	1	−1	1	1	−1	T_x	—	IR	4	3

[a] After Shirakawa and Ikeda (1971).
[b] N and n denote the number of normal modes and of internal vibrations, respectively.

increase with temperature of polymerization are attributed to the trans structure; the bands that behave oppositely are attributed to the cis units. The frequency ratios of the 3013, 1292, and 1015 cm^{-1} bands in *trans*-(CH)$_x$ to the 2231, 916, and 752 cm^{-1} bands in *trans*-(CD)$_x$ are 1.35, 1.41, and 1.35, respectively. The theoretical ratio calculated from the masses of hydro-

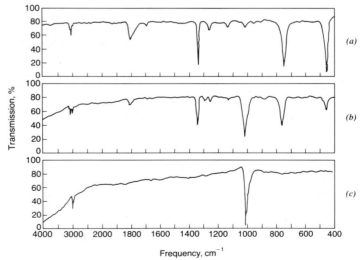

Fig. 6.2 Infrared spectra of (CH)$_x$ prepared at (a) 195, (b) 293, and (c) 423 K. [After Shirakawa and Ikeda (1971).]

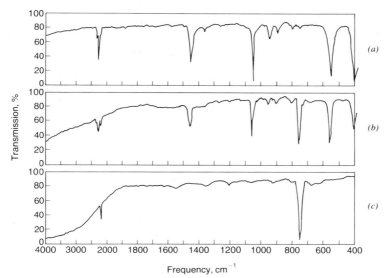

Fig. 6.3   Infrared spectra of $(CD)_x$ prepared at (a) 195, (b) 297, and (c) 423 K. [After Shirakawa and Ikeda (1971).]

gen and deuterium is 1.36. These bands correspond to the C—H stretching and deformation modes; their assignments given in Table 6.3 are based on vibrations in low molecular weight analogs. By the same reasoning, the 3057, 3044, 1329, 1249, and 740 cm^{-1} bands (2275, 2255, 1050, 947, and 548 cm^{-1} bands) in *cis*-$(CH)_x$ (*cis*-$(CD)_x$) are designated as C—H (C—D) stretching and C—H (C—D) deformation modes. The 446 cm^{-1} band in *cis*-$(CH)_x$ with a corresponding band at 402 cm^{-1} in poly(acetylene-d_2) has $\nu_H/\nu_D = 1.11$; thus the vibration does not involve hydrogen atoms. The band is assigned to B_{1u} C—C—C deformation.

It was not possible to determine from an analysis of the infrared spectra of cis-rich polyacetylene whether the polymer has predominantly the trans-cisoid or the cis-transoid structure. The C=C stretching should be active in the cis-transoid structure, whereas the C—C stretching should be active in the trans-cisoid. The C=C stretching vibration occurs in the range of 1680–1620 cm^{-1} and is lowered toward ~ 1600 cm^{-1} for conjugated systems. The C—C stretching vibrations occur between 1200–1000 cm^{-1} and are raised for conjugated systems. In *cis*-$(CH)_x$ there is no band in the vicinity of 1600 cm^{-1}, but there is a very weak band at 1118 cm^{-1}, which may be assigned to the C—C stretching mode for the trans-cisoid form. Unfortunately, there is no corresponding absorption in *cis*-$(CD)_x$ between

TABLE 6.3

Assignments of the Infrared Spectra of $(CH)_x$ and $(CD)_x$[a]

Observed frequency (cm^{-1}) and intensity								Assignment	
$(CH)_x$				$(CD)_x$					
3047	vw[b]	vw[c]	—	2275	vw[b]	vw[d]	—	B_{1u}	C—H stretching in cis
3044	w	w	—	2255	m	w	—	B_{3u}	C—H stretching in cis
3013	vvw	w	m	2231	w	w	m	B_{3u}	C—H stretching in trans
1800	m	w	—						1329 + 466 = 1785?
				1448	s	m	—		1050 + 402 = 1452?
1690	vw	vvw	—						1249 + 466 = 1695?
				1360	vw	—	—		947 + 402 = 1349?
1329	s	m	—	1050	s	m	—	B_{1u}	C—H in-plane deformation in cis
1292	—	vw	vvw	975	vvw	vvw	—	B_{1u}	C—H in-plane deformation in trans
1249	w	vw	—	916	—	vvw	vw	B_{3u}	C—H in-plane deformation in cis
1118	vw	vvw	—	947	w	vw	—	B_{3u}	C—C stretching in cis?
1015	vw	s	vs	892	w	vw	—	B_{2u}	C—H out-of-plane deformation in trans
				800	vw	vw	—		In -(trans-CH=CH$_2$)-
980	vvw	—	—	752	vvw	s	vs		between -(cis-CH=CH)$_n$-
940	vw	—	—	735	vvw	vvw	—		In -(trans-CH=CH)$_t$-
				720	vvw	—	—		between -(cis-CH=CH)$_n$-
740	vs	s	—	548	vs	s	—	B_{3h}	C—H out-of-plane deformation in cis
446	vs	m	—	402	vs	s	—	B_{1u}	C—C deformation in cis

[a] Shirakawa and Ikeda (1971).
[b] Intensity in the polymer prepared at 195 K.
[c] 293 K.
[e] 297 K.
[d] 423 K.

$1000-1100$ cm^{-1}, expected for the same vibration, which has a small isotopic shift.

The IR spectra of $(C_2D_2\text{-}co\text{-}C_2D_2)_x$ show differences from superimposed spectra of the homopolymers especially in the region of C—H and C—D out-of-plane deformation modes. They moved to lower frequencies in the trans configuration and to higher frequencies in the cis configuration and split into several bands. Thus the trans C—H and C—D deformation bands appeared at $745-715$ cm^{-1}, the cis C—H bands at $810-1743$ cm^{-1}, and the *cis*-C—D bands at $663-552$ cm^{-1}. In the case of terpolymer $(C_2H_2\text{-}co\text{-}C_2D_2\text{-}co\text{-}C_2HD)_x$ the shifts of the out-of-plane deformation vibrations with respect to the homopolymer are even greater. Tentative assignments have been made for some of them as shown below:

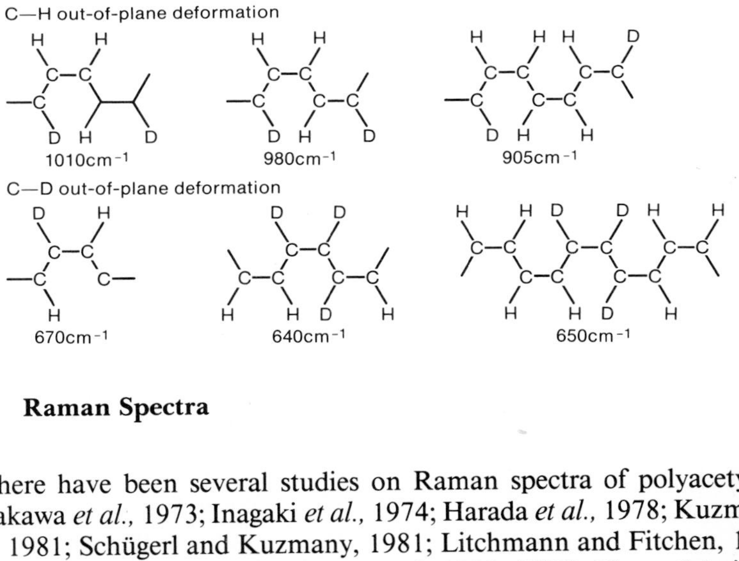

(6.3)

6.1.3 Raman Spectra

There have been several studies on Raman spectra of polyacetylene (Shirakawa *et al.*, 1973; Inagaki *et al.*, 1974; Harada *et al.*, 1978; Kuzmany, 1980, 1981; Schügerl and Kuzmany, 1981; Litchmann and Fitchen, 1979, 1981; Litchtmann, 1981; Litchmann *et al.*, 1980, 1981). Figure 6.4 gives a set of typical spectra for polyacetylene and poly(acetylene-d_2); the weaker lines are not seen clearly. In addition to the frequency reduction expected for the deuterated polymer, there are other differences between the two isotopic polyacetylenes. In the case of $(CH)_x$, there are two strong Raman bands for the trans polymer and three for the cis polymer. It is reversed for $(CD)_x$, which has three bands for the trans isomer and two bands for the cis polymer. Tables 6.4 and 6.5 give the assignments for the Raman lines.

The intense Raman bands are due to the resonance-enhancement effect when the exciting frequency approaches or enters the electronic absorption bands. Because the visible absorption of polyacetylene belongs to $\pi \rightarrow \pi^*$

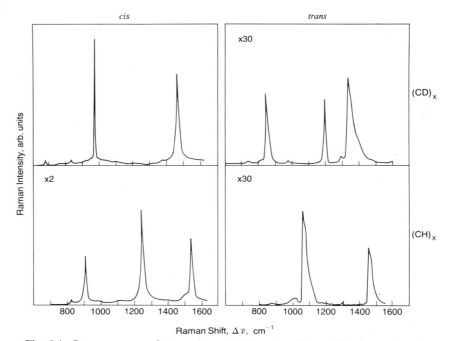

Fig. 6.4 Raman spectra of *cis*- and *trans*-(CH)$_x$ and (CD)$_x$ at 77 K for 600-nm laser excitation. [After Lichtmann *et al.* (1980).]

transition, the resonance-enhanced Raman modes depend on the contribution of C—C and C=C stretching vibrations. This consideration enables the elucidation of the structure of *cis*-polyacetylene. In the cis-transoid configuration there are one A_g C=C stretching and two C—C stretching modes of A_g and B_{2g} symmetry. On the other hand the trans-cisoid configuration has two C=C stretching modes (A_g and B_{2g}) and one A_g C—C stretching mode. Table 6.4 shows that there are in the Raman spectra of *cis*-(CH)$_x$ (*cis*-(CD)$_x$) one A_g C=C stretching line at 1544 (1460) cm^{-1}, an A_g C—C stretching mode at 1252 (978) cm^{-1}, and a B_{2g} C—C stretching vibration at 915 (870) cm^{-1}. Thus it can be concluded that *cis*-polyacetylene has the cis-transoid configuration. This conclusion supports the mechanism of Ziegler–Natta polymerization involving cis opening of the triple bond of acetylene in the propagation step (Section 2.5.2).

6.1.4 Vibrational Analysis

Force-constant calculations had been carried out by Shirakawa and Ikeda (1971) for the C—H out-of-plane vibrations. They made the simpli-

TABLE 6.4

Raman Lines of *cis*-(CH)$_x$ and *cis*-(CD)$_x$

$\nu(CH)_x$ (cm^{-1})	$\nu(CD)_x$ (cm^{-1})	$\dfrac{\nu(CH)_x}{\nu(CD)_x}$		Assignment
295 vw	265 vw	1.11	B_{3g}	C—C—C deformation, out-of-plane
445 w	406 vw	1.09	B_{2g}	C—C—C deformation, in-plane
820 vw	830 vw			not identified
915 s	870 s	1.06	B_{2g}	C—C stretch plus contribution from trans in the deuterated polymer
985 m	767 w	1.28	B_{3g}	C—H deformation out-of-plane
1008 m	745 m	1.35	B_{1g}	C—H deformation in-plane
$\begin{bmatrix}1070\\1130\ \text{s}\end{bmatrix}$	870 m			from residual *trans*-(CH)$_x$
1170 w	930 w	1.26	B_{2g}	C—H deformation in-plane
$\begin{bmatrix}1230\ \text{sh}\\1252\ \text{vs}\end{bmatrix}$	$\begin{bmatrix}965\ \text{sh}\\987\ \text{s}\end{bmatrix}$	1.28	A_g	C—C stretch with considerable addition of C—H deformation in-plane
1292 w	1205 s	1.07	A_g	C—H deformation in-plane with considerable addition of C—C stretch in the deuterated polymer
1490 sh	1390 sh			from residual *trans*-(CH)$_x$
1544 vs	1460 vs	1.05	A_g	C=C stretch
3030 m	2260 m	1.34	B_{2g}	C—H stretch
3090 m	2315 w	1.34	A_g	C—H stretch

TABLE 6.5

Raman Lines of *trans*-(CH)$_x$ and *trans*-(CD)$_x$

$\nu(CH)_x$ (cm^{-1})	$\nu(CD)_x$ (cm^{-1})	$\dfrac{\nu(CH)_x}{\nu(CD)_x}$		Assignment
878 w	808 w			Not identified
1008 s	745 m	1.35	B_g	CH deformation out-of-plane
1090–1120 s	856–870 s	1.28	A_g	C—C stretch with considerable addition of CH deformation in-plane
1170 w	960 w			Not identified
1292 w	1207 w	1.07	A_g	C—C stretch
$\begin{bmatrix}1470\\1520\end{bmatrix}$ vs	$\begin{bmatrix}1370\\1450\end{bmatrix}$ vs	1.07	A_g	C=C stretch
2990 w	2230 m	1.34	A_g	C—H stretch

TABLE 6.6

Force Constants for C—H Out-of-Plane Deformations in Polyacetylenes[a]

Mode[b]	Force constant mdyn Å
	0.427
	0.057
	−0.080
	0.059
	−0.036
	0.002
	0.019
	0.002

[a]Shirakawa and Ikeda (1971).
[b]Broken line indicates the two coordinates involved in the interaction.

TABLE 6.7

Calculated and Observed C—H and C—D Out-of-Plane Deformation Frequencies

	Frequency (cm⁻¹)	
	Obsd	Calcd
cis-CH	740	740.5
cis-CD	548	543.7
trans-CH	1015	1014.8
trans-CD	752	745.2

TABLE 6.8

Raman Spectra of Polyacetylene

cis-(CH)$_x$			*cis*-(CD)$_x$			*trans*-(CH)$_x$		*trans*-(CD)$_x$		
obsd		calcd	obsd		calcd	obsd	calcd	obsd	calcd	
a	b	c	a	d	c	b,e,f	g	d	g	f
3090		3090	2315		2301	1474–1500[j]	1900	1340	1358	1420
1544	1541	1527	1460	1470	1486			1296		
1292		1209	1205			1080–1100	1190	1198	1226	1185
1252	1247	906	978	972	951	1016–1020	1120	848	850	949
3030		3049	2260		791			816		
1170		1575	1170	1102	1423			744	749	818
915	908	1013	960	835,1042	858			625		
445		508	406	403	445			539		

[a] Kuzmany (1980).
[b] Lefrant *et al.* (1982).
[c] Galtier *et al,* (1982).
[d] Lichtmann *et al.* (1980).
[e] Shirakawa *et al.* (1973).
[f] Hsu *et al.* (1978).
[g] Mele and Rice (1980a).
[h] Inagaki *et al.* (1974).
[j] Underlined are strong peaks.

fying assumptions that all C—C lengths were 1.40 Å, that all C—H lengths were 1.08 Å, and that all C—C—H and C—C—C angles were 120°. The results are summarized in Table 6.6. There is good agreement between calculated and observed infrared-active out-of-plane deformation frequencies, as shown in Table 6.7.

An early analysis of Raman vibrations was made by Inagaki *et al.* (1975) for *trans*-polyacetylene and a later one by Schügerl and Kuzmany (1981). In

Table 6.8 the calculated and observed Raman line frequencies from these and other sources are summarized.

6.1.5   Raman Line Dispersion

An interesting but somewhat controversial aspect of resonance Raman spectroscopy of polyacetylene is the dispersion effect of excitation wavelength first observed by Shirakawa *et al.* (1973) for the 1470 cm^{-1} trans band in cis-rich sample. The dispersion behaviors are very different for the cis and trans polymers. The effect of excitation wavelength on the resonance Raman spectra of *cis*-polyacetylene is small, but large for *trans*-polyacetylene.

In the case of *cis*-(CD)$_x$, Fig. 6.5 shows the effect of excitation wavelength on the 978 cm^{-1} A_g band of the C—C stretching mode and the 1460 cm^{-1}

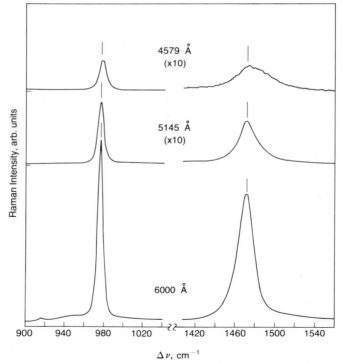

Fig. 6.5   Resonance Raman dispersion of *cis*-(CD)$_x$ at 77 K excited by three differrent frequencies. [After Lichtmann *et al.* (1980).]

A_g band of the C=C stretching mode. The band profile is symmetric for the 978 cm^{-1} band at all three excitation wavelengths and for the 1460 cm^{-1} band, using 514.5- and 600-nm excitation. When excited by 457.9-nm radiation, the C—C band was broadened on the higher frequency side. Very small shifts of approximately 2 and 5 cm^{-1} were seen for the 976 cm^{-1} and 1470 cm^{-1} modes, respectively. Because some workers had attributed much larger dispersion effects in *trans*-polyacetylene to conjugation length-dependent modes and distribution in conjugation length, the results on *cis*-(CD)$_x$ had been said to indicate either that, as grown, cis film has a narrow distribution in conjugation lengths or that the modes of the cis isomer are not dependent on conjugation length. Similar effects of excitation wavelength on resonance-line intensity and width were observed for *cis*-(CH)$_x$.

In comparison, the long-wavelength dispersion anomalies are very marked for *trans*-polyacetylene (Kuzmany, 1980, 1981; Schügerl and Kuzmany, 1981; Lichtmann *et al.,* 1980, 1981). The effect of excitation wavelength on the 1100 and 1500 cm^{-1} bands in *trans*-(CH)$_x$ for the C—C and C=C stretching modes, respectively, and on the 860, 1200, and 1400 cm^{-1} bands in *trans*-(CH)$_x$ for the two C—C stretching modes and C=C stretching mode, respectively, are shown in Fig. 6.6 (Schügerl and Kuzmany, 1981). There are shifts in the long wavelength edge of the 1100 and 1500 cm^{-1} bands in *trans*-(CH)$_x$ and of the 860 and 1400 cm^{-1} bands in *trans*-(CD)$_x$ with increase in excitation frequency. There are concommitent increases in bandwidths, and the bands show two or maybe even three maxima. Lichtmann *et al.* (1980, 1981) reported different dispersion results for these bands. The long-wavelength band edges were sharp and not affected by changes in excitation wavelength. Increases in excitation frequency resulted in the development of a broad band to the high frequency with one or two maxima. The integrated intensity of each band does not vary much as the laser is tuned through the visible wavelengths. The two studies agreed in that the 1200 cm^{-1} band in *trans*-(CD)$_x$ does not show dispersion anomaly.

The excitation profile also shows the dispersion phenomenon (Fig. 6.7). The top curve shows the variation in intensity at the low-frequency edge of the band, which is strongest for excitation in the red. The bottom curve is for the variation in intensity at the highest-frequency edge, which is strongest in the blue. For intermediate vibrational frequencies, the peak in the excitation curves lies between the two above.

The dispersion anomaly is apparently sample dependent; samples of *trans*-polyacetylene subjected to longer exposure to air or prepared under less than ideal isomerization conditions show Raman profiles that are shifted increasingly to higher frequencies (Fig. 6.8).

One interpretation given to the dispersion anomalies for *trans*-polyacet-

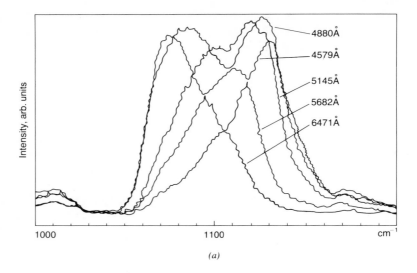

(a)

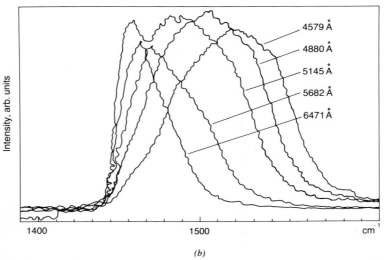

(b)

Fig. 6.6   The resonance-enhanced Raman lines of backbone vibrations excited with various laser lines: (a) 1100 cm⁻¹ band in *trans*-(CH)$_x$; (b) 1500 cm⁻¹ band in *trans*-(CH)$_x$; (c) 860 cm⁻¹ band in *trans*-(CD)$_x$; (d) 1200 cm⁻¹ band in *trans*-(CD)$_x$; (e) 1400 cm⁻¹ band in *trans*-(CD)$_x$. [After Schügerl and Kuzmany (1981).]

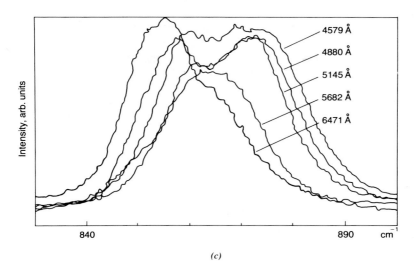

(c)

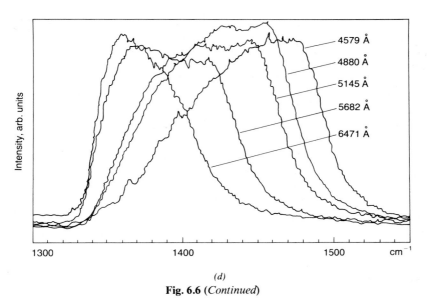

(d)

Fig. 6.6 (*Continued*)

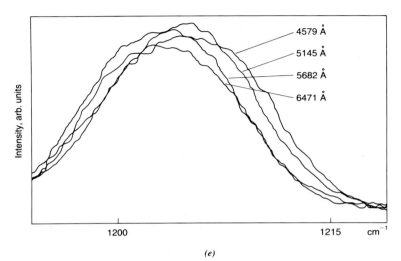

(e)

Fig. 6.6 (*Continued*)

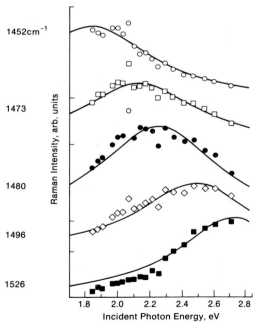

Fig. 6.7 Excitation profiles for the Raman intensity at five different frequencies within the C=C band. [After Lichtmann (1981).]

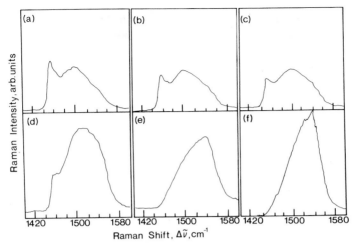

Fig. 6.8 Sample dependence of Raman dispersion effect excited at 457.9 nm in *trans*-$(CH)_x$. Curves (a) to (f) show the effect of increasing exposure to air and less than ideal isomerization conditions: (a) is the best sample; (f) is the worst sample. [After Fitchen (1982).]

ylene is that shorter conjugation lengths have higher vibration frequencies and are excited at higher photon energies. The nature of interruption of backbone conjugation is not understood but is thought to be due to cross-links, bond defects, impurities, etc. Kuzmany (1981) and Schügerl and Kuzmany (1981) have considered the effect of the C=C stretching force constant on the frequencies of the Raman lines (Fig. 6.9). The results showed that the C—C stretching force constant does not have a strong influence on Raman mode frequency. On the other hand, the C=C stretching force constant has a strong effect on Raman mode frequency.

The observed Raman line profile was considered to be the sum of contributions from polyacetylene segments of different conjugation lengths, each having a Lorentzian line shape and assumed to have a width of 20 cm^{-1} corresponding to the width of the narrowest line observed in *cis*-$(CH)_x$. Fitchen (1982) assumed a distribution function $P(n)$ for the number of chains of conjugation length n, so that the total absorption coefficient is

$$\alpha(v) = \sum_{n=1}^{\infty} P(n)\alpha_n(v), \tag{6.4}$$

and the Raman profile at laser frequency v_L is

$$\sigma(v_L, \omega) = \sum_{n=1}^{\infty} P(n)\sigma_n(v_L, \omega). \tag{6.5}$$

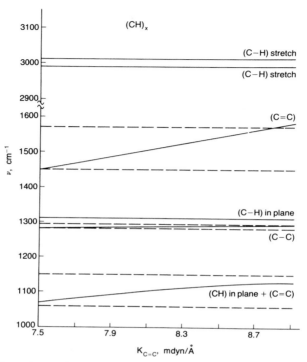

Fig. 6.9 Dependence of Raman-active molecular mode on the C=C stretching force constant (——) calculated (----) range of observed dispersion [After Schügerl and Kuzmany (1981).]

A distribution of $P(n) \alpha n^{-2.3}$ was found to fit the experimental profile. In a similar vein, Kuzmany (1980) chose a distribution that was the sum of two log normal distributions with maxima at $n = 5$ and 40. In another estimate, Kuzmany (1981) used a modified Gaussian distribution function of the form

$$\phi(n) = n^2 \exp[-(n - n_0)^2/2\Delta^2] \qquad (6.6)$$

and found that $n_0 = 30$ and $\Delta = 20$ afforded a good fit with the observed spectral profile.

Mele and Rice (1980a) treated the anomalous systematic decrease of the frequency of bond stretching vibrations in another way. The premise is that *trans*-polyacetylene, as a Peierls insulator, has a phonon spectrum screened by extended π-electron states. As a consequence, the intrinsic lattice vibrations should possess long-wavelength dispersion anomalies. They added to the usual Born–Oppenheimer Hamiltonian a phonon–electron interaction term:

$$H = H_e + H_p + H_{ep}, \tag{6.7}$$

where H_e explicitly includes the nearest-neighbor coupling of the π orbitals with a bandwidth of ~ 12 eV (Grant and Batra, 1979a), and H_p describes the unscreened phonon frequency, using a Born model and bond-stretching and bond-rotation force constants of similar vibrations in small organic molecules. The bond-stretching force constants are scaled linearly with bond length. The electron–phonon coupling term H_{ep} is defined by the chain conformation and represents the modulation of the nearest-neighbor electronic interaction integral due to atomic vibrations. The only parameter in H_{ep} is $\beta = \partial t_{n,n+1}/\partial x$ where $t_{n,n+1}$ is the nearest-neighbor electronic interaction integral and x is the bond length. When the unscreened force field model, i.e., $\beta = 0$, is applied to *trans*-polyacetylene, there is no agreement between calculated and observed Raman bands. The introduction of β improves agreement, and variation of β resulted in best agreement between zone-center frequency and experimental values for $\beta = 8$ eV Å^{-1}.

The general Raman-active modes were calculated at 1487, 1180, and 1030 cm^{-1}. The H_{ep} causes softening of optical phonons and dispersions of the modes near the zone center. For the two most strongly softened phonon bands the long-wavelength dispersions for $q < 0.10$ Å^{-1}, where q is the wave vector, are: $\omega_\lambda^2(q) = \omega_\lambda^2(0) + c^2 q^2$. For the $\omega_\lambda(0) = 1487$ cm^{-1} band, $c = 1.8 \times 10^6$ cm sec^{-1}; for the $\omega_\lambda(0) = 1030$ cm^{-1} band, $c = 1.05 \times 10^6$ cm sec^{-1}.

If the dispersion of the Raman profile is simply due to the presence of distribution in conjugation length, then it would seem that the 1100 cm^{-1} and 1500 cm^{-1} bands in *trans*-$(CH)_x$ should display similar monotonic conjugate length dependence and profile changes for the two bands as the trans content of the polymer changes. Lefrant et al. (1982) studied the dispersion behavior of polyacetylene that has been isomerized from film having a cis content $> 90\%$ at 413 K under vacuum for periods of $1 – 10$ min to obtain specimens having various trans contents. They found that major differences in the profiles of the trans bands exist between pure *trans*-polyacetylene and the slightly isomerized cis polymer. The Raman bands of pure *trans*-$(CH)_x$ broaden and develop a double-peak structure as the laser excitation is changed from 676.4 nm to 457.9 nm. The opposite behavior was observed for the trans bands in the cis-rich polymer. Figure 6.10 shows the variation in the profile of the trans Raman bands with changing trans content for excitation wavelengths 647.1 and 676.4 nm. The band profiles for 20% *trans*-$(CH)_x$ have very different line shapes, depending on excitation frequencies. When excited at 647.1 nm, the low-frequency band has two maxima and the high-frequency band has three maxima, but at 676.4-nm excitation this is reversed. The behavior for the high-frequency band in the sample with a 10% trans sample is opposite to the 20% trans

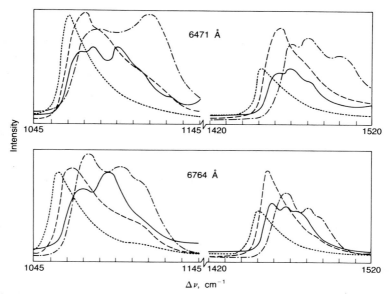

Fig. 6.10   Raman spectra of *trans*-(CH)$_x$ bands in: (· · · ·) 100% trans; (----) 50% trans; (- · - · - · -) 20% trans; (——) 10% trans. [After Lichtmann (1981).]

sample, whereas the low-frequency band is not affected by the excitation frequency. Moreover, the Raman bands from the 10% *trans*-(CH)$_x$ film show shifts to lower frequency as compared to the 20% *trans*-(CH)$_x$ sample. These results cast doubt on the simple interpretation of the Raman band profile relation with conjugation length. But rather there seems to be a critical dependence of the response of the trans segments on the sample composition and excitation energy upon the injection of the e–h pair and the dynamics of the lattice distortion involved in the scattering process.

Mele (1982) proposed that the frequency-dependent Raman line shapes are due to hot luminescence in very long *trans*-polyacetylene chains. Following laser excitation the polymer is unstable to a relaxation that will lower the one-electron eigenvalue of the final state and raise the energy of the initial state connected in the absorption. The various symmetry-lowering distortions of the lattice vary as a function of the exciting radiation. The short-time (10^{-14} sec) lattice dynamics is highly specific to the frequency of the absorbed photon. For instance, for the four lowest allowed optical transitions of a 48-atom polyene chain, the lowest optical excitation between band-edge states resulted in a strong depression of the bond alternation amplitude at the center of the chain. For the second excitation level,

Mele obtained a uniform suppression of the bond-alternation amplitude and a shorter-wavelength modulation of this amplitude. For the higher-lying excitations the shorter-wavelength modulation has the form of a standing wave with wavelength $\lambda_n = L/(n - 1)$ for the nth excitation in the polyene chain of length L. The excess population of the excited intermediate states decays by the adiabatic radiative channel corresponding in this picture to inelastic scattering by the hot luminescence channel.

Thus in addition to the normal Raman scattering, the photoinduced relaxation should be manifested as emission of vibrational quanta seen as a "satellite" line in the Raman spectrum. The deviation of the "satellite" line from the primary line depends on the incident photon energy. This dependence increases in the order of $(CD)_x$, 1198 cm^{-1}; $(CD)_x$, 840 cm^{-1}; $(CH)_x$, 1450 cm^{-1}; $(CH)_x$, 1055 cm^{-1}; $(CD)_x$ 1330 cm^{-1}, bands all for the trans polymer.

Finally, there are also nonadiabatic channels for the relaxation of the hot carrier, which is important for describing the long-time (10^{-12} sec) processes and may be responsible for part of the width of the secondary emission and the high-frequency tail on the observed Raman line.

It seems that much more careful and thorough investigation of the resonance Raman dispersion is needed especially with regard to sample preparation, isomerization, and handling. Conductivity, EPR, IR, and other measurements should be made on the same specimen used for Raman study for correlation. If the conjugation lengths are as short in *trans*-polyacetylene as suggested by the analysis of Raman dispersion, can the soliton waves be sustained without displaying boundary effects? Other properties to be discussed in this chapter are also seemingly incongruent with the hypothesis of very short conjugation lengths.

6.2 OPTICAL ABSORPTION AND REFLECTIONS

The absorption spectra of polyacetylenes are shown in Fig. 6.11. For *cis*-polyacetylene the absorption begins gradually, rising sharply between 1.8 and 1.9 eV. The first absorption maximum occurs at 2.1 eV, followed by two others at 2.3 and 2.4 eV. The peak absorption coefficient is $\sim 4 \times 10^5$ cm^{-1}, comparable to 10^6 for typical direct-gap semiconductors. The low-energy tail structure may be due to partial isomerization to some trans structures in the polymer. The intrinsic optical gap is about 1.8 eV.

The intrinsic absorption edge of *trans*-polyacetylene is shifted to lower energies with respect to the cis polymer. The absorption begins slowly at

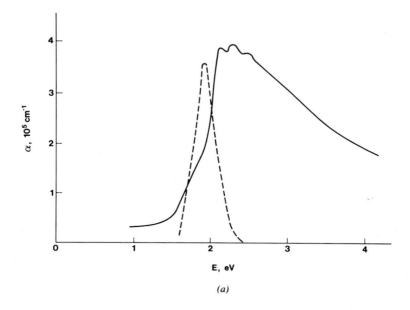

(a)

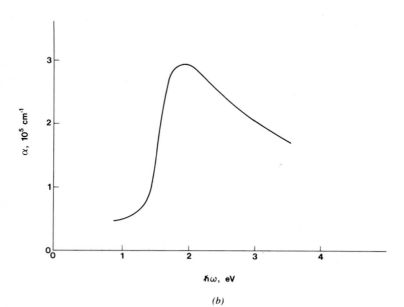

(b)

Fig. 6.11   Optical absorption curves for (a) *cis*-polyacetylene and (b) *trans*-polyacetylene. [After Fincher *et al.* (1979a).]

1.0 eV, rising abruptly at 1.4 eV to a peak at about 1.9 eV, which is red-shifted by 0.3 eV from the cis polymer. The absorption maximum is without structure and has an absorption coefficient of $\sim 3 \times 10^5$ cm^{-1}. The low-energy absorption tail may be attributed to the low dimensionality of the polyacetylene. In a typical three-dimensional semiconductor, the energy surface may be written

$$E_{cv}(k) = E_0 + t_{0x}k_x^2 + t_{0y}k_y^2 + t_{0z}k_z^2, \tag{6.8}$$

where the t_0s are the transfer integrals. The density of states in the valence band or the conduction band is

$$\rho(E_{cv}) = \frac{(E - E_G)^{1/2}}{4\pi^2(t_{0x}t_{0y}t_{0z})^{1/2}}, \tag{6.9}$$

and the band edge should rise as $(E - E_G)^{1/2}$. This is shown in Fig. 6.11a as the broken curve. However, as the dimensionality decreases, t_{0y} and t_{0z} decrease. In the case of polyacetylene the transverse bandwidth is only ~ 0.4 eV as compared to the longitudinal bandwidth of ~ 10 eV. Thus the density of states near the gap edge varies as

$$\rho(E) = \rho(0)[1 - E/E_G]^{-1/2}, \tag{6.10}$$

where $\rho(0)$ denotes the density of states at the band center in the absence of a band gap. This gives rise to the absorption tail shown as a dotted curve in Fig. 6.12b. Figure 6.12a is a schematic diagram of the energy band structure of polyacetylene. The direct gap results from the bond alternation. Interchain coupling is shown here to give the E_G^{ind}, where b is one of the transverse lattice vectors. The direct-gap transition occurs at $(\pi/2a, 0, 0)$ and an indirect gap from $(\pi/2a, \pi/b, 0)$. Optical absorption begins when the photon energy is greater than the indirect gap and then increases rapidly as $h\nu > E_G^{dir}$. Absorption following maximum decreases for large-energy photons.

Reflectance spectra have been measured on stretch-aligned ($l/l_0 = 2.94$) polyacetylene films by Fincher *et al.* (1978, 1979a). The results are shown in Fig. 6.13. There is a broad maximum for $R_\parallel$ at ~ 2 eV, corresponding to the absorption peak. The reflectance perpendicular to the stretch direction is rather insensitive to photon energy, and there is a small maximum at ~ 1.7 eV. The largest value of anisotrophy is $R_\parallel/R_\perp = 10$ at 2.5 eV.

The reflectivity is given by

$$R = \frac{(n-1)^2 + \alpha^2}{(n+1)^2 + \alpha^2}, \tag{6.11}$$

where n is the refractive index and α the extinction coefficient, which are related to the dielectric function of nonmagnetic materials by

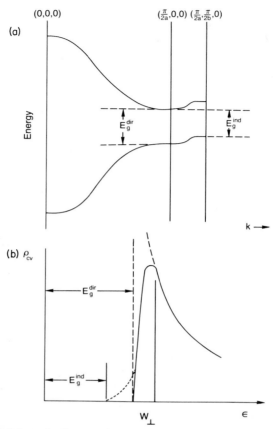

Fig. 6.12 (a) Schematic diagram of energy-band structure of polyacetylene showing the direct gap due to bond alternation and indirect gap from interchain interaction. (b) The joint optical density of states for band structure in (a). [After Fincher *et al.* (1979a).]

$$n = \{\tfrac{1}{2}[(\epsilon_1^2 + \epsilon_2^2)^{1/2} + \epsilon_1]\}^{1/2} \tag{6.12}$$

$$\alpha = \{\tfrac{1}{2}[(\epsilon_1^2 + \epsilon_2^2)^{1/2} - \epsilon_1]\}^{1/2}. \tag{6.13}$$

ϵ_1 (ϵ_2) is the real (imaginary) part of the complex dielectric function. The frequency dependence of the dielectric function is given by the Kramers–Kronig dispersion relations

$$\epsilon_1(\omega) - 1 = \frac{2}{\pi} P \int_0^\infty \frac{\omega' \epsilon_2(\omega')}{(\omega')^2 - \omega^2} \, d\omega', \tag{6.14}$$

$$\epsilon_2(\omega) = -\frac{2\omega}{\pi} P \int_0^\infty \frac{[\epsilon_1(\omega') - 1]}{(\omega')^2 - \omega^2} \, d\omega'. \tag{6.15}$$

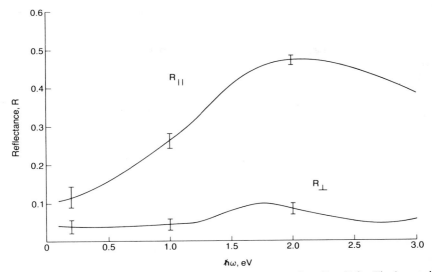

Fig. 6.13   Reflectance spectra of partially oriented polyacetylene film. [After Fincher *et al.* (1978,1979a).]

Analysis of the parallel reflectance spectrum of Fig. 6.13 gave the variation of ϵ_1 with frequency (Fig. 6.14). In Eqs. (6.14) and (6.15) P represents the principal value. The low-frequency limiting value for $\epsilon_{||}(0) \simeq 4$.

   The optical conductivity σ can be obtained from the absorption coefficient α by

$$\sigma = nc\alpha/4\pi, \tag{6.16}$$

where c is the speed of light. Figure 6.15 showed $\sigma_{||}(\omega)$ as a function of energy, which is zero at low frequency, increasing to a maximum value of 4×10^3 $(\Omega \text{ cm})^{-1}$ at about 2 eV.

   The oscillation strength sum rule for solids is

$$\int_0^\infty \omega\epsilon_2(\omega)\,d\omega = \frac{\pi\omega_p^2}{2} = \frac{\pi(\omega_p^\circ)^2 m}{2m^*}, \tag{6.17}$$

where ω_p° is the plasma frequency for π electrons, which for free electron mass is

$$(\omega_p^\circ)^2 = 4\pi Ne^2/m \simeq 6 \times 10^{31}, \tag{6.18}$$

and N is the number of π electrons per unit volume. N is about 2×10^{22} cm^{-3} for polyacetylene with a density of 0.4 g cm^{-3}. Now

$$\epsilon_2 = 4\pi\sigma/\omega, \tag{6.19}$$

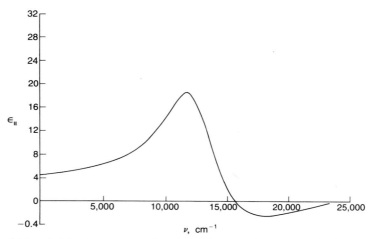

Fig. 6.14 $\epsilon_\parallel(\omega)$ for *trans*-polyacetylene from Kramers–Kronig analysis of the $R_\parallel$ reflectance spectrum. [After Fincher *et al.* (1979a).]

and we can rewrite Eq. (6.16)

$$8 \int_0^\infty \sigma(\omega)\, d\omega = (\omega_p{}^\circ)^2 \frac{m}{m^*} \simeq 3 \times 10^{31} \qquad (6.20)$$

Integration of Fig. 6.15 gives $(\omega_p{}^\circ)^2 m/m^* \simeq 3 \times 10^{31}$. Comparison with the $(\omega_p{}^\circ)^2$ value of 6×10^{31} implies an effective electron mass of nearly unity, i.e., $1.0 < m^*/m < 1.5$.

The effect of hydrostatic pressure on the absorption spectra of polyacetylene was studied by Moses *et al.* (1982a) from ambient pressure to 8.8 kbar. There was a large "red shift" of the absorption edge of the trans polymer. At a pressure of 8.8 kbar the energy gap is reduced by about 0.1 eV. In the case of the cis polymer, there is a similar effect above the low-energy absorption knee. The latter is due to absorption by the unavoidable trans isomeric units in the cis-rich polymer. When at the end of a pressure cycle the absorption was remeasured at ambient pressure, it was found to be identical with those obtained before the application of high pressure. Thus the shift is real and reversible.

The above results were used by Moses *et al.* (1982a) to estimate the magnitude of interchain transfer integral $t_\perp = t_x = t_y$. Equation (6.8) can be rewritten

$$E_\pm(k) = E_0 \pm [2t_0^2 + 2t_1^2 + 2(t_0^2 - t_1^2)\cos 2kc]^{1/2}$$
$$\doteq E_0 + E(k_z), \qquad (6.21)$$

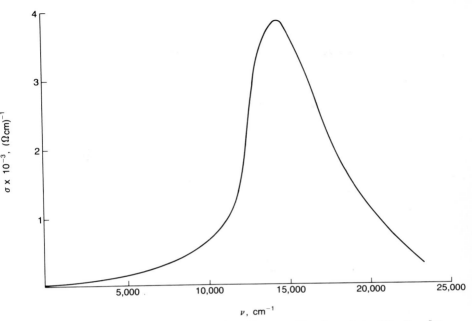

Fig. 6.15 $\sigma_\parallel(\omega)$ for *trans*-polyacetylene from Kramers–Kronig analysis of the $R_\parallel$ reflectance spectrum. [After Fincher *et al.* (1979a).]

where c is the average interchain separation between adjacent carbon atoms, $t_0 + t_1$ is the transfer integral for the short C—C bond and $t_0 - t_1$ is the transfer integral for the long C—C bond. For a tetragonal array of polyacetylene chains, the approximate solution for the energy bands with nearest-chain coupling only has a minimum gap energy given by

$$E_G = 4t_1 - 8t_\perp. \tag{6.22}$$

The magnitude of the "red shift" suggests $t_\perp \simeq 0.025 - 0.05$ eV at ambient pressure and $t_\perp/t_0 \simeq 10^{-2}$.

6.3 PHOTOEXCITATION, LUMINESCENCE, AND PHOTOCONDUCTIVITY

Absorption of photons by traditional semiconductors generates free carriers (electron and hole), resulting in photoconductivity. Recombination of electron and hole results in luminescence. Therefore, the two phenomena

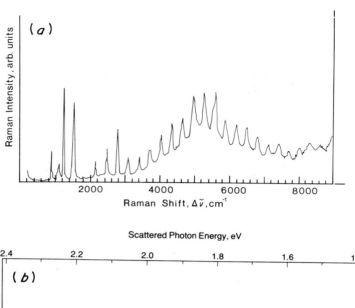

Fig. 6.16 Photoluminescence and multiple-overtone Raman structure from (a) *cis*-(CH)$_x$ and (b) *cis*-(CD)$_x$. [After Lichtmann *et al.* (1981); Lichtmann and Fitchen (1981)].

are intimately related. Although the recombination process can be either radiative or nonradiative at room temperature, luminescence will always be observed at low temperatures inasmuch as the carriers were confined at the site of their formation until recombination occurs. However, the photoexcitation properties of polyacetylenes are markedly different from those of

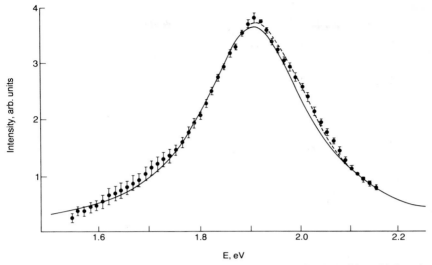

Fig. 6.17 The luminescence spectrum of *cis*-polyacetylene (Fig. 6.16 with multiple-order Raman lines subtracted): (———) fit of a Lorentzian curve with a half width of 0.13 eV, (----) fit of two Lorentzian functions. [After Lauchlan *et al.* (1981).]

typical semiconductors. In fact, the two properties are mutually exclusive. These anomalies have been explained by the soliton mechanism.

The luminescence from *cis*-polyacetylene can be seen from its Raman spectra (Fig. 6.16). In addition to the backbone fundamental Raman-active vibrations at 909, 1250, and 1541 cm^{-1} in *cis*-$(CH)_x$, and 976 and 1470 cm^{-1} in *cis*-$(CD)_x$, there was a long progression of overtones and combinations of the fundamental modes. The progression has the appearance of a single series, but, in fact, the peaks are in most cases superpositions of two or more lines because the frequencies of the fundamental modes are approximately multiples of a common frequency. Superimposed on these lines are two broad luminescence bands at ~ 1.9 and ~ 1.3 eV (~ 1.85 and ~ 1.4 eV) for *cis*-$(CH)_x$ and $(CD)_x$. When the excitation wavelength is changed, the Raman overtone features show dispersion effects. However, the Raman fundamental bands show no appreciable dispersion in profile (see above) and the luminescence bands always appear at the same energy. The main luminescence band is actually quite sharp and lies just below the sharp increase in the absorption coefficient (Fig. 6.11b). Lauchlan *et al.* (1981) had analyzed the main luminescence spectral region by subtracting the multiple-order Raman lines to obtain Fig. 6.17. It was fitted with Lorentzian lines; the major one is centered at 1.9 eV (half width at half maximum is 0.13 eV) and a weak one at 2.0 eV with half width of about

0.04 eV. There remained a component near 2.2 eV that is not Lorentzian. Neither the photoluminescence bandwidth nor its intensity is sensitive to temperature. Above 150 K there were indications of cis–trans isomerization during the scattering measurement. If one assumes the luminescence to be due to the cis structure only and corrects the data for the decrease of cis structure with increasing temperature, then the corrected luminescence linewidth and intensity remain relatively constant to ±20% between 7 and 300 K.

The overtone progressions are reminiscent of multiphonon scattering series observed in inorganic semiconductors. There, scattering of light with incident energy greater than E_G gives rise to multiphonon processes, reflecting the strong coupling of electron–hole intermediate states to longitudinal optical phonons. This coupling is reflected in the excitation spectrum (Fig. 6.18). The excitation threshold is 2.05 eV, whereas the luminescence energy

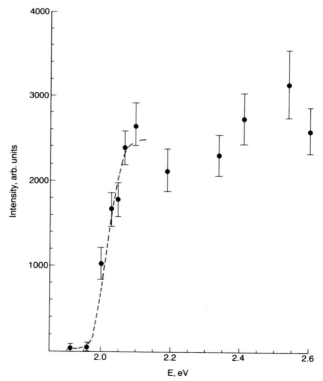

Fig. 6.18   Excitation spectrum for the luminescence from *cis*-polyacetylene. [After Lauchlan *et al.* (1981).]

is 1.9 eV. The excess energy may be interpreted as a Stokes shift, presumably due to lattice distortion around the photoexcited electron–hole pair. Above the threshold, the luminescence intensity is insensitive to the increase in excitation frequency even though the multiple-order Raman lines shift and change in intensity with the laser frequency.

In sharp contrast to *cis*-polyacetylene, Lauchlan *et al.* (1981) found the trans polymer to be virtually devoid of luminescence within instrumental sensitivity. However, Imhoff *et al.* (1982) observed a very weak luminescence at 1.2 eV in *trans*-polyacetylene, which was suggested to arise from perturbed residual cis units or unquenched ^1A_g luminescence from the trans segments. On the other hand, the latter is photoconducting, whereas the cis isomer does not show photoconductivity above the dark current. Photocurrent begins to be detectable when *trans*-polyacetyelene is excited at ~ 0.7 eV, referred to as the knee; it increases steeply above 1 eV (Fig. 6.19). Therefore, photoconductivity occurs at photon energies lower than in the absorption process. This difference is shown more clearly in Fig. 6.20, which is a replot of the photocurrent spectrum in Fig. 6.19. The variation of absorption coefficient with photon energy is also included for comparison. The free-carrier generation efficiency rises exponentially above the threshold and changes to a slow increase above the onset of interband transition.

The principal behavior of the photoconductivity of *trans*-polyacetylene

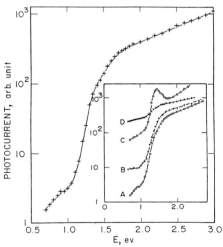

Fig. 6.19   Photocurrent versus photon energy for *trans*-(CH)$_x$ for sample A. The inset shows the photoconductivity responses of inferior samples B, C, and D. [After Etemad *et al.* (1981b,c).]

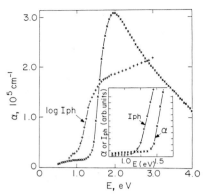

Fig. 6.20 Comparison of photocurrent I_{ph} (+) and absorption coefficient α (x) as a function of photon energy. Inset is a plot of both quantities linearly. [After Etemad *et al.* (1981b,c).]

is reproducible. However, the absolute amount of photoconductivity and photoresponse below 1.0 eV varied from sample to sample (Fig. 6.19, inset). Those specimens that showed significant photoresponse below 1 eV were of low quality, and it may be that impurities are giving dark conductivity. For instance, the worst sample D upon compensation with ammonia decreases the low-energy photoresponse and converts the photocurrent spectrum to one like that of the best sample A. The fact that the light-to-dark current ratio can be varied by doping and/or compensation by more than five orders of magnitude indicated that the photoconductivity is not related to the dark conductivity.

In summary, the two isomeric polyacetylenes display unusual photoexcitation properties different from traditional semiconductors; the latter show both photoconductivity and recombination luminescence upon excitation. *cis*-Polyacetylene emits band-edge luminescence but is without photoconductivity, whereas *trans*-polyacetylene is a photoconductor that does not luminesce. This fundamental difference between polyacetylene and traditional semiconductors was explained by the soliton mechanism (Lauchlan *et al.*, 1981; Etemad *et al.*, 1981b,c).

Let us consider first the case of *trans*-polyacetylene. The absorption of a photon of energy $h\omega \geqslant 2\Delta$, creates an e–h pair within the rigid polyacetylene chain. In time less than 10^{-13} sec the latter evolve into a charged soliton pair (S^+, S^-), i.e.,

$$e + h \rightarrow S^+ + S^-. \tag{6.23}$$

Flood *et al.* (1982) performed EPR measurements during photoexcitation using pulsed radiation of 10^{19} photons/pulse and double modulation to improve sensitivity of detection. The quantum efficiency for photoproduction of spins was found to be less than 2×10^{-7}. The result implies that the photoinduced excited states and the photogenerated carriers are spinless and are to be associated with charged solitons. The two charged solitons are free to move in an applied electric field to give rise to photoconductivity.

That a charged soliton pair is not directly photogenerated can be understood in the following way. The minimum energy for this process (the threshold energy) is only $2(2\Delta/\pi) < 2\Delta$ (Section 10.2). But this transition would require large lattice distortion and thus has low quantum efficiency. This may be regarded as the solid-state version of the Franck–Condon principle. This quantum efficiency increases exponentially as the photon energy approaches 2Δ. For $\hbar\omega \simeq 2\Delta$ the excitation is an interband transition to produce an e–h pair with quantum efficiency of near unity. Thus the photocurrent rises rapidly above 1 eV $\simeq 2(2\Delta/\pi)$ and more slowly above 1.5 eV $\simeq 2\Delta$.

The free charged solitons do not recombine to emit radiation but rather undergo charge neutralization to produce a neutral soliton–antisoliton pair.

$$S^+ + S^- \rightarrow S \cdot + \bar{S} \cdot . \tag{6.24}$$

The neutral solitons may relax to the ground state by multiphonon emission or nonradiative chemical processes. The decay of photocurrent has been measured after laser pulse excitation (Fig. 6.21). The log–log plot is a straight line; $I_{ph}(k) \sim t^{-\beta}$ where $0.55 \leqslant \beta \leqslant 0.65$ for different specimens. This result shows that the carrier recombination time must be greater than 2 msec. The low decay may be attributed to the topological requirement that the photoexcitation in *trans*-polyacetylene must decay in pairs and the finding that the photogenerated solitons are in fact pinned (Section 9.4.2).

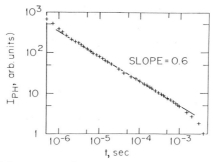

Fig. 6.21 Decay of photocurrent after laser-pulse excitation of *trans*-(CH)$_x$. [After Lauchlan (1981).]

Let us now turn to discuss the photoexcitation behavior of *cis*-polyacety-lene. It differs fundamentally from the trans polymer in that the ground state configurations of the cis polymer are the cis-transoid and trans-cisoid, which are not degenerate in energy but differ by a small amount ΔE_c. Photoexcitation of a charged soliton – antisoliton pair required an energy for the photon of

$$\hbar \omega = 2(2\Delta/\pi) + n \, \Delta E_c, \tag{6.25}$$

where $2\Delta \simeq 2.0$ eV for *cis*-polyacetylene and n is the number of CH units separating the two π-amplitude kinks. The important point is that the greater the separation of the charged solitons, the greater the energy. Conse-quently, as the charged solitons are being formed, they would be confined; they have the characteristics of a bound polaron. Band-edge luminescence takes place either from the confined e–h pair, charged soliton pair, or the polaronlike intermediate. That free solitons or free polarons are not pro-duced by photoexcitation can be concluded from the Stokes shift of about 0.15 eV. A shift of about 0.6 eV would be expected to be associated with the production of free solitons. The observed Stokes shift together with the multiple-order Raman lines indicate strong electron–phonon coupling in *cis*-polyacetylene.

6.4 HEAT CAPACITY

The heat capacities of polyacetylene have been measured by Moses *et al.* (1980) using a computerized adiabatic calorimeter calibrated at the Na-tional Bureau of Standards with potassium chrome alum between 0.3 and 7 K. The results are shown in Fig. 6.22. The specific heats of *cis*-polyacety-lene have a T^3 dependence at higher temperatures similar to other semicrys-talline polymers. The behavior of *trans*-polyacetylene is similar but exhibits smaller slopes, as expected, due to its more stable conformation.

The lattice dynamic contribution to the specific heats of polymers originates from vibrations along the backbone and transverse vibrations. The former corresponds to a quasi-one-dimensional phonon mode with interchain forces playing a role only at very low temperature. The phonon modes of the latter are essentially three-dimensional with the interchain forces playing a central role. Therefore, the lattice specific heat of polyacety-lene may be considered as the sum of two contributions, C_l of longitudinal vibrations and C_t of the transverse vibrations. The transverse vibration has T^3 temperature dependence of Debye theory, i.e., $C_t \propto (T/\theta)^3$, where θ is

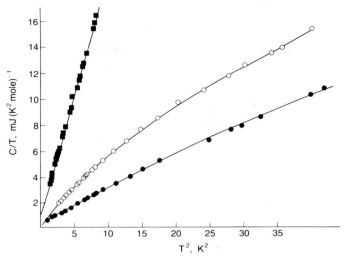

Fig. 6.22 Specific heats of (○) *cis*-polyacetylene; (●) *trans*-polyacetylene; (■) [CH(AsF$_5$)$_{0.12}$]$_x$ plotted as C/T versus T^2. [After Moses *et al.* (1980).]

the Debye temperature. For C_l one assumes that the longitudinal phonon density of state of the coupled chain structure is three-dimensional at low temperatures and one-dimensional at higher temperatures. The total specific heat is $C = C_l + C_t$. The heat capacity data can be fitted with these three parameters.

6.5   THERMAL CONDUCTIVITY

Anomalous thermal conductivity κ behavior has been reported by Guckelsberger *et al.* (1981), as shown in Fig. 6.23 for both undoped and saturation-doped polyacetylene (with iodine). The (log κ)-versus-T plot has a linear T dependence from 2 to 50 K, then the slope increased to a T^3 dependence. From κ, C_v (specific heat per unit volume), and v (average sound velocity), the phonon mean free path X is given by $X = 3\kappa/C_v v$. The variation of X with T (Fig. 6.23b) showed a minimum at 40 K. Under this condition, X is comparable to the phonon wavelength and the phonons are severely damped. The results were interpreted as follows: heat is mainly transported by the fast longitudinal phonons along the polyacetylene chains, and the phonons have frequencies up to 500 cm^{-1}. On the other hand, because of the weak interchain interaction, the frequencies of the transverse

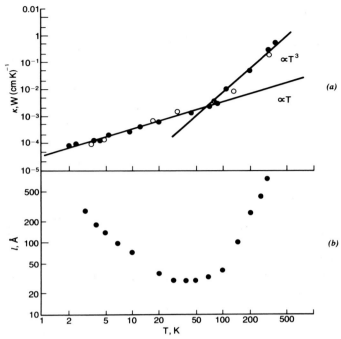

Fig. 6.23 (a) Thermal conductivity and (b) phonon mean free path as a function of temperature for (●) $(CH)_x$, (□) $(CHI_{0.137})_x$. [After Guckelsberger *et al.* (1981).]

vibration of one chain against another is less than 50 cm^{-1}. Although these slow transverse phonons are not effective in heat transport, they can strongly scatter the longitudinal phonons at low temperatures. Above 50 K the mean free paths of the longitudinal phonons increase rapidly with temperature, and scattering by the transverse phonons become insignificant.

Thermal conductivity has also been measured by Newman *et al.* (1981) with the heat flow both parallel and perpendicular to the film. The latter has the very low value of < 3.5 mW (cm K)$^{-1}$. On the other hand, $\kappa_\parallel$ increases with increasing temperature as described above, reaching values of about 50 mW (cm K)$^{-1}$ at room temperature. These results are expected for weak interchain interactions.

Both studies showed very little effect of heavy doping with either iodine or AsF_5. This is understandable because the dopants act to separate the polymer chains, thus separating the lattice and electronic contributions. First, the dopants do not form a continuous lattice. Second, according to the Weidman–Franz law

$$\kappa = LT\sigma, \qquad (6.26)$$

where L is the Lorentz number $\simeq 2.45 \times 10^{-8}$ W Ω K^{-2}. For [CH (AsF$_5$)$_{0.15}$]$_x$ at room temperature $\sigma \simeq 200$ (Ω cm)$^{-1}$ and $\kappa_{\parallel}^e \simeq 1.6$ mW (cm K)$^{-1}$ for the electronic contribution. This is only 4% of the observed thermal conductivity and is experimentally unobservable.

6.6   TENSILE PROPERTIES

Tensile and mechanical properties of polyacetylene has already been mentioned in Section 4.6.1 in connection with partial alignment of poly-acetylene films. The main objective is to increase the conductivity in the draw direction. Shirakawa and Ikeda (1979) showed that extension ratios up to 4.0 can be obtained for freshly prepared *cis*-polyacetylene. Druy *et al.* (1980) found that specimens of *cis*-polyacetylene mounted in an Instron machine did not exhibit an elastic region upon initial loading nor a yield point (the stress above which a permanent strain is produced). Even a very small stress induced a permanent deformation, and the polymer exhibits work hardening.

Prestraining of pristine *cis*-polyacetylene introduced elastic properties. In Fig. 6.24 the polymer is extended to point A. Upon removal of the load the polymer relaxes via path ABC. Stretching returns the sample through D to point A. These loading–relaxation cycles are nearly linear and elastically reversible. However, a stress greater than the point A results in plastic

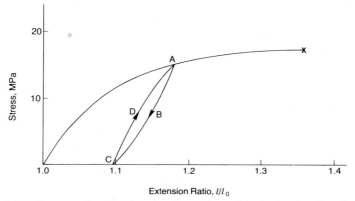

Fig. 6.24   Stress–strain curve for *cis*-polyacetylene. X is the break point. See text for explanation. [After Druy *et al.* (1980).]

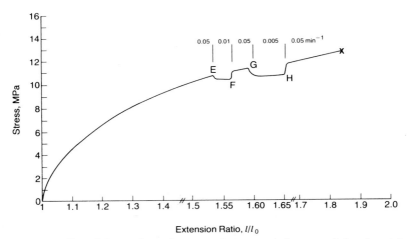

Fig. 6.25   Effect of changes in strain rate on the stress–strain curve of *cis*-polyacetylene. [After Druy *et al.* (1980).]

deformation. So a stress of about 15 MPa (1 MPa $= 1$ kg mm^{-2} g, where $g = 9.8$ m sec^{-2}) and a strain of about 1.17 corresponds to the yield point. The Young's modulus in the elastic region is about 200 MPa (the slope of the line joining points A and C in Fig. 6.24). The yield point depends upon the amount of prior strain, and the shape of the stress–strain curve is not very sensitive to the strain rate. Figure 6.25 shows the effect of suddenly changing the strain rate. At point E, the strain rate was decreased to 0.01 min^{-1}, causing plastic deformation to continue at a lower level of

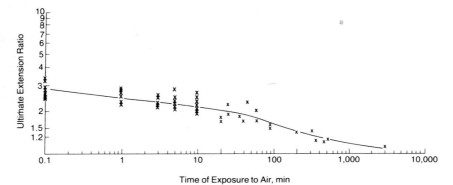

Fig. 6.26   Variation of ultimate extension ratio of *cis*-polyacetylene as a function of time of exposure to air before stretching. [After Druy *et al.* (1980).]

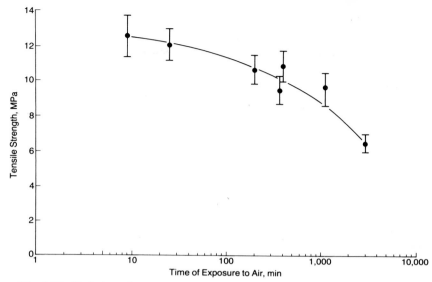

Fig. 6.27   Variation of *cis*-polyacetylene tensile strength with time of exposure to air. [After Druy *et al.* (1980).]

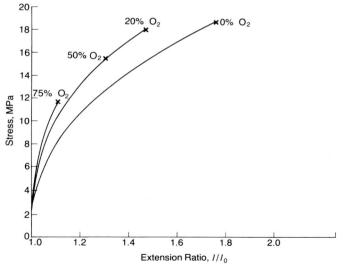

Fig. 6.28   Stress–strain curves and fracture points of *cis*-polyacetylene previously exposed to indicated O_2–Ar mixture for 3 hr. [After Druy *et al.*(1980).]

TABLE 6.9

Dependence of Tensile Properties on Density of Polyacetylene

Initial monomer pressure (atm)	Tensile strength (MPa)	Density (g cm^{-3})
0.4	21	0.4 ± 0.08
0.4	21	0.4 ± 0.08
0.8	33	0.5 ± 0.07
0.8	30	0.5 ± 0.07

stress. At point F the strain rate was restored to the original 0.05 min^{-1} as does the stress level, etc.

Air is deleterious to the extensibility of *cis*-polyacetylene. The ultimate extensibility is less, the longer the specimen is exposed to air, as shown in Fig. 6.26. Exposure to air also tends to reduce the polymer's tensile strength, as shown in Fig. 6.27. The effect of oxygen increases with increase of its partial pressure. When *cis*-polyacetylene films from one single polymerization are exposed to air for 35 min during cutting and sample preparation and then exposed to O_2–Ar atmospheres of different compositions for 3 hr, the ultimate extension ratio decreases with increases in the oxygen partial pressure (Fig. 6.28).

It is to be expected that with the fibril morphology, the tensile properties should increase with density. Druy *et al.* (1980) prepared polyacetylene of different densities by varying the monomer pressure. Higher-density polyacetylene films have more fibrils per unit volume, resulting in higher tensile strength, as given in Table 6.9.

6.7 DIELECTRIC CONSTANTS

Pristine polyacetylene is a poor semiconductor and therefore its dielectric properties are of interest. Devreux *et al.* (1981a) had measured the dielectric constants ϵ of the polymer by the cavity perturbation method at 9.1 GHz. Neglecting the depolarization factor, ϵ is given by

$$\epsilon - 1 = \delta/f, \qquad (6.27)$$

where δ is the relative frequency shift of the cavity resonance and f is a filling factor, which is determined from

$$f = \gamma m/\rho V_c, \qquad (6.28)$$

where $\gamma = 2.1$, determined by the geometry of the cavity, V_c is its volume, m is the sample weight, and the density ρ was taken to be 1.18 g cm^{-3}. For an isotropic sample,

$$\epsilon_{av} = \tfrac{2}{3}\epsilon_\perp + \tfrac{1}{3}\epsilon_\| \qquad (6.29)$$

The samples were made on two roll mills, which reduced a 100–200-μm-thick polyacetylene film to less than 20 μm in thickness. The ϵ_{av} for these partially oriented specimens gave ϵ_{av} of 5.0 and 5.5, respectively, for *cis-* and *trans*-polyacetylene. The anisotropy in the plane of *trans*-polyacetylene was small: $\epsilon_\| = 5.7$ and $\epsilon_\perp = 4.0$, as might be expected.

6.8 PROTON NUCLEAR MAGNETIC RESONANCE

There were several ^{1}H NMR studies on polyacetylene; the results are either in poor agreement or in contradiction. The quantity measured in wide-line NMR is the peak-to-peak linewidth ΔH_{pp}. For precisely Gaussian line shape, the second moment M_2 is related to ΔH_{pp} by

$$M_2 = (\tfrac{1}{2}\Delta H_{pp}). \qquad (6.30)$$

It is related to the chemical structure of the polymer by

$$M_2 = 358 \sum_j (r_{ij})^{-6}, \qquad (6.31)$$

where r_{ij} is the distance between the ith and jth protons, which are assumed to be fixed in the solid. The various results are summarized in Table 6.10.

Mihály *et al.* (1979a) recorded ^{1}H NMR of polyacetylene in a nitrogen atmosphere at room temperature and 11 MHz. The cis polymer has a maximum second moment of 14.6 G^2; it is 8.9 G^2 for the trans polymer. Ikehata *et al.* (1981) found $M_2 = 6.3$ G^2 for undoped *trans*-polyacetylene at

TABLE 6.10

Values of ^{1}H NMR Second Moments of Polyacetylenes (in G^2)

Source temperature	Mihály *et al.* (1979a) Ambient	Ikehata *et al.* (1981) Low	High	Meuer *et al.* (1982) 100 K	300 K
cis-(CH)$_x$	14.6 ± 0.5	—	—	10.3 ± 0.5	9 ± 0.5
trans-(CH)$_x$	8.9	19 ± 2	6 ± 2	8.9 ± 0.5	8.1 ± 0.5

ambient temperature. The values at 300 K were found to be 9 ± 0.5 and 8.1 ± 0.5 G^2 for *cis*-(CH)$_x$ and *trans*-(CH)$_x$, respectively, by Meuer *et al.* (1982).

The single-chain contribution to M_2, calculated by Eq. (6.31), is 9.2, 7.6, and 4.1 G^2, respectively, for the trans-cisoid, cis-transoid, and trans-transoid forms. Therefore, M_2 is much smaller for *trans*-polyacetylene than *cis*-polyacetylene. In addition, there are the interchain contributions that can be calculated from the crystal structure models. Table 6.11 gives the calculated M_2 values for the three models of *trans*-polyacetylene. Examination of Table 6.11 shows that the second-moment results on *trans*-(CH)$_x$ are in best agreement with the *trans*-polyacetylene crystal structure of Shimamura *et al.* (1981). Meuer *et al.* (1982) concluded that these results seem to rule out the structure proposed by Lieser *et al.* (1980b).

With regard to *cis*-polyacetylene, Mihály *et al.* (1979a) calculated the interchain contribution to M_2 to be 4.9 G^2 for a total value of $M_2 = 12.5$ G^2. They assumed the crystal structure of Baughman *et al.* (1978) for the cis-transoid form. The experimental values of Meuer *et al.* (1982) are much smaller and those of Mihály *et al.* (1979a) and larger than the calculated values. It would be desirable to calculate the second moment for the correct crystal structure of Chien *et al.* (1982e).

Mihály *et al.* (1979a) reported that there was a very narrow resonance of less than 0.2 G in width superimposed on the broad Gaussian line. The intensity of this narrow line varies from sample to sample and gradually increases during doping. The integrated intensity of this narrow resonance in doped material represents about 2% of the protons. The origin of this resonance was proposed to be protons on mobile methyl or ethyl groups at chain ends. However, Ikehata *et al.* (1981) did not observe this narrow component in their ^{1}H NMR spectra of doped polyacetylene. Also no sp^3 type of carbon was seen in the ^{13}C NMR spectra.

The second moment of polyacetylene decreases with doping, as shown in Fig. 6.29. There was a large decrease in M_2 for *cis*-polyacetylene with only 0.5% iodine. This was taken to be due to cis–trans isomerization induced by

TABLE 6.11

Calculated and Observed Second Moments of *trans*-Polyacetylene at Room Temperature

Crystal structure	Method	M_2 interchain	M_2 total
Baughman *et al.* (1978)	X ray	3.1	7.2
Lieser *et al.* (1980b)	ED	7.1	11.2
Shimamura *et al.* (1981)	ED	3.5	7.6

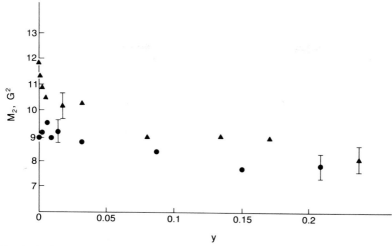

Fig. 6.29   ¹H NMR second moment versus iodine concentration: (▲) *cis*-polyacetylene; (●) *trans*-polyacetylene. [After Mihály *et al.* (1979a).]

doping and to the transformation of 80% of the polymer to the trans structure at the 0.5%-doping level. However, this is contrary to the more recent IR results (Tanaka *et al.,* 1983) showing very little isomerization even at 1.3%-iodine doping. The slow decrease of M_2 with doping of trans and cis polymer, the latter after the initial rapid decline, was interpreted to be the result of lattice expansion when iodine is intercalated. Two processes of intercalation were envisioned: (a) the lattice planes are strongly separated by a small concentration of iodine and the remaining space gradually filled as doping proceeds and (b) the dopant only affects the chain segments immediately surrounding the dopant. The gradual decrease of M_2 with doping seems to support the second possibility. The decrease in M_2 for *trans*-polyacetylene by doping is about 1.5 G², whereas the experimental error bar is approximately 1.1 G². In the other ¹H NMR study, doping with AsF₅ did not decrease the line width of *trans*-polyacetylene.

Ikehata *et al.* (1981) observed an unusually sharp decrease in the linewidth at 155 K; below and above this transitional temperature T_n, the linewidth is essentially constant (Fig. 6.30). This temperature decreases with doping. T_n is 110 K for $[CH(AsF_5)_{0.026}]_x$ and 95 K for $[CH(AsF_5)_{0.053}]_x$. There is no corresponding transition by differential scanning calorimetry. It was suggested that the motional narrowing may be due to a twisting motion of the π kinks. As the kinks diffuse by, a proton is said to undergo a large amplitude motion corresponding to full rotation of the polyacetylene chain. When $\Delta\omega\tau_c < 1$, where $\Delta\omega$ is the static proton linewidth and τ_c is the

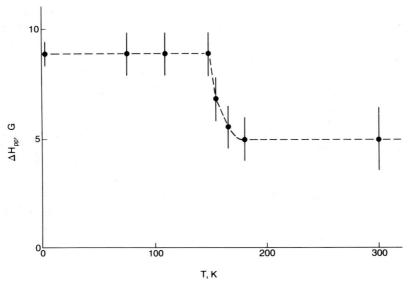

Fig. 6.30   Variaton of ^{1}H NMR linewidth with temperature. [After Ikehata *et al.*(1981).]

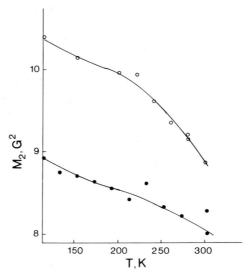

Fig. 6.31   Variation of second moment of ^{1}H NMR for (○) *cis-* and (●) *trans*-polyacetylene with temperature. [After Meuer *et al.*(1982).]

correlation time for a given interchain proton pair, then motional narrowing occurs. An alternative interpretation offered is the loss of long-range order at T_n within the *trans*-polyacetylene chain due to translational fluctuations strong enough to disrupt interchain structural coherence. The differences in the results of the two ^{1}H NMR investigations suggest the necessity of detailing the method of preparation and history of the polyacetylene samples.

A more recent ^{1}H NMR study by Meuer *et al.* (1982) gave results in complete disagreement with those of Ikehata *et al.* (1981). The second moment of *cis*-polyacetylene decreased monotonically from 10.3 ± 0.5 G^2 to 9 ± 0.5 G^2 from 100 to 300 K (Fig. 6.31). In the case of *trans*-polyacetylene, the second moment decreased smoothly from 8.9 ± 0.5 G^2 to 8.1 ± 0.5 G^2 between 100 and 300 K. There was no indication of the sharp drop in the vicinity of 150 K reported by Ikehata *et al.* (1981), and the overall decrease of M_2 was much smaller. This casts doubt upon the line-narrowing transition and the proposed mechanism for it.

6.9   CARBON-13 NUCLEAR MAGNETIC RESONANCE

Natural abundance ^{13}C NMR spectra have been obtained with cross polarization and magic angle spinning by several groups (Clarke *et al.*, 1981; Peo *et al.*, 1981a; Maricq *et al.*, 1978; Gibson *et al.*, 1981; Bernier *et al.*, 1981a). The *trans*-polyacetylene chemical shift is 136 to 139 ppm upfield from TMS; the shift for *cis*-polyacetylene is 126 to 129 ppm. The exact values differ slightly, probably because of referencing. All reports showed a chemical-shift difference of 10 ppm between the backbone ^{13}C NMR of the cis and trans polymers. The chemical shifts are believed to be purely of chemical origin and not from contributions of paramagnetic effects. Only in the earliest report of Maricq *et al.* (1978) was there found a small resonance of $\sim 5\%$ attributable to sp^3 carbon resonance. Inasmuch as later research did not find such resonance, the early measurements might have been made on inferior-quality polymer. The findings of a large fraction of sp^3 carbon resonance in polyacetylenes prepared by catalysts other than the Ti(OBu)$_4$–AlEt$_3$ had been given in Section 2.8.

Chapter 7

Chemical Reactions

Polyacetylene with its conjugated backbone is expected to be highly susceptible to some chemical reactions and may even have unique ones of its own. The insolubility of the polymer makes it difficult to characterize the reaction products and complicates the reaction kinetics. Since poly(methylacetylene) is soluble and is a close chemical analog of polyacetylene, results on the reactions of the former would further the understanding of the chemical properties of the latter. In this chapter most of the reported chemical reactions are discussed.

The thermolysis of polyacetylene has been investigated by pyrolysis gas chromatography–mass spectroscopy (GC–MS). More than two dozen products have been identified, their yields measured quantitatively, and the activation energies for their formation obtained. It is surprising to find large amounts of proton-enriched products in addition to the major product benzene. The results suggest very facile electron–proton exchange processes. Similar studies of poly(methylacetylene) show that electron–methyl exchange reactions are also rapid.

In Section 7.2 the reactions involving hydrogen are described. These include the equilibration of H_2 and D_2 over n-doped polyacetylene, exchange of D_2 with the protons on n-doped polyacetylene, and the hydrogenation of n-doped polyacetylene. The last process converts about half of the polyacetylene to incompletely hydrogenated, partially crosslinked polyethylenelike products. Reactions with bromine and ozone are discussed in Sections 7.3 and 7.4, respectively.

The autoxidations of polyacetylene and poly(methylacetylene) are described in Section 7.5, including the kinetics and activation energies. Oxygen is found to be involved directly in the initiation processes. Poly(methylacetylene) can be very efficiently stabilized by a combination of hydroperoxide decomposers and scavengers for alkyl and peroxy radicals. The mechanisms for autoxidation and stabilization are presented.

The electrochemistry of polyacetylene in aqeous and nonaqeous electrolytes has been investigated by MacDiarmid and co-workers. The results, described in Section 6, show the unique properties of carbanions and carbonium ions of polyacetylene.

Finally, the question of cross-linking is critically addressed in Section 7.7. Present results indicate that chain scission, elimination of aromatic products, and cross-linking may occur by oxidation at elevated temperatures and under pyrolysis conditions. However, there is little or no solid evidence for spontaneous and extensive cross-linking reactions for polyacetylene and poly(methylacetylene) at ambient temperatures.

7.1   THERMOCHEMISTRY

7.1.1   Polyacetylene

7.1.1.1   Pyrolysis – GC – MS

Thermogravimetric analysis of polyacetylene at a helium flow rate of 40 ml min^{-1} and a heating rate of 3 K min^{-1} showed that pyrolysis occurs in two stages (Fig. 7.1). The onset of the first stage occurs at 594 K; the second stage has a threshold temperature of ~ 770 K. These results can be compared with the differential scanning calorimetry thermograms reported by Ito *et al.* (1975), which showed (Fig. 5.3) an exotherm at 598 K and a broad endotherm at about 713 K.

An early brief study (Ito *et al.,* 1975) identified benzene as the major product and small amounts of hydrogen, ethane, ethylene, propylene, and butane as minor products of pyrolysis of polyacetylene at 598 K. The brown high-boiling products were not characterized.

We have (Chien *et al.,* 1982c) performed pyrolysis experiments using a Chemical Data Systems Pyroprobe 100 instrument. A weighed solid polymer sample was placed in a quartz tube (25 $\times$ 2 mm inside diameter), which was then inserted into a heated interface connected to a gas chromatograph. The sample was heated to a final temperature limit preset at a value up to 1273 K. The probe heating rate and the duration of pyrolysis were preset and

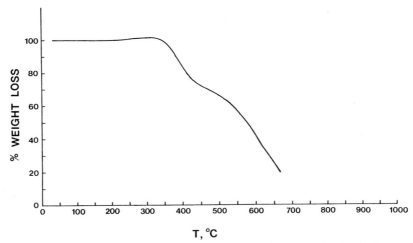

Fig. 7.1   TGA curve of polyacetylene at a heating rate of 3 K min⁻¹ and a He flow rate of 40 ml min⁻¹. Sample size 2.07 mg.

were highly reproducible. Because the pyroprobe was interfaced directly to the chromatograph, trapping of the pyrolyzates and reinjection was not needed.

The gas chromatograph was a Varian Instruments Model 3700 using helium carrier gas at a flow rate of 40 ml min⁻¹. For higher-boiling products a SE-30 silicone gum, fused-silica capillary column (13 m × 0.32 mm inner diameter) was employed, programmed from 313 to 498 K at 5 K min⁻¹, with a helium flow rate of 2 ml min⁻¹ and the inlet splitting ratio was 30:1. Pyrolysis products were detected by flame ionization. About 100 μg of polyacetylene was weighed into the quartz tube of the pyrolyzer and purged with helium for 10 min; the temperature of the pyrolysis interface was kept at ∼ 503 K to avoid the condensation of pyrolyzates. Tenax GC (60/80 mesh) was packed in the glass sleeve inside the injector to keep any very high molecular weight products from the column.

The eluted pyrolysis peaks were identified by a Hewlett–Packard Model 5985A mass spectrometer coupled with a 5840A gas chromatograph by a jet separator for the packed column or by a platinum–iridium transfer line for the capillary column. For this part of the study, only total-ion chromatograms were collected, using the mass spectrometer as the sole detector. This ensured optimal transfer of pyrolyzates to the ion source of the mass spectrometer. The electron-impact ionization source operated at an ionizing voltage of 70 eV, the ion source temperature being 473 K. A Hewlett–Packard Model 2648A data system was used to aid in the interpretation of mass

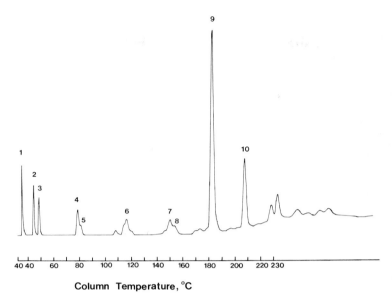

Column Temperature, °C

Fig. 7.2   Low-boiling pyrogram of polyacetylene: 60-μg sample, pyrolysis temperature 923 K in He, 1.8 m × 32 mm inside diameter chromosorb 102 column, carrier gas 40 ml min⁻¹, temperature programmed at 10 K min⁻¹ up to 503 K.

spectra. Figures 7.2 and 7.3 show typical low- and high-boiling pyrograms, respectively. Because the products are all hydrocarbons, they have the same sensitivity to flame ionization detectors. It is possible to obtain quantitative comparisons of the relative yields. These are given in Tables 7.1 and 7.2.

The unexpected aspects of the above results are the large amounts of C_1 and C_2 alkanes, C_2 and C_3 alkenes, and butadiene, which totaled in number of moles 1.34 times that of benzene at 873 K, and the ratio is 2.4 at 1073 K. Acetylene was not detected as a product. In addition, significant quantities of alkyl-substituted aromatic compounds such as toluene and xylenes were produced. These proton-enriched products must gain hydrogen atoms from the proton-depleted products, chief among which is the carbonaceous char, which amounted to about 25%. There is relatively good material balance. For instance, at 1073 K (Tables 7.1 and 7.2) the proton-enriched products gained a total of 9.7 μmole of hydrogen atoms. Whereas the proton-depleted products and char lost a total of 7.8 μmole of hydrogen atoms. Therefore, there is no formation of hydrogen during pyrolysis, based on proton balance, and none was observed. At 873 K the hydrogen-atom gain is about 5.9 μmole by the proton-enriched products, and the loss is about 6.9 μmole for the char and proton-depleted products. Therefore, over this broad temperature range the material balance is within 20%.

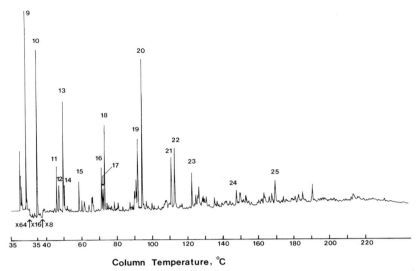

Fig. 7.3   High-boiling pyrogram of polyacetylene: 100-μg sample, pyrolysis temperature 923 K in He, 15 m × 0.32 mm inside diameter SE-30 fused-silica capillary column, carrier gas 2 ml min⁻¹, temperature programmed at 5 K min⁻¹ up to 498 K.

7.1.1.2   Kinetics

The yield of benzene increases very slightly with increase of temperature. Figure 7.4 shows this variation, from which an activation energy of about 3.1 kcal mole⁻¹ was obtained. At 925 K the rate is about 5×10^{-5} mole of benezene per mole of monomer per sec. The yields of other products, relative to benzene, as a function of pyrolysis temperature are given in Fig. 7.5. From these plots the apparent activation energies were obtained and summarized in Table 7.3. The proton-rich products have an activation energy of formation larger than that for benzene; the higher the value, the greater the proton enrichment. Methane has the highest ΔE of ~ 11 kcal mole⁻¹. Most products with molecular weights greater than that of benzene have activation energies comparable to or less than that for benzene. Only *p*-xylene and indene seem to be somewhat out of line. All these activation energies are very small in comparison with those for the pyrolysis of other hydrocarbon polymers such as polypropylene (Kiang *et al.,* 1980) and polyisoprene (Chien and Kiang, 1979a).

The rate of formation of benzene at 573 K was measured indirectly. This temperature was selected because it is the lowest temperature for measurable reaction. Polyacetylene was heated at this temperature in vacuum from

TABLE 7.1

Low-Boiling Pyrolyzates of Polyacetylene

Peak number	Product[a]	Pyrolysis temperature (K)									
		873		923		973		1023		1073	
		Weight[b]	Mole[c]	Weight[b]	Mole[c]	Weight[b]	Mole[c]	Weight[b]	Mole[c]	Weight[b]	Mole[c]
1	CH_4	9.0 ± 0.6	0.44	11.5 ± 0.1	0.56	14.7 ± 0.6	0.72	17.4 ± 1.7	0.85	21.9 ± 0.3	1.07
2	C_2H_4	8.8 ± 0.3	0.25	9.8 ± 0.4	0.27	12.5 ± 0.3	0.35	16.1 ± 2.0	0.45	18.1 ± 0.6	0.50
3	C_2H_6	9.1 ± 0.5	0.24	9.5 ± 0.1	0.25	10.6 ± 0.4	0.28	10.6 ± 1.0	0.28	11.5 ± 0.1	0.30
4	C_3H_6	8.8 ± 0.3	0.16	9.3 ± 0.2	0.17	11.1 ± 0.2	0.27	13.0 ± 1.4	0.24	13.9 ± 0.4	0.26
5	C_3H_8	4.2 ± 0.3	0.074	3.8 ± 0.1	0.067	3.8 ± 0.3	0.067	3.2 ± 0.3	0.057	3.2 ± 0.1	0.057
6	C=C—C=C	12.6 ± 0.3	0.18	12.4 ± 0.6	0.18	14.3 ± 0.3	0.27	16.0 ± 1.5	0.23	15.8 ± 0.4	0.23
7	(cyclopentadiene)	8.0 ± 0.2	0.095	8.0 ± 0.3	0.095	8.7 ± 0.2	0.10	9.2 ± 0.5	0.11	9.5 ± 0.2	0.11
8	C=C—C=C—C	6.3 ± 0.1	0.072	6.2 ± 0.1	0.071	6.5 ± 0.1	0.075	6.4 ± 0.5	0.073	5.6 ± 0.2	0.064
9	(benzene)	100.	1.0	100.	1.0	100.	1.0	100.	1.0	100.	1.0

[a] Average of three pyrolyses.
[b] Relative weight normalized to benzene.
[c] Relative mole normalized to benzene.

TABLE 7.2

Peak number	Product[a]	Pyrolysis temperature (K)			
		723		873	
		Weight[b]	Mole[c]	Weight[b]	Mole[c]
9	(benzene ring)	100.	1.0	100.	1.0
10	(ring)—CH$_3$	21.6 ± 1.3	0.18	20.8 ± 1.8	0.17
11	(ring)—C$_2$H$_5$	2.9 ± 0.1	0.021	2.5 ± 0.1	0.018
12	(structure)	1.4 ± 0.1	0.010	1.4 ± 0.1	0.010
13	(ring)—C=C	6.8 ± 0.2	0.051	6.2 ± 0.3	0.047
14	(structure)	2.0 ± 0.2	0.015	1.6 ± 0.1	0.012
15	(ring)—C=C	1.7 ± 0.1	0.011	1.8 ± 0.1	0.012
16	(ring)—C=C—C	2.8 ± 0.2	0.019	2.5 ± 0.2	0.017
17	(indane)	1.6 ± 0.1	0.011	1.4 ± 0.1	0.009
18	(indene)	4.6 ± 0.2	0.031	4.8 ± 0.2	0.032
19	(indene)—CH$_3$	4.1 ± 0.1	0.025	4.2 ± 0.1	0.025
20	(naphthalene)	8.8 ± 0.2	0.054	8.7 ± 0.5	0.053
21	(methylnaphthalene)	3.8 ± 0.2	0.021	3.4 ± 0.3	0.019
22	(methylnaphthalene)	5.4 ± 0.1	0.030	4.5 ± 0.6	0.025
23	(biphenyl)	2.2 ± 0.1	0.011	2.3 ± 0.2	0.012
24	(fluorene)	—		—	
25	(anthracene)	—		—	

[a] Average of three pyrolyses.
[b] Relative weight normalized to benzene.
[c] Relative mole normalized to benzene.

Pyrolysis temperature (K)							
923		973		1023		1073	
Weight[b]	Mole[c]	Weight[b]	Mole[c]	Weight[b]	Mole[c]	Weight[b]	Mole[c]
100.	1.0	100.	1.0	100.	1.0	100.	1.0
21.9 ± 0.6	0.19	20.5 ± 1.2	0.17	22.8 ± 0.6	0.19	23.1 ± 1.1	0.20
3.1 ± 0.2	0.023	2.6 ± 0.3	0.019	2.9 ± 0.1	0.021	2.8 ± 0.1	0.021
1.8 ± 0.1	0.013	1.8 ± 0.3	0.013	2.1 ± 0.2	0.015	2.1 ± 0.1	0.015
6.4 ± 0.3	0.048	5.8 ± 0.5	0.044	6.3 ± 0.4	0.047	6.1 ± 0.4	0.046
1.8 ± 0.1	0.013	1.6 ± 0.2	0.012	1.8 ± 0.1	0.013	1.7 ± 0.1	0.013
1.8 ± 0.1	0.012	1.6 ± 0.1	0.011	1.7 ± 0.1	0.011	1.5 ± 0.1	0.010
2.6 ± 0.2	0.017	2.0 ± 0.1	0.013	2.3 ± 0.1	0.015	1.7 ± 0.1	0.011
1.4 ± 0.1	0.009	1.1 ± 0.1	0.007	1.2 ± 0.1	0.008	1.1 ± 0.1	0.007
5.1 ± 0.3	0.034	4.7 ± 0.1	0.032	5.6 ± 0.3	0.038	6.2 ± 0.5	0.042
4.2 ± 0.3	0.025	4.1 ± 0.1	0.025	4.2 ± 0.2	0.025	4.0 ± 0.1	0.024
8.6 ± 0.4	0.052	8.0 ± 0.5	0.049	8.8 ± 0.5	0.054	9.0 ± 0.2	0.055
3.2 ± 0.3	0.018	2.9 ± 0.1	0.016	3.4 ± 0.1	0.019	3.3 ± 0.1	0.018
3.9 ± 0.3	0.021	3.4 ± 0.2	0.019	3.6 ± 0.1	0.020	3.5 ± 0.1	0.019
2.2 ± 0.2	0.012	1.9 ± 0.2	0.010	2.2 ± 0.1	0.011	2.3 ± 0.1	0.012
1.0 ± 0.1	0.005	0.9 ± 0.1	0.004	1.0 ± 0.1	0.005	1.1 ± 0.1	0.005
1.4 ± 0.1	0.006	1.3 ± 0.1	0.006	1.5 ± 0.1	0.007	1.6 ± 0.1	0.007

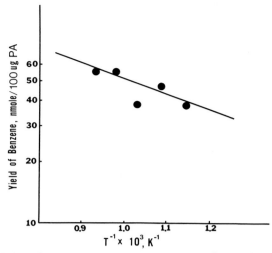

Fig. 7.4 Variation of yield of benzene with temperature of pyrolysis of polyacetylene.

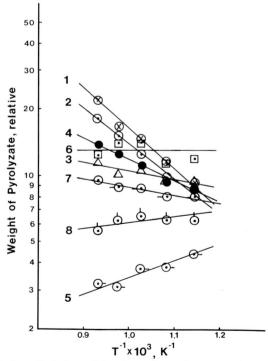

Fig. 7.5 Variation of relative weight of pyrolyzates from polyacetylene as a function of temperature. The numbers correspond to the peak numbers in Tables 7.1 and 7.2.

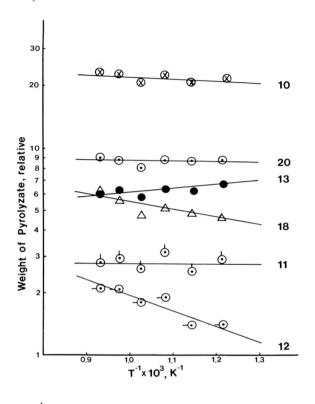

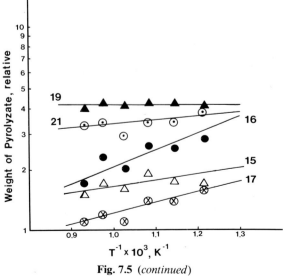

Fig. 7.5 (*continued*)

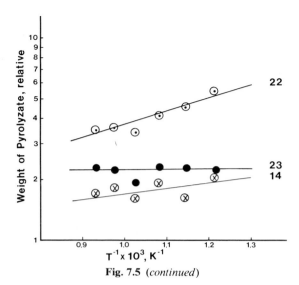

Fig. 7.5 (*continued*)

a few minutes to 2 hr. The samples were subsequently pyrolyzed for 20 sec at 923 K in He, and the products were analyzed by GC–MS. In the case of benzene, the amount liberated at 923 K decreases with increasing time of preheating at 573 K. In other words, a portion of the benzene was produced during the preheating so that lesser amounts can be formed by the subsequent pyrolysis. Figure 7.6 is a first-order plot of these experimental results. The rate constant for the process is 1.8×10^{-4} sec^{-1}. The amount of benzene

TABLE 7.3

Activation Energies for the Formation of Pyrolysis Products from Polyacetylene

Product	ΔE (kcal mole^{-1})	Product	ΔE (kcal mole^{-1})
Methane	11.0	*p*-Xylene	7.2
Ethylene	10.4	*o*-Xylene	1.9
Propylene	7.8	Styrene	2.5
Ethane	5.3	Methylstyrene	1.8
Cyclopentadiene	4.9	α-Methylstyrene	0
1,3-Butadiene	3.1	Indane	0.3
Propane	0.6	Naphthalene	3.1
1,3-Pentadiene	1.9	2-Methylnaphthalene	2.3
Benzene	3.1	1-Methylnaphthalene	0.2
Toluene	3.7	Indene	5.1
Ethylbenzene	3.1	Methylindene	3.1
m-Xylene	3.5	Biphenyl	3.1

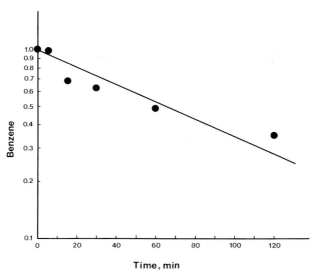

Time, min

Fig. 7.6   First-order rate plot of the formation of benzene at 573 K from polyacetylene.

given in the figure for each time of preheating is the average of six pyrolyses. The standard deviation is about ±5%.

The effects of preheating on the yields of other products upon pyrolysis are summarized in Table 7.4. All the results are the averages of six experiments; those of benzene were plotted in Fig. 7.6. The effect seems to depend

TABLE 7.4

Yield of Pyrolyzate in nmole/100 μg of Polyacetylene Pyrolyzed after Preheating at 572 K

Product	Time of preheating (min)					
	0	5	15	30	60	120
Methane	27.7	26.8	24.0	34.1	41.1	43.6
Ethylene	15.4	14.1	11.1	13.2	14.1	13.8
Ethane	12.2	12.1	11.1	15.2	18.2	19.1
Propylene	9.95	9.22	7.48	9.01	9.79	9.37
Propane	2.86	3.03	2.95	4.10	5.10	5.51
Butadiene	10.6	10.5	8.28	6.64	7.68	7.01
Cyclopentadiene	5.92	5.30	4.07	4.52	4.62	4.20
Cyclopentene	3.29	3.24	2.73	3.20	3.34	3.02
Toluene	16.5	14.3	10.8	11.9	11.2	9.54
Benzene	46.1	45.1	31.4	29.0	22.5	15.5
Char (%)[a]	25.0	25.1	28.0	21.5	26.2	23.3

[a] Weight % of initial sample weight.

on the H/C ratio of the particular product. For the products with low H/C ratios, toluene (1.14) and cyclopentadiene (1.2), preheating tends to lower the subsequent pyrolysis yields, as for benzene. However, whereas the effect on benzene is a threefold decrease in pyrolysis yield after 120 min of preheating, the decrease is only 1.7 times for toluene and 1.4 times for cyclopentadiene. That is, the effect is smaller for large H/C ratio. Second, preheating has negligible effect on products with H/C ratios between 1.4 and 2.0. These include ethylene, propylene, butadiene, and cyclopentene. Third, the highly proton-enriched products have increased pyrolysis yields by preheat treatment. Methane, ethane, and propane all increase in pyrolysis yields following preheating at 592 K. Finally, the absolute yield of char was not affected by preheating.

The main results of heating at 592 K are the degradation of polyacetylene to low H/C products so that their yields are decreased in subsequent pyrolysis. Proton migration occurs during the treatment so that greater amounts of high H/C products are produced by pyrolysis. There is no significant effect on either the products with intermediate H/C ratio or the char.

7.1.1.3   *Effect of Dopant*

The pyrograms for $[CH(AsF_5)_{0.14}]_x$ are shown in Figs. 7.7 and 7.8 and those for $(CHI_{0.06})_x$ are shown in Figs. 7.9 and 7.10, all for pyrolysis at 923 K. Most of the peaks can be identified with those products from undoped

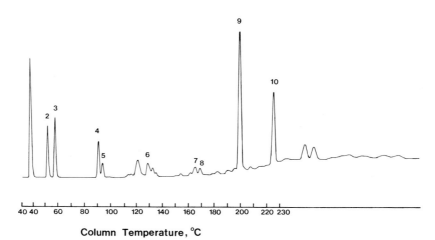

Column Temperature, °C

Fig. 7.7   Low-boiling pyrogram of $[CH(AsF_5)_{0.14}]_x$: 336 μg. Conditions same as in Fig. 7.2.

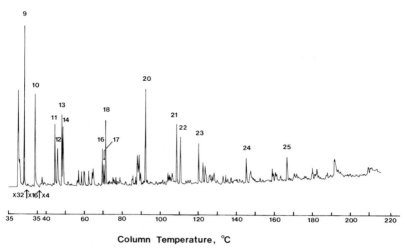

Fig. 7.8 High-boiling pyrogram of $[CH(AsF_5)_{0.14}]_x$: 440 μg. Conditions same as in Fig. 7.3.

polyacetylene. The relative yields in weight are summarized in Tables 7.5 – 7.8 for results obtained at 923, 1023, and 1123 K.

Doping greatly enhances the formation of proton-enriched products as can be seen by the comparison of 923-K pyrolysis yields in Table 7.9. This effect is greater for iodine doping than it is for AsF$_5$-doped polymers. In contrast, products that are multiples of CH units or that are proton depleted are not significantly affected by doping. For instance, compare methyl-

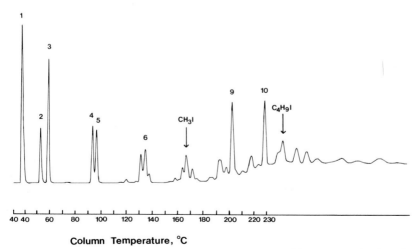

Fig. 7.9 Low-boiling pyrogram of $[CHI_{0.06}]_x$: 223 μg. Conditions same as in Fig. 7.2.

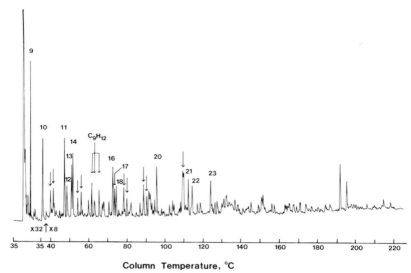

Fig. 7.10   High-boiling pyrogram of $[CHI_{0.06}]_x$: 390 μg. Conditions same as in Fig. 7.3.

styrene with styrene and indane with indene. Furthermore, the yield enhancement is greater for the more saturated hydrocarbons, i.e., methane, ethane, and propane as compared to ethylene, propylene, and xylenes and ethylbenzene over toluene and styrene.

TABLE 7.5

Distribution of Low-Boiling Products of Pyrolysis of AsF_5-Doped Polyacetylene as a Function of Temperature[a]

Peak number	Product	Temperature (K)		
		923	1023	1123
1	Methane	70.7 ± 1.1	127 ± 4.5	149 ± 2.7
2	Ethylene	26.0 ± 0.9	28.2 ± 0.5	37.7 ± 1.7
3	Ethane	27.8 ± 1.2	31.6 ± 2.2	34.3 ± 1.0
4	Propylene	19.5 ± 0.5	19.9 ± 0.7	23.0 ± 1.2
5	Propane	7.2 ± 0.5	7.5 ± 0.5	8.4 ± 0.6
6	1,3-Butadiene	13.7 ± 1.0	13.4 ± 0.4	18.2 ± 0.9
7	Cyclopentadiene	7.7 ± 0.2	7.8 ± 0.3	7.9 ± 0.2
8	1,3-Pentadiene	5.7 ± 0.1	5.1 ± 0.1	3.7 ± 0.2
9	Benzene	100.	100.	100.

[a] $[CH(AsF_5)_{0.14}]_x$ pyrolyzed in He for 20 sec and products separated on a Chromosorb 102 column (6 ft × $\frac{1}{8}$ in.). Products are given in relative weight with benzene taken to be 100. All values are averages of three runs.

TABLE 7.6

Distribution of High-Boiling Products of Pyrolysis of AsF$_5$-Doped Polyacetylene as a Function of Temperature[a]

Peak number	Products	Temperature (K)		
		923	1023	1123
9	Benzene	100.	100.	100.
10	Toluene	32.6 ± 0.6	42.5 ± 1.5	46.4 ± 3.6
11	Ethylbenzene	4.9 ± 0.5	5.4 ± 0.4	5.5 ± 0.7
12	p-Xylene	3.2 ± 0.1	4.4 ± 0.3	4.6 ± 0.7
13	Styrene	6.0 ± 0.3	6.2 ± 0.2	6.1 ± 0.6
14	o-Xylene	4.8 ± 0.2	5.0 ± 0.3	4.6 ± 0.5
16	trans-α-Methylstyrene	3.1 ± 0.1	2.7 ± 0.3	2.2 ± 0.3
17	Indane	1.7 ± 0.1	1.7 ± 0.2	1.2 ± 0.3
18	Indene	5.5 ± 0.3	5.9 ± 0.1	6.2 ± 0.4
20	Naphthalene	7.1 ± 0.3	7.3 ± 0.7	8.4 ± 0.7
21	2-Methylnaphthalene	4.9 ± 0.2	5.1 ± 0.4	4.8 ± 0.6
22	1-Methylnaphthalene	4.1 ± 0.3	3.9 ± 0.2	3.8 ± 0.3
23	Biphenyl	3.1 ± 0.1	2.8 ± 0.1	2.9 ± 0.2
24	Fluorene	1.8 ± 0.1	2.0 ± 0.1	2.1 ± 0.3
25	Anthracene	1.8 ± 0.1	1.9 ± 0.1	2.0 ± 0.2
	Char (% of total sample)	35.5	34.4	35.4

[a] $[CH(AsF_5)_{0.14}]_x$ pyrolyzed in He for 20 sec and products separated on a SE-30 fused-silica column (15 m × 0.324 mm). Products are given in relative weight with benzene taken to be 100. All values are averages of three runs.

TABLE 7.7

Distribution of Low-Boiling Products of Pyrolysis of Iodine-Doped Polyacetylene as a Function of Temperature[a]

Peak number	Product	Temperature (K)		
		923	1023	1123
1	Methane	156.7 ± 10.6	237.8 ± 5.3	217.9 ± 8.2
2	Ethylene	48.4 ± 1.9	69.1 ± 1.4	81.7 ± 6.5
3	Ethane	109.3 ± 5.6	124.4 ± 3.8	102.8 ± 3.8
4	Propylene	56.2 ± 2.0	71.2 ± 2.5	75.0 ± 5.0
5	Propane	53.8 ± 2.3	54.3 ± 2.7	37.3 ± 1.8
6a	C$_4$	34.7 ± 0.9	40.2 ± 1.6	28.4 ± 0.8
6b	C=C—C=C	45.5 ± 1.8	49.7 ± 3.7	34.7 ± 4.2
6c	C$_4$	9.3 ± 0.3	13.5 ± 1.7	10.6 ± 1.5
	CH$_3$I	41.4 ± 2.0	46.8 ± 4.0	32.4 ± 2.3
8	C=C—C=C—C	18.2 ± 0.9	22.7 ± 0.7	14.6 ± 1.1
9	⬡	100	100	100

[a] Same footnote as in Table 7.5, except that the polymer is $(CHI_{0.066})_x$.

TABLE 7.8

Distribution of High-Boiling Products of Pyrolysis of Iodine-Doped Polyacetylene
as a Function of Temperature[a]

Peak number	Products	Temperature (K)		
		923	1023	1123
9	Benzene	100	100	100
10	Toluene	59.5 ± 2.2	74.6 ± 1.8	72.9 ± 0.4
10a		5.2 ± 0.2	5.3 ± 0.6	4.2 ± 0.6
10b		6.1 ± 3.3	6.1 ± 0.6	4.2 ± 0.4
11	Ethylbenzene	15.1 ± 1.0	19.2 ± 0.8	14.3 ± 0.2
12	*p*-Xylene	6.3 ± 0.1	8.6 ± 0.6	7.7 ± 0.5
13	Styrene	8.6 ± 0.1	9.3 ± 0.4	10.4 ± 0.3
14	*o*-Xylene	12.9 ± 0.2	14.5 ± 0.9	12.6 ± 0.9
14a		3.8 ± 0.3	2.9 ± 0.5	2.6 ± 0.1
14b		5.1 ± 0.2	4.3 ± 0.6	3.1 ± 0.2
14c	C_9H_{12}	6.7 ± 0.8	6.5 ± 0.4	4.4 ± 0.3
14d	C_9H_{12}	3.5 ± 0.2	3.8 ± 0.3	2.8 ± 0.1
14e	C_9H_{12}	5.4 ± 0.4	5.8 ± 0.4	4.1 ± 0.2
16	*trans*-α-Methylstyrene	9.4 ± 0.2	7.6 ± 0.9	5.9 ± 0.6
17	Indane	5.9 ± 0.2	5.7 ± 0.3	4.0 ± 0.2
18	Indene	5.6 ± 0.2	7.9 ± 0.5	10.5 ± 0.9
18a		5.6 ± 0.3	4.9 ± 0.6	3.3 ± 0.5
18b		3.8 ± 0.2	3.1 ± 0.2	2.2 ± 0.1
18c		6.7 ± 0.2	5.1 ± 0.7	3.4 ± 0.3
18d		4.1 ± 0.3	3.3 ± 0.4	2.9 ± 0.1
20	Naphthalene	9.2 ± 0.2	9.7 ± 0.6	9.7 ± 0.3
21	2-Methylnaphthalene	6.6 ± 0.1	6.4 ± 0.6	6.2 ± 0.4
22	1-Methylnaphthalene	5.2 ± 0.3	4.8 ± 0.5	4.9 ± 0.3
23	Biphenyl	6.3 ± 0.3	5.2 ± 0.5	4.4 ± 0.2
	Char (% of total sample)	34.6	34.7	—

[a] Same footnote as in Table 7.6, except that the polymer is $(CHI_{0.066})_x$.

Doping also affects the yield of carbonaceous char. The samples listed in Tables 7.6 and 7.8 were first used in TGA measurements up to 573 K. In the case of $(CHI_{0.06})_x$ the iodine content in the polymer was 4.8 wt %. Consequently, the average char yield corresponds to 37% of the polyacetylene. Similarly, the TGA run left $[CH(AsF_5)_{0.14}]_x$ 35 wt % of AsF_5. The average char yield after pyrolysis is 52% of the polymer. The relative yields of char from $(CH)_x:[CH(AsF_5)_{0.14}]_x:(CHI_{0.06})_x = 1:2.1:1.5$. Therefore, the effect of doping on proton-depleted char yield is smaller than it is on proton enrichment.

In the case of AsF_5-doped polyacetylene, the pyrograms (Figs. 7.7 and 7.8) resemble closely those for the undoped polymer (Figs. 7.2 and 7.3). The major difference appears to be the resolution of some of the products into two or more peaks with very similar retention times, which may indicate

TABLE 7.9

Comparison of Pyrolyzate Yields of Undoped and Doped Polyacetylenes

Products	$(CH)_x:[CH(AsF_5)_{0.14}]_x$ $:(CHI_{0.066})_x$[a]	Products	$(CH)_x:[CH(AsF_5)_{0.14}]_x$ $:(CHI_{0.066})_x$
Methane	1:7.3:13.6	Styrene	1:1:1.5
Ethylene	1:1.7:4.3	o-Xylene	1:2.8:8.1
Ethane	1:3.0:11.7	Methylstyrene	1:1.6:3.8
Propylene	1:1.5:5.5	Indane	1:1.4:4.8
Propane	1:2.3:17	Indene	1:1.1:1.4
Butadiene	1:0.8:6.4	Naphthalene	1:0.9:1.1
Pentadiene	1:0.8:1.5	2-Methylnaphthalene	1:1.2:1.9
Toluene	1:1.9:3.3	1-Methylnaphthalene	1:1.4:1.3
Ethylbenzene	1:1.9:6.6	Biphenyl	1:1.3:2.4
p-Xylene	1:2.1:4.1		

[a] Ratio of yields of undoped and doped polyacetylenes pyrolyzed at 1023 K under identical conditions.

isomers. If this is the case, then AsF_5 may be catalyzing rearrangement reactions. There were no new products that may be attributed to fluorinated compounds. On the other hand, new products were definitely produced in the pyrolysis of $(CHI_{0.06})_x$. In Fig 7.9, the peaks due to CH_3I and C_4H_4I are indicated. In Fig. 7.10 there are many new peaks that were not present in the pyrograms of $(CH)_x$ and $[CH(AsF_5)]_x$. These are shown by arrows. The structures of these products have not been determined because they are not formed in significant quantities.

In conclusion, the major effect of doping is to facilitate further the electron–proton exchange process discussed in Section 7.1.1.4.

7.1.1.4 Mechanism — Electron – Proton Exchange

A thermolysis mechanism for polyacetylene must explain some very unusual characteristics of the processes. The activation energies for the formation of products are very low (Table 7.3). In comparison, the activation energies for the thermolysis of polypropylene is $51-56$ kcal mole^{-1} (Kiang *et al.,* 1980; Chien and Kiang, 1978) and it is 41 kcal mole^{-1} for polyisoprene (Chien and Kiang, 1979a). Whereas monomer is usually one of the main products in the pyrolysis of polymers, acetylene was not detected here. Hydrogen was also apparently not produced. Instead, large quantities of proton-enriched products were obtained upon pyrolysis of polyacetylene. They cannot be the result of hydrogen abstraction by methylene, or other intermediates because the yields of methane are the same at 1023 K from pyrolysis of polyacetylene and poly(methylacetylene) (Section 7.1.2.1). We propose a mechanism involving a series of facile proton–electron exchange migrations (Chien *et al.,* 1982c), accompanied by the shifting of alternate short- and long-backbone carbon bonds.

We envision the following thermolysis mechanism for polyacetylene. The initiation process is the scission of a backbone bond, with or without formation of a solition.

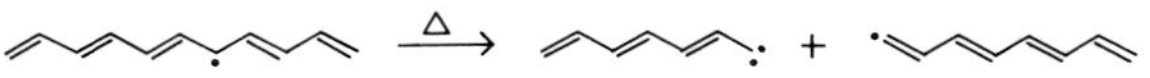

(7.1)

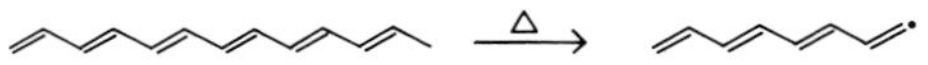

(7.2)

Intramolecular electron migration and ring closure leads immediately to benzene, which is the most abundant product:

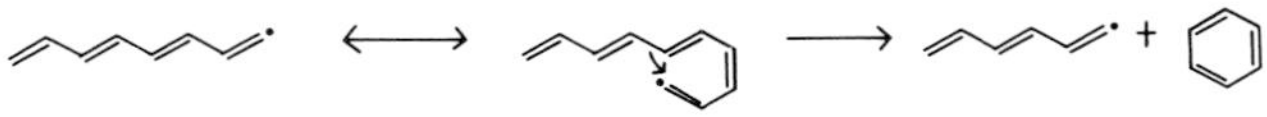

(7.3)

To account for the production of methane and ethane, we have already discounted the hydrogen abstraction by CH, CH_2, or CH_3 fragments. It is also not viable to invoke their formation from the ethyl end groups, which are present in amounts of the order of 0.1% of the polymer, which has a molecular weight of about 11,000. Instead, we take cognizance of the fact that the solitons in polyacetylene are extremely diffusive. At elevated temperatures one may conceive similarly a high mobility for protons. We propose that as π bonds are broken and the π electrons move to another part of the chain to reform π bonds, protons migrate in the opposite direction for chemical balance. The two may be thought of as occurring in a countercurrent concerted manner and may even be quantum-mechanical tunneling in nature. This can be depicted, albeit clumsily, by a series of classical chemical events. Consider the radical produced in the chain scission. The terminus gains a proton from somewhere down the polymer chain in exchange for its π electron as follows:

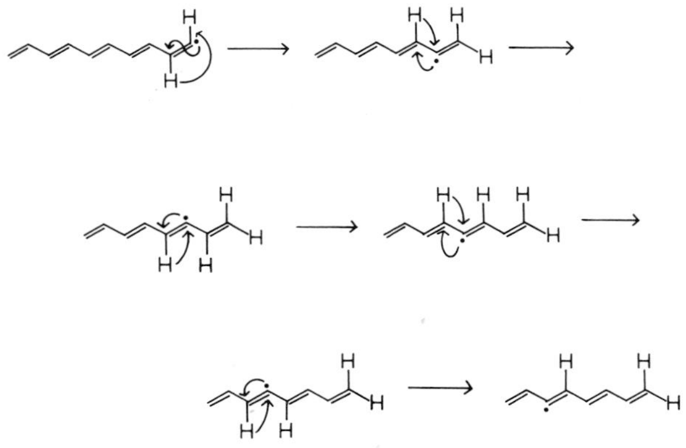

(7.4)

Therefore, we can write an abbreviated representation for the above process as

$$\text{(7.5)}$$

To produce methane, the events are

$$\text{(7.6)}$$

and it is

$$\text{(7.7)}$$

for the formation of ethylene. Ethane is obtained in a similar manner

$$\text{(7.8)}$$

To depict the formation of toluene, we arrange the chain end to resemble the product before proton–electron exchange for illustrative purposes:

$$\text{(7.9)}$$

Finally, for naphthalene, which was produced in low yield, we write

Here the proton migration is distal, but the electron migrates locally.

7.1.2   Poly(methylacetylene)

7.1.2.1   *Pyrolysis – GC – MS*

Thermogravimetric analysis was performed as described above for poly-acetylene. Thermal decomposition of poly(methylacetylene) commences at about 423 K, reaching completion at about 703 K and leaves 7% of the polymer as carbonaceous char (Fig. 7.11). Therefore, poly(methylacetylene) is thermally more unstable than polyacetylene (compare Figs. 7.11 and 7.1). At least two factors probably contribute toward the greater thermal stability of polyacetylene. Polyacetylene is a more conjugated molecule than the methyl derivative; the former has identical backbone carbon atoms but they are different in poly(methylacetylene) by virtue of the substitutent. The nonbonded repulsion between the pendant methyl groups may cause devia-tion of the backbone from planarity. Finally, the cooperative π interaction between polyacetylene molecules is strong, as evidenced by its insolubility and infusibility. This interaction is small in poly(methylacetylene), and the polymer is readily dissolved in common solvents.

The pyrolysis – GC – MS conditions used above for polyacetylene when applied to poly(methylacetylene) yielded nonreproducible pyrograms. Ap-parently, the polymer is unstable at the heated interface temperature of 503 K. This was remedied by constructing an aluminum block cooled by running water, which lowered the temperature of the exterior of the inter-face as well as the sample prior to pyrolysis. In this experiment, the pyro-probe containing poly(methylacetylene) was inserted into the outermost region of the pyrolysis interface and purged with helium for 10 min. The

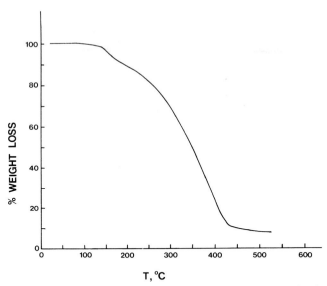

Fig. 7.11   TGA curve of poly(methylacetylene). Conditions same as in Fig. 7.1.

pyroprobe was then inserted into the heated interface with 3-sec, delay, which allows for the heating of the stainless-steel support for the platinum coil in the pyroprobe.

Figures 7.12 and 7.13 show typical low-boiling and high-boiling pyrograms of poly(methylacetylene). The numbers adjacent to each peak refer to the peak numbers in Tables 7.10 and 7.11. Table 7.10 contains the relative abundance of the low-boiling compounds and Table 7.11 the values for the high-boiling products. All the values were normalized with respect to mesitylene.

There are several very interesting points about the pyrolysis data. Mesitylene is the major product, as expected. Significant amounts of alkanes and alkenes are formed as well; their total yields are 0.23 mole at 723 K as compared to one mole of mesitylene, which increased to 1.4 times greater than mesitylene at 923 K. The high molecular weight pyrolysates are 0.7 and 1.1 times that of mesitylene at 723 and 923 K, respectively. Second, very little benzene is formed from poly(methylacetylene), i.e., about 1% of the mesitylene yield. The yields of alkanes and alkenes from polyacetylene and poly(methylacetylene) (given in parentheses) are compared at the same pyrolysis temperature of 923 K normalized to benzene (mesitylene). They are methane 0.72 (0.8), ethylene 0.35 (0.14), ethane 0.28 (0.13), propylene 0.21 (0.17), propane 0.067 (0.056), butadiene 0.21 (0.087), and pentadiene 0.10 (0.064). In other words, the relative yields of these products are within a

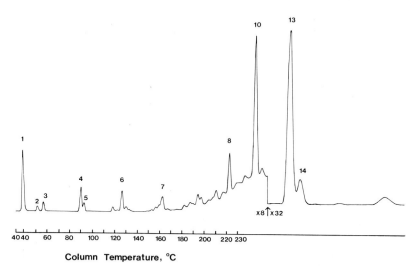

Fig. 7.12   Low-boiling pyrogram of poly(methylacetylene): 200-μg sample, pyrolysis temperature 773 K in He, 1.8 m $\times$ 32 mm inside diameter chromosorb 102 column, carrier gas 40 ml min^{-1}, temperature programmed at 10 K min^{-1} to 503 K.

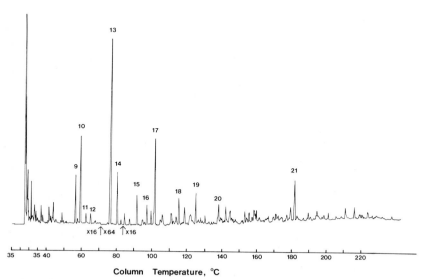

Fig. 7.13   High-boiling pyrogram of poly(methylacetylene): 200-μg sample, pyrolysis temperature 773 K in He, 15 $\times$ 0.32 mm inside diameter SE-30 fused-silica capillary column, carrier gas 2 ml min^{-1}, temperature programmed at 5 K min^{-1} to 498 K.

TABLE 7.10

Low-Boiling Pyrolyzates of Poly(methylacetylene)

Peak number	Product[a]	Pyrolysis temperature (K)									
		773		823		873		923		1023	
		Weight[b]	Mole[c]	Weight[b]	Mole[c]	Weight[b]	Mole[c]	Weight[b]	Mole[c]	Weight[b]	Mole[c]
1	CH_4	2.1 ± 0.1	0.16	3.4 ± 0.1	0.25	5.0 ± 0.2	0.36	7.2 ± 0.1	0.54	10.7 ± 0.3	0.80
2	C_2H_4	0.12 ± 0.01	0.005	0.25 ± 0.02	0.011	0.54 ± 0.01	0.023	1.33 ± 0.03	0.057	3.2 ± 0.1	0.14
3	C_2H_6	0.24 ± 0.01	0.010	0.47 ± 0.02	0.019	0.93 ± 0.04	0.037	1.8 ± 0.1	0.070	3.2 ± 0.1	0.13
4	C_3H_6	0.71 ± 0.01	0.020	1.23 ± 0.04	0.035	2.1 ± 0.1	0.060	3.49 ± 0.04	0.10	5.8 ± 0.2	0.17
5	C_3H_8	0.19 ± 0.01	0.005	0.42 ± 0.02	0.011	0.9 ± 0.1	0.023	1.4 ± 0.1	0.038	2.0 ± 0.1	0.056
6	$C{=}C$	0.9 ± 0.1	0.018	1.2 ± 0.1	0.026	1.85 ± 0.05	0.040	2.6 ± 0.1	0.046	3.63 ± 0.04	0.064
7	$C{=}C{-}C{=}C{-}C$	0.78 ± 0.02	0.014	1.6 ± 0.1	0.027	1.9 ± 0.1	0.033	2.6 ± 0.1	0.046	3.63 ± 0.04	0.064
8	(dimethylbenzene structure)	4.1 ± 0.4	0.054	3.2 ± 0.2	0.042	3.1 ± 0.3	0.041	3.9 ± 0.2	0.051	6.2 ± 0.1	0.082
13	(trimethylbenzene structure)	100.	1.0	100.	1.0	100.	1.0	100.	1.0	100.	1.0

[a] Average of three pyrolyses.
[b] Weight normalized to mesitylene.
[c] Mole normalized to mesitylene.

TABLE 7.11

High-Boiling Pyrolyzates of Poly(methylacetylene)

Peak number	Product[a]	Pyrolysis temperature (K)							
		723		773		873		1023	
		Weight[b]	Mole[c]	Weight	Mole	Weight	Mole	Weight	Mole
9	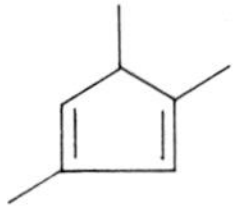	7.5 ± 0.2	0.083	8.4 ± 0.5	0.093	5.9 ± 0.2	0.066	4.2 ± 0.5	0.047
10		9.3 ± 0.1	0.105	12.2 ± 0.7	0.14	16.6 ± 0.4	0.19	27.5 ± 3.7	0.31
11	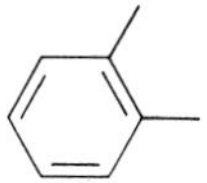	1.2 ± 0.1	0.014	1.7 ± 0.1	0.019	3.2 ± 0.1	0.036	6.4 ± 0.9	0.072
12	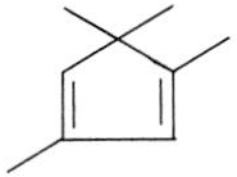	1.1 ± 0.1	0.011	1.9 ± 0.1	0.019	2.7 ± 0.1	0.027	1.8 ± 0.3	0.018
13	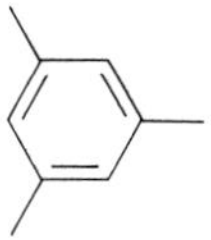	100	1.0	100	1.0	100	1.0	100	1.0
14	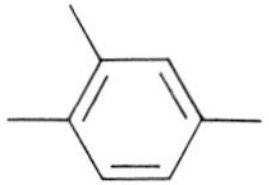	24.8 ± 0.3	0.25	25.7 ± 1.3	0.26	23.0 ± 0.5	0.23	27.7 ± 2.0	0.28
15	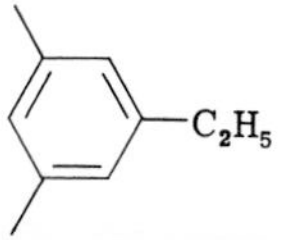	2.8 ± 0.1	0.025	3.2 ± 0.2	0.029	4.5 ± 0.2	0.04	9.0 ± 2.5	0.081

274

16	1.2 ± 0.1	0.011	1.8 ± 0.2	0.016	2.7 ± 0.1	0.025	5.7 ± 0.8	0.052
17	12.7 ± 0.6	0.11	12.8 ± 0.3	0.12	9.6 ± 0.3	0.086	11.1 ± 1.4	0.099
18	3.2 ± 0.1	0.024	4.1 ± 0.2	0.031	3.7 ± 0.3	0.028	3.2 ± 0.6	0.024
19	3.7 ± 0.1	0.024	4.4 ± 0.2	0.033	3.3 ± 0.1	0.025	3.4 ± 0.7	0.026
20	1.4 ± 0.1	0.011	2.2 ± 0.2	0.017	3.7 ± 0.1	0.028	4.4 ± 0.6	0.033
21	5.8 ± 0.1	0.031	6.4 ± 0.2	0.034	4.5 ± 0.4	0.024	4.8 ± 0.8	0.026

[a] Average of six pyrolyses.
[b] Wt normalized to mesitylene.
[c] Mole normalized by mesitylene.

factor of 2 for the two polymers. Finally, there are a number of aromatic pyrolyzates that do not have methyl groups on alternating carbon atoms and others have methyl groups on adjacent carbon atoms. These results suggest strongly that electron–proton exchange is facile and electron–methyl exchange also occurs with relative ease for poly(methylacetylene).

7.1.2.2 Kinetics

The temperature dependence of the formation of mesitylene is given in Fig. 7.14. At 925 K the rate for this process is 8×10^{-5} mole per monomer per second. The corresponding rate for the formation of benzene from polyacetylene at this temperature is 5×10^{-5}. However, the activation energy for the former process is only 0.8 kcal mole^{-1} as compared to 3.5 kcal mole^{-1} for the latter. Therefore, the production of mesitylene from poly(methylacetylene) is not slowed appreciably by decreases in temperature.

The temperature dependence of pyrolyzate yields is shown by Arrhenius plots in Fig. 7.15. There are two types of behavior. The first is a linear Arrhenius plot, and some of the products belong to this class. The activation energies for the formation of these products are given in Table 7.12. It is interesting to compare these values with those for polyacetylene pyrolysis

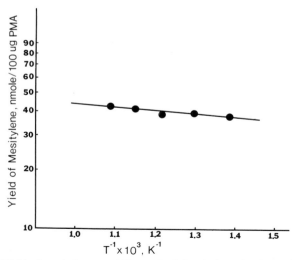

Fig. 7.14 Yield of mesitylene per 100 μg of poly(methylacetylene) pyrolyzed at various temperatures.

(Table 7.3). For the two polymers the value of ΔE for methane formation are nearly identical. On the other hand, the activation energies of formation for the other low-boiling products are anywhere from 5 to 10 kcal mole^{-1} higher from poly(methlacetylene) than those for polyacetylene. The results indicate that methyl-group migration requires considerable activation energy even though proton migration occurs with comparable ease in the two polymers.

The second type of behavior gives nonlinear Arrhenius plots. Two products, notably toluene and 1,2,4-trimethylbenzene, have concave plots. There are others: 1,3,5-trimethylcyclopentadiene, 1,3,5,5-tetramethylcyclopentadiene, and 1,3,7-trimethylindene, all of which have convex plots. It may be speculated that increase of temperature at first favors the formation of the latter group but at high temperature they may decompose to give products of the former group.

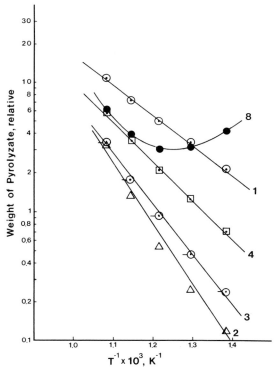

Fig. 7.15   Variation of relative weight of pyrolyzates from poly(methylacetylene) as a function of temperature. The numbers correspond to the peak numbers in Tables 7.11 and 7.12.

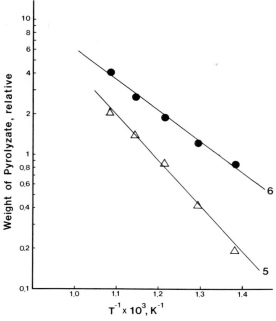

Fig. 7.15 (*continued*)

7.1.2.3 Mechanism

The thermolysis mechanism for poly(methylacetylene) probably involves facile migrations of electrons, protons, and methyl groups. The initiation process is the random chain scission:

$$\text{(7.11)}$$

(I)   (II)

The nature of the products suggests that both electron–proton and electron–methyl exchange can occur by means of either 1,2 or 1,3 shifts. These are illustrated for 1,2 exchanges

$$\text{I (II)} \longrightarrow \qquad\qquad \left(\qquad\qquad \right) \qquad\qquad \text{(7.12)}$$

and for the 1,3 exchanges

$$\text{I (II)} \longrightarrow \qquad\qquad \left(\qquad\qquad \right) \qquad\qquad \text{(7.13)}$$

TABLE 7.12

Activation Energies for the Formation of Pyrolysis Products from Poly(methylacetylene)

Product	ΔE (kcal mole^{-1})	Product	ΔE (kcal mole^{-1})
CH_4	11.6	(1,4-disubstituted benzene)	5.2
C_2H_4	22.3	(1,2-dimethylbenzene)	9.0
C_2H_6	18.9	(1,3-dimethyl disubstituted benzene)	0.8
C_3H_6	14.7	(1,2,4-trisubstituted benzene)	1.4
C_3H_8	16.4	(dimethylphenyl)$-C_2H_5$	6.8
$C{=}C$	11.3	(dimethylphenyl)$-C{=}C$	8.2
$C{=}C{-}C{=}C{-}C$	10.9	(trimethyl disubstituted benzene)	0

The proton or methyl group can originate from more distant carbon atoms and can be written

$$\qquad\qquad (7.14)$$

$$\qquad\qquad (7.15)$$

Based on these migrations, one can account for all the observed pyrolysis products of poly(methylacetylene). In the case of mesitylene, we write

$$\text{(7.16)}$$

$$\text{(7.17)}$$

The aliphatic hydrocarbons and olefins can be produced by

$$\text{(7.18)}$$

$$\text{(7.19)}$$

$$\text{(7.20)}$$

The following pathway to the formation of toluene can be extended to account for all the other aromatic products:

$$\text{(7.21)}$$

7.2   REACTIONS WITH HYDROGEN

Undoped or *p*-doped polyacetylene apparently does not react with hydrogen at elevated temperature and pressure even in the presence of various hydrogenation catalysts. Shirakawa *et al.* (1980a,b) found, however, that *n*-doped polyacetylene will react with hydrogen. Both hydrogen–deuterium exchange and hydrogenation have been observed.

7.2.1   Hydrogen – Deuterium Equilibration

There are two types of H – D exchange reactions. One is the equilibration of a mixture of isotopic molecules H_2 and D_2 in the presence of n-doped polyacetylene. The other is the exchange reaction between D_2 and the hydrogen atoms in the polymer. When a gaseous mixture of H_2 and D_2 is exposed to $(CHNa_y)_x$ or $(CHK_y)_x$, equilibration of the isotopes occurs at room temperature (Shirakawa *et al.*, 1980a).

It is probable that the alkali metal-doped polyacetylene acts as a low work-function metal. H_2 – D_2 equilibration may proceed as follows:

$$(CHNa_y)_x + H_2 \rightleftharpoons (CHNa_y)_x^+ + H_2^-$$
$$H_2^- + D_2 \rightleftharpoons HD^- + HD \qquad (7.22)$$
$$(CHNa_y)_x^+ + HD^- \rightleftharpoons (CHNa_y)_x + HD$$

The rate constant for equilibration k_{eq} is given by

$$k_{eq} = \frac{1}{t} \ln\left(\frac{[P_{HD}]_0 - [P_{HD}]_{eq}}{[P_{HD}]_t - [P_{HD}]_{eq}}\right), \qquad (7.23)$$

where $[P_{HD}]_0$, $[P_{HD}]_t$, and $[P_{HD}]_{eq}$ denote the pressure of HD at $t = 0, t$, and ∞, respectively. From the values of k_{eq} obtained between 333 and 403 K an activation energy of about 13 kcal mole^{-1} was obtained for H_2 – D_2 equilibration promoted by metallic $(CHNa_y)_x$.

7.2.2   H – D Exchange with Polyacetylene

The exchange reaction with the hydrogen atoms in the n-doped polyacetylene requires temperatures higher than 423 K. For instance, when 85 torr of D_2 gas was introduced over $[(CH)Na_{0.094}]_x$ and heated to 473 K, HD and H_2 appeared in the gas phase after an induction period of a few hours. A constant ratio of $H_2 : D_2 : HD$ was obtained. Figure 7.16 shows the kinetics of the reaction. The rate constant for exchange k_{ex} is obtained from

$$k_{ex} = (1/t) \ln([P_{D_2}]_0 / [P_{D_2}]_t), \qquad (7.24)$$

and the equilibrium constant K is

$$K = [P_{HD}]^2 / [P_{H_2}][P_{D_2}]. \qquad (7.25)$$

The composition of HD, D_2, and H_2 reached equilibrium after 250 hr, which corresponds closely to all the H and D atoms in the system. This suggests that all the H atoms of the doped polyacetylene can exchange with D_2 in the gas phase. From the results of k_{ex} determined from 433 to 503 K, a value of 24 kcal mole^{-1} was obtained for the activation energy of this

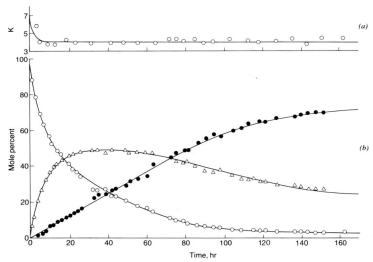

Fig. 7.16 Exchange between D_2 and H in $(CHNa_{0.064})_x$ at 473 K and an initial $P_{D_2} = 85$ torr. (a) Equilibrium constant K $(=[HD]^2/[H_2][D_2])$ versus time; (b) variation of isotopic composition of the gas with time: (●) H_2; (○) D_2; (△) HD. [After Shirakawa *et al.* (1980a).]

exchange process. A possible mechanism is the following:

$$(7.26)$$

The fact that all the hydrogen atoms on the polymer can be exchanged is probably attributable to the nonlocalized nature of the negative carrier, i.e.,

$$(7.27)$$

7.2.3   Hydrogenation of Polyacetylene

The observation of exchange reactions led Shirakawa *et al.* (1980b) to attempt hydrogenation of polyacetylene. Sodium-doped polyacetylene film was allowed to react with H_2 at 550 torr and 473 K for 7 hr. The original silvery film turned to a transparent pale brown, which upon cooling to room temperature became opaque. After the Na^+ ion in the polymer was removed by aqueous methanol, the product was flexible and had thermoplastic properties. It exhibited the typical IR spectrum for polyethylene with the exception of a band at 965 cm^{-1} that may be attributed to C—H out-of-plane vibration of a residual isolated double bond or a short sequence of conjugated double bonds.

Even though the hydrogenation was carried out at 473 K, which is some 70 K above the melting point of polyethylene, the polymer film retained its initial shape, suggesting that the product is crosslinked. Extraction of the material with boiling tetralin for 5 hr gave soluble polyethylene in approximately 60% yield. This polyethylene has IR absorption at 720 and 730 cm^{-1}, indicating a crystallinity of $\sim 50\%$. The melting transition of the polymer occurred between 385 and 394 K. GPC analysis of this polyethylene gave a $\overline{M}_n$ of 6200 and a polydispersity of 3.44. The GPC curve (Fig. 7.17) is bimodal with a considerable amount of higher molecular weight products. Because the initial $\overline{M}_n$ of polyacetylene is $\sim 11,000$, the hydrogenation

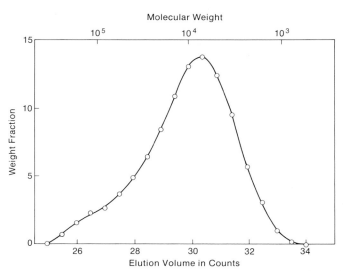

Fig. 7.17   GPC curve of the soluble fraction of hydrogenated (CH)$_x$. [After Shirakawa *et al.* (1980b).]

results may be interpreted in two ways. One possible explanation is that the low molecular weight fraction of polyacetylene is more readily hydrogenated and solubilized than the higher molecular weight chains. An alternative interpretation is that chain scission can occur during the severe hydrogenation conditions such as

$$(7.28)$$

About 40% of the polyacetylene remained insoluble after hydrogenation. These products may represent either incompletely hydrogenated material or cross-linked polymers. Cross-links can be formed by Diels–Alder reactions. Another possibility is that during the course of hydrogenation unpaired spins become localized by aliphatic CH_2 segments that would react like ordinary radicals. This is shown schematically by

$$(7.29)$$

The mechanism of hydrogenation is likely to be the same as that proposed for H–D exchange between D_2 and polyacetylene [Eq. (7.26)]. This is strongly supported by the observation that Na-doped polyacetylene after hydrogenation, upon immersion in an evacuated methanol–water mixture, liberates H_2, as expected, if sodium hydride is present.

7.3 REACTION WITH BROMINE

Doping of polyacetylene with halogen was discussed in Section 8.2.1. In the case of iodine, iodination of the C=C bonds does not seem to occur at ambient temperature. Alkyl iodides were, however, detected in the pyrolysis of $(CHI_{0.06})_x$. Doping with iodine increases conductivity monotonically until saturation, and the excess iodine can be removed by evacuation. On the other hand, doping with bromine and chlorine increases conductivity at first, followed by decline, suggesting addition of halogens to the C=C bonds. The reaction of polyacetylene with bromine was studied in detail by Kletter *et al.* (1980).

When *cis*-polyacetylene film is exposed to bromine vapor at $\leqslant 10$ torr, using a 250-K CCl_4 slush bath, film undergoes a series of changes in color and resistivity corresponding to changes in polymer composition, as shown schematically in Fig. 7.18. The lightly doped polymer appears golden; resistivity decreases with increasing doping. The material goes through red and violet till minimum resistivity is attained when the polymer is blue. Additional reaction causes an increase in resistivity. The green film is still shiny and flexible. Excessively brominated polyacetylene is yellow to white and is dull and brittle. A sample of composition $(CHBr_{0.16})_x$ has a metallic conductivity of 20.8 $(\Omega\ cm)^{-1}$ and a thermoelectric power of 26.9 $\mu V\ K^{-1}$. This material is therefore a *p*-type metal.

An ingenious synthesis of a semiconducting derivative of polyacetylene makes use of the bromination and dehydrobromination reactions. When the matallic $(CHBr_y)_x$ film is heated in a sealed container in vacuo between 393 and 423 K, HBr and Br_2 are first evolved. The reactor was returned to room temperature, allowing the bromine to react with the polymer. Repetition of this procedure resulted in complete reaction of Br_2, leaving only HBr in the gas phase. The following reaction sequence may be proposed. The polymer was first brominated by the dopant:

$$(7.30)$$

Heating leads to dehydrobromination:

$$(7.31)$$

Subsequent bromination can occur at both double and triple bonds,

$$(7.32)$$

The above dehydrobomination, although shown as an intramolecular process in Eq. (7.31), can also occur by intermolecular pathways. The reactions almost certainly include cis–trans isomerization. Based on the amount of HBr evolved and elemental analysis for C, H, and Br, the material has a composition of $(CH_{0.82}Br_{0.14})_x$. Its transport properties are $\sigma = 2.1 \times 10^{-6}$

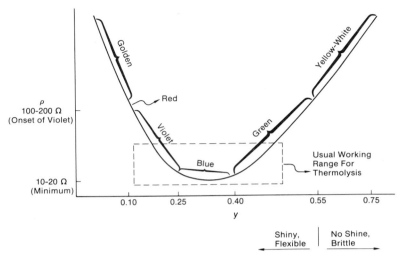

Fig. 7.18   Schematic drawing for the color change and resistivity of polyacetylene upon doping with bromine. $P_{Br_2} \leqslant 10$ torr, temperature 250 K, doping time 5 – 15 min. At minimum $\rho : \sigma_{RT} = 20.8 \ (\Omega \text{ cm})^{-1}$, $S = +269 \ \mu V \ K^{-1}$, $(CHBr_{0.16})_x$. [After MacDiarmid (1982).]

$(\Omega \text{ cm})^{-1}$ and $S = 650 \ \mu V \ K^{-1}$, characteristic of a p-type semiconductor. These properties can be finely adjusted by varying the composition of the polymer through the extent of original doping, temperature, and duration of heating.

These p-type semiconductors can be converted to p-type metal. Doping with iodine produces silvery-black materials such as $(CH_{0.89}Br_{0.09}I_{o.12})_x$ with $\sigma = 10 \ (\Omega \text{ cm})^{-1}$ and $S = 18 \ \mu V \ K^{-1}$. AsF_5 doping gave a shiny golden film having transport of $\sigma = 12 \ (\Omega \text{cm})^{-1}$ and $S = 13.2 \ \mu V \ K^{-1}$ for a composition of $[CH_{0.82}Br_{0.13}(AsF_5)_{0.12}]_x$.

7.4   REACTION WITH OZONE

Complete ozonlysis of defect-free polyacetylene should yield only glyoxal aside from very small amounts of propionaldehyde due to the ethyl end groups. Products other than these would signify the presence of defects such as crosslinks. Li *et al.* (1982) treated polyacetylene with 6% O_3 in O_2 until complete disappearance of the polymer particles (12 – 36 hr). In order to perform GC – MS analysis of the products, the aldehydes were converted to the corresponding methyl esters by reaction with H_2O_2 followed by

TABLE 7.13

Polyacetylene Ozonolysis Products[a]

Sample	% Yield		
	Dimethyl oxalate	Dimethyl succinate	Dimethyl malonate
A *cis*-(CH)$_x$	99.52	0.10	0.38
B *cis*-(CH)$_x$	99.49	0.28	0.23
C *trans*-(CH)$_x$	98.79	0.50	0.71
D *trans*-(CH)$_x$	98.55	0.41	1.04

[a] After Li *et al.* (1982).

acidification to pH 4, removing the excess H_2O_2 by decomposition with
PtO_2, and treating the reaction mixture with CH_2N_2. The esters were
extracted with $Et_2O–CH_2Cl_2$, dried, and concentrated. Because the products, especially dimethyl oxalate, are volatile and have low partition coefficients between H_2O and CH_2Cl_2, the material balance was quite poor
($\leq 10\%$). Therefore, standard aqueous solutions of the carboxylic acids were
esterified and extracted as above. These were used as references to compensate for product loss during extraction and for flame-ionization detector
response. The results are summarized in Table 7.13.

If the polyacetylene backbone is without defect, only dimethyl oxalate
will be formed. This is indeed the major product, as shown in Table 7.13.
The succinate and malonate are apparently derived from defects on the
polymer backbone, but the mechanism of their formation is not clear. The
appearance of dimethyl succinate was not due to the copolymerization of a
trace of ethylene [from reactions between $Ti(OBu)_4$ and $AlEt_3$] and ozonolysis of the double bonds flanking the ethylene unit. Acetylene polymerized
in the presence of 25 mole % of ethylene-d_2 did not yield more dimethyl
succinate or deuterated derivative after ozonlysis. The authors proposed
that if there are crosslinks, ozonolysis would produce a tetracarboxylic acid
that would undergo rapid double decarboxylation to yield the succinate,

$$
\begin{array}{c}
\text{(structure)} \xrightarrow[\text{work up}]{O_3} \\
\begin{array}{c}
\text{HOOC} \quad \text{COOH} \\
\backslash \quad / \\
\text{CH} \\
| \\
\text{CH} \\
/ \quad \backslash \\
\text{HOOC} \quad \text{COOH}
\end{array}
\xrightarrow{-2\ CO_2} \\
\text{HOOC–CH}_2\text{–CH}_2\text{–COOH} \xrightarrow{CH_2N_2} \\
\text{MeOOC–CH}_2\text{–CH}_2\text{–COOMe}
\end{array}
\tag{7.33}
$$

The formation of malonic esters may proceed by more than one pathway. The reaction of ozone with an olefin is considered to involve electrophilic attack by one of the oxygen atoms of ozone followed by ring closure with the incipient carbocation leading to the "initial" ozonide. If there is a proton–hole exchange and the addition of ozone is slightly reversible, a CH_2 unit can be formed:

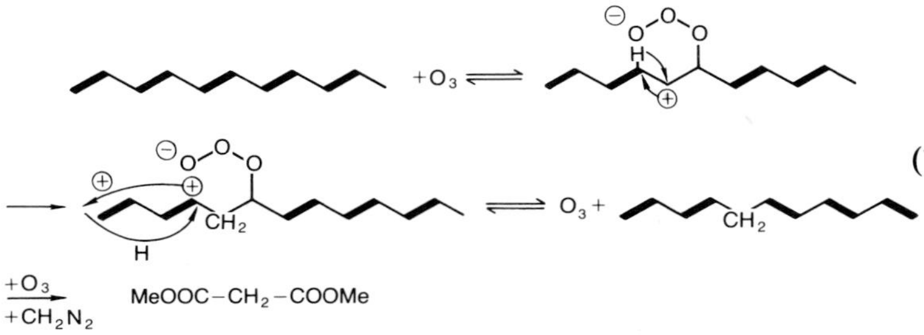

(7.34)

Another possible mechanism is the electron–proton exchange process proposed above for the pyrolysis of polyacetylene,

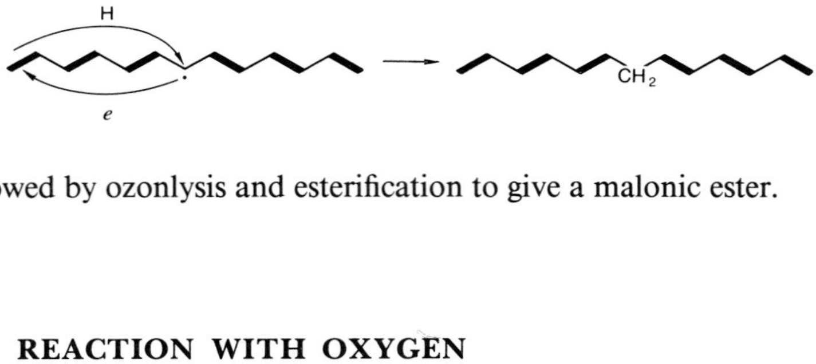

(7.35)

followed by ozonlysis and esterification to give a malonic ester.

7.5   REACTION WITH OXYGEN

The interactions of oxygen with polyacetylene are very complicated. They cause changes in physical, chemical, and mechanical properties; some are reversible or partially reversible and others are irreversible. Much depends on the experimental conditions and the properties being investigated. The effects of oxygen in catalyzing cis–trans isomerization, on reversible broadening of EPR, on ^1H NMR T_1, and on dynamic nuclear polarization have been given above. In this section we concentrate on autoxidation and stabilizations.

7.5.1 Autoxidation of Polyacetylene

Autoxidation of polyacetylene was measured by Chien *et al.* (1983f) with an automatic oxygen uptake apparatus (Chien and Boss, 1967a,b; Chien and Wang, 1975). Measurements were made at either 760 or 150 torr of oxygen, which was kept constant while decrease in the volume of oxygen reacted was recorded.

The oxidation curves for polyacetylene resemble those for polyolefins. There was a short induction period which becomes shorter as the temperature is increased. However, at elevated temperature, the induction period approaches a constant value. The time for the sample to reach thermal equilibrium may contribute to this period. This is followed by an extended period of a relatively constant rate of oxidation. At 343 K the rate did not diminish even after 170 ml of oxygen was absorbed per gram of polyacetylene. This corresponds to one O_2 per 10 CH units. The oxygen uptake curves for polyacetylene are shown in Fig. 7.19.

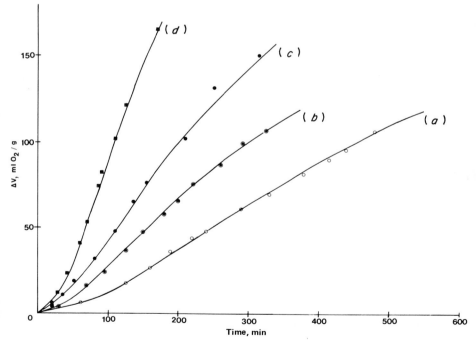

Fig. 7.19 Oxygen uptake for polyacetylene at 760 torr: (a) 313 K; (b) 323 K; (c) 333 K; (d) 343 K.

TABLE 7.14

Rate of Autoxidation of Polyacetylene[a]

Temperature (K)	Rate $\times$ 10^7 mole (g sec)$^{-1}$
313	1.95
323	3.0
333	4.42
343	9.16

[a] At 760 torr oxygen.

The rates of oxidation at various temperatures are summarized in Table 7.14. The Arrhenius plot gives an activation energy of 9.6 kcal mole^{-1} (Fig. 7.20). This value is much lower than those for the autoxidation of other polymers. For instance, $\Delta E = 16$ kcal mole^{-1} for polypropylene (Chien and Kiang 1979c) and it is 27 kcal mole^{-1} for the autoxidation of polyisoprene (Chien and Kiang, 1979b).

Autoxidation of polyacetylene had also been followed by elemental analysis and weight uptake (Gibson and Pochan, 1982) and by infrared analysis (Gibson and Pochan, 1982; Yen *et al.*, 1980). Autoxidation of polyacetylene in dry air in the absence of UV light (screened by Plexiglas) at room temperature has a 22-hr induction period. For very long experiment times, the data were analyzed in terms of two first-order processes, and it was said that the oxidation had no effect on cis–trans isomerization, which is inconceivable in view of our results described in Section 5.3.2. When the

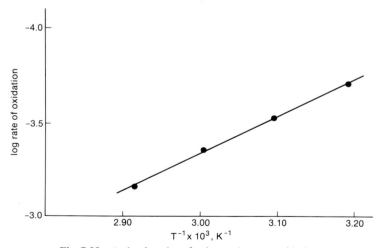

Fig. 7.20 Arrhenius plot of polyacetylene autoxidation.

sample was exposed to ambient air, humidity, and room fluorescent light, the induction period disappeared and the kinetics of autoxidation was found to be simple first order, which was sample-thickness independent. A sample initially 53% in cis content absorbed 46% oxygen after 5 months. This corresponds to one O_2 molecule per three CH units.

The autoxidation in the presence of ambient light, when monitored by IR and weight uptake, gave different kinetic results (Gibson and Pochan, 1982). The oxidation kinetics are said to comprised two pseudo-first-order domains; cis–trans isomerization also can be analyzed by two first-order equations. Several new IR bands were formed during autoxidation. There are two carbonyl bands, at 1665 and 1720 cm^{-1}, attributed by Shirakawa and Ikeda (1971) to α,β-unsaturated and isolated ketone chromophores, respectively. There is an absorption at 3450 cm^{-1} that corresponds to the O—H stretching mode of a hydrogen-bonded hydroperoxy group and/or hydroxy moiety. Peaks also appear at 1140 cm^{-1} (C—O stretch) and 887 cm^{-1} (peroxide, epoxide).

The effect of oxygen on the conductivy of *trans*-(CH)$_x$ at elevated temperature is not unlike that caused by bromine but only more dramatic (Pochan *et al.,* 1980, 1981). There is an initial increase in conductivity followed by a precipitious decline (Fig. 7.21). Similar behavior was observed for polyacetylene with various trans contents. The results may be interpreted as initial doping of polyacetylene by oxygen, which is partially reversible. This is followed by autoxidation leading to reduction in conductivity. This is supported by the results in Fig. 7.21, which showed that the drop in conductivity occurs when rapid buildup of carbonyl groups begins. The processes can be represented by

$$(CH)_x + O_2 \rightleftharpoons [CH(O_2)_y]_x \longrightarrow \text{oxidized product} \qquad (7.36)$$

$$\begin{array}{ccc} \text{undoped} & \text{oxygen } p\text{-doped} & \text{nonconducting} \\ \text{polymer} & \text{polymer} & \text{polymer} \end{array}$$

The electrical property change accompanying these reactions was written (Gibson and Pochan, 1982)

$$\frac{1}{R_t} = \frac{N_u}{R_u} + \frac{N_d}{R_d} + \frac{N_i}{R_i}, \qquad (7.37)$$

where subscripts t, u, d, and i denote the total, undoped, doped, and nonconducting polymers, respectively, and N represents the fraction of each of the species present. R_d may be assumed to be at least three orders of magnitude less than R_u, and R_i is at least four orders of magnitude lower than R_u. Therefore, Eq. (7.37) is controlled by the term for the doped species. In these experiments, the current i is monitored with an applied

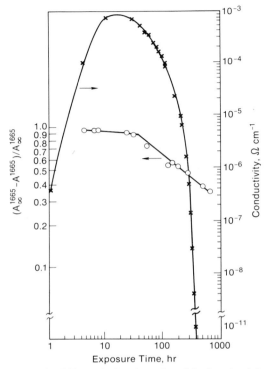

Fig. 7.21 Log(reduced 1665 cm⁻¹ absorbance) and log(conductivity of polyacetylene initially 53% *cis,* 60 μm thick) versus exposure to air in the absence of UV light at 298 K. [After Gibson and Pochan (1982).]

voltage V. Equation (7.37) can be approximated and recast as

$$i = V/R_t \simeq V N_d/R_d. \tag{7.38}$$

In view of the fact that the conversion from the doped polymer to an insulator product is first order, Eq. (7.38) can also be written

$$i = (V/R_d)N_d^\circ \, e^{-t/\tau}. \tag{7.39}$$

The experimental results on polymers with initial trans content of 60% showed that there may be two first-order processes with different rate constants. However, only one decay process was observed for a 30% *trans*-polyacetylene sample. The rate constants for the second faster process were determined as a function of temperature, and the results gave apparent activation energies ranging from 9.4 to 13.8 kcal mole⁻¹.

7.5.2   Autoxidation of Poly(methylacetylene)

Poly(methylacetylene) is a good substrate for autoxidation. The polymer is noncrystalline and soluble in common solvents so that reactions can be performed under homogeneous conditions either in solution or in the solid state. Effects of diffusion are minimized. It is the closest analog to polyacetylene, and the kinetics and mechanism of autoxidation obtained for poly(methylacetylene) will be relevant to those for polyacetylene. Finally, stabilizers can be introduced into poly(methylacetylene) uniformly, whereas it is difficult to assure homogeneous distribution of any stabilizer introduced into polyacetylene.

The autoxidation of poly(methylacetylene), as measured by oxygen uptake, is shown in curve 1 of Fig. 7.22 (Chien *et al.,* 1983d). There was virtually no induction period; oxygen consumption begins immediately though at a slightly slower rate, which may be due to thermal equilibration. The maximum rate of oxidation at 343 K is about 9.9×10^{-7} mole g^{-1} sec^{-1}. There is only a short period during which the rate of oxidation remains constant, which slows down somewhat more markedly than in the autoxidation of polyacetylene. The oxidation curve is hyperbolic in shape in the

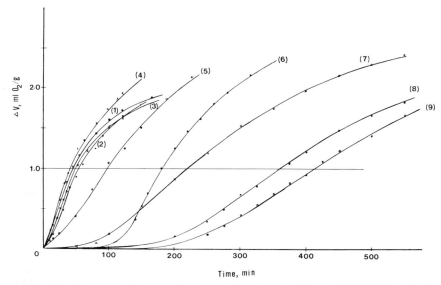

Fig. 7.22   Oxygen-uptake curves for poly(methylacetylene) films at 343 K, 760 torr O$_2$: (1) without additive; (2) 1% TNP; (3) 1% DSTDP; (4) 1% DMA; (b) 1% Tinuvin 770; (6) 1% 4010; (7) 1% BHT; (8) 1% BHT + 1% TNP; (9) 1% BHT + 1% DSTDP.

absence of an effective inhibitor. In Curve 1 shown in Fig. 7.22, 100 ml of oxygen was taken up by the polymer, corresponding to 0.18 mole of oxygen per backbone carbon atom, comparable to the polyacetylene oxidation.

Another way to monitor the autoxidation is through infrared spectroscopy. Figure 7.23a shows a spectrum of pristine poly(methylacetylene); the spectrum after 14.7 hr of exposure to O_2 at 333 K is given in Fig. 7.23b. The spectra were obtained on film cast from a concentrated solution of the polymer. The two bands at 1375 and 1460 cm^{-1} are assigned to the methyl-group symmetric and antisymmetric bending vibrations, respectively. The other bands in the order of increasing frequencies are tentatively assigned as follows: 910 cm^{-1} to the $=CH$ rocking mode, 1620 cm^{-1} to $C=C$, and 2960, 2920, and 2875 cm^{-1} to the CH_3 stretching vibrations.

As poly(methylacetylene) was oxidized, the following spectral changes were observed. There is produced a broad band with v_{max} at 3420 cm^{-1} and a bandwidth of about 300 cm^{-1}, which may be attributed to the OH stretching vibration of a hydroperoxy or hydroxyl group hydrogen bonded to a carbonyl group, hydroperoxy, or hydroxyl group. In the carbonyl region the v_{max} at 1720 and 1680 cm^{-1} are probably the bands for carbonyl and $C=C-C=O$ groups. The CH_3 stretching modes decrease in intensity with oxidation, as expected. There is absorption intensity increase beneath the prominent 1375 and 1460 cm^{-1} bands, which is believed to be related to the bending vibration of methylene groups (connected to OH). Finally, there is increasing absorption in the 1000–1300 cm^{-1} region, which could be associated with vibrations of peroxy groups.

It was realized that the total area between 1300 and 1500 cm^{-1} remained constant as methyl groups were oxidized to methylene groups. Apparently, the decrease of absorbance of the methyl-group vibrations was compensated by the absorption of the methylene group connected to OH in this frequency region. Therefore, it is possible to monitor the rate of autoxidation from the increase in absorption of the OH band at 3420 cm^{-1} or of the carbonyl band at 1720 cm^{-1} with respect to the absorbance between 1300 and 1500 cm^{-1}. Figure 7.24 gives the variation of $A(\geq C=O)/A(-CH_3 + =CH_2)$ and $A(-OH)/A(-CH_3 + =CH_2)$, and the time it takes to reach a ratio of 1.0 and 0.6, respectively, was used to obtain the Arrhenius plot (Fig. 7.25). From the results obtained between 298 and 343 K, the apparent activation energy of autoxidation was found to be 14.6 kcal $mole^{-1}$.

Oxidation of poly(methylacethylene) results in the decrease of conjugation length. This is shown by the electronic spectra (Fig. 7.26); after 60 hr of exposure to air at 298 K, the absorption is blue shifted by ~ 30 nm.

Autoxidation of poly(methylacetylene) was also performed in air (Fig. 7.27). Table 7.15 compares this with that obtained in pure oxygen (Curve 1, Fig. 7.22). The comparison for the time to reach certain levels of oxidation

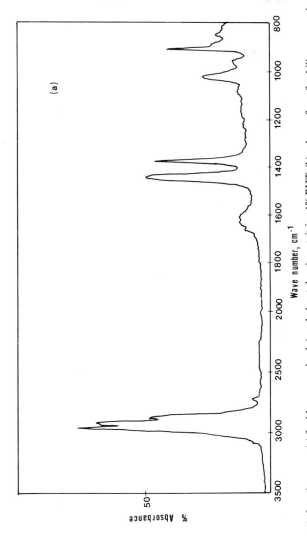

Fig. 7.23 Infrared spectra: (a) freshly prepared poly(methylacetylene) containing 1% BHT; (b) polymer free of stabilizer exposed to O₂ at 333 K for 880 min.

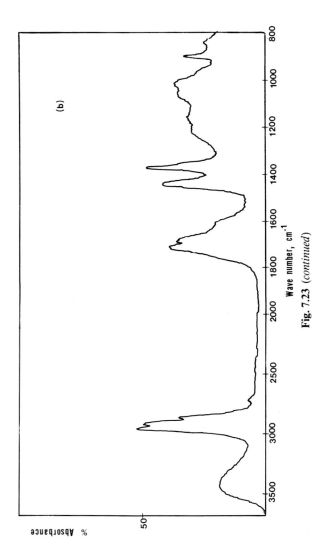

Fig. 7.23 (*continued*)

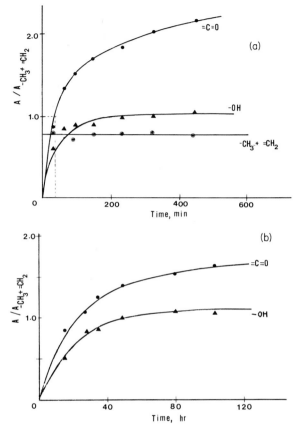

Fig. 7.24   Variation   of $A(\!\!>\!\!C\!\!=\!\!O)/A(\!\!-\!\!CH_3 + =\!CH_2)$ and $A(\!-\!OH)/A(\!-\!CH_3 + =\!CH_2)$ with time of oxidation: (a) 343 K; (b) 298 K; 760 torr O_2.

(Column 2) is 4.9 times longer in air than in oxygen; the slope for oxidation in air, both initial and final, is about 0.21. These results indicate that autoxidation of poly(methylacetylene) is linearly dependent on the oxygen partial pressure. The slopes from the $C\!\!=\!\!O$ band directly differ by a smaller than expected ratio because the band is not completely resolved.

7.5.3   Inhibition

Oxidation of hydrocarbon polymers can be inhibited by various additives and their synergistic combinations (Blumberg *et al.*, 1965; Chien and Boss, 1967a, 1972; Chien and Kiang, 1980; Kiang and Chien, 1980; Chien

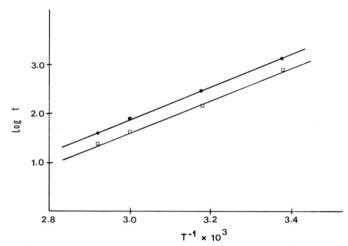

Fig. 7.25   Arrhenius plot for autoxidation of poly(methylacetylene): (●) for $\diagup$C$=$O formation, (□) for —OH formation.

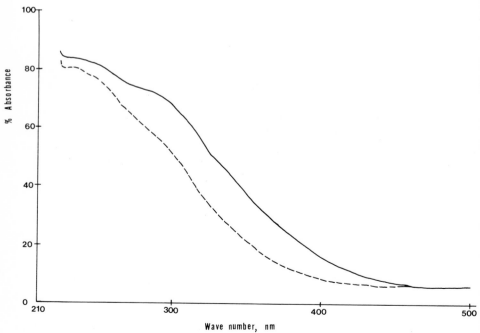

Fig. 7.26   Electronic spectra of poly(methylacetylene): (——) before oxidation, (----) after exposure to air for 60 hr at 298 K. 2.5 mg polymer in 100 ml of CH_2Cl_2.

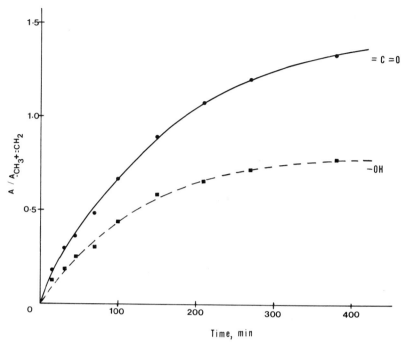

Fig. 7.27   Autoxidation of poly(methylacetylene) in an atmosphere of air at 343 K.

and Connor, 1968). To inhibit thermal autoxidation, the stabilizers are hindered phenols, which scavenge peroxy radicals, and phosphites and sulfides, which act to decompose hydroperoxides. To protect polyolefins against photo-oxidation, nickel chelates are used to quench excited chromophores. Hindered amines are very efficient stabilizers, although the mechanism of its action is not yet understood. We have studied the effects of these additives on the oxidations of polyacetylene and poly(methylacetylene) (Chien *et al.*, 1983d). The stabilizers with their structures and abbreviated names are contained in Table 7.16.

TABLE 7.15

Comparison of Autoxidation of Poly(methylacetylene) at 343 K in Air and Oxygen

	Time (min) to $\dfrac{A(\gtrsim C=O)}{A(-CH_3 + CH_2)} = 1.0$	Initial slope		Final slope	
		C=O band	OH band	C=O band	OH band
Air	182	0.009	0.005	0.0008	0.0002
Oxygen	37	0.028	0.023	0.002	0.0009
Ratio (air/O$_2$)	4.9	0.32	0.21	0.4	0.22

TABLE 7.16

Stabilizers

Abbreviation	Name		Structure	Mode of action
	Chemical			
BHT	Di-*tert*-butyl-*p*-cresol			Peroxy radical scavenger
4010	*N*-Phenyl-*N'*-cyclohexyl-*p*-phenylenediamine			Peroxy radical scavenger
Inganox 1010	Tetrakis[methylene 3-(3,5-di-*tert*-butyl-4-hydroxyhydrocinnamate]methane			Peroxy radical scavenger

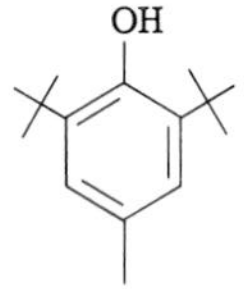

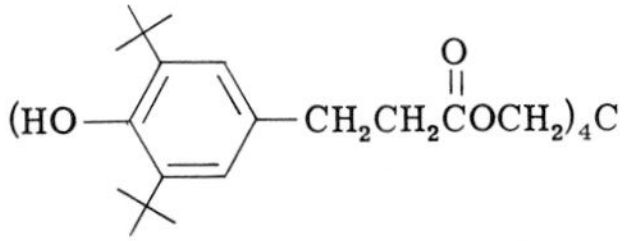

		Structure	Function
Tinuvin 770	Bis(2,2,6,6-tetramethylpiperidinyl) sebacate	$\left[\overset{(HN)}{\underset{}{}}-O-\overset{O}{\overset{\|}{C}}-(CH_2)_4\right]_2$	Multiple functions
TNP	Tris(nonylphenyl)phosphite	$P(O-\!\!\!\!\bigcirc\!\!\!\!-C_9H_{19})_3$	Hydroperoxide decomposers
DSTDP	Disterarylthiodipropionate	$S[CH_2CH_2\overset{O}{\overset{\|}{C}}O(CH_2)_{16}CH_3]_2$	Hydroperoxide decomposers
DMA	9,10-Dimethylanthracene	(9,10-dimethylanthracene structure with CH$_3$ groups)	Quencher
BPN	N-tert-butyl-2-phenylnitrone	$\bigcirc\!\!-CH=\overset{O}{\overset{\uparrow}{N}}-C_4H_9$	Spin trap

The incorporation of 4% BHT into polyacetylene foam resulted in a significant inhibition of polyacetylene autoxidation at 323 K (Fig. 7.28). There is an induction period of 460 min, whereas it was approximately 10 min without the additive. The rate of oxidation was decreased from 3×10^{-7} mole (g sec)$^{-1}$ to 4.9×10^{-1} mole (g sec)$^{-1}$.

More extensive study was made by Chien *et al.* (1983d) on the effect of stabilizers in autoxidation of poly(methylacetylene). Some of the results are shown in Fig. 7.22 by oxygen uptake and in Fig. 7.29 by IR. Tables 7.17 and 7.18 give the induction periods and rates of oxidation. Within experimental accuracy, it was determined that autoxidation is not affected by DMA. The hydroperoxide decomposers alone (TNP and DSTDP) have no effect on the oxidation. This is as expected because phosphites and sulfides are secondary stabilizers, and they are effective only in the presence of a peroxy radical scavenger. Tinuvin 770 slightly retards the oxidation of poly(methylacety-lene). The presence of radical scavengers BHT and 4010 elicited a distinct induction period. The induction period was significantly increased, and the rates of subsequent autoxidation were retarded when BHT was used with a synergist such as TNP and DSTDP. The superior performance of BHT over Irganox 1010 is in strong contrast to the latter, which is one of the best available antioxidants for polypropylene. The reversal of performance is simple to rationalize. Irganox 1010 is a high molecular weight compound,

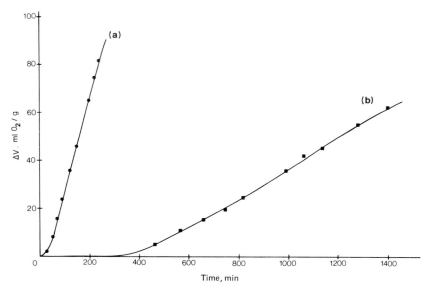

Fig. 7.28 Autoxidation of polyacetylene at 323 K and 760 torr: (a) no inhibitor, (b) inhibited by 4% of BHT.

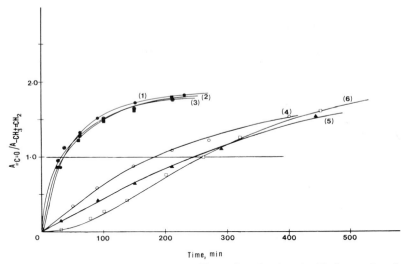

Fig. 7.29   Effect of stabilizer on the rate of formation of carbonyl oxidation products by IR at 343 K, 760 torr oxygen: (1) no additive, (2) 1% DSTDP, (3) 1% DMA, (4) 1% Irganox 1010, (5) 1% BHT, (6) 1% BHT + 1% DSTDP.

compatible with polypropylene, and has very low volatility. In 423-K circulating-oven tests for aging lifetime of polypropylene, Irganox 1010 is extremely effective, whereas the low molecular weight BHT was volatilized in a few hours (Blumberg *et al.,* 1965). On the other hand, for the stabilization of poly(methylacetyelene), which oxidizes very readily, the higher

TABLE 7.17

Effect of Stabilizers on the Autoxidation of Poly(methylacetylene) by Oxygen Uptake[a]

Stabilizer	t_{ind} (min)	Maximum rate of oxidation $\times 10^7$ (mole g^{-1} sec^{-1})
None	None	9.9
TNP	None	9.9
DSTDP	None	9.9
DMA	None	11.0
Phenylenediamine	None	17.0
Hydroquinone	23	7.9
Tinuvin 770	9	11.9
4010	117	5.9
BHT	75	2.5
BHT + TNP	175	2.1
BHT + DSTDP	225	1.9

[a] At 343 K, 760 torr oxygen, with 1 wt % of stabilizer.

TABLE 7.18

Effect of Stabilizers on the Autoxidation of Poly(methylacetylene) by IR Spectra[a]

	Time in min to reach	
Stabilizer	$A(\geq C{=}O)/$ $A(-CH_3 + {=}CH_2) = 1.0$	$A(-OH)/$ $A(-CH_3 + {=}CH_2) = 0.6$
None	37	32
TNP	35	30
DMA	42	38
DSTDP	35	35
Irganox 1010	178	110
BHT	250	250
Irganox 1010 + DSTDP	175	115
BHT + TNP	200	190
BHT + DSTDP	260	280

[a] At 343 K, 760 torr oxygen, with 1 wt % of stabilizer.

diffusivity of BHT over Irganox 1010 is an advantage and volatility is of no concern.

Figure 7.30 shows the autoxidation of poly(methylacetylene) containing 1% BHT from 293 to 343 K. From the Arrhenius plot (Fig. 7.31) a value of 18.9 kcal mole^{-1} was obtained.

The above results suggest that the autoxidation of poly(methylacetylene) is a free-radical process. Therefore, we looked for the presence of radicals by EPR. It will be recalled that freshly prepared poly(methylacetylene) is devoid of the $g = 2$ signal, which was evident in polyacetylene and copolymers of acetylene and methylacetylene. However, when poly(methylacetylene) is exposed to oxygen at 343 K for a few minutes, a weak EPR signal developed with $g = 2.0022$ and a linewidth of 16 G. The intensity of this resonance did not alter appreciably with time of oxidation. At 323 K the resonance doubles its intensity in 60 min. The unpaired spin is apparently very stable; dissolution of the oxidized polymer film in tetrachloroethylene changes neither the linewidth nor the intensity of the EPR spectra. It seems that the unpaired spin is related to the delocalized neutral defects with restricted mobility, and may or may not play a role in the free-radical chain process.

In none of the oxidized poly(methylacetylene) was there a detectable EPR signal characteristic of peroxy radicals (Chien and Boss, 1967a,b,c). Therefore, apparently the peroxy radicals of poly(methylacetylene) are too unstable to be present in a high enough steady-state concentration to be observed by EPR.

A spin trap, *N-tert*-butyl-2-phenylnitrone (BPN) was used to detect the concentration of carbon radicals indirectly. A toluene slution of 5%

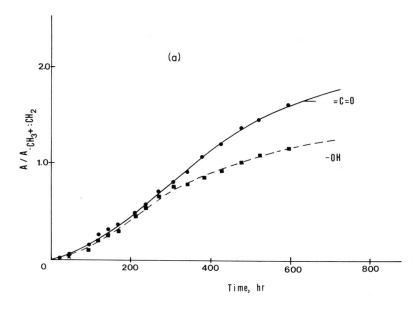

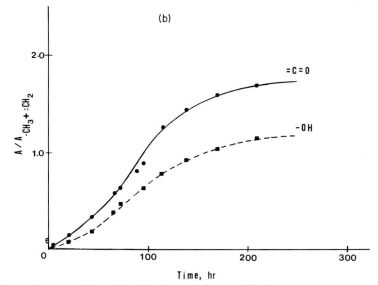

Fig. 7.30   Autoxidation of poly(methylacetylene) stabilized by 1% BHT, $P_{O_2} \pm 760$ torr: (a) 293 K; (b) 313 K; (c) 333 K; (d) 343 K.

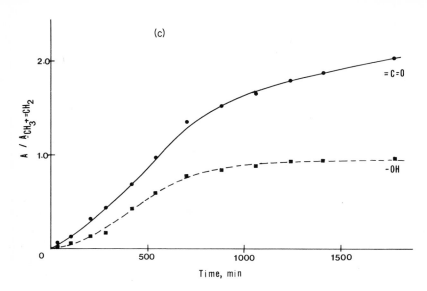

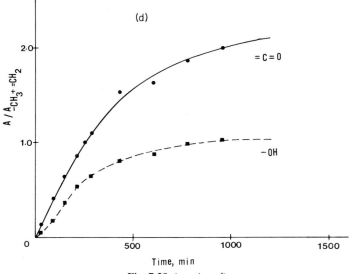

Fig. 7.30 (*continued*)

poly(methylacetylene), 2.5% BHT, and 2.5% BPN oxidized with 760 torr O_2 at 343 K for 40 min gave the EPR spectrum shown in Fig. 7.32a with nitrogen and proton hyperfine coupling constants of 13.8 and 2 G, respectively. The spectrum can be attributed to a nitroxide radical

$$R\cdot \; + \; \langle \bigcirc \rangle \!\!-\!\! \underset{H}{\overset{+}{C}} \!\!-\!\! \underset{\underset{O_-}{|}}{N} \!\!-\!\! CMe_3 \; \longrightarrow \; \langle \bigcirc \rangle \!\!-\!\! \underset{H}{\overset{R}{C}} \!\!-\!\! \underset{\underset{O}{|}}{N} \!\!-\!\! CMe_3 \qquad (7.40)$$

where R· is a carbon radical of poly(methylacetylene). Figure 7.32b shows the EPR spectrum of the same radical produced in a polymer film having the characteristic feature of a nitroxide radical in an immobilized state.

If the carbon radical is the principal chain carrier, then BPN should show inhibitory effect on the autoxidation of poly(methylacetylene). Curve 1 of Fig. 7.33 shows that BPN by itself has no effect whatsoever; there is no induction period, and the maximum rate of oxidation is 11.6×10^{-7} mole g^{-1} sec^{-1}, as in an unstabilized polymer. This result implies that the carbon radical is unstable and/or reactive and that BPN only traps a small fraction of it. However, in a synergistic combination with BHT or with BHT and DSTDP, BPN enhances the oxidative stability of poly(methylacetylene). The rates of oxidation were reduced by about a factor of 2 as compared with stabilization by BHT alone. Similar indications of stabilization by the IR results are shown in Fig. 7.34. The induction period of oxidation of BHT-stabilized poly(methylacetylene) was increased from 75 to 130 min by the presence of BPN. The spin trap increased inhibition time of the BHT- and DSTDP-stabilized polymer from 225 to 280 min. The effectiveness of the spin trap as a synergist is shown in Table 7.19.

The nitroxide radical concentration as a function of time of oxidation was followed at 343 K, 1 atm O_2, and in the presence of 1% BPN and 1% BHT. As soon as the sample reached 343 K, $\sim 10^{-5}$ mole of spins per liter were observed. During the induction period the concentration remained unchanged. The radical concentration increases as the polymer begins to autoxidize and appears to reach a constant value of 4×10^{-5} mole liter^{-1} during the period of steady autoxidation (Fig. 7.35).

We have also made efforts to stabilize poly(methylacetylene) against photooxidation. A 450-W medium-pressure mercury lamp with water cooling was used to irradiate poly(methylacetylene) films under 760 torr of oxygen. The pyrex glass container effectively filtered out radiation with wavelengths shorter than 280 nm. The distance separating the light source and the specimen was maintained at 8 cm. The results of photooxidation in the presence of a number of stabilizers are shown in Fig. 7.36. The changes in the IR spectra are shown in Fig. 7.37. The results were generally disappointing. None gave an induction period. Only Irganox 1010 and Tinuvin 770 showed some retardation effects.

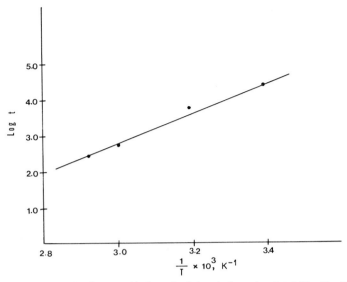

Fig. 7.31   Arrhenius plot for autoxidation of poly(methylacetylene) stabilized by 1% BHT.

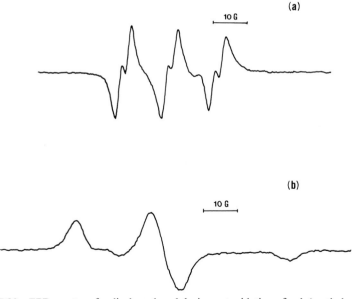

Fig. 7.32   EPR spectra of radical produced during autoxidation of poly(methylacetylene) in the presence of BPN: (a) in toluene solution, (b) in solid polymer film.

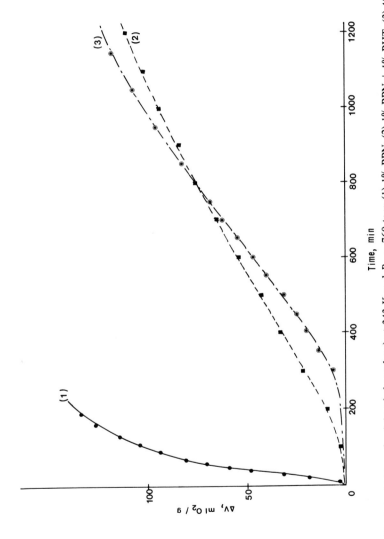

Fig. 7.33 Autoxidation of poly(methylacetylene) at 343 K and $P_{O_2} = 760$ torr: (1) 1% BPN; (2) 1% BPN + 1% BHT; (3) 1% BPN + 1% BHT + 1% DSTDP.

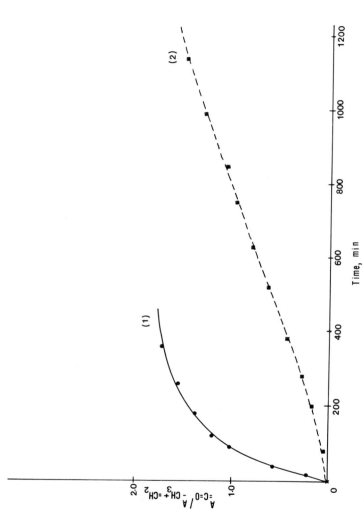

Fig. 7.34 Infrared spectra of autoxidizing poly(methylacetylene) at 343 K and $P_{O_2} = 760$ torr: (1) 1% BPN; (2) 1% BPN + 1% BHT.

TABLE 7.19

Effect of Spin Trap on the Autoxidation of Poly(methylacetylene)

Antioxidant	t_{ind} (min)	Rate of oxidation (mmole g^{-1} sec^{-1})
1% BPN	11	11.6×10^{-4}
1% BPN + 1% BHT	130	0.86×10^{-4}
1% BPN + 1% BHT + DSTDP	280	1.1×10^{-4}

7.5.4   Mechanisms of Autoxidation and Stabilization

Poly(methylacetylene) has three allylic protons per monomer unit. One might think that they would play an important role in autoxidation. This is found not to be the case by direct comparison with the autoxidation of polyacetylene. At 343 K polyacetylene and poly(methylacetylene) have

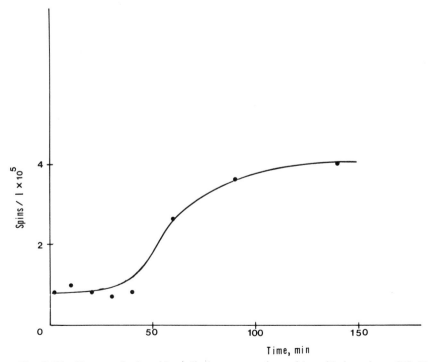

Fig. 7.35   Change of nitroxide radical concentration with oxidation time: 343 K, $P_{O_2} = 760$ torr, 1% BPN + 1% BHT.

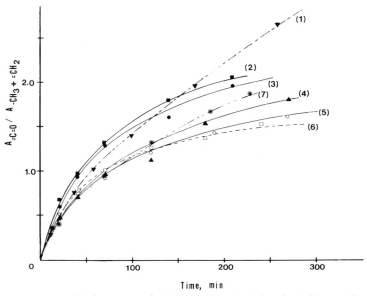

Fig. 7.36 Photooxidation curves of poly(methylacetylene) irradiatd with a medium-pressure mercury lamp for 260 min at 298 K and 760 torr O_2: (1) no additive; (2) 1% 4010; (3) 1% BHT; (4) 1% BHT + 1% DSTDP; (5) 1% Irganox 1010; (6) 1% Tinuvin 770, (7) 1% DMA.

nearly the identical autoxidation rates of $\sim 9 \times 10^{-7}$ mole g^{-1} sec^{-1}. However, the activation energy of polyacetylene oxidation is only about 9.6 kcal $mole^{-1}$. This shows that the allylic hydrogens, although certainly involved in the autoxidation of poly(methylacetylene), do not play a dominant role in the autoxidation of acetylenic polymers. As a result, one may assume that the mechanisms of autoxidation and stabilization of polyacetylene and poly(methylacetylene) share some similarities.

Initiation.   Because poly(methylacetylene) autoxidizes without an induction and the rate is first order with respect to p_{O_2}, it may be assumed that oxygen is directly involved in the initiation processes. One such process is the catalysis by oxygen to form neutral defects as has been demonstrated by EPR studies:

$$+ \; O_2 \; \xrightarrow{\; k_1 \;}$$

$$+ \tag{7.41}$$

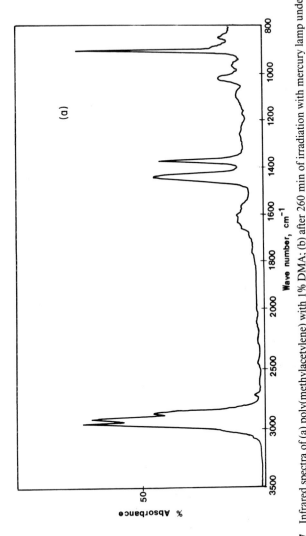

Fig. 7.37 Infrared spectra of (a) poly(methylacetylene) with 1% DMA; (b) after 260 min of irradiation with mercury lamp under oxygen.

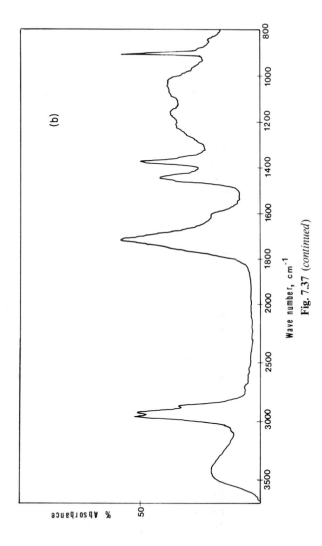

Fig. 7.37 (*continued*)

The spins were observed as the weak $g \approx 2$ EPR signal and the more intense nitroxide signal upon reaction with the spin trap. Another possible direct process is

$$\text{(structure)} + O_2 \xrightarrow{k_2} \text{(structure with OOH)} \qquad (7.42)$$

this hydroperoxide is probably unstable and may rearrange by

$$\text{(structure with OOH)} \xrightarrow{k_3} \text{(structure with O)} + H_2O \qquad (7.43)$$

The reactions shown in Eqs. (7.42) and (7.43) are probably necessary to account for the results shown in Fig. 7.38. The rates of formation of carbonyl and hydroxyl groups in the absence of antioxidant are the greatest at the beginning of the oxidation. Even in the presence of an antioxidant (Fig. 7.38b), the rates were high during the induction period. The intensity of the OH band reaches a constant value while the carbonyl band intensity continues to increase with time.

With the low-lying electronic states of poly(methylacetylene) and the presence of unpaired spins, there is a possibility that the polymer may promote ground-state oxygen to a singlet state, i.e., $^1O_2 (\Delta g)$, especially when illuminated with UV radiation. The singlet oxygen can undergo "ene" reaction with the polymer to produce hydroperoxides. However, this possibility is probably unimportant because DMA is ineffective as a stabilizer in either thermal or photooxidation of poly(methylacetylene).

The hydroperoxides produced in the reaction of Eq. (7.42) and in the subsequent propagation steps will provide the source for secondary initiation, i.e.,

$$\text{(structure with OOH)} \xrightarrow{k_4} \text{(structure with O)} + HO\cdot \qquad (7.44)$$

This is the principal initiation reaction in the autoxidation of polyolefins and is responsible for the continuation of radical–chain processes. On the other hand, this process does not seem to play a similarly significant role in the autoxidation of poly(methylacetylene) because it is without the characteristic rate acceleration, or the role is a very slight one. Nevertheless the participation of the reaction of Eq. (7.44) is supported by the beneficial effect of TNP and DSTDP as synergistic hydroperoxide decomposers.

Propagation.   The propagation reactions involve $R\cdot$, $RO\cdot$, and $RO_2\cdot$

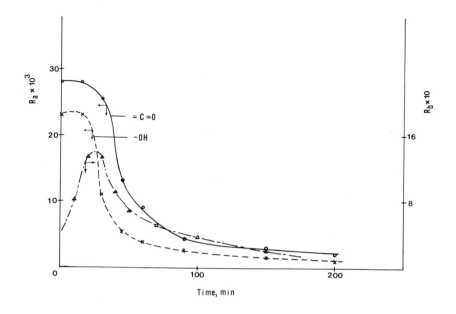

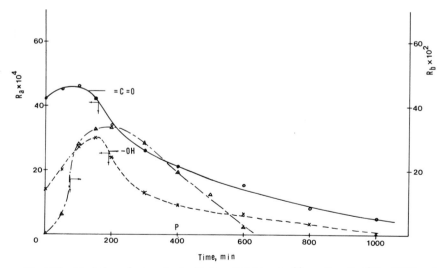

Fig. 7.38   Variation of rates of autoxidation, R_b (▲) and of formation R_a of C=O (●) and —OH (×) with time of reaction at 343 K, P_{O_2} = 760 torr: (a) poly(methylacetylene); (b) with 1% BHT.

radicals. The main reaction for the R· radical is the one with oxygen:

$$\text{(7.45)}$$

In the case of autoxidation of polyolefins the alkyl radicals react with oxygen at diffusion-limited rates. This is probably not true for poly(methylacetylene) because the spin trap was found to be able to compete with oxygen, i.e., the reactions of Eqs. (7.40) and (7.45) occur simultaneously. If we designate the rate constants for the spin-trap reaction [Eq. (7.40)] to be k_5, then $k_5[\text{BPN}] > k_6[\text{O}_2]$. The only possible explanation for this competition is that the R· radical of polymethylacetylene is stabilized by delocalization.

Several reactions for the RO_2· radical can be envisioned. The hydrogen abstraction is supported by the formation of the $=\text{CH}_2$ group whose vibration was observed between 1300 and 1500 cm^{-1}. This reaction can be represented for the intramolecular process as

$$\text{(7.46)}$$

and analogously for intermolecular processes. Alternate reactions are

$$\text{(7.47)}$$

and

$$\text{(7.48)}$$

Apparently $k_8 + k_9 \gg k_7$ because autoxidation of polyacetylene occurs at rates comparable to, and with an activation energy smaller than, those for

poly(methylacetylene). A set of reactions equivalent to those shown in Eqs.
(7.46 – 7.48) are available for the RO· radical.

Dissociation of hydroperoxide yields RO· radicals. The RO· radical is
most likely to be immediately converted to a carbonyl or hydroxyl product
with the formation of the stable unpaired spin on the conjugated backbone:

$$\text{(7.49)}$$

Scission of poly(methylacetylene) can occur by

$$\text{(7.50)}$$

Terminations.   The most probable termination reaction is that between
two RO_2· radicals:

$$\text{(7.51)}$$

Although a similar reaction is possible for the RO· radicals:

$$\text{(7.52)}$$

however, it will probably be superseded by the reactions of Eqs. (7.49) and
(7.50).

7.5.5   Kinetics

Oxygen-pressure dependence in autoxidation can arise from several
causes. At low oxygen pressure, when not all terminations occur between

two $RO_2 \cdot$ radicals (k_{t_1}) but when there are also terminations between two $R\cdot$ radicals (k_{t_2}) and cross reactions between $RO_2 \cdot$ and $R\cdot$ radicals ($k_{t_{12}}$), the rate of oxidation is as given by Walling (1957),

$$-\frac{d[O_2]}{dt} = \frac{k_7 k_6 [RH][O_2] R_i^{1/2}}{(k_7 k_6 k_{t_{12}}[O_2][RH] + 2k_6^2 k_{t_1}[O_2]^2 + 2k_7^2 k_{t_2}[RH]^2)^{1/2}} \quad (7.53)$$

where [RH] is the polymer concentration and R_i is the rate of initiation. The kinetic order dependence on $[O_2]$ will lie between zero and one, depending on the relative magnitude of the rate constants and oxygen pressure. Another source of oxygen-pressure dependence is the diffusion limitation for semicrystalline polymers at elevated temperatures and thick specimens (Chien, 1968). This is not believed to be the case for the very thin films of poly(methylacetylene) oxidized at mild temperatures.

The observed dependence of poly(methylacetylene) autoxidation rate on oxygen partial pressure is best attributed to initiation by oxygen, i.e., the reactions of Eqs. (7.41) and (7.42). Assuming steady-state approximation, one finds

$$\frac{-d[O_2]}{dt} = k_2[RH][O_2] + (k_7 + k_8 + k_9)[RO_2\cdot] + k_{12}[RO_2\cdot]^2. \quad (7.54)$$

The observed results are consistent with initiation predominantly by the reaction of Eq. (7.42), and very low $[RO_2\cdot]$ (not detected by EPR).

The stabilization of poly(methylacetylene) is most effective when a combination of synergistic stabilizers is used. The $RO\cdot$ and $RO_2\cdot$ radicals can be scavenged by hindered phenols:

$$(7.55)$$

The hydroperoxides were decomposed by phosphorus and sulfur compounds:

$$\begin{aligned} ROOH + R_2S &\rightarrow ROH + R_2SO \\ ROOH + R_3'P &\rightarrow ROH + R_3'PO, \end{aligned} \quad (7.56)$$

and the R· radicals are scavenged by the spin trap [Eq. (7.40)].

7.6   CROSS-LINKING

Here we devote a section to a critical examination of the question of cross-linking. There seem to exist many pathways by which polyacetylene can cross-link readily. In fact, formation of cross-links had been often invoked to explain insolubilization of the polymer, resistance to chlorination upon standing (Wegner, 1981), failure to convert all the polyacetylene to a soluble polyethylenelike product by hydrogenation (Shirakawa *et al.*, 1980b), rapid loss of mechanical properties, etc. In fact, Wegner (1981) concluded that "polyacetylene undergoes spontaneous crosslinking reactions and moreover crosslinking even occurs during the polymerization if the Shirakawa technique is used."

Possible reactions leading to cross-links are the following:

(a)   Cycloaddition of the Diels–Alter type:

$$\text{(7.57)}$$

$$\text{(7.58)}$$

$$\text{(7.59)}$$

(b)   Addition assisted by fragments of the Ziegler–Natta catalyst:

$$2 \quad \underline{\hspace{2cm}} + \text{AlEt}_3 + \text{HX} \longrightarrow$$

$$\text{(7.60)}$$

where $X = OH$, Cl, OBu, etc.

(c) Intermolecular annihilation of solitons:

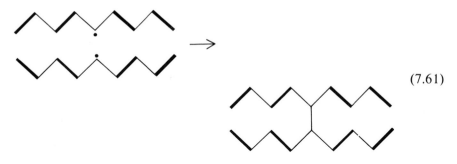

(7.61)

(d) Autoxidation by the reaction of Eq. (5.55) or by

$$2 \quad \text{(polymer chain with } O_2^=\text{)} \quad \rightarrow$$

(7.62)

$$2 \quad \text{(polymer chain)} + O_2 \rightarrow$$

(7.63)

The above reactions serve to illustrate the possibilities but not to exclude others.

The question as to whether or not a polymer becomes cross-linked is usually simply settled by the fact that a cross-linked polymer is insoluble in any solvent and may be swollen to a gel. This criterion is inapplicable to polyacetylene because of its intrinsic insolubility and infusibility. Some may argue that this is because the polymer is already cross-linked; others would attribute these characteristics to a high degree of crystallinity and strong and cooperative interchain $\pi-\pi$ interactions. The next question is concerned with the number of crosslinks per chain. The failure to detect sp^3 carbon atoms by solid high resolution ^{13}C NMR in highly crystalline polyacetylene and to find ozonolysis products attributable to cross-linked units imply an upper limit of a few crosslinks per chain. We sought for evidence that would either support or refute the occurrence of spontaneous and extensive cross-linking reactions for polyacetylene and poly(methylacetylene).

We have taken low molecular weight poly(methylacetylene), selected for reasons of both high degree of chain mobility and ease of characterization. The initial polymer has a $\overline{M}_n = 1394$ and $\overline{M}_w/\overline{M}_n = 5.01$, and is soluble in most common solvents (Fig. 7.39). It was cast into film and oxidized for 27 hr to 100 ml O_2 g^{-1} at 298 K. The resulting polymer is completely soluble in more polar solvents, such as methyl ethyl ketone or tetrahydrofuran, because oxidation produced polar functional groups. GPC gave $\overline{M}_n = 1091$ and $\overline{M}_w/\overline{M}_n = 5.27$ (Fig. 7.39b). Thus there was no evidence of cross-linking under oxidative conditions. In another experiment, the poly(methylacety-lene) was cast into film and heated in vacuo at 453 K for 15 min. This is the

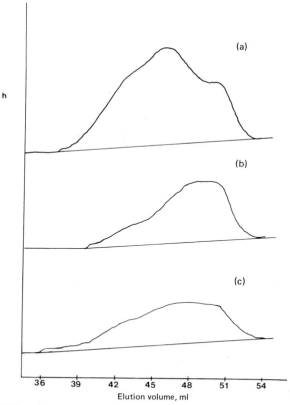

Fig. 7.39   GPC curves: (a) initial poly(methylacetylene); (b) after oxidized at 298 K for 27 hr; (c) after heating at 453 K for 15 min. in vacuo.

condition often used in cis–trans isomerization of polyacetylene. After the heating, the polymer was found to be completely soluble and to have $\overline{M}_n = 860$ and $\overline{M}_w/\overline{M}_n = 3.35$ (Fig. 7.39c).

Therefore, during the heat treatment poly(methylacetylene) was degraded to a polymer of lower molecular weight. Furthermore, the molecular weight distribution was narrowed; cross-linking would tend to broaden the distribution. Thus, at least in the case of poly(methylacetylene), chain scission and degradation occurred in preference to cross-linking. In view of the above discussions of similarities between poly(methylacetylene) and polyacetylene, we infer that excessive cross-linking does not occur for polyacetylene by oxidation or heating.

We now consider the effect of preheating polyacetylene at 573 K on the product yields and distributions of subsequent flash pyrolysis at high temperatures (Table 7.4). During preheating three types of chemical changes could occur. First, the dominant products requiring low energy of activation would be formed and volatilized so that their yields are decreased in the subsequent pyrolysis. Such is the case for benzene; preheating for 120 min at 572 K lowered its yield threefold for flash pyrolysis. Second, the major products, such as methane and propane, needing high energy of activation and multiple electron–proton exchanges would be produced in increased yields in the subsequent pyrolysis. The important implication is that those reactions that may occur during the preheating treatment do not in any way impede the facile electron–proton exchange postulated for the formation of proton-enriched products for the flash pyrolysis afterwards. Third, if extensive cross-linking reactions of the types shown in Eqs. (7.57–7.59) occurred during the preheating, then substituted cyclohexene, cyclopentene, cyclohexadiene, and cyclopentadiene would be produced. Based on the sensitivity of the GC–MS technique, they were not produced in amounts exceeding the level of detection of <0.11 wt %. As electron–proton exchange must take place with the preheating treatment, there would also be proton depletion elsewhere. Yet the proton depletion does not inhibit further electron–proton exchange processes. This would mean either the formation of $C \equiv C$ bonds or cross-linking by aromatization, but not reactions like those of Eqs. (7.57)–(7.62), which would produce barriers for facile electron–proton exchange.

We have noted that under optimum isomerization conditions an annealing effect was observed so that the crystalline order was enhanced. This would be hardly possible if polyacetylene is highly cross-linked.

Finally, at least for the polyacetylene molecules in the crystalline domain, one can calculate the closest approach of carbon atoms on adjacent chains from the crystal structure (Sections 3.5 and 3.6). The distances are ~4.4 Å in both *cis*- and *trans*-polyacetylene. If the cross-linking reactions of

Eqs. (7.57)–(7.61) were to occur, the carbon atoms must be brought to a nearness of less than 1.5 Å and bond angles must be decreased in most instances. These events are unlikely to occur with the constraints of the crystalline order. However, oxidative cross-linking [Eq. (7.62)] may overcome these restrictions.

In conclusion, many of the physical chemical changes attributed to the cross-linking reaction may be attributed to other causes such as crystallization, annealing, and changes in crystalline structure. The principal result of heating and autoxidation under moderate conditions is chain scission. At elevated temperatures, formation of benzene, electron–proton exchange to enrich chain ends with protons, and formation of $C\equiv C$ (catacondensed) or aromatic units become the dominant processes.

Chapter 8

Doping

8.1 INTRODUCTION

Pristine *cis-* and *trans-*polyacetylenes have room-temperature conductivities of 10^{-10} and 10^{-5} $(\Omega\ cm)^{-1}$, respectively. The conductivity can be raised to about 10^3 $(\Omega\ cm)^{-1}$ by proper doping. Consequently, doping constitutes an important part of the technology associated with polyacetylene.

The various methods used to dope polyacetylene are described in Section 8.2. Doping with vapor is the earliest and also the most common method used for *p*-type dopants. The solution method is the one of choice for *n*-type doping. Polyacetylene can be *p*-doped with salts containing oxidizing cations such as NO_2^+, NO^+, Ag^+, and Fe^{3+}. An interesting photoinitiated method was developed at IBM, which photolyzes triarylsulphonium or diaryliodonium salts to oxidize and dope the polyacetylene. The most elegant method of doping is through electrolysis. The precise rate and level of doping can be controlled by the applied voltage and current passed.

The chemical structures of dopant species are discussed in Section 8.3. Iodine was shown to exist as I_3^- and I_5^- in polyacetylene by Mössbauer and Raman spectroscopies. The relative amounts of the two species depend on the level of doping; more I_5^- is found in heavily doped polymer and more I_3^- is present in less heavily doped polyacetylene. The total electric charges on the I_3^- and I_5^- species are -0.83 and -0.77, respectively. There is some disagreement about the nature of AsF_5 dopant in polyacetylene. Depending on experimental conditions and subsequent treatment, it can be $As_2F_{10}^{2-}$, AsF_6^-, $HAsF_6^-$, or even $HAsF_5OH$. The nature of the protonic-acid dopant is yet to be elucidated; the subject is complicated by the fact that the common protonic acid dopants are also strong oxidizing agents. Tables 8.1 and 8.2 list the known dopants and the room-temperature conductivities of the doped polyacetylenes.

Attempts have been made to study the kinetics of doping and the diffusion of dopant as described in Sections 8.4 and 8.5. The efforts were by and large unfruitful because of the complexity of the system. Compensation is discussed in Section 8.6, and the effects of doping on polyacetylene morphology are presented in Section 8.7. The mechanisms of doping are proposed in Section 8.8.

8.2 DOPING METHODS

Both *cis-* and *trans*-polyacetylene can be doped. There are reasons favoring doping of the cis because it is more flexible and mechanically stronger and attains higher ultimate conductivity than the trans polymer. In any event, doping should be performed on freshly prepared polymer. When a new dopant is investigated, the composition of the doped polymer must be precisely determined by elemental analysis and the structure of the dopant species must be elucidated by all available means.

8.2.1 Gaseous Doping

Polyacetylene was first doped by exposure to vapors of electron-attracting compounds such as iodine, bromine, AsF_5, and SbF_5 (Chiang *et al.,* 1977; Shirakawa *et al.,* 1977). Depending on the amount of the polymer, the thickness, the desired rate, and the level of doping, the vapor pressure of the dopant is maintained accordingly. To find the best condition for doping or to dope the sample to a certain intermediate level, we recommend the four-probe apparatus as shown in Fig. 8.1. A strip of polyacetylene film is mounted on the four-probe with the aid of an electrodag. Another larger piece of preweighed film was placed in the bottom of the apparatus.

After the apparatus is thoroughly evacuated and the dopant reaches thermal equilibrium with the bath in which it is immersed, the dopant is admitted into the chamber containing the polymer. In the case of standard

(*a*)

(*b*)

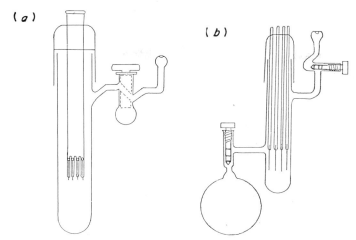

Fig. 8.1 Four-probe apparatus for doping and simultaneous conductivity measurements: (a) type I, (b) type II.

polyacetylene film, doping with iodine requires about 24 hr to approach a constant conductivity value. By careful control of the vapor pressure, polyacetylene can be doped to any specified conductivity, as seen in Fig. 8.2 (Chiang *et al.,* 1978a). It is a general rule that doping should be performed as slowly as possible to avoid inhomogeneous doping. For instance, the vapor pressure of iodine should be kept below 0.1 torr for overnight doping of polyacetylene to the metallic state. To lightly doped in situ polymerized polyacetylene film of less than 1000 Å thick, an iodine vapor pressure of $\sim 10^{-3}$ torr and exposure time of a few seconds was employed.

The larger piece of film in the reactor was used both to determine the doping level and to be used for other measurements. For dopant concentration of $y > 10^{-4}$, the level of doping can be determined by weighing. For very low levels of doping, radioactive isotopes can be used, i.e., $^{125}I_2$. The doped specimen was usually pumped to constant conductivity before handling. An assumption is made here that the two specimens in the chamber are doped to identical contents. This has not been experimentally demonstrated because the electrodag on the film attached to the four-probe prevented the analysis of the doped polymer. The assumption is more likely to be valid the slower the doping process.

Bromine or chlorine doping is complicated by halogenation reactions. Low levels of doping increases conductivity, but conductivity decreases at high levels (Fig. 8.3). For bromine the maximum conductivity corresponds to $y \sim 0.05$. The bromination of polyacetylene and dehydrobromination reactions were described in Section 7.3.

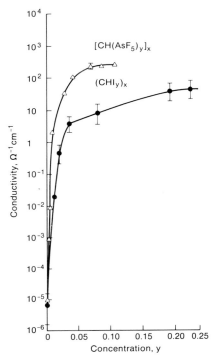

Fig. 8.2   Electrical conductivity of halogen-doped polyacetylene as a function of halogen concentration: (●) iodine, (△) AsF$_5$. [After Chiang *et al.* (1978a).]

Arsenic pentafluoride is a very powerful dopant. It is not easy to dope polyacetylene with AsF$_5$ to a desired concentration and conductivity. Rapid doping tends to result in inhomogeneously doped material and probably causes the discordant reports on its properties. We have until recently used the "slow" doping method. A type-I apparatus is connected to a vacuum line with a 1-liter storage vessel filled to 300–500 torr with AsF$_5$. The cold finger of the vessel was cooled to ∼ 171–175 K, using a methanol–dry-ice slush bath at the temperature of which the AsF$_5$ vapor pressure is about 20–30 torr. With the stopcock connecting the storage vessel to the manifold closed, the stopcock of the vessel was opened for 1–3 sec, allowing AsF$_5$ to fill the small volume between the two stopcocks. This gas was then expanded into the manifold and the doping apparatus. The pressure of AsF$_5$ under these conditions is very low, usually <0.01 torr. This process may be repeated to achieve the desired doping while monitoring the increase in the conductivity of the specimen. Stopcock grease is to be avoided in any part of the vacuum line exposed to AsF$_5$, and the apparatus must be flamed out completely. For

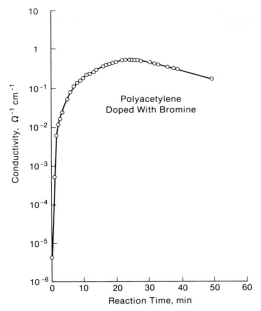

Fig. 8.3 Electrical conductivity of *trans*-polyacetylene during bromine doping. [After Chiang *et al.* (1978b).]

doping with higher than atmospheric pressure of AsF_5, a completely stainless-steel vacuum system is mandatory. Even with this slow doping technique the polymer is doped apparently nonuniformly.

It is now known that the difficulty of uniform doping with AsF_5 is due to gaseous products formed in the process [Section 8.3.3, Eqs. (8.14) and 8.15)], such as AsF_3, which can act to deter infusion of the dopant. Consequently, a better method than the above slow-doping technique is the following. With the polyacetylene film mounted on the four-probe apparatus, AsF_5 at <0.01 torr is admitted for 15 sec and the system is evacuated for 2 min. This process is repeated as many times as is necessary to reach the desired dopant concentration. Evidence for nearly uniform doping according to electrical conductivity and thermopower measurements will be given in Chapter 11. It is our opinion that this "cyclic" doping technique should be used if there is any reason to believe that gaseous products may be formed by the doping reactions.

Doping with H_2SO_4 and other similar substances of low volatility is accomplished in the following manner (Gau *et al.*, 1979). After mounting the sample in either a type-I or type-II four-probe apparatus and pumping to dry the electrodag contacts, the apparatus is opened to argon and ~ 0.5 ml

H_2SO_4 (Fisher Reagent A.C.S., 98%) is introduced in the bottom of the doping chamber. The apparatus is connected to the vacuum line and pumped dynamically for several hours, allowing the H_2SO_4 vapor to contact and dope the sample to the desired concentration. The same technique was used for $HClO_4$ doping. As in the case of AsF_5 doping, we recommend the cyclic doping – evacuation method for the same reasons.

8.2.2 Solution Doping

Solution doping is used mostly to obtain *n*-doped polyacetylene though not limited to it. For instance, iodine doping can be carried out in a toluene or pentane solution of iodine or an aqueous solution of KI_3. A procedure for doping polyacetylene with sodium is described below.

About 50 ml of dry THF, which was refluxed over CaH_2 in a solvent still, was transferred to a Schlenk tube under argon. A piece of sodium metal (coated with paraffin oil) was dipped into pentane to remove the oil and cut into a cube (~ 1.28 g). The sodium was flattened into a sheet with a hammer and, while holding the sheet above the mouth of the Schlenk tube containing the THF, small pieces were cut with scissors and dropped into the tube under argon flow. The suspension was stirred with a magnetic bar, and ~ 6.4 g of naphthalene was introduced into the tube under argon. The solution became deep green-black, characteristic of the sodium naphthalide radical anion, and was stirred overnight to complete solubilization of the components.

The sample to be doped was mounted on the probes of a type-II apparatus. The apparatus was pumped overnight on a vacuum line. Argon was introduced through the ball joint of the four-probe and the Teflon plug of the stopcock attached to the bulb was removed. The sodium naphthalide – THF solution was transferred into the bulb under argon, using a syringe. A small magnetic bar was dropped into the bulb, the stopcock was replaced, and the four-probe was closed off from the argon supply.

The apparatus was connected to a vacuum line, and the doping chamber was pumped for at least 1 hr. The sodium naphthalide – THF solution was subjected to several freeze – pump – thaw cycles. After disconnecting the apparatus from the vacuum line, the bottom of the doping chamber was cooled to 77 K for several seconds and then the stopcock to the bulb was opened. The sodium naphthalide – THF solution was then immediately poured into the doping chamber to cover the sample connected to the probe wires. The initial cooling of the bottom of the chamber was done in order to facilitate the transfer of the solution. The conductivity was monitored initially by a simple ohmmeter (two-probe) and finally by using the four-probe method described previously. For samples studied in this work, the conductivity usually reached saturation within a few minutes.

The sample-doping vessel was washed free of excess sodium naphthalide as follows. After reaching the saturation conductivity of the sample, the bulb was cooled to 77 K, and the solution in the doping chamber was poured into the bulb. The solution in the bulb was stirred vigorously, and pure THF was distilled into the doping chamber by cooling the bottom of the chamber with liquid nitrogen. After warming the THF to room temperature, it was poured back into the bulb and this process was repeated until the THF in the chamber was colorless. The stopcock to the bulb was then closed. The solution could be reused as long as it was still deep green. A light brown solution suggested significant decomposition due to interaction with moisture and/or air, and in this case the solution was discarded by decomposing it in a large quantity of *tert*-butyl alcohol.

In principle the above technique can be used for *n*-doping to low levels by employing a dopant solution of low concentration. This has not been reported because of instability of the *n*-doped polyacetylene. Lightly *n*-doped material would be difficult to characterize because of the effect of slightest contamination.

8.2.3 Doping Assisted by Oxidizing Cation

Salts containing NO^+ or NO_2^+ ions have been demonstrated to be convenient reagents for *p*-type doping. The cation of the salt oxidizes polyacetylene and at the same time introduces anions that stabilize the polycarbocation to produce *p*-doped $(CH)_x$. In the process the oxidizing cation is liberated as the reduced neutral gaseous molecule. For example, SbF_6^- can be introduced into polyacetylene simply by treating a cis polymer film with a $CH_3NO_2-CH_2Cl_2$ solution of appropriate salts. Thus $NO_2^+SbF_6^-$ yields golden, flexible, highly conducting films (Gau *et al.*, 1979). The process can be written

$$(CH)_x + 0.05x NO_2^+ SbF_6^- \rightarrow [CH(SbF_6)_{0.05}]_x + 0.05x NO_2 \qquad (8.1)$$

HSO_4^- ions can be introduced into polyacetylene from a CH_3NO_2 suspension of $NO^+HSO_4^-$ (Rachdi *et al.*, 1981). The stoichometry for the reaction is represented by

$$(CH)_x + 0.12x NO^+HSO_4^- + 0.02x CH_3NO_2 \rightarrow$$
$$[CH(HSO_4)_{0.12}(CH_3NO_2)_{0.02}]_x + 0.12x NO \qquad (8.2)$$

In this instance not all the solvent nitromethane was removable.

The doping reaction is more complicated in the case of $NO^+SbCl_6^-$. Immersion of a polyacetylene film in a solution of $NO^+SbCl_6^-$ rendered the polymer highly conducting. Analysis indicated the composition to be either $[CH(SbCl_8)_{0.0095}]_x$ or $[CH(SbCl_6)_{0.009}]_x$, depending on slight variations in experimental procedure. The $SbCl_8^-$ species may be formulated as consisting of a Cl_3^- ion coordinated to $SbCl_5$. In the vapor-phase doping at room temperature, the product has a composition of $[CH(SbCl_5)_{0.022}]_x$. Apparently, there is predissociation of $NO^+SbCl_6^-$

$$NO^+SbCl_6^- \rightarrow NOCl + SbCl_5 \qquad (8.3)$$

and the doping by $SbCl_5$ proceeded analogously to AsF_5.

A complex reaction occurs between polyacetylene and a solution of $NO_2^+BF_4^-$ in neat CH_3NO_2 to yield highly conducting films, when analyzed were found to have the composition $[CH(BF_2)_{0.09}]_x$. The other products are presumably CH_3F, NO_2F, and NO_2. The dopant ion is postulated as BF_2^-.

p-Type doping can also be accomplished with a silver salt. Clark *et al.* (1978) obtained $[(CH)^{y+}(ClO_4^-)_y]_x$ by dipping a polyacetylene film into solutions of silver perchlorate to obtain a material with a conductivity of 3 $(\Omega \text{ cm})^{-1}$ and a thermopower of 20 $\mu V/K$. The reaction is believed to be

$$(CH)_x + yAgClO_4 \rightarrow [(CH)^{y+}(ClO_4^-)_y]_x + yAg^0 \qquad (8.4)$$

where $y = 0.006-0.018$. This method is inferior to others because of the difficulty of removing the metallic silver.

The cation can also be a transition-metal ion in its higher oxidation state. We found that polyacetylene is readily doped by immersing the polymer in a saturated solution of anhydrous $Fe(ClO_4)_3$ in dry toluene with stirring. After doping to the desired conductivity, the films were removed, washed thoroughly with toluene and CH_2Cl_2, then dried in vacuo overnight (Reynolds *et al.*, 1982). The room temperature conductivity for $(CH)_x$ films, doped to saturation with $Fe(ClO_4)_3$, is greater than 500 $(\Omega \text{ cm})^{-1}$. The thermopower coefficient was found to be $+30 \mu V/K$. The undoped films have a silvery appearance and upon doping turn a brilliant gold. The conductivity increase can be monitored with time, reaching its limiting value after 2–3 hr, when the dopant has diffused completely into the film. Interestingly, experiments carried out in either CH_3OH or H_2O as the solvent were unsuccessful. Presumably the solvated ferric ion does not have the redox potential needed to oxidize polyacetylene. Figure 8.4 shows the dependence of conductivity on dopant level from 2–6 mole percent as determined by weight uptake experiments. The conductivity can be controlled by controlling the dopant level, though at low levels the dopant distribution is probably inhomogeneous due to incomplete diffusion. The reproducibility of the data is shown for different experiments in the figure.

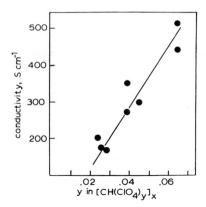

Fig. 8.4   Variation of σ with y for polyacetylene doped with $Fe(ClO_4)_3$.

Elemental analysis of a $Fe(ClO_4)_3$-doped film is C, 76.56%; H, 6.62%; Cl, 7.38%; O, 9.44%; Fe, <0.05%. The theoretical values are: C, 77.48%; H, 6.46%; Cl, 5.73%; O, 10.32% for $C_{1.00}H_{1.04}Cl_{0.032}O_{0.092}$. The Fe content is less than 6.5×10^{-5} mole Fe per mole carbon, indicating that the dopant species is not a transition-metal complex. The doping reaction probably occurs by the following mechanism:

$$(CH)_x + xyFe(ClO_4)_3 \rightarrow [(CH)^{y+}(ClO_4)_y^-]_x + xyFe(ClO_4)_2 \qquad (8.5)$$

A sample that by analysis was found to have the composition $[CH(ClO_4)_{0.045}]_x$ has a Dysonian EPR spectrum with a A/B ratio of 8/1, indicative of metallic electrons. The calculated skin depth of 29 μm for films with a conductivity of 300 $(\Omega\ cm)^{-1}$ is less than the 100-μm thickness of our films, consistent with these results.

8.2.4   Photoinitiated Doping

Triarylsulphonium and diaryliodonium salts are known to decompose when exposed to ultraviolet radiation to produce the corresponding protonic acids. Because some protonic acids have been shown to be effective dopants for polyacetylene, Clarke et al. (1981) impregnated polyacetylene films with triphenylsulfonium hexafluoroarsenate and exposed the sample to UV irradiation. The polymer rapidly gained conductivity by ten orders of magnitude to $\sigma \simeq 2\ (\Omega\ cm)^{-1}$ in a few minutes of irradiation. The conductivity depends on the concentration of the salt in CH_2Cl_2 introduced by the impregnation step. Also, the unchanged salt can be subsequently removed by solvent extraction. This method affords a convenient way to prepare samples of a controlled level of doping. Furthermore, a mask can be used to effect doping only in the exposed area, and the salt under the masked area can be removed after irradiation. This photoinitiated doping has obvious implications for lithographic applications in microelectronics.

8.2.5   Electrochemical Doping

Electrochemical doping is both simple and elegant. A *cis*-polyacetylene film ($1 \times 3 \times 0.01$ cm) can be used as the anode in the electrolysis of aqueous 0.5 M KI solution. At a potential of 9 V, the initial current was only 1 mA because of the low conductivity of the pristine polymer. As the polymer became doped, its conductivity increased. In 30 min the current increased to 43 mA and the polymer was doped to the metallic state with the composition $(CHI_{0.07})_x$ and $\sigma_{RT} = 9.7$ $(\Omega$ cm$)^{-1}$.

In the electrolysis of 0.5 M $Bu_4N^+ClO_4^-$ in CH_2Cl_2 at 9 V, 1 hr of doping gave $[CH(ClO_4)_{0.0645}]_x$. The current increased from 0.95 to 3.4 mA; the material has a room temperature conductivity of 970 $(\Omega$ cm$)^{-1}$. Similar results were obtained by the electrolysis of $Bu_4N^+SO_3CF_3^-$, the doped polymer is assumed to contain $SO_3CF_3^-$ ions. In the electrolysis of a CH_2Cl_2 solution of $(n - Pr)_3NH^+AsF_6^-$, the product was found by analysis to be $[CH(AsF_4)_{0.077}]_x$.

Spontaneous *n*-doping of polyacetylene occurs in the electrolysis of a 1-M solution of $Bu_4N^+ClO_4^-$ in THF with the polymer film as the cathode. Levels of doping can be controlled by the applied voltage. $[CH(Bu_4N)_y]_x$ materials of various levels of doping were obtained at voltages of 1.50 ($y \approx$ 1%), 1.35 ($y \approx$ 3%), 1.30 ($y \approx$ 5%), and 1.25 V ($y \approx$ 8%).

Electrochemical doping offers some distinct advantages. Doping level can be precisely controlled by the number of coulombs of current passed. The voltage versus charge ($V-Q$) curve can be determined independently. A polyacetylene film of known weight can be electrochemically doped at a given voltage and allowed to stand for one day to permit diffusion of the counterion from the surface to the interior of the polyacetylene films while the potential is recorded. The doped film is electrochemically returned to 2.5 V, characteristic of neutral polyacetylene, and the charge liberated is measured. From such $V-Q$ curves the relationship between V and y is known. Second, the approach to equilibrium of doping can be ascertained by the current level. Third, electrochemical undoping is cleaner than chemical compensation, forming no chemical products requiring removal, and the sample is returned to a neutral though not pristine state. The method opens up possibilities of doping with species that cannot be introduced by any obvious conventional chemical means. Both *p*- or *n*-type doping can be achieved by the electrochemical process.

The electrochemical technique has not yet been used to obtain lightly doped polyacetylene. The technique is certainly capable of doping polyacetylene to any level and should be further investigated in this regard.

8.3   DOPANTS AND THE NATURE OF DOPANT SPECIES

8.3.1   Dopants for Polyacetylene

Polyacetylene has been *p*-doped by a large variety of compounds and *n*-doped by a smaller number of ions (MacDiarmid and Heeger, 1979). It is common to report only the limiting room-temperature conductivity of samples doped to the maximum level. For a few systems, conductivity as a function of dopant concentration was determined and sometimes the temperature dependency as well. In Tables 8.1 and 8.2 the types of doped polyacetylene reported in the literature are summarized.

8.3.2   Nature of Iodine in Doped Polyacetylene

The compositions of the doped polyacetylene given in Table 8.1 cannot be taken literally in all cases as if the structure of the dopant species were known. They represent, in some instances, merely the empirical formula based on elemental analysis. In many cases only the increase in weight after doping was determined, and the dopant was assumed to have its original structure except for a change in its oxidation state.

Among the dopants for polyacetylene, only the nature of iodine in the polymer is known with some certainty. It exists largely as I_3^- and I_5^-. Quantitative estimates were obtained by Matsuyama *et al.* (1981) and by Kaindl *et al.* (1982) using ^{129}I-Mössbauer spectroscopy. The gamma source used as $^{66}Zn^{129m}Te$ or $^{119}Sn^{129m}Te$ ($t_{1/2}$ for ^{129m}Te is 33.5 days). The absorber is the ^{129}I-doped polyacetylene. The 27.7-keV nuclear–gamma rays of the $\frac{5}{2}-\frac{7}{2}$ transition were detected by a large-area intrinsic Ge diode. The sample temperature was maintained at 4.2 K. A typical Mössbauer spectrum is shown in Fig. 8.5, which contains three subspectra. The spectra of the

TABLE 8.1

p-Type Dopants for Polyacetylene[a,b]

Sample	Conductivity [$(\Omega \text{ cm})^{-1}$ at 298 K]
cis-[CH(HF)$_{0.115}$]$_x$[c]	4.9
trans-[CH(HBr)$_{0.04}$]$_x$	7×10^{-4}
trans-(CHCl$_{0.02}$)$_x$[d]	1×10^{-4}
trans-(CHBr$_{0.23}$)$_x$	4×10^{-1}
cis-[CH(ICl)$_{0.14}$]$_x$	5×10^{1}
cis-(CHI$_{0.30}$)$_x$	5.5×10^{2}
trans-(CHI$_{0.20}$)$_x$	1.6×10^{2}
cis-[CH(IBr)$_{0.15}$]$_x$	4.0×10^{2}
trans-[CH(AsF$_5$)$_{0.10}$]$_x$	4.0×10^{2}
cis-[CH(AsF$_5$)$_{0.10}$]$_x$[c]	1.2×10^{3}
cis-[CH$_{1.1}$(AsF$_6$)$_{0.10}$]$_x$	$\sim 7 \times 10^{2}$
cis-[CH(AsF$_4$)$_{0.077}$]$_x$	2.0×10^{2}
cis-[CH(SbF$_6$)$_{0.05}$]$_x$	4.0×10^{2}
cis-[CH(SbF$_5$)$_{0.06}$]$_x$[c]	5×10^{1}
cis-[CH(SbCl$_6$)$_{0.009}$]$_x$	1×10^{-1}
bis-[CH(SbCl$_8$)$_{0.0095}$]$_x$	1×10^{1}
cis-[CH(SbCl$_5$)$_{0.022}$]$_x$	2
cis-[CH(BF$_2$)$_{0.09}$]$_x$	1×10^{2}
cis-[CH(IF$_{5.63}$)$_{0.096}$]$_x$	1.5×10^{2}
cis-[CH(SO$_3$F)$_y$]$_x$	7×10^{2}
cis-[CH(ClO$_4$)$_{0.0645}$]$_x$	9.7×10^{2}
cis-[CH$_{1.11}$(AsF$_5$OH)$_{0.011}$]$_x$	$\sim 7 \times 10^{2}$
cis-[CH$_{1.058}$(PF$_5$OH)$_{0.058}$]$_x$[e]	$\sim 3 \times 10^{1}$
cis-[CH(HSO$_4$)$_{0.12}$(CH$_3$NO$_2$)$_{0.02}$]$_x$	4.3×10^{2}
cis-[CH(H$_2$SO$_4$)$_{0.106}$(H$_2$O)$_{0.070}$]$_x$	1.2×10^{3}
cis-[CH(HClO$_4$)$_{0.127}$(H$_2$O)$_{0.297}$]$_x$	1.2×10^{3}
cis-[CH(CF$_3$SO$_3$H)$_y$)$_x$[f]	—
cis-[CH(XeOF$_4$)$_{0.025}$]$_x$	5×10^{1}

[a]Cis or trans refers to the principal isomeric composition before doping. [After Chiang *et al.* (1978c).]

[b]Composition by elemental analysis unless stated otherwise.

[c]Composition by weight uptake.

[d]Dopant was (SO$_3$F)$_2$; no composition or analysis was given. Anderson *et al.* (1978).

[e]By electrochemical doping.

[f]Resistivity as a function of *y* and temperature given, but not conductivity. Rolland *et al.* (1980c).

TABLE 8.2

n-Type Dopants for Polyacetylene[a,b]

Sample	Conductivity [$(\Omega$ cm$)^{-1}$ at 298 K]
cis-$[Li_{0.30}(CH)]_x$	2.0×10^2
cis-$[Na_{0.21}(CH)]_x$	2.5×10^1
cis-$[K_{0.16}(CH)]_x$	5.0×10^1
trans-$[Na_{0.28}(CH)]_x$	8.0×10^1
cis-$[CH(Bu_4N)_{0.08}]_x{}^c$	—

[a]Cis or trans refers to principal isomeric condition before doping. [After Chiang *et al.* (1978c).]
[b]Composition by weight uptake.
[c]Electrochemical doping.

heavily doped polymer show no evidence of I^- ions, which would, if present, give rise to a sharp singlet line at around -0.5 mm sec^{-1}.

Subspectra 1, 2, and 3 have isomer shift (and quadrupole splitting) values of 1.39 ± 0.04 mm sec^{-1} (-1731 ± 19 MHz), 0.345 ± 0.025 mm sec^{-1} (-885 ± 19 MHz), 0.34 ± 0.04 mm sec^{-1} (-1244 ± 16 MHz), respectively. For a linear I_3^- ion the central atom is neutral and each terminal

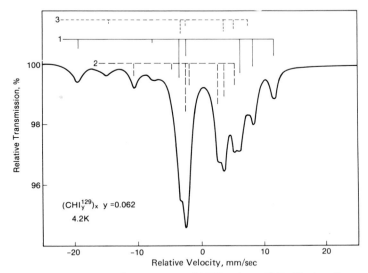

Fig. 8.5 Mössbauer absorption spectrum of $(CHI_{0.062})_x$ at 42 K. The bar diagrams show the positions and intensities of the lines of the three subspectra (see text). [After Kaindl *et al.* (1982).]

atom has a charge of 0.5. Therefore, I_3^- is expected to contribute toward subspectra 1 and 2 with relative intensities $1:2$. The species is labeled as

$$I(\alpha)-I(\beta)-I(\alpha) \tag{8.6}$$

The linear $D_{\infty h}$ I_5^- ion is

$$I(a)-I(b)-I(c)-I(b)-I(a) \tag{8.7}$$

The charges have been estimated as $-\frac{1}{3} \leqslant Z_a \leqslant 0$, $Z_b = 0$ and $-\frac{1}{3} \leqslant Z_c \leqslant -\frac{2}{3}$. Thus, I(b), I(c), and I(a) contribute to subspectra 1, 2, and 3, respectively, with relative intensities $2:1:2$. In other words, the Mössbauer parameters of $I(\beta)$ are very similar to $I(b)$ and those of $I(\alpha)$ are nearly the same as $I(c)$. Consequently, the subspectra intensities A_i are related to relative amounts of the polyiodide ions. If we use braces to denote mole fractions, then

$$A_1 : A_2 : A_3 = (\{I_3^-\} + 2\{I_5^-\}):(2\{I_3^-\} + \{I_5^-\}):(2\{I_5^-\}). \tag{8.8}$$

Furthermore, it is assumed that $\{I_3^-\} + \{I_5^-\} = 1$. Equation (8.8) is thus simplified to

$$A_1 : A_2 : A_3 = (2 - \{I_3^-\}):(1 + \{I_3^-\}):(2 - 2\{I_3^-\}). \tag{8.9}$$

This enables a determination of the relative amounts of I_3^- and I_5^- present in $(CHI_y)_x$, as shown in Fig. 8.6.

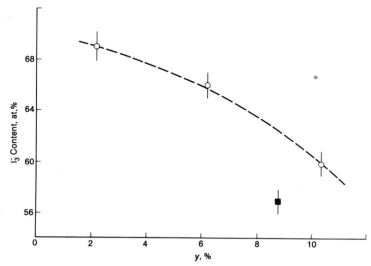

Fig. 8.6 I_3^- content (in at. %) as a function of y for $(CHI_y)_x$: (○) *cis*-polyacetylene, (■) *trans*-polyacetylene. [After Kaindl *et al.* (1982).]

TABLE 8.3

Distribution of I_3^- and I_5^- Ions in $[CHI_y]_x$

Sample	Mole fraction		References
	I_3^-	I_5^-	
cis-$[CHI_{0.022}]_x$	0.69	0.31	a
cis-$[CHI_{0.062}]_x$	0.67	0.33	a
cis-$[HCI_{0.103}]_x$	0.60	0.40	a
$cis[CHI_{0.20}]_x$	0.21	0.69	b
cis-$[CHI_{0.23}]_x$	0.07	0.93	b
$trans$-$[CHI_{0.088}]_x$	0.57	0.43	a

[a] Kaindl *et al.* (1982).
[b] Matsayama *et al.* (1981).

Table 8.3 summarizes the Mössbauer results. The I_5^- ions increase rapidly with y reaching 93 mole % in cis-$[CHI_{0.23}]_x$. Evacuation removes I_2 to give cis-$[CHI_{0.20}]_x$, which causes large reduction in I_5^- ions and increases I_3^- ions to 21 mole %. In cis-$[CHI_{0.022}]_x$, 69 mole % of the dopant ions are I_3^-. There is no I_2 present in the most heavily doped sample. I_2 would contribute toward subspectrum 1 and causes $A_1/A_2 > 2$, which is contrary to the observed ratio, which is less than 2.

Intense Raman bands were observed in the low-frequency region for iodine-doped polyacetylene (Hsu *et al.*, 1978, Lefrant *et al.*, 1979). One band located at $105-107$ cm^{-1} was assigned to the symmetric stretching vibration in I_3^-, based on similar bands in other known I_3^--containing substances. The other band at $150-160$ cm^{-1} is close to that expected for asymmetric stretching vibration of I_3^-. However, vibrations of I_5^- should also appear in this frequency region; Me_4NI_5 shows a band at $143-164$ cm^{-1}. Assignment of the $150-160$-cm^{-1} band to I_5^- was favored by these authors. However, Raman spectra cannot be used to determine the relative amounts of I_3^- and I_5^- in the polymer because the ratio of the intensities of the two bands is very sensitive to the excitation wavelength.

X-ray photoelectron spectroscopy (XPS) has been used to study iodine-doped polyacetylene. In these measurements the monochromatic x-ray beam from the AlK_α radiation was incident at 40° to the sample surface and the ejected photoelectron was detected normal to the sample surface. Because of the low energy of the electrons, the technique probes only the near surface region of the specimen. Thus about 63% of the XPS signal for 1.4-keV electrons originates within the top 21 Å of the sample, 82% from about 42 Å, etc. The iodine $3d_{3/2}-3d_{5/2}$ core level spectrum of $(CHI_y)_x$ is shown in Fig. 8.7 (Hsu *et al.*, 1978). The iodine $3d_{5/2}$ levels can be deconvoluted into two components at 620.6 and 619 eV with a half width of 1.4 eV.

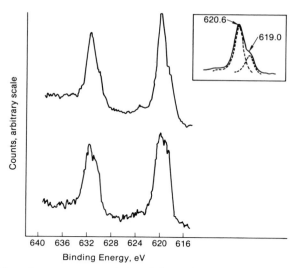

Fig. 8.7 Photoelectron spectra of the $I3d_{3/2}$–$I3d_{5/2}$ core level: upper curve, *trans*-$(CHI_{0.22})_x$; lower curve, *cis*-$[CHI_{0.22}]_x$; insert, deconvolution of the $I3d_{5/2}$ level for *trans*-$(CH)_x$. [After Hsu *et al.* (1978).]

The lower binding-energy component was assigned to I_3^-, whereas the higher binding-energy component appears to be associated with a less negatively charged iodine species such as I_5^-. Salaneck *et al.* (1980) found that the 620.6-eV peak is about twice the height of the 619-eV peak in $(CHI_{0.28})_x$. Upon pumping, the two peaks become comparable in intensity.

Further support for the presence of I_3^- and I_5^- came from the mass spectrometric study of Allen *et al.* (1979). *trans*-Polyacetylene was saturation doped with iodine, and the sample was inserted directly into the ion source of a Hewlett–Packard 5985 A quadrupole mass spectrometer. Figure 8.8a shows the amount of I_2^+ detected as the specimen was heated from 303 to 643 K at a rate of 20 K min^{-1}. There was a sharp peak of I_2^+ at ~ 343 K and a very broad peak at higher temperatures. This doublet indicates the presence of two iodine species. When the iodine-doped polyacetylene was heated at 303 K for five min and then examined as above, the low-temperature I_2^+ peak was not observed (Fig. 8.8b). The latter reappears for a sample heated at 303 K for 5 min and redoped with iodine (Fig. 8.8c). The results may be interpreted as the occurence at 303 K of the following equilibrium:

$$[CH(I_5^-)_y]_x \overset{\Delta}{\rightleftharpoons} [CH(I_3^-)_y]_x + yxI_2 \tag{8.10}$$

At higher temperatures the I_3^- ions were eliminated. Similar results have been observed for bromine- and chlorine-doped polyacetylene.

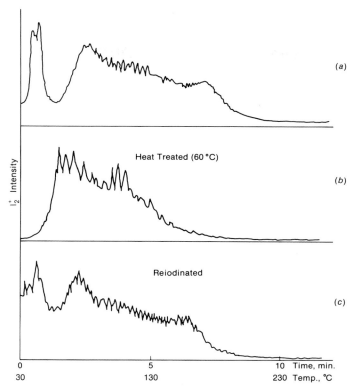

Fig. 8.8 Profiles of I_2^+ intensities versus temperature for heavily doped $(CHI_y)_x$. (a) Fresh doped sample, (b) heated at 303 K for 5 min, (c) sample b after redoping. [After Allen *et al.* (1979).]

The charge on each iodine atom in a polyiodide ion can be related to the quadrupole coupling constant from Mössbauer spectroscopy for pure σ bonding,

$$eq^{mol} = h_{p_z} eq^{at} \qquad (8.11)$$

where h_{p_z} denotes the number of holes in the 5_{p_z} orbital of iodine, the charge is $z = h_{p_z} - 1$, and eq^{at} is the electric-field gradient due to one p_z hole (for a free iodine atom with $5s^2 5p^5$ configuration). For ^{129}I, $e^2 q^{at} Q(7/2) = -1608$ MHz. From those observed quadrupole coupling parameters cited above, Kaindl *et al.* (1982) estimated the charges on $I(\alpha)$ and $I(\beta)$ in I_3^- to be -0.45, and $+0.01$, respectively. The total charge on I_3^- is -0.83. Similarly, the charges on $I(a)$, $I(b)$, and $I(c)$ in I_5^- were found to be -0.23, $+0.07$, and -0.45, respectively, for a total charge of -0.77. In other words, the efficiency of charge transfer is high and each dopant injects about 0.8 charged carriers into polyacetylene.

Charge transfer is expected to shift the carbon $1s$ core binding energy in XPS. Hsu *et al.* (1978) and Salaneck *et al.* (1979a,b, 1980) found a chemical shift for the C($1s$) peak energy of 0.2 ± 0.2 eV. From this they estimate a charge transfer of about 0.03 electrons per carbon atom for $[CH(I_3^-)_{0.03}]_x$, which corresponds to one charge carrier per I_3^- dopant. However, if one takes into consideration that there are approximately equal numbers of I_3^- and I_5^- ions in these specimens (Fig. 8.6), then the sample is approximately $[CH(I_3^-)_{0.025}(I_5^-)_{0.015}]_x$ and there is about 80% charge transfer per dopant, in good agreement with the Mössbauer results.

8.3.3   Nature of AsF$_5$ in Doped Polyacetylene

Because AsF_5 is the most commonly employed dopant, there has been intense interest in the nature of this dopant in polyacetylene. This knowledge would also be helpful in understanding the doping reactions of similar compounds, such as BF_3, SbF_5, and $SbCl_5$. There have been some disagreements about the form of AsF_5 that exists in doped polyacetylene. This is not unexpected because arsenic has $+3$ and $+5$ stable oxidation states and AsF_5 is extremely reactive. Disproportionation of AsF_5 and its reaction with H_2O to form HF are some of the complications.

The group at the University of Pennsylvania had prepared $[CH(AsF_5)_y]_x$ samples with y ranging from 0.086 to 0.104. Elemental analysis was quantitative for $As:F = 1:5.0$. Therefore, the first step of the doping is thought to be

$$(CH)_x + yAsF_5 \rightarrow [CH^{y+}(AsF_5^- \cdot)_y]_x \qquad (8.12)$$

However, EPR does not show any paramagnetism due to the $AsF_5^- \cdot$ radical anion. The authors proposed that the dopant species is the diamagnetic $As_2F_{10}^{2-}$ by dimerization of the $AsF_5^- \cdot$:

$$[CH^{y+}(AsF_5^- \cdot)_y]_x \rightarrow [CH^{y+}(As_2F_{10})_{y/2}^{2-}]_x \qquad (8.13)$$

Even though $As_2F_{10}^{2-}$ has not been reported before, postulation of its existence was deemed plausible because of the stabilization influence of the large anion by the polycation of $(CH)_x$, as in the case of halogen-doped polyacetylene. MacDiarmid and Heeger (1979) supported this hypothesis by cryogenic pumping of AsF_5-doped polyacetylene at 298 K or below. Only AsF_5 was liberated, with traces of HF and AsF_3, consistent with the elemental analyses of the samples.

However, when the above specimen was warmed to 373 K, its composition changed to $[CH(AsF_6)_{0.09}]_x$. During the heating, mostly AsF_3 and HF were evolved between 333 and 373 K. A reversible equilibrium was proposed:

$$[(CH)^{+y}(As_2F_{10})^{2-}_{y/2}]_x \rightleftharpoons [(CH)^{+2/3y}(AsF_6)_{2/3y}]_x + \frac{xy}{3}AsF_3 \qquad (8.14)$$

During this heat treatment the room temperature conductivity was unchanged because the number of positive carriers was reduced only by $\frac{1}{3}$, i.e., from those associated with $(CH)^{+y}$ lowered to $(CH)^{+2/3y}$.

Clarke *et al.* (1979) and Clarke and Street (1979) used X-ray absorption of synchrotron radiation by AsF_5-doped polyacetylene to show that the K-shell preabsorption edge in the specimen is shifted by 1.4 eV with respect to AsF_5 vapor. This shift is identical to that observed by Bartlett *et al.* (1978) for AsF_6^- in graphite. In addition, there is a very weak shoulder located 6.7 eV below the main absorption in the doped polyacetylene that corresponds to the peak identified by Bartlett *et al.* as AsF_3. Based on this evidence, it was proposed that the dopant species is AsF_6^- (Clarke *et al.*, 1979):

$$2(CH)_x + 3xyAsF_5 \rightarrow 2[CH^+(AsF_6)^-_y]_x + xyAsF_3 \qquad (8.15)$$

They attributed the weak signal of AsF_3 absorption of the hard X-ray to the low affinity of polyacetylene for AsF_3 and loss of a fraction of AsF_3 during evacuation.

Weber *et al.* (1981) doped many samples of polyacetylene with AsF_5 and found the As:F ratios to vary from 1:5.0 to 1:6.0. The mass spectra of the volatile species that could be pumped from these samples was mainly AsF_5, and the very small amount of AsF_3 observed can be attributed to the decomposition of AsF_5 by means of wall reactions in the inlet system of the mass spectrometer. The samples apparently contain an admixture of $As_2F_{10}^{2-}$ and AsF_6^- species. Fortunately, the electrical properties of AsF_5-doped polyacetylene are not sensitive to the As:F ratio. In other words, in the metallic regime, mobility of the positive carriers is not much affected by the nature of the anion, whether it is $As_2F_{10}^{2-}$ or AsF_6^-.

Other complications may be encountered in AsF_5 doping. Any moisture present in the system would produce HF, which could alter the nature of the dopant species. For instance, when $[CH(AsF_5)_y]_x$ is treated with $\sim 50\%$ aqueous HF or with AsF_5 containing HF, the following transformation apparently occurs:

$$[CH(AsF_5)_y]_x + xyHF \rightarrow [CH(HAsF_6)_y]_x \qquad (8.16)$$

This form of doped polyacetylene was apparently obtained when the polymer film was very slowly doped at an AsF_5 pressure of less than 0.1 torr at room temperature for 2–4 days. The dopant species is apparently $HAsF_6$ inasmuch as the material slowly evolves HF and AsF_3 at room temperature, namely,

$$[CH(AsF_5)_y]_x \rightarrow 2xyHF + xyAsF_3 + C_xH_{x-2xy} \qquad (8.17)$$

The HF produced would then react according to Eq. (8.16). In this process hydrogens are abstracted from the polymer and the carbon–carbon single bonds are converted to carbon–carbon double bonds or the carbon–carbon double bonds changed to carbon–carbon triple bonds.

In the presence of moisture, another possible dopant $HAsF_5OH$ is formed:

$$AsF_5 + H_2O \rightarrow HAsF_5OH \qquad (8.18)$$

This occurs even when $[CH^{y+}(As_2F_{10})^{2-}{}_{y/2}]_x$ is pumped dynamically at room temperature in the vacuum system for 24–40 hr. The resulting substance is found by analysis to be $[CH(HAsF_5OH)_y]_x$ ($y = 0.018$–0.11). Apparently, the minute quantities of water present in the vacuum system is sufficient to react with the $[CH^{y+}(As_2F_{10})^{-}_{y/2}]_x$ to give $HAsF_5OH$.

The degree of charge transfer upon AsF_5 doping was estimated from the shift in the C(1s) core level XPS spectrum. Relative to the undoped polyacetylene, the C(1s) peak moved to higher binding energies by 0.6 ± 0.2 eV. The direction of the shift corresponds to charge transfers from the polymer to AsF_5 dopant. From Siegbahn's experimental tabulations, there is a 5.6 eV/electron chemical shift corresponding to 0.11 ± 0.01 electron per carbon atom in the $[CH(AsF_5)_{0.11}]_x$ sample. This result implies the transfer of one electron from polyacetylene to each AsF_5 moiety upon doping.

8.3.4 Nature of Protonic Acids in Polyacetylene

Polyacetylene doped with protonic acids attains conductivity as high as by any other type of dopant ($\sim 10^3 \ \Omega^{-1} \ cm^{-1}$). The exact nature of the dopant species is as yet unknown. However, the following addition reaction does not occur:

$$(8.19)$$

This reaction would interrupt the backbone π system with saturated units. Also, treatment with an amine would result in the formation of a quaternary ammonium compound on the polyacetylene, and the parent compound cannot be regenerated by this compensation,

$$+ \text{Et}_3\text{N} \longrightarrow \qquad\qquad + \text{H}_2\text{O} \qquad (8.20)$$

which is contrary to experimental observations. Finally, the reactions shown in Eqs. (8.19) and (8.20) for $(CD)_x$ would result in isotopic exchange, which was not observed.

The following observations made by MacDiarmid (1982) showed that there is only very weak complexing of the hydronium ion with the polymer. A free-standing cis-rich $(CD)_x$ was doped with $HClO_4$ vapor for ~ 10 hr at 298 K to a metallic state of composition $[(CD)(H_3O^+ClO_4^-)_{0.094}]_x$. Compensation with triethylamine vapor for six days converted the material to $[(CD)(Et_3NH^+ClO_4^-)_{0.094}]_x$. The conductivity decreased about eight orders of magnitude to $\sim 10^{-5}$ $(\Omega\ cm)^{-1}$. More than 80% of the $Et_3NH^+ClO_4^-$ can be removed from the polymer by extraction with ether and was found not to contain any $Et_3ND^+ClO_4^-$. Therefore, there is no isotopic exchange. Infrared showed the compensated polymer to have the trans structure; protonic-acid doping followed by chemical compensation results in isomerization of *cis*-polyacetylene to *trans*-polyacetylene. Possible structures for the $HClO_4$-doped polyacetylene may be

$$\text{or} \qquad\qquad\qquad\qquad (8.21)$$

One unfortunate aspect of protonic-acid doping so far is that these acids are also strong oxidizing agents. MacDiarmid (1982) now believes that with perchloric acid the doping reaction involves oxidation of polyacetylene

$$9HClO_4 + 8(CH) \longrightarrow 8(CHClO_4) + HCl + 4H_2O \qquad (8.22)$$

Furthermore, we found that $HClO_4$-doped polyacetylene often gave analysis with O/Cl between 5 and 6. Thus, true protonic acid doping not complicated by *p*-type doping has not yet been established.

8.4   DOPING OF *cis*- AND *trans*-POLYACETYLENES

It was established in Section 5.3 that chemical or electrochemical *p*- or *n*-doping of *cis*-polyacetylene results in isomerization. Essentially the same metallic material was obtained upon heavy doping of either the cis or the trans polymer. Because the latter process does not involve structural isomerization, doping of *trans*-polyacetylene is expected to proceed more rapidly and uniformly than the doping of *cis*-polyacetylene.

We have compared the kinetics of AsF_5 doping of *cis*- and *trans*-polyacetylene, the latter obtained by heating the former at 383 K for 5.5 hr. The results are shown in Fig. 8.9. The polymers were mounted on four probes; resistivity was monitored as the specimens were exposed to AsF_5. The AsF_5 reservoir was kept at 143 K so that the doping proceeded as slowly as possible. The figure shows that the increase in conductivity of trans material

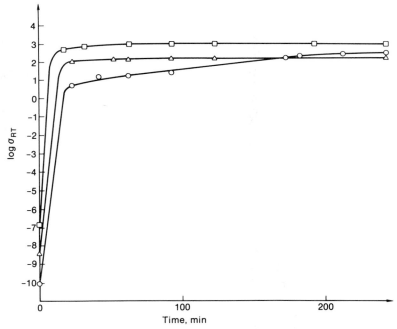

Fig. 8.9   Comparison of the rate of AsF_5 doping as measured by the change of σ_{RT} with time of doping: (o) *cis*-$[CH(AsF_5)_{0.08}]_x$; (△) $\sigma_\perp$ for stretched *trans*-$[CH(AsF_5)_{0.14}]_x$, $l/l_0 = 1.8$; (□) $\sigma_\parallel$ for stretched *trans*-$[CH(AsF_5)_{0.14}]_x$, $l/l_0 = 1.8$.

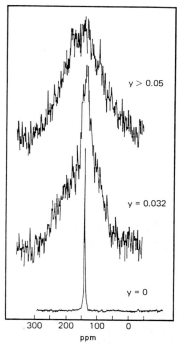

y > 0.05

y = 0.032

y = 0

300   200   100   0
ppm

Fig. 8.10 ^{13}C NMR spectra of *trans*-$[CH(AsF_5)_y]_x$. [After Clarke *et al.* (1983).]

occurs smoothly and rapidly. In contradistinction, the cis polymer was doped more slowly and the kinetics are more complicated. This can be interpreted by the dual processes of isomerization and doping in the latter case.

More striking differences between doping of the two isomers are shown by the natural abundance ^{13}C NMR spectra obtained, using cross polarization and magic-angle spinning. Three effects of the ^{13}C NMR spectra are expected with doping. The removal of electrons from the π system of polyacetylene by dopant should result in a downfield shift of the resonance. There should be a Knight shift associated with the semiconductor-to-metal transition. Third, the increase in Pauli susceptibility as the polymer becomes metallic would result in significant line broadening. Figure 8.10 shows that the ^{13}C NMR spectrum of the AsF_5-doped trans polymer ($y = 0.032$) is different from that of the undoped material. It appears to consist of a broadened peak at the original trans resonance position superimposed on an even broader downfield-shifted signal characteristic of the metallic regimes.

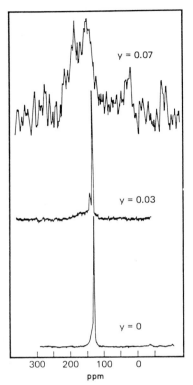

Fig. 8.11 ^{13}C NMR spectra of *cis*-[CH(AsF$_5$)$_y$]$_x$. [After Clarke *et al.* (1983).]

At $y > 0.05$, the ^{13}C NMR spectrum is that of a uniform metallic substance.

The ^{13}C NMR spectrum of the pristine *cis*-polyacetylene is shown in the bottom of Fig. 8.11, which shows a shoulder to the low-field side of the main resonance, indicating the presence of a small amount of trans structures. Upon doping with AsF$_5$ to $y = 0.01$ the trans CH resonance is resolved, and a broad weak resonance at ~ 170 ppm can be discerned. These latter features become more prominent at $y = 0.03$ (Figs. 8.11 middle, and 8.12). Finally, the spectrum of *cis*-[CH(AsF$_5$)$_{0.07}$]$_x$ (Fig. 8.11 top) is essentially the same as that of the metallic AsF$_5$-doped *trans*-polyacetylene.

From the above results it may be inferred that doping of *cis*-polyacety-lene is complicated. There is isomerization of cis to trans segments as well as paramagnetic broadening. The doped polymer does not become uniform on the NMR time scale until $y = 0.07$. The results on *trans*-polyacetylene suggest that doping is relatively uniform.

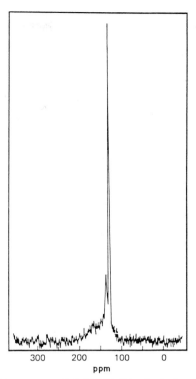

Fig. 8.12 ^{13}C NMR spectrum of *cis*-[CH(AsF$_5$)$_{0.01}$]$_x$. [After Clarke *et al.* (1983).]

8.5   DIFFUSION OF DOPANTS

Diffusion of small molecules in semicrystalline polymers has been much studied. These polymers are continuous materials. Polyacetylene is full of void space having a texture like a piece of paper or cloth. There have been several reports on the rates of absorption and desorption and of diffusion. However, not all the investigators seem to be aware of the complications.

(a)   *Morphological effect.*   Polyacetylene is composed of fibrils that fill only about 35% of the volume of a film specimen. Exposure of the film to dopant should result in a very rapid initial diffusion of the dopant into the interfibrillar space, followed by dissolution of the dopant in the amorphous phase. Diffusion then occurs within the amorphous phase from the surface into the center of the fibril. Dissolution and diffusion of dopant into the crystalline region can be orders of magnitude slower than in the amorphous phase.

(b)   *Structural effect.*   Doping of *trans*-polyacetylene is accompanied by a change in the crystal structure from the undoped to the doped structure. In the case of *cis*-polyacetylene, a larger lattice modification is involved from the cis polymer crystal structure to the doped trans structure.

(c)   *Dopant fixation.*   The diffusion of dopant in polyacetylene is not a reversible process. The dopant undergoes oxidation or reduction to an ionic species that becomes fixed by coulombic interaction with the charge on the polyacetylene chain. Only when the polymer is doped beyond its saturation limit is the surplus dopant weakly adsorbed and may be removable by evacuation or extraction.

Because of the above complications the diffusion of dopant in polyacetylene cannot be treated as a single Fickian process. Kiess *et al.* (1980) had doped polyacetylene with iodine at low pressure ($\leqslant 5 \times 10^{-2}$ torr), and the dopant could be removed by one day of evacuation at $10^{-5} – 10^{-6}$ torr. Reexposure of this specimen to iodine gave exactly the previously doped conductivity; however, the doping time required is considerably shortened. Under their experimental conditions, it took 10^4 sec for the initial doping of the pristine polyacetylene. The time was reduced to $1 – 5 \times 10^2$ sec for the third and successive doping. This is attributable to the structural changes that occurred during the first, doping which made it easier for redoping.

For the initial iodine doping Kiess *et al.* (1980) noted very rapid weight uptake in the first 10 sec. For $t > 10$ sec the process can be described by $d[I_2]dt \propto t^n$ ($\frac{1}{3} < n < \frac{1}{2}$). They estimated a diffusion constant of $\sim 3 \times 10^{-14}$ cm^2 sec^{-1} at 298 K. The rate is slower than that of the diffusion of much larger molecules in semicrystalline polymers.

Bernier *et al.* (1981a) studied the adsorption and desorption of $^{125}I_2$ in polyacetylene and found the diffusion constant to be a strong function of the concentration of the iodine solution in which the polymer sample is immersed. Long doping time gave a diffusion constant of $\sim 1 – 2 \times 10^{-9}$ cm^2 sec^{-1}, which is independent of the external iodine concentration. In the desorption experiment, the diffusion constant unexpectedly is proportional to iodine concentration. That is, a lower concentration of iodine in pentane resulted in a smaller coefficient of diffusion for the desorption of the dopant species in $(CHI_y)_x$. The results summarized in Table 8.4 showed the data should not have been treated by the simple Fick's law. The concentration of ^{125}I in the polyacetylene surface C_s is several orders of magnitude greater than the concentration in pentane, C_l. Furthermore, C_s is not strongly dependent on C_l; a two hundredfold increase in C_l caused less than sevenfold increase in C_s. The author nevertheless applied Fick's law, assuming the

TABLE 8.4

Diffusion of Iodine in Polyacetylene[a]

C_I(mg cm^{-3})	5.4×10^{-3}	10^{-2}	4.9×10^{-2}	10^{-1}	1
C_s (mg cm^{-3})	22	27	48	72	145
y	5.6×10^{-3}	7×10^{-3}	1.2×10^{-2}	1.8×10^{-2}	3.9×10^{-2}
D (cm^2 sec^{-1}) [absorption, long time, Eq. (8.23)]	—	1.2×10^{-9}	—	1.5×10^{-9}	1.7×10^{-9}
D (cm^2 sec^{-1}) [absorption, short time, Eq. (8.24)]	—	1.4×10^{-9}	—	7.3×10^{-9}	2.0×10^{-8}
D (cm^2 sec^{-1}) [desorption, short time, Eq. (8.25)]	—	2.1×10^{-10}	—	1.1×10^{-9}	1.5×10^{-9}

[a] Bernier *et al.* (1981a).

polyacetylene film to be a homogeneous slab of material. They gave approximate expressions for the diffusion coefficient to be

$$\text{Absorption, long time } \overline{C} \simeq C_s\left[1 - \frac{8}{\pi}\exp\left(\frac{\pi^2 Dt}{d^2}\right)\right], \qquad (8.23)$$

$$\text{Absorption, short time } \overline{C} \simeq \frac{4C_s}{d}\left(\frac{Dt}{\pi}\right)^{1/2}, \qquad (8.24)$$

$$\text{Desorption, short time } \overline{C} \simeq C_s\left[1 - \frac{1}{d}\left(\frac{Dt}{\pi}\right)^{1/2}\right], \qquad (8.25)$$

where $\overline{C}$ is the average concentration of ^{125}I in the polymer and d the thickness of the film. The value of D given in Table 8.4 showed that it depends upon the duration of the measurement and C_I. Absorption and desorption measurements gave different results.

In the investigation by Benoit *et al.* (1981) on the diffusion of I$_2$ and SbF$_5$ vapor into polyacetylene, the chemical binding of the dopant species mentioned above was recognized. They imbedded the 200-μm polymer film after doping in epoxy resin, cured and polished. Casting microprobe analysis was done with a 1-μm diameter electron beam and a 5-μm ionization sphere. The typical profiles for I$_2$- and SbF$_5$-doped polyacetylene are shown in Fig. 8.13. As one might expect, the dopant concentration is high at the surface and low in the interior of the specimen for short doping time. Long doping times for I$_2$ (>515 min) resulted in nearly uniform dopant profiles. How-

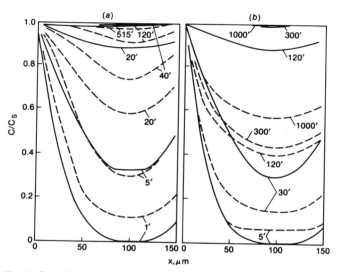

Fig. 8.13   Profile of doping in polyacetylene by casting microprobe analysis with time of doping, in min, given next to each curve (solid lines, theoretical; broken lines, experimental): (a) iodine doping, $D = 7.89 \times 10^{-8}$ cm^2 sec^{-1}; (b) SbF$_5$ doping, $D = 1.31 \times 10^{-8}$ cm^2 sec^{-1}. X is distance from sample surface. [After Benoit *et al.* (1981).]

ever, in the case of SbF$_5$ doping even after 1000 min the dopant concentration at 100 μm beneath the surface is still only about 55% of C_s. If at any given time doping was stopped by pumping away the dopant, the dopant-concentration profile in the polyacetylene specimen remains unchanged after a very long time either under vacuum or in a nitrogen atomosphere. Also C_s is very much greater than C_l in the vapor phase, as it was in the case of solution doping. Therefore, Fick's law cannot be valid for these systems. Benoit *et al.* (1981) introduced a chemical-binding term to the diffusion equation and let the apparent diffusion coefficient be the sum of an intrinsic diffusion coefficient and a concentration-dependent contribution. They found modest agreement between theory and experiment only for very short doping time but no agreement otherwise.

Louboutin and Beniere (1982) diffused ^{125}I$_2$ from pentane solution into polyacetylene film at temperatures from 213 to 293 K for 2–4 hr. The sample was then embedded in hard resin and microtomed sections for radioassay. Typical results are shown in Fig. 8.14. They applied Fick's second law modified with a chemical fixation term, which is first order (rate constant k) with respect to iodine concentration in pentane in the interfibril space, and obtained

$$D = 1.73 \times 10^{-3} \exp[-2880/RT] \quad \text{cm}^2 \text{ sec}^{-1} \qquad (8.26)$$

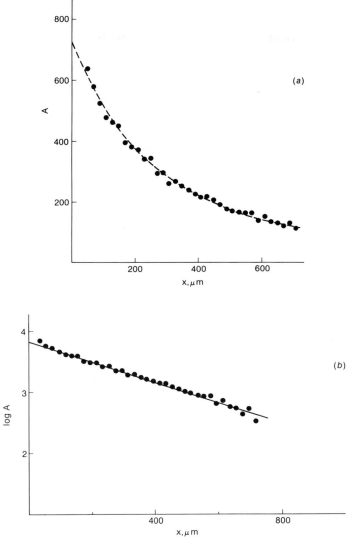

Fig. 8.14   Distribution profile of iodine in polyacetylene immersed in 0.01 g $^{125}I_2$/l pentane solution for 4 hr: (a) specific activity versus x, (b) first order plot of (a). X is the distance from sample surface. [After Louboutin and Beniere (1982).]

and

$$k = 0.754 \exp[-2046/RT] \quad \text{sec}^{-1}. \tag{8.27}$$

At room temperature this corresponds to $D = 1.3 \times 10^{-5}$ cm^2 sec^{-1} and $k = 0.023$ sec^{-1}.

It was possible to monitor directly the approach to diffusion equilibrium by optoelectrochemical spectroscopy (Section 9.1). In these experiments, after a spectrum was recorded and the voltage increased to a new level, the monochromator was set at 0.8 eV for the midgap transition. Figure 8.15 shows the approach to diffusion equilibrium of the ClO$_4^-$ ions in the forms of time dependence of the cell current $i_c(t)$ and midgap absorption $\delta\alpha$ (0.8 eV) on a semilog plot. After the initial transient, both i_c and $\delta\alpha$ decay exponentially:

$$i_c(t) = t_0 \exp(-t/\tau) \tag{8.28}$$

$$(\alpha_f - \alpha_i) = \delta\alpha_0 \exp(-t/\tau) \tag{8.29}$$

where α_f and α_i are final and initial absorbance. The time constant τ ranges from about 6 to 10 hr.

In the electrochemical doping, simple theory gave diffusion constants that are orders of magnitude too small. Kaufman *et al.* (1983) showed that there is an electric field enhancement of the diffusion of dopant ion. Under conditions of current flow, the ion current density in the polyacetylene electrode $\mathbf{J}_I$ is given by

$$\mathbf{J}_I = -D\,\nabla^2 C_I + \bar{\sigma}_I\mathbf{E}/e, \tag{8.30}$$

where C_I is the ion concentration, which will in general vary across the

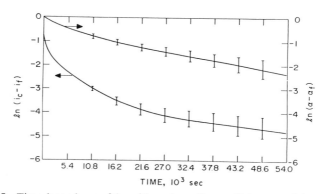

Fig. 8.15   Time dependence of the midgap absorption coefficient α, and the cell current i_c, after a step increase in applied voltage. [After Feldblum *et al.* (1982a).]

thickness of a fibril, and $\bar{\sigma}_I$ is the average ionic conductivity in the absence of an electric field E. Assuming that ion movement into the polyacetylene fibril limits the kinetics, the local ion concentration will differ from that of electrons while current is flowing resulting in the electric field term in Eq. (8.30). Differentiation of the equation and using the continuity equation for the ion current $(\nabla \cdot \mathbf{J}_I = \partial C_I / \partial t)$ and the Maxwell equation $[\nabla \cdot \mathbf{E} = 4\pi e(C_I - C_e)]$ leads to a "forced" transport equation

$$\frac{\partial C_I}{\partial t} = D \nabla^2 C_I + 4\pi \bar{\sigma}_I (C_I - C_e) \tag{8.31}$$

where the net charge density is $e(C_I - C_e)$, e is the electronic charge, and C_e is the local electron concentration (or holes in the case of p doping).

From the above discussion one must come to the conclusion that the determination of the diffusion constant for the chemical doping of polyacetylene is a futile exercise. The results are almost without significance. The "diffusion constant" obtained is at best that value specific for the cis–trans content of the sample, whether doping is from vapor or soliton. The final level of doping is dependent on whether or not the polymer has been doped previously and whether it is for the interfibril diffusion or diffusion in the amorphous or the crystalline phase. This is not to say that it is not important to know the conditions required for uniform doping, which needs to be determined empirically for every new dopant.

8.6 COMPENSATION

An interesting and useful property of polyacetylene, characteristic of classic semiconductors, is that the n- and p-type dopants can compensate one another. This property is illustrated by the following experimens of Chiang *et al.* (1978a). Na-doped $(CH)_x$ films were prepared by treating the cis polymer with a solution of sodium naphthalide, $Na^+C_{10}H_8^-$. The material is metallic with an initial composition of $(CHNa_{0.27})_x$. It was then exposed to AsF_5 (Fig. 8.16) and I_2 (Fig. 8.17). The compensation proceeds more slowly than the original doping; the electrical conductivity of the n-type sample gradually decreases and reaches a minimum. Continued doping results in conversion to p-type material with an associated increase in conductivity. Similar results have been reported by Francois *et al.* (1981a).

Compensation is a stoichiometric process. The compensation point with

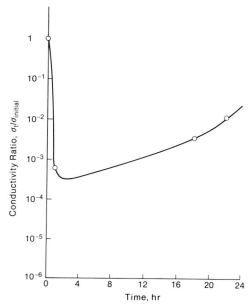

Fig. 8.16 Compensation curve for Na-doped $(CH)_x$; $\sigma_t/\sigma_{initial}$ versus time exposure to AsF_5. [After Chiang *et al.* (1978a).]

iodine, as determined by weight uptake, occurs in the above experiment at $(CHNa_{0.27}I_{0.28})_x$. According to theory, the Na-doped polyacetylene has conduction electrons in the midgap state, and the gap itself was closed following incommensurate–commensurate transitions induced by the dopant Na^+ ions. Upon exposure to an electron-accepting substance, conduction electrons are removed to form A^-. The A^- ion modifies the coulombic interaction of the D^+ ions with the electrons and the polymer becomes once again a Peierls-unstable system and a semiconductor. Due to the relatively high concentration of ionic species in the sample, the conductivity of the fully compensated sample at minimum remained higher than that of the original polymer prior to *n*-doping. Another contributing factor is the extreme sensitivity of conductivity to low levels of doping, and precise compensation would be difficult to accomplish.

Because the Na^+ ions in the *n*-doped polyacetylene were retained in the polymer, after compensation there is a half mole of NaI per CH unit. This would account for the low conductivities attainable after the second doping passed the compensation, as shown in Fig. 8.17. The conductivity will be limited by the amount of new and uncompensated dopant the sample is able to accommodate.

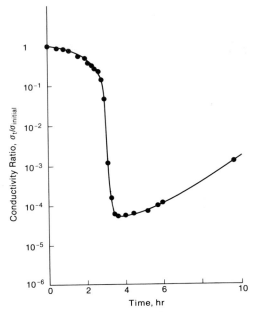

Fig. 8.17   Compensation curve for Na-doped $[CH]_x$; $\sigma_t/\sigma_{initial}$ versus time of exposure to iodine. [After Chiang *et al.* (1978a).]

8.7   EFFECT OF DOPING ON MORPHOLOGY

The effect of iodine doping on the crystal structure of *cis*-polyacetylene is discussed in Section 9.2. That study showed that polyiodide ions were intercalated in the polymer; other dopant ions may be assumed to behave similarly. Ion intercalation is expected to cause an expansion of the lattice. Francois *et al.* (1981b) used a microtensile apparatus to measure simultaneously the conductivity and precise length of the polyacetylene specimen being doped by iodine or sodium. The doping was done by immersing the apparatus in (1) iodine saturated heptane solution, (2) THF solution of naphthalene and sodium, or (3) THF solution of sodium and benzophenone. The conductivity was measured by two pressed contacts, and the weight increase of reference samples was determined for dopant concentration. The volume expansion is assumed to be proportional to the length increase,

$$\Delta V/V_0 = 3\Delta L/L_0 \tag{8.32}$$

This is borne out by the observation (Fig. 8.18) that shows that at low levels

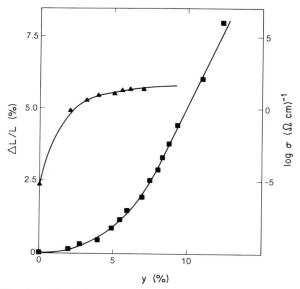

Fig. 8.18   The elongation $\Delta L/L_0$ (■) and log σ (▲) as a function of y in $(CHI_y)_x$. [After Francois *et al.* (1981b).]

of iodine doping there is only a small change (0.5%) in the sample length while the conductivity increases 10^6-fold. When $y > 0.07$, $\Delta L/L$ increases linearly with y without any great affect on σ. In other words, a 1% increase in y leads to a 3% increment in volume.

Like other properties of polyacetylene, the lattice expansion with doping is an irreversible process, as seen in Fig. 8.19. An iodine-doped $(CH)_x$ was compensated by sodium. Compensation caused a large decrease in conductivity but was not accompanied by any significant change in sample dimension. When the sample is redoped with iodine, the volume change resumed without apparent discontinuity and the effects are cumulative. These observations can be accounted for by the fact that even though compensation eliminated the positive soliton, the dopant ion is not easily removed because of ion–dipolar interaction. We can write

$$\text{(8.33)}$$

Rommelmann *et al.*(1981) examined the morphology of doped polyacetylene with SEM. Doping with iodine to low levels causes small, if any, observable changes. At doping level $y > 0.15$, the fibril diameter increased

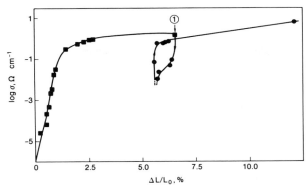

Fig. 8.19   Variation of σ versus elongation during iodine doping of $(CH)_x$ (■) followed by compensation indicated by an arrow and redoping with Na (●). [After Francois *et al.* (1981b).]

about twofold. AsF_5 doping causes more dramatic changes. The fibril diameter increases from 20 to 40 nm at $y = 0.01$, 75 nm for $y = 0.039$, and 100 nm at $y = 0.06$. The difference between iodine and AsF_5 can be readily accounted for by the fact that the polyiodide ion is linear and cylindrical in shape whereas AsF_6^-, or $As_2F_{10}^{2-}$, is a spherical or obloidal ion of large diameter. Examination of the fractured surface showed that the morphological changes are uniform through the thickness of the polyacetylene film.

We have doped ultrathin in situ polymerized polyacetylene film and obtained direct-transmission electron microscopy. The results are very informative except for the disadvantage that the level of doping cannot be determined on such specimens. For *cis*-polyacetylene doped with iodine to intermediate level, Fig. 8.20a shows fusion of the large-diameter fibrils (compare with undoped micrograph Fig. 3.9). This is especially noticeable where the density of fibrils is high. The small-diameter microfibrils showed local and irregular swelling along their lengths. Figure 8.20b shows that when polyacetylene is doped to maximum level, greatly swollen and fused fibrils appear. All the microfibrils had disappeared unless the smallest entity in this figure represents the swollen microfibrils. This is improbable because it would represent more than a tenfold increase in diameter. More likely, microfibrils had fused with others by the dopant. In both Figs. 8.20a and b there were very dark regions that may be due to high concentrations of dopant. Therefore, in these instances the distribution of dopant is not uniform, at least not at this microscopic level.

Figure 8.20c is particularly interesting. The ultrathin polyacetylene was doped extremely slowly. Except for a few dark spots that may be due to catalyst residue because they do not correspond to the fibrils, the remainder of the micrographs show very uniform doping. Where fibrils are aggregated,

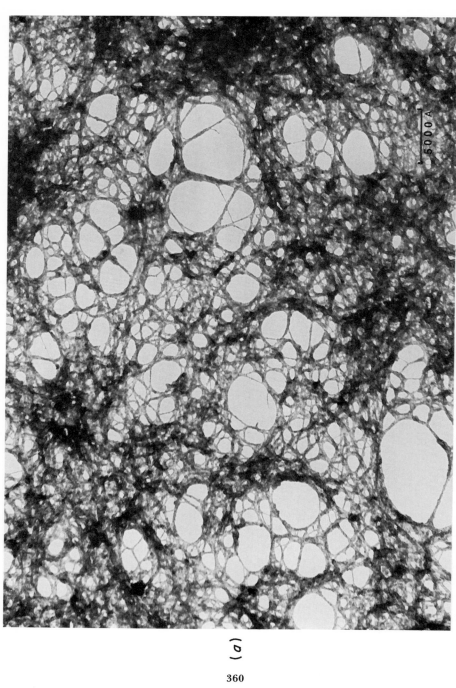

5000 Å

Fig. 8.20 Electron micrographs of *cis*-polyacetylene doped with iodine: (a) doped to intermediate level, observed at 298 K; (b) doped to maximum level, observed at 123 K; (c) doped very slowly with iodine kept at 195 K for 8 days, observed at 298 K.

(*a*)

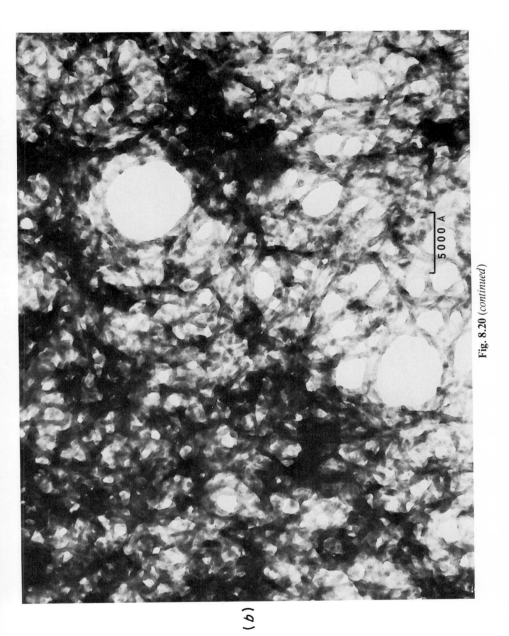

(b)

Fig. 8.20 (continued)

5000 Å

361

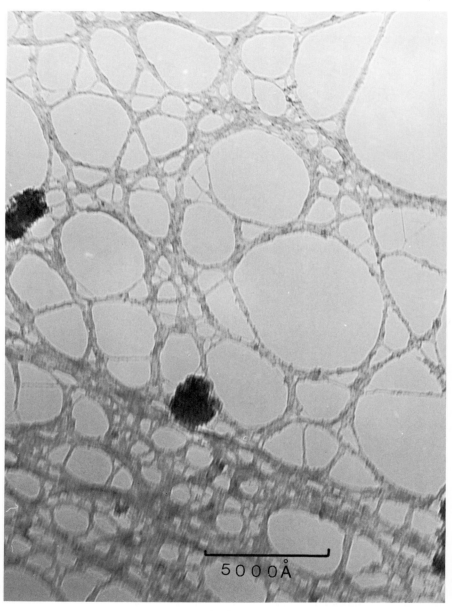

(*c*)

Fig. 8.20 (*continued*)

they appear to be fused together. However, the microfibrils remained; some of them have uniform appearance throughout their lengths whereas others do not.

The results of the effect of iodine doping on the morphology of ultrathin film of *trans*-polyacetylene (Fig. 8.21) are essentially similar to those described above for the cis polymer.

Figure 8.22 shows the morphology of AsF₅-doped *cis*-polyacetylene. At intermediate levels of doping there is swelling and fusion of fibrils and the doping appears to be nonuniform (Fig. 8.22a). At heavy levels of AsF₅ doping (Fig. 8.22b) the polymer fibrils appear to be fused into a more or less

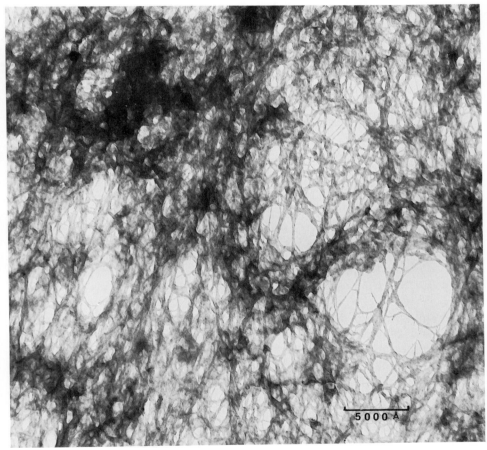

Fig. 8.21a Electron micrograph of *trans*-polyacetylene lightly doped with iodine for 5 sec at 296 K.

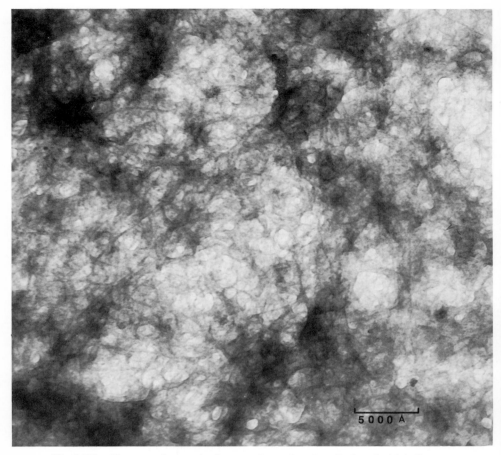

Fig. 8.21b   Electron micrograph of *trans*-polyacetylene heavily doped with iodine.

continuous chain. The dark spots here and there may be the side-reaction products of doping. Except for a few microfibrils and fibrils, the remainder of the specimen no longer has a discernible fibrillar morphology.

8.8   MECHANISMS OF DOPING

The fundamental effect of doping is to introduce free carriers and the net result of doping is the conversion of the nearly insulating polyacetylene to a metallic, conducting material. Two processes can be envisioned. The first is

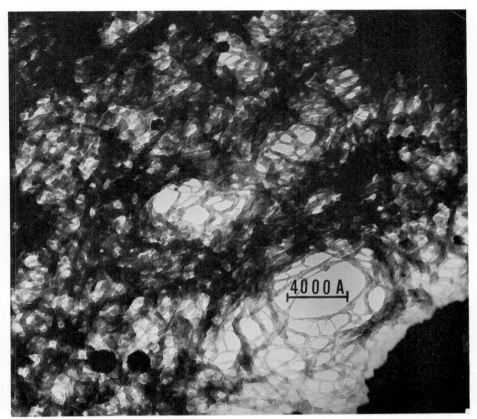

Fig. 8.22a Electron micrograph of *cis*-polyacetylene doped with AsF_5: intermediate doping level.

the interaction of a dopant with an existing neutral soliton. The soliton state is a nonbonding MO (Fig. 8.23). It is a state derived from one-half a state each from the valence band (π bonding MOs) and the conduction band (π-antibonding MOs). In fact, two neutral solitons are created simultaneously, as indicated in Eqs. (5.15) and (5.16), thus depleting one state from each of the bands. The planar zig-zag structure showing an unpaired spin is a simplified representation. However, the neutral solitons have none of the usual free-radical characteristics. At the bottom of Fig. 8.23 is shown the projection of the backbone carbon atoms with the p orbitals perpendicular to the plane of zig-zag chain and the π-phase kink nature of the species. Even though only four carbon atoms are shown as having the π wave functions opposite in phase to their neighboring atoms, there should be 2ξ of the atoms involved. p-Type doping converts a S $\cdot$ to a positive soliton. Mele and Rice

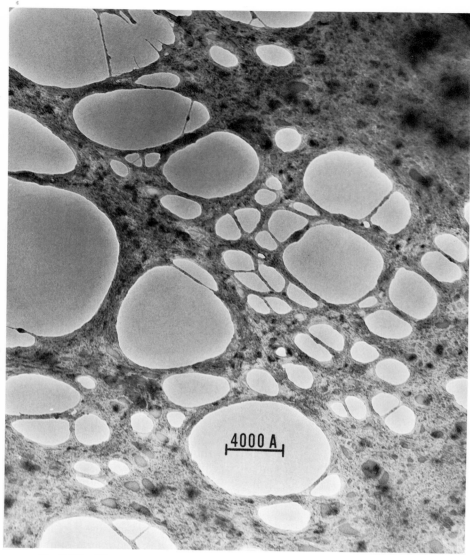

Fig. 8.22b   Electron micrograph of *cis*-polyacetylene heavily doped with AsF_5.

(1980a) calculated that the energy of a bare S^+ is about 0.4 eV higher in energy than $S\cdot$ and much larger domain width. However, the attractive Coulomb interaction with the dopant ion lower the energy of S^+ to only about 0.08 eV above $S\cdot$ and correspondingly somewhat greater delocalization. Conversely, n-doping can convert $S\cdot$ to S^-. The same energetic and

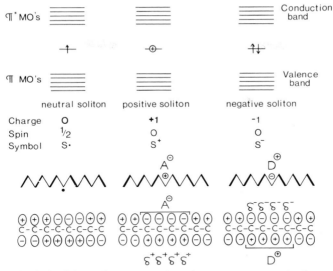

Fig. 8.23   Schematic representation of neutral and charged solitons

delocalization due to the pinning potential apply to S⁻ as in S⁺. This mechanism of doping is more probable for *trans*-polyacetylene than for *cis*-polyacetylene because the trans polymer has larger number of defects that are highly diffusive.

A second mchanism of doping is the direct oxidation of polyacetylene by *p*-type dopant resulting in the formation of positive polaron P⁺ or a delocalized radical cation. It comprises a neutral defect and a charge defect. There is attractive interaction between the two defects, and the energy is smallest, 0.65 eV, when the two defects are in proximity, corresponding to the polaron state (Bredás *et al.*, 1982b). With respect to the midgap level, at 0.7 eV, the polaron binding energy, which is related to the deformation of the lattice around it, is 0.05 eV. The optimized value for ξ is about 7 as it is for neutral soliton. It is shown in Fig. 8.24 that the polaron introduces two defect levels, 0.4 eV above and below the Fermi level. For P⁺, the bonding polaron state is half occupied. Direct reduction of polyacetylene by *n*-type dopant produces negative polaron P⁻ or a delocalized radical anion. The bonding polaron state is fully occupied but only single occupancy in the antibonding polaron state.

It requires energy to separate the two defects of a polaron, as shown in Fig. 8.25. The figure also shows that two S · tend to recombine leaving no deformation on the polyacetylene chain.

At low doping levels S · is converted to S± by the first process or P± is created by the second process. The EPR data to be described in Section 9.3

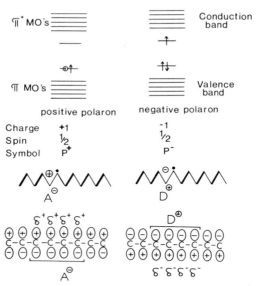

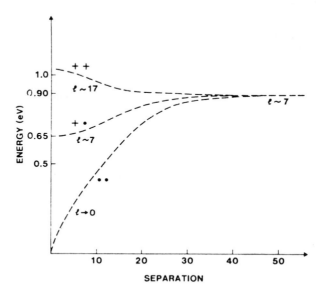

Fig. 8.24 Schematic representations of polarons.

Fig. 8.25 Energetics of separation of two defects (*l* corresponds to ξ in the text): + + indicating two S⁺, + · indicating a radical cation for large separation and a polaron for small separation, and · · indicating two neutral solitons. [After Bredás *et al.* (1982b).]

showed that the EPR intensity of iodine and AsF$_5$-doped *cis*-polyacetylene show fairly significant sample-to-sample scattering for $0 < y < 10^{-3}$, implying that both processes occur randomly. In the case of *trans*-polyacetylene, which contains about one-half of a neutral soliton per chain, the EPR intensity is very constant for $y \leqslant 10^{-3}$. The first process decreases a spin. There are two possibilities for the second process. If a polaron is created on a chain of molecular weight of $\sim 11,000$ some chains will begin to have two hand, creation of a polaron on a chain already having a neutral soliton would result in the annihilation of the two neutral defects. Even though there is attractive interaction between the charge and neutral defect of the polaron, the other neutral defect is highly diffusive and is not under the influence of a pinning field.

As the doping level exceeds 10^{-3} or about one dopant per polyacetylene chain of molecular weight of $\sim 11,000$ some chains will begin to have two polarons on the same chain. Bredás *et al.* (1982b) estimated the interactions between two polarons. Figure 8.26 shows that at low doping an appreciable energy is required to separate the neutral defects from the charge defects. It was proposed that the polaron states form bands. The second process is that

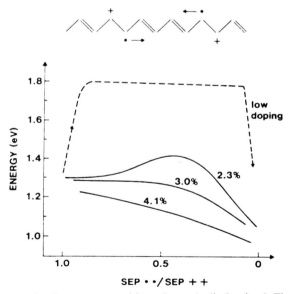

Fig. 8.26 Interaction between two positive polarons (radical cations). The horizontal axis is the ratio of the radical separations to the charge separations. Going from extreme left to the extreme right of the figure represents conversion of two positive polarons to two positive solitons. Doping levels as mole percent are indicated. [After Bredás *et al.* (1982b).]

if the two polarons are close to one another, they will be converted to two positive solitons. That homogeneous doping of polyacetylene above 10^{-3} for y resulted in rapid decrease in EPR intensity until none can be detected for $y(I_3^-) \approx 10^{-2}$ suggests that the process occurs more readily than Fig. 8.25 indicates. In this concentration range positive solitons begin to form a band.

Bredás *et al.* (1982b) thought that there are coexisting polaron and soliton bands in the intermediate doping regime up to a level of 3–4%. A different possibility exists; this may occur in heavily doped polyacetylene. Consider the case of two positive polarons separated by two positive solitons. The latter pair oscillates between the two polarons while the neutral defect tends to oscillate and more less confined by the positive charges of the two positive defects. In such heavily doped polymer this interaction may result in a contribution toward conductivity by carriers with spin.

In the case of electrochemical doping, two analogous processes can be postulated:

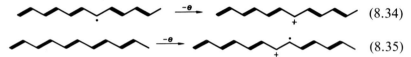

$$ (8.34) $$

$$ (8.35) $$

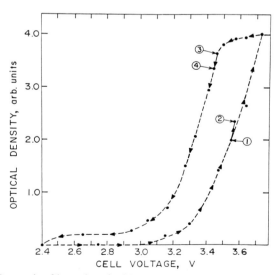

Fig. 8.27 Hysteresis of intensity of midgap absorption during electrochemical charging and discharging cycle. [After Feldblum *et al.* (1982a).]

and also for the equivalent electrode reductions. It is understood that a counterion from the electrolyte will become associated with the charged soliton or polaron.

The most important consideration for the choice of a dopant is based on the redox potential of polyacetylene,

$$(CH)_x^+ + e \to (CH)_x; \qquad E_0 = 1.0 \text{ V} \tag{8.36}$$

For *p*-type processes, the dopant must have $E_0 < +1$ V, and E_0 should be greater than 1 V for *n*-dopant.

In electrochemical doping, the results suggest the main mechanism to be polaron injection, in Eq. (8.35). In the optoelectrochemical spectroscopy investigation by Feldblum *et al.* (1982), it was found that the intensity of the midgap transition shows hysteresis on the charging and discharging cycle (Fig. 8.27). Detailed discussion of the midgap transition is given in Section 9.4.1. For instance, at 3.4 V the transmitted light intensity is an order of magnitude larger on doping than undoping. This hysteresis is not due to a diffusion artifact because the measurements were made after equilibrium was established. For instance, point 1 at 3.56 V was taken $1\frac{3}{4}$ hr after the voltage was set. Point 2 was taken after the cell was allowed to set for 17 hr. A similar procedure was followed at 3.46 V on the discharge cycle. Thus the approach to equilibrium is along the curve defining the hysteresis.

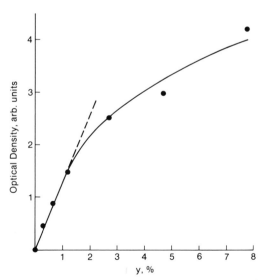

Fig. 8.28 Variation of absorption coefficient at 0.8 eV with *y*. [After Feldblum *et al.* (1982a).]

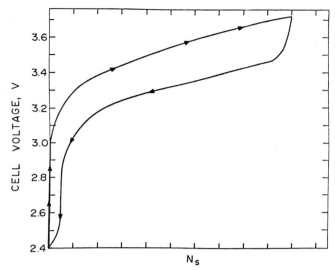

Fig. 8.29   Number of solitons N_s as a function of cell voltage. [After Feldblum *et al.* (1982a).]

The observed hysteresis implies that charge is injected at higher voltages and taken out at lower voltages, which may be explained by a mechanism of charge injection by means of polarons and charge removal by means of solitons. If one assumes that the midgap absorption is proportional to the soliton and dopant, although this is strictly true only for $y \leqslant 0.01$ (Fig. 8.28), one obtains a plot of the number of solitons N_s as a function of cell voltage (Fig. 8.29). Deviation from the assumed linearity would not change the curve qualitatively. What is important is that there is a threshold for charge injection near 3 V but not all the charges are removed until the voltage is reduced below about 2.5 V. This must mean that the species involved during charging and discharging are not the same, otherwise the processes will occur at the identical voltage.

Injection of a single charge into *trans*-polyacetylene causes lattice distortion due to electron–phonon coupling to a polaron. Solitons are not formed because they can only be produced in pairs, which means simultaneous injection of two charges—an improbable process. Continued introduction of polarons results in a reaction to convert them to charged solitons as discussed above. The injection of polaron occurs at about 3.0 V, which implies that undoped *trans*-polyacetylene is positive with respect to Li metal by about 2.3 V because $\Delta \approx 0.7$ eV. As the charge is withdrawn, neutral solitons are formed

$$S^+ + e \rightarrow S \cdot \tag{8.37}$$

which rapidly disappear. Therefore the charge is removed from the midgap state. The midgap absorption would not disappear until the applied voltage is lowered to the electrochemical potential of the neutral *trans*-polyacetylene, which is about 2.2 – 2.6 V.

In situ optical studies (Feldblum *et al.*, 1982b) showed that electrochemical isomerization of *cis*-polyacetylene commences at $y \approx 0.01$.

Chapter 9

Conducting Polyacetylene

9.1 INTRODUCTION

The most interesting property of polyacetylene, both from a theoretical and a technological viewpoint, is the 10^{13}- and 10^8-fold increase in conductivity upon doping of the cis and trans polymers, respectively. If we are to understand this new type of material, it will be inadequate just to investigate the properties of the polymer in its insulating and metallic limits. Instead, it is necessary to monitor the evolutionary changes over the widest possible doping range in small increments. As will be seen, such results are scarce. Those available data will be described. The transport properties, however, will be presented separately in Chapter 11.

The crystal structure of polyacetylene undergoes distinct changes upon doping. Section 9.2 describes the results of iodine doping of the polymer, from which it is deduced that the dopant ions are intercalated between polyacetylene chains in a random arrangement. The variation of magnetic properties with doping is both interesting and controversial. The paramagnetic susceptibility can be obtained by EPR, which showed that it remained constant up to $y \sim 10^{-3}$. Above this concentration, the spin susceptibility decreases rapidly. Observations of high Curie susceptibility in lightly doped specimens, especially if it has a Dysonian line shape, is probably associated with heterogeneous doping forming discrete metallic islands (Section 9.3).

Convincing evidence supporting the soliton concept are the doping-induced midgap transition and new infrared-active vibrational modes. However, there are alternative interpretations, as discussed in Section 9.4. Several other properties of conductive polyacetylenes are given in Sections 9.5–9.7. The stability of doped polymers is the topic of Section 9.8.

9.2 CRYSTAL STRUCTURE

Baughman *et al.* (1978) and Hsu *et al.* (1978) first examined iodine-doped polyacetylene with X-ray diffraction. A sample of $(CHI_{0.16})_x$ had diffraction lines at positions similar to those of the pristine polymer. However, a low-angle reflection at $2\theta = 11.6°$, corresponding to d spacing of 7.6–7.9 Å, appeared in the doped polymer. The authors interpreted the results with a model in which the iodine intercalates between the (100) planes of polyacetylene. The 7.6–7.9-Å spacing is reasonably close to the sum of the separation of the (100) planes of polyacetylene, which gives rise to the most intense diffraction lines in the precursor polymers (3.7–3.8 Å) and the van der Waals diameter of iodine, which is 3.96 Å. The authors postulated that, assuming a linear array of iodine ions (having a linear density approximated by that for an I_3^- ion) are incorporated into one-half of the chain sites (such as by intercalation between adjacent close-packed planes), a composition of $(CHI_{0.33})_x$ is calculated. Mössbauer spectroscopy showed that for $y > 0.2$ the dopants are mostly in the form of I_5^- ions.

Monkenbusch *et al.* (1982) had obtained powder X-ray diffraction patterns of polyacetylene doped with AsF_5, SbF_5, I_2, and PF_5. Their results on iodine-doped polymer are not identical to the earlier ones. Whereas the scattering related to undoped polyacetylene at $2\theta = 24°$ ($d = 3.71$ Å) and 25° ($= 3.56$ Å) remained, there are one or possibly two peaks located at about $2\theta = 22°$, and a third peak is observed at about 31°. The new low-angle reflection was found at $2\theta = 10.6° \pm 0.5°$, corresponding to d-spacing of 8.35 ± 0.04 Å as compared to the earlier value of 7.6–7.9 Å. In the case of polyacetylene doped with AsF_5 and SbF_5, the strong reflection at $2\theta = 23.2°$ of the pristine polyacetylene completely disappeared. Sharp reflection at $2\theta = 10.6°$ appeared in those materials, as was seen in the iodine-doped polymer, as well as the rather broad peak having a 2θ value ranging from about 15 to 21°. The broad maximum around $2\theta = 20°$ was shown to consist of at least two individual reflections at 2θ of about 16° and 22° in PF_5-doped polyacetylene. Such a sample does not display a lower angle reflection. The authors attributed this absence to an unfavorable structure factor for the lattice planes containing the PF_6^- ions as compared to those containing anions of much heavier elements.

We have obtained electron-diffraction patterns of partially aligned *cis*-polyacetylene doped with iodine to different levels (Chien *et al.,* 1982a,e; Shimamura *et al.,* 1982). *cis*-Polyacetylene, in situ polymerized onto the EM grid, was doped with iodine, the latter being kept at 250 K for 4 hr to reach an intermediate semiconducting material (sample A). The sample was kept at 195 K for a short while before transferring it into the electron microscope with a cold stage maintained at 123 K.

The electron diffraction pattern of the aligned fibrils in sample A is shown in Fig. 9.1, and Fig. 9.2 is a schematic representation thereof. The pattern is more complicated than that of undoped *cis*-polyacetylene. All the equatorial reflections can be assigned to the pristine cis polymer. Therefore, most of the sample A retains its cis crystal structure; the mole fraction of iodine in the sample is between 10^{-3} and 10^{-2}. Along the meridian there are

Fig. 9.1   Electron-diffraction pattern of aligned fibrils in sample A (*cis*-polyacetylene doped with iodine to semiconducting state).

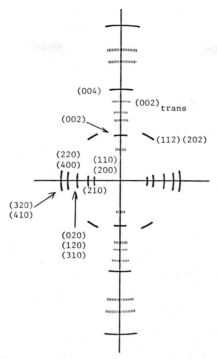

Fig. 9.2 Schematic representation of the electron diffraction pattern in Fig. 9.1.

reflections corresponding to (002) and (004) of *cis*-polyacetylene. However, there are also reflections attributable to the (002) reflection of *trans*-polyacetylene. In addition, there are new reflections that do not correspond to any spacing of either of the undoped polyacetylene structures. These are apparently the reflections from the new doped structure. The results suggest that doping can promote isomerization of polyacetylene even at low temperatures.

    cis-Polyacetylene was heavily doped with iodine to the metallic state. Sample B was obtained by exposing the cis polymer overnight to iodine kept at 263 K ($p_{I_2} = 8.1 \times 10^{-2}$ torr). Figure 9.3 gives the electron-diffraction pattern. All the reflections attributable to pristine *cis*-polyacetylene have disappeared. The diffraction data are summarized in Table 9.1 and Fig. 9.4. There are only two broad equatorial reflections with spacings 3.8 and 2.1 Å. On the meridian, eight discrete peaks were observed. In addition, we found very diffuse scattering around $s = 2 (\sin \theta)/\lambda = 0.95$ Å^{-1}. The intensity of the peak at $s = 0.81$ Å^{-1} (d spacing $= 1.23$ Å) suggests that the observed fifth peak is due to the repetition of the zig-zag conformation of the

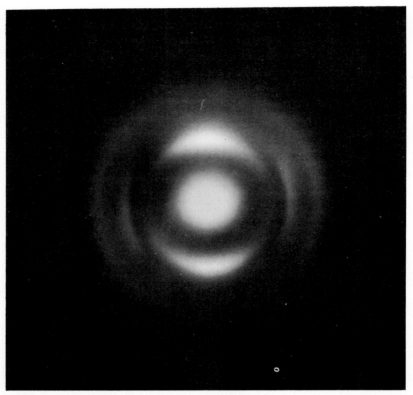

Fig. 9.3 Electron-diffraction pattern of aligned fibrils in sample B (*cis*-polyacetylene saturation doped with iodine to metallic state).

polyacetylene chain. The absence or undetectably weak intensity at $s = 0.41$ $Å^{-1}$ (half of $s = 0.81$ $Å^{-1}$) implies that the carbon–carbon bond has nearly the same length.

Attempts to index the other meridional-intensity peaks has been tried but without success. Diffraction theory implies that the broadness of these peaks produces virtually no interference between the polyiodide ions, i.e., the intensity distribution corresponds to scattering from independent polyiodide ions without correlation between them. According to theory, polyacetylene becomes a metal only when the chains are Coulomb-coupled to random potential of dopant ions in three dimensions. Periodic potential of dopant ions cannot remove the Peierls instability, and bond alternation persists.

Since the polyiodide ions are present as either I_3^- or I_5^-, we have calculated

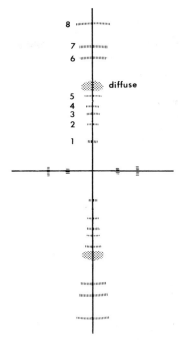

Fig. 9.4   Schematic representation of diffraction in Fig. 9.3.

the intensity distribution of the electron diffraction along the meridian as a function of the I–I distance d and the neighboring C–I distance along the chain (Δ). Figure 9.5a shows the results for the 30-CH-long polyacetylene chain. The intensity peaks at $s = 0.81$ and 1.62 Å^{-1} correspond to the observed fifth and eighth peaks, respectively. Interference with the polyiodide ions only are shown in Fig. 9.5b. Comparison of peak positions for the large s values between the observed and calculated intensities gave the I–I distance as 3.14 Å. Finally to introduce interference between the polyacetylene and the dopant ions, the distance between reference CH units and I atoms along the polyion chain was varied. The observed meridional-intensity distribution represents the structure factor of a segment composed of several tens of CH units and one polyiodide ion. An extra peak at about $s = 0.95$ Å^{-1} in the calculated curve may correspond to the very diffuse scattering maximum in the experimental electron-diffraction data.

The relative intensities in Fig. 9.5c are sensitive to the choice of Δ. In the case of I$_3^-$, a value of $\Delta = 0$ Å gave a good fit with the observed intensity profile, whereas if Δ is taken to be 0.615 Å, for example, it increased the

TABLE 9.1

Electron Diffraction of Polyacetylene Heavily Doped with Iodine

Reflection number		$s_i = \sin\theta_i/\lambda^a$	s_i/s_1	
			Observed	Calculated for $d = 3.3$ Å[b]
Meridional	1	0.16 ($=s_1$)	1.0 (vs)	1.0
	2	0.26	1.63	1.54
	3	0.32	1.98 (m)	2.03
	4	0.37	2.27	—
	5	0.41	2.50 (s)	2.51
				2.86
	6	0.53	3.28	3.17
	7	0.63	3.88	3.63
	8	0.69	4.29	4.11
Equatorial	9	0.13	—	—
	10	0.23	—	—

[a] s_i corresponds to distance of ith peak from center of electron-diffraction pattern.
[b] Here d is the I–I distance.

$s = 0.95$ Å^{-1} intensity, contrary to observation. On the other hand, the best fit for I_5^- was found with $\Delta = 0.615$ Å.

Taking into consideration the charge distribution in polyiodide ions determined by Mössbauer spectroscopy (Yamaoka *et al.*, 1981) the following structures were proposed (Figs. 9.6 and 9.7). For $[CH(I_3^-)_y]_x$, the positively charged central iodine atom ($+0.01e$) is located between two CH units, and the terminal iodine atoms with $-0.5e$ charge each are situated next to the CH units. This implies fluctuation of electron-density distribution along the polyacetylene chain. From the atomic diameters of iodine and carbon atoms, the interchain distance should be $3.96 + 3.60 \approx 7.6$ Å, which is twice the observed spacing of 3.8 Å ($s = 0.25$ Å^{-1}). In the model for $[CH(I_5^-)_y]_x$, the negatively charged ($-0.5e$) central iodine atom and ($-0.25e$) terminal iodine atoms are positioned next to CH units, whereas the other two nearly neutral iodine atoms are situated between two CH units. This I_5^- ion would also induce fluctuation of the positive-distribution in the polyacetylene chain.

Neutron-diffraction patterns have been obtained for AsF$_5$-doped *cis*-(CD)$_x$ and *trans*-(CD)$_x$ (Riekel *et al.*, 1982). The specimens were found to be highly crystalline with apparent crystallite dimensions 7 and 13 nm perpendicular and parallel to the chain direction, respectively. From the

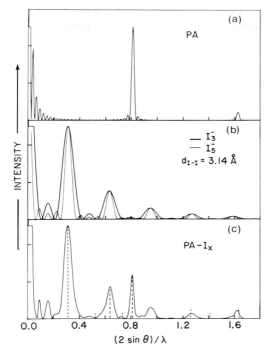

Fig. 9.5   Calculated scattering intensity profiles for: (a) polyacetylene of 30 CH units; (b) I_3^- and I_5^-; (c) iodine-doped polyacetylene where dotted lines correspond to observed peaks whose heights represent relative intensity. Fracton of $I_3^-/I_5^- = 1$, $\Delta = 0$ for I_3^- and 0.615 Å for I_5^-.

reflection positions, s_i values of 0.13, 0.26, 0.31, and 0.34 Å^{-1} were obtained. They correspond closely to the first to fourth meridional reflections of iodine-doped polyacetylene by electron diffraction (Table 9.1).

9.3   MAGNETIC SUSCEPTIBILITIES

The total spin susceptibilities for doped polyacetylenes can be described by

$$\chi(T) = \chi_d + \chi_p + \chi_s(T) \tag{9.1}$$

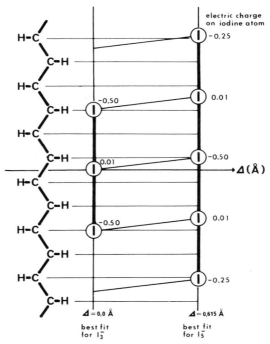

Fig. 9.6   Dependence of iodine atom positions relative to CH units as a function of the distance between reference CH and I atoms Δ. Electric charge distribution on the polyiodide ions are those proposed by Yamaoka *et al.* (1981).

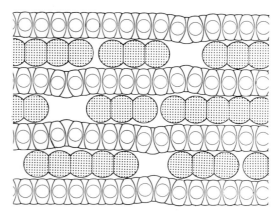

Fig. 9.7   Structural model for iodine-doped polyacetylene.

where χ_d is the temperature-independent diamagnetic core susceptibility, χ_p is the temperature-independent Pauli susceptibility, and

$$\chi_s(T) = n_i \mu_B^2 / k_B T, \qquad (9.2)$$

is the Curie susceptibility of neutral $S = \frac{1}{2}$ soliton. Equation (9.1) shows clearly that the $S = 0$ charged solitons $S^\pm$ do not contribute toward the magnetic susceptibility. Fortunately, their presence can be shown by visible and IR absorptions and their concentrations estimated from the spectral data. χ_d is very important in direct measurement of magnetic susceptibility; it is of no concern in EPR and optical spectroscopic studies.

$\chi(T)$ can be measured by a field-gradient Faraday balance. Samples were placed inside a gelatin capsule and suspended on the balance; then the sample chamber was evacuated to be filled with helium. $H(dH/dz)$ was measured, where H is the field and dH/dz the field gradient, which had been calibrated with a standard. The $\chi(T)$ thus obtained includes all contributions to Eq. (9.1). The diamagnetic susceptibility can be approximately estimated from Pascal's constants. However, χ_d is large and can contribute significant error to χ_p, which is obtained by the difference [Eq. (9.1)]. This error becomes less important when polyacetylene is doped to the metallic state for which there is no Curie susceptibility, and all the paramagnetic susceptibility arises from temperature-independent χ_p. The susceptibility from electron spin only can be determined with standard EPR technique, usually at X-band frequency. Only a few milligrams of sample are needed for this measurement. Comparison of the EPR intensity with a standard yields χ_s. Pauli susceptibility should be absent for a lightly doped polymer unless there is gross doping inhomogeneity, resulting in small metallic domains.

The spin susceptibility of heavily doped and highly conducting polyacetylene cannot be measured at microwave frequencies because of the skin effect (Section 9.3.3). Measurements must be done at rf frequencies with a NMR reference to normalize out any field nonuniformity resulting from the skin effect. Weinberger *et al.* (1979) used the Schumacher–Slichter technique with a Q-meter continuous-wave spectrometer operating at 37 MHz. Both the electron-spin and reference nuclear resonances were observed without changing any rf circuit parameters. The proton in the polyacetylene has a susceptibility given by

$$\chi_H = N I (I + 1)(\gamma_H \hbar)^2 / 3 k T, \qquad (9.3)$$

where I and γ_H are the nuclear spin and gyromagnetic ratio of the proton and N is the number of protons per unit volume.

9.3.1 Curie Susceptibility and EPR Saturation Characteristics

EPR is the best technique for the measurement of low concentrations of paramagnetic species. Early EPR studies of AsF$_5$-doped polyacetylene by Goldberg *et al.* (1979) found that the EPR intensity increased steadily with doping at 60 torr of AsF$_5$ pressure. At a pressure of 120 torr, the EPR signal decreased slightly followed by large increases with doping times and the spectra developed a Dysonian line shape. This is now recognized to be the result of doping too rapidly, resulting in the formation of small metallic domains.

We have conducted a very thorough and careful study of the effect of very slow iodine doping on the Curie spin over a range of $10^{-6} < y < 10^{-1}$, the results of which were given in Section 5.5.1. A similar study has also been made on AsF$_5$ doping.

9.3.1.1 Iodine-Doped Polyacetylene

Doping with iodine does not change the EPR line shape of *trans*-poly-acetylene, which remains Lorentzian with slightly broadened width up to $y \simeq 10^{-2}$. There is a slight asymmetry at higher levels of doping. The A/B ratio is 1.2 at $y = 0.015$ and 1.5 at $y = 0.022$. Since Dysonian theory for metallic electrons does not apply to $A/B \leq 2.7$, the observed value suggests stationary paramagnetic impurities situated pseudouniformly in the metal-lic domain. The definition of the A/B ratio is given in Section 9.3.3 on Dysonian line shape. When the EPR signal is symmetric, its line width is independent of microwave power. This is in contrast to undoped *trans*-poly-acetylene whose line width increases with increasing microwave power (Fig. 5.23).

The EPR intensity is unchanged by light doping up to $y \approx 10^{-3}$, above which level the intensity decreased rapidly. No EPR was observable for y greater than 0.022. The results are shown in Fig. 9.8. Doping in the range of y from 3×10^{-6} to 3×10^{-3} was done with $^{125}I_2$; ordinary iodine was used for $y \geq 7 \times 10^{-4}$. The dopant concentrations measured by radioassay and weight uptake between $y = 7 \times 10^{-4}$ and 3×10^{-3} are in good agreement. The EPR can be saturated for all doped $(CHI_y)_x$ at 298 K; the microwave power required to saturate EPR increases with increase of y (Fig. 9.9). However, at 77 K, heavily doped samples cannot be saturated (Fig. 9.10). This may be due to excessive inhomogeneous broadening or may be caused by a nonuniform microwave field in the sample due to a skin effect.

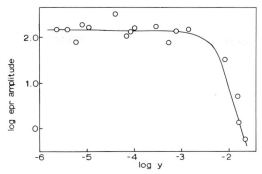

Fig. 9.8 Variation of EPR intensity versus doping for *trans*-$(CHI_y)_x$.

Iodine-doped *cis*-polyacetylene has EPR intensity about an order of magnitude less than that in *trans*-polyacetylene (Fig. 9.11). The variability of [S·] is about a factor of 3. Thus unlike the case of *trans*-polyacetylene, iodine doping did not eliminate the scattering of EPR measurements in *cis*-polyacetylene. The EPR intensity starts to decrease at about $y = 10^{-3}$, but because of the initial low number of the unpaired spin and its broad linewidth the decrease can neither be followed with accuracy nor to a high level of doping. The EPR signal disappears at $y > 10^{-3}$. This change can be better observed by following the variation of EPR amplitude with time of doping a sample directly in the EPR cavity (Fig. 9.12). The amplitude was

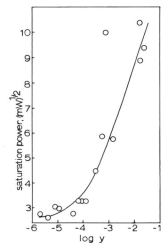

Fig. 9.9 Variation of EPR saturation power for *trans*-$(CHI_y)_x$.

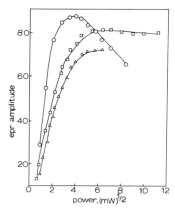

Fig. 9.10   EPR saturation curves at 77 K for *trans*-$(CHI_y)_x$; (⊙) $y = 7.8 \times 10^{-3}$; (△) $y = 1.6 \times 10^{-2}$; (□) $y = 2.2 \times 10^{-2}$.

constant but decreases rapidly after 2 min of doping. It is estimated that under these particular conditions the polymer was doped to $y \approx 10^{-3}$.

The EPR spectra of *cis*-$(CHI_y)_x$ all have a Lorentzian line shape. Furthermore, the line width is constant at 6–7 G for all levels of doping (Table 5.4). This shows that no extensive cis–trans isomerization occurred in the vicinity of the neutral defects at $y'(I_3^-) = 7 \times 10^{-3}$ or $y''(I_5^-) = 4 \times 10^{-3}$. Alternatively the neutral defects introduced by polaron doping of *cis*-polyacetylene

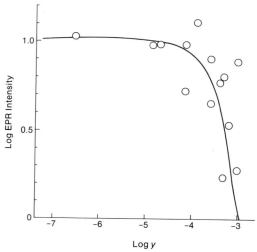

Fig. 9.11   Variation of EPR intensity with y for iodine-doped *cis*-polyacetylene.

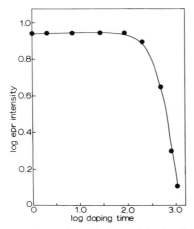

Fig. 9.12 Variation of EPR intensity with time of doping (in min) of *cis*-polyacetylene with iodine.

is also confined. The saturation results for iodine-doped polyacetylenes are discussed in Section 5.5.1.

9.3.1.2 *AsF₅-Doped Polyacetylene*

Results in this and the next sections and on conductivity of $[CH(AsF_5)_y]_x$ (Section 11.3) suggest strongly that AsF_5 doping is more prone to be inhomogeneous than iodine doping. It is likely that AsF_5 dopes the polymer on contact and the concentration gradient of dopant ion from the surface of the polyacetylene fibril to the center is greater than it was for iodine doping. This has been demonstrated for doping with SbF_5, which resembles AsF_5 chemically (Section 8.5). Another complication may be the formation of AsF_3, which can hinder the diffusion of AsF_5 into the specimen.

The variation of absolute [S·] with doping for *trans*-$[CH(AsF_5)_y]_x$ is shown in Fig. 9.13. [S·] remained constant up to $y \approx 10^{-3}$, beyond which [S·] decreases rapidly, as in iodine doping. There is a great deal of scatter in this and other results on saturation and relaxation times, suggesting sample-to-sample variability. In the case of *trans*-$(CHI_y)_x$, iodine eliminated this variability. This is attributed to the oxidation of Ti^{3+} ions to Ti^{4+} ions. Even though AsF_5 is perfectly capable of doing the same, probably its rapid reaction with the polymer prevents uniform interaction with the small number of metal ions present.

The EPR spectrum of *trans*-$[CH(AsF_5)_y]_x$ is symmetric for y up to 10^{-2}, above which it becomes strongly Dysonian (Fig. 9.14). It is recalled from

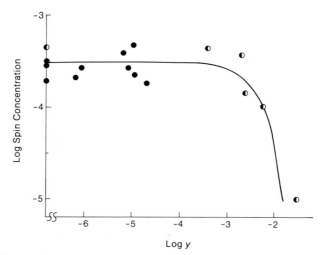

Fig. 9.13   Variation of EPR intensity versus doping with AsF$_5$ of *trans*-polyacetylene: (●) data of Chien *et al.* (1982b), (◐) data Ikehata *et al.* (1980).

Fig. 9.13 that the symmetrical signal started to decrease rapidly at $y \geqslant 10^{-3}$. The Dysonian line probably arises from spins in the metallic domain.

Figure 9.15 contains some examples of EPR saturation curves for *trans*-[CH(AsF$_5$)$_y$]$_x$. The EPR linewidth remains relatively constant for all levels of doping until the spectra assumed a Dysonian line shape (Fig. 9.16). Figure 9.17 showed that T_1 is only slightly affected by AsF$_5$ doping. This is in marked contrast to the large decrease of T_1 with y in iodine doping of the same polymer. Furthermore, T_1 starts to decrease rapidly with increasing y

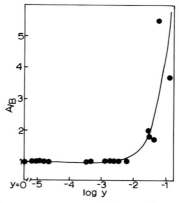

Fig. 9.14   EPR symmetry of *trans*-[CH(AsF$_5$)$_y$]$_x$ as a function of y.

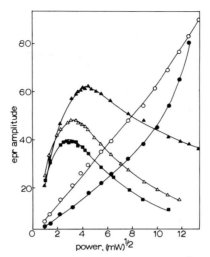

Fig. 9.15 EPR saturation curves at room temperature for *trans*-[CH(AsF$_5$)$_y$]$_x$: ($\triangle$) $y = 3.6 \times 10^{-4}$; ($\blacksquare$) $y = 1.5 \times 10^{-3}$; ($\blacktriangle$) $y = 3.7 \times 10^{-3}$; ($\bigcirc$) $y = 2.8 \times 10^{-2}$; ($\bullet$) $y = 5.9 \times 10^{-2}$.

above a value of $\sim 10^{-3}$ in *trans*-[CH(AsF$_5$)$_y$]$_x$, whereas for *trans*-[CH(I$_y$)]$_x$ the change of T_1 versus y becomes moderated for $y > 10^{-3}$. The T_2 dependence on AsF$_5$ concentration (Fig. 9.18) parallels that for T_1, the value of T_1 is always 200–300 times larger than T_2 at all levels of doping. The average values of T_1 and T_2 for $y < 10^{-3}$ are nearly the same as for pristine *trans*-polyacetylene, implying that the majority of unpaired spins do not feel

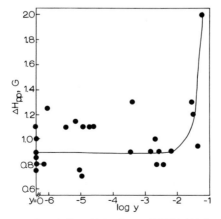

Fig. 9.16 Variation of EPR linewidth of *trans*-[CH(AsF$_5$)$_y$]$_x$ with doping level.

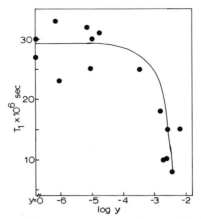

Fig. 9.17   Variation of T_1 versus log y for *trans*-[CH(AsF$_5$)$_y$]$_x$.

the presence of AsF$_5$ at low concentrations. These samples were prepared by the "slow" doping procedure (Section 8.2.1), which will be shown to be nonuniform in dopant distribution (Section 11.6).

That doping of *cis*-polyacetylene by AsF$_5$ is inhomogeneous can be demonstrated definitively. When doping is done rapidly, one observes four general types of EPR line shapes, depending on the dopant concentration. The spectra in Fig. 9.19 are representative of (a) undoped *cis*-(CH)$_x$; (b) $y \simeq 0.005$, which contains both a broad and narrow component; (c) samples

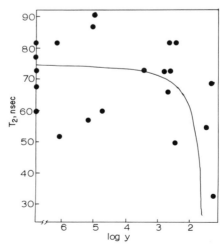

Fig. 9.18   Variation of T_2 versus log y for *trans*-[CH(AsF$_5$)$_y$]$_x$.

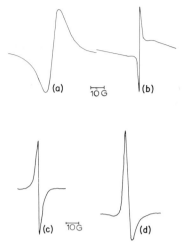

Fig. 9.19 EPR spectra of *cis*-$[CH(AsF_5)_y]_x$ at room temperature: (a) $y = 0$; (b) $y = 5 \times 10^{-3}$; (c) $8 \times 10^{-3} \leqslant y \leqslant 0.02$; (d) $y \approx 0.08$.

doped in the range $0.008 < y < 0.02$, which is like the diffusive neutral defects in *trans*-polyacetylene; and (d) heavily doped "metallic" material displaying Dysonian line shape. Very different results were obtained with very slow doping. Figure 9.20 shows that ΔH_{pp} is ~ 6 G for $y < 10^{-4}$ but narrowed dramatically to ~ 0.5 G at $y \sim 10^{-3}$. The line width increases

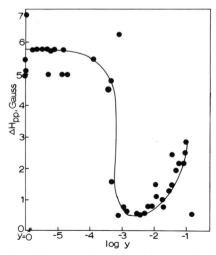

Fig. 9.20 Variation of EPR linewidth of *cis*-$[CH(AsF_5)_y]_x$ with doping level.

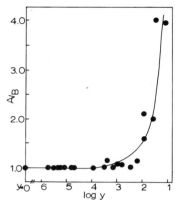

Fig. 9.21   EPR asymmetry of *cis*-[CH(AsF$_5$)$_y$]$_x$ with doping level.

with further doping due to development of a Dysonian line shape (Fig. 9.21). These results can be interpreted to mean that a small fraction of the polyacetylene chains were extensively or completely isomerized by AsF$_5$ doping at $y = 10^{-3}$. According to ^{13}C NMR results (Figs. 8.11 and 8.12), there may be about 15% trans structure in *cis*-[CH(AsF$_5$)$_{0.03}$]$_x$. The failure to observe the broad EPR of the spins in the cis structure at a much higher doping level is due to the following. Isomerization creates more neutral defects in the trans segments; the neutral solitons in the trans chains have EPR linewidths less than a tenth of that in the cis chains. Take together, the latter signal became too weak to be discerned from the baseline as the instrument gain was reduced to recorded the growth of the narrow component. If we assume that at $y \sim 10^{-3}$ about 10% of the *cis*-polyacetylene chain is isomerized to the trans structure, then based on the $\overline{M}_n$ of the polymer each AsF$_5$ dopant isomerizes about 17 CH units. This corresponds well with the soliton domain width. For complete isomerization, it will require $y \sim 0.06$, consistent with the ^{13}C NMR results of Clarke *et al.* (1983). It should be noted that iodine doping of *cis*-polyacetylene, if done slowly, never yielded a product whose EPR spectrum displayed a narrow component. The variation of absolute spin concentration with doping for *cis*-[CH(AsF$_5$)$_y$]$_x$ (Fig. 9.22) demonstated further that doping is inhomogeneous even when done very slowly. In this case [S·] increases when $y > 10^{-3}$ and reaches a concentration comparable to [S·] in undoped *trans*-polyacetylene at $y \sim 10^{-2}$.

A few representative EPR saturation curves for *cis*-[CH(AsF$_5$)$_y$]$_x$ are shown in Fig. 9.23; the EPR cannot be saturated for $y \geqslant 0.018$. Table 9.2 gave the relaxation results for those samples that could be saturated. Based on the T_1 and T_2 values, the EPR spectra probably arise from the solitons in

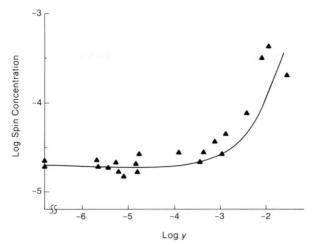

Fig. 9.22 Variation of unpaired spin per CH unit versus doping with AsF$_5$ for *cis*-polyacetylene.

the trans regions of the sample. The values of T_1 are much smaller compared to those of *trans*-$[CH(AsF_5)_y]_x$ at a similar doping level (Fig. 9.17). This suggests high local concentrations of AsF$_5$ in the cis polymer and possibly more nonuniform doping than for the trans polymer.

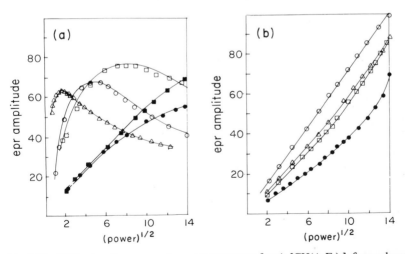

Fig. 9.23 EPR saturation curves at room temperature for *cis*-$[CH(AsF_5)_y]_x$ for y values: (a) ($\triangle$) 0; ($\bigcirc$) 5×10^{-3}, ($\square$) 8×10^{-3}; ($\bullet$) 1.8×10^{-2}; (X) 3.3×10^{-2}; (b) ($\bigcirc$) 4.8×10^{-2}; ($\triangle$) 6.5×10^{-2}; ($\square$) 7.9×10^{-2}; ($\bullet$) 9.4×10^{-2}.

TABLE 9.2

EPR Relaxation Data for *cis*-[CH(AsF$_5$)$_y$]$_x$

y	ΔH^c_{pp} (G)	T_1 (sec)	T_2 (sec)
0	~ 7	5.6×10^{-6}	9.3×10^{-9}
2×10^{-3}	~ 0.5	7.7×10^{-8}	1.3×10^{-8}
5×10^{-3}	~ 0.8	7.9×10^{-8}	8.2×10^{-8}
8×10^{-3}	~ 0.8	3.0×10^{-8}	8.2×10^{-8}

9.3.2 Pauli Susceptibility

Pauli susceptibility is due to the electrons in the Fermi gas. In principle χ_p is directly proportional to the Fermi density of states [Eq. (9.2)]. However, χ_p is only a small part of the total susceptibility [Eq. (9.1)]. Furthermore, even though χ_p is said to be temperature independent in Eq. (9.1), it has a slight positive temperature dependence for common metals. A large contribution of core diamagnetic susceptibility and any Curie susceptibility have to be deleted from the total observed susceptibility to obtain χ_p. Yet the matter is crucial to the physics of polyacetylene; a detailed account of data available to date is given. The susceptibility of conducting materials is usually measured by complementary techniques such as Faraday balance, Schumacher–Slichter magnetic resonance, and EPR.

Ikehata *et al.* (1980) has measured the temperature-independent Pauli Susceptibility of [CH(AsF$_5$)$_y$]$_x$ as a function of y. At concentrations below $y < 0.053$ the accuracy is limited by the residual Curie-law contribution and χ_p is less than 5×10^{-8} emu/mole. At higher concentrations, e.g., at $y = 0.053$, $\chi_p = (5 \pm 3) \times 10^{-8}$ emu mole^{-1}. The Pauli contribution increases rapidly above $y = 0.07$ to a value of about 3×10^{-6} emu mole^{-1}. Figure 9.24 shows the variation of χ_p versus y. For $0 < y < 5 \times 10^{-2}$, χ_p is very small and particularly sensitive to the doping procedure. For instance, χ_p may be smaller for a more highly doped polymer than a lightly doped sample. The variation of χ with T for three [CH(AsF$_5$)$_y$]$_x$ samples is shown in Fig. 9.25. At low values of y of 5×10^{-3} and 1.6×10^{-2} there are temperature-dependent contributions to $\chi(T)$. However, for the metallic specimen with $y \geqslant 4 \times 10^{-2}$, the magnetic susceptibility is temperature independent (Table 9.3).

The Pauli susceptibility corresponds to the existence of a degenerate sea of metallic charge carriers. Form the Pauli formula the Fermi-surface density of states $N(E_F)$ may be obtained. But for polyacetylene the total

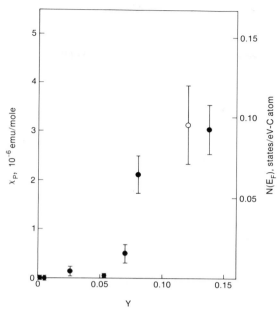

Fig. 9.24 Pauli susceptibility χ_p and implied density of states at Fermi level versus y for *trans*-$[CH(AsF_5)_y]_x$. [After Ikehata *et al.* (1980).]

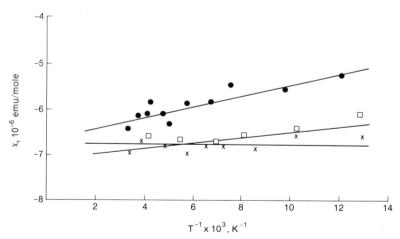

Fig. 9.25 Susceptibility as a function of temperature for three sample of $[CH(AsF_5)_y]_x$: (●) $y = 0.005$; (□) $y = 0.016$; (X) $y = 0.07$. [After Weinberger *et al.* (1979).]

TABLE 9.3

Pauli Susceptibility of Heavily Doped $[CH(AsF_5)_y]_x{}^a$

y	$\chi_{ob}{}^b$	$\chi_d{}^{b,c}$	$\chi_p{}^b$	$N(E_F)$ [states (eV C)$^{-1}$]
0.016	−6.7	−7.3	0.6	0.014
0.04	−5.6	−9.2	3.6	0.11
0.07	−6.8	−11.4	4.6	0.14
0.10	−6.2	−13.6	7.4	0.23
0.13	−9.8	−15.6	5.8	0.18

[a] After Weinberger *et al.* (1979).
[b] Value at 298 K in units emu/mole carbon $\times 10^{-6}$.
[c] Estimated from Parcal constants.

susceptibility χ_{ob}, measured by a Faraday balance, is negative for doped and undoped materials, indicating dominance of diamagnetic atomic-core contributions. This contribution was estimated from Pascal's constants. Therefore,

$$\chi_p = \chi_{ob} - \chi_d \text{ (Pascal)} = \mu_B^2\, N(E_F). \qquad (9.4)$$

Table 9.3 shows that the magnitude of correction is much greater than the measured susceptibility. Thus the calculated χ_p and $N(E_F)$ values have low accuracy.

The regime with $0.005 < y < 0.05$ is of particular interest because both Curie and Pauli susceptibilities are absent. Yet the polymer possesses high electronic conduction. The charge carrier must be spinless, but an independent soliton description is inappropriate.

Tomkiewicz *et al.* (1979, 1981) reported EPR spin susceptibility data different from those obtained by Ikehata *et al.* (1980). The authors elected to study AsF$_5$-doped *cis*-polyacetylene for the following reasons. The susceptibility of the undoped *cis*-polyacetylene is much lower than that of undoped *trans*-polyacetylene, and so the effects of doping would be more easily distinguished from the background. Second, doping produces nearly complete isomerization, and the authors contended that they will have a doped trans polymer that is less imperfect than that obtained by doping after isomerization. An early sample of *cis*-$[CH(AsF_5)_{0.009}]_x$ showed complicated $\chi(T)$ behavior. Above 35 K, the susceptibility showed a slight increase with increasing temperature and this was attributed to a degenerate Fermi gas of spin-carrying holes. Below 35 K, the susceptibility increased rapidly with decreasing temperature, which may be due to a few remaining localized electrons. The conductivity of this specimen was found to vary with $\exp[-(T_0/T)^{1/2}]$. The authors concluded that conduction occurs by means of tunneling between metallic droplets and that the sample is very inhomo-

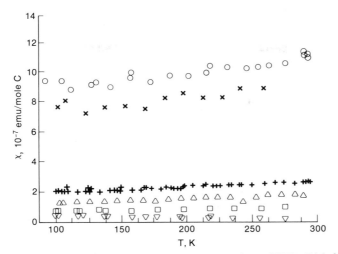

Fig. 9.26   Spin susceptibility as a function of temperature for *cis*-[CH(AsF$_5$)$_y$]$_x$ for y values: ($\bigtriangledown$) 0; ($\square$) 3 × 10^{-3}; ($\triangle$) 7.7 × 10^{-3}; (+) 9.5 × 10^{-3}; (X) 3.2 × 10^{-2}; ($\bigcirc$) 4 × 10^{-2}. [After Tomkiewicz *et al.* (1981).]

geneously doped. Subsequently, Tomkiewicz *et al.* (1981) reported that five *cis*-[CH(AsF$_5$)$_y$]$_x$ samples exhibit temperature, independent susceptibility (Fig. 9.26) in the temperature range $100 < T < 300$ K. The spin susceptibility at room temperature was found to vary linearly with y from 0.3 to 12 mole % (Fig. 9.27). Even a sample of *trans*-[CH(AsF$_5$)$_{0.003}$]$_x$, in which the

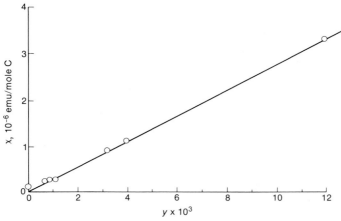

Fig. 9.27   Room-temperature spin susceptibility of *cis*-[CH(AsF$_5$)$_y$]$_x$ as a function of y. [After Tomkiewicz *et al.* (1981).]

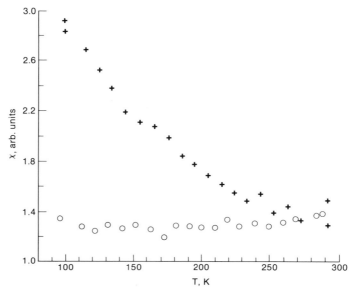

Fig. 9.28 Spin susceptibility as a function of temperature for: (+) pristine *trans*-polyacety-lene, (○) *trans*-[CH(AsF$_5$)$_{0.005}$]$_x$. [After Tomkiewicz *et al.* (1981).]

dopant concentration is below that said to be required for the semiconduc-tor–metal transition, was shown to have only Pauli susceptibility (Fig. 9.28). The authors reaffirmed their earlier conclusions and cited other evidence to support inhomogeneous doping and the presence of metallic islands. It will be shown in Section 11.6 that *cis*-polyacetylene is nonuni-formly doped even by the cyclic procedure.

Epstein *et al.* (1981a) found iodine doping to be nonuniform, especially if it is done rapidly by exposing samples to iodine vapor carried by a flow of dry N$_2$ gas. A specimen of *trans*-(CHI$_{0.135}$)$_x$ doped in this manner was found to have a Curie-spin susceptibility of 200 ppm, which is within a factor of 2 of undoped *trans*-polyacetylene. They also used a slower doping procedure to obtain a specimen of nearly the same concentration and found a much lower Curie susceptibility of 70 ppm. Even this is much greater than our very uniformly doped *trans*-(CHI$_y$)$_x$, which has a Curie susceptibility of 2.5 ppm at $y = 0.02$ and no detectable spin susceptibility for higher dopant concen-tration. Also for the low-Curie-susceptibility sample, Epstein *et al.* (1981a) observed that above 120 K there was a positive dependence of χ on T (Fig. 9.29). This behavior is reminiscent of that found by Tomkiewicz *et al.* (1981) for inhomogeneously doped *cis*-[CH(AsF$_5$)$_y$]$_x$. Figure 9.30 presents the Pauli and Curie susceptibilities derived from low-temperature (77 K)

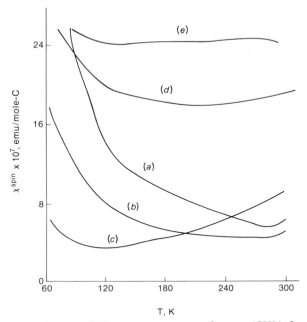

Fig. 9.29 Total spin susceptibility versus temperature for *trans*-(CHI$_y$)$_x$ for y values: (a) 0; (b) 6.9 × 10^{-3}; (c) 0.132; (d) 0.135; (e) 0.159. [After Epstein *et al.* (1981a), who expressed dopant concentration as I$_3^-$ ions.]

data; the core susceptibility was determined experimentally as $-(6.60 \pm 0.15) \times 10^{-6}$ emu (mole C)$^{-1}$ for polyacetylene and $-(43.55 \pm 0.2) \times 10^{-6}$ emu (mole I)$^{-1}$ for the dopant ion. Again, as in the study of AsF$_5$-doped polyacetylene, the diamagnetic corrections are greater than the total spin susceptibility.

Another Faraday-balance study was made on doped polyacetylenes by Peo *et al.* (1981b) from 10 to 300 K. χ_c was calculated from the diamagnetism data of Haberditzl (1968). Figure 9.31 is a plot of $\chi_{ob} - \chi_c$ versus T^{-1} for [CH(AsF$_5$)$_{0.07}$]$_x$ and (CHI$_{0.18}$)$_x$ along with undoped polyacetylene, which is presumably the cis polymer. This is not clearly stated in the paper under discussion but in the paper on the Knight shift (Peo *et al.,* 1981a) the authors wrote "our (CH)$_x$ films were prepared at $-78°C$ following the procedure described by Ito *et al.* and doped by exposure to AsF$_5$ vapor." From the slopes and intercepts of Fig. 9.31, the results of χ_p in 10^{-6} emu (mole C)$^{-1}$ and (N_c in spins per 10^6 C atoms) are $+0.08$ (630), $+0.15$ (380), and $+1.9$ (800) for *cis*-(CH)$_x$, (CHI$_{0.185}$)$_x$, and [CH(AsF$_5$)$_{0.07}$]$_x$, respectively.

The considerable disagreement regarding the magnetic properties of

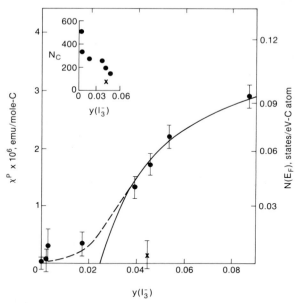

Fig. 9.30   Pauli susceptibility versus I_3^- concentration for $[CH(I_3)_y]_x$. The inset shows the Curie spin susceptibility in ppm. [After Epstein *et al.* (1981a).]

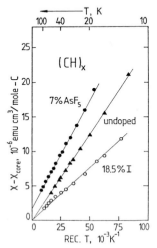

Fig. 9.31 $\chi_{total} - \chi_{core}$ for polyacetylenes versus T^{-1}. [After Peo *et al.* (1981b).]

doped polyacetylene may be attributed to differences in the polymers, whether cis or trans polymer was used, the dopant, and the doping procedure resulting in nonuniform distribution of dopant ion to various extent.

9.3.3   Dysonian Line Shape

The EPR of conduction electrons in metals was first observed by Feher and Kip (1955). The line shape and intensity of the observed resonances are decisively affected by the diffusion of electrons in and out of the skin of the metal. This is because the radio frequency (rf) field creates a certain macroscopic magnetization in the material as a result of flipping the magnetic moment of the conduction electrons. The difference between metallic and nonmetallic substances is that the rf field in the former in turn depends on the magnetization because the penetration of the rf field into the metal is changed by the magnetization. Because the magnetization is carried by the electrons, their diffusion through samples with different thicknesses and relaxation times determines the EPR features.

Dyson (1955) obtained closed theoretical solutions for the EPR of metals with classic and anomalous skin effects. From the analysis of the line shape one can deduce the size of the conducting domain, relaxation time, electron mean free path, and other properties of the material. Heavily doped polyacetylene often exhibits Dysonian EPR (Fig. 9.32). Analysis of the spectra should be very informative. The limiting cases for Dyson's theory are discussed below.

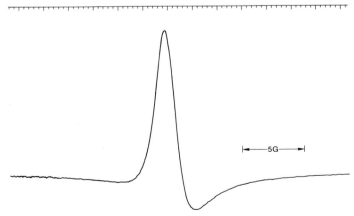

Fig. 9.32   EPR spectrum of a metallic perchalorate-doped *trans*-polyacetylene.

The main parameters in Dyson's theory are T_1, the electron-spin–lattice relaxation time, which is equal to the spin–spin relaxation time T_2 for metal; T_T, the time it takes the electron to traverse the sample; T_D, the time it takes an electron to diffuse through the skin depth δ with $\delta = c(4\pi^2\sigma f)^{-1/2}$. For X-band EPR, $f = 9 \times 10^9$ Hz and σ is the conductivity. In the anomalous case, the skin depth is equal to the mean free path Λ where $\Lambda = \tau v$ and τ is the electron mean collision time and v is the velocity of the electron. The average time an electron will diffuse across the skin depth is $T_D = (3\delta^2/2v\Lambda)$.

When the metallic domain is very small and $T_T \ll T_D$, then the EPR is simply a symmetric Lorentz line with the natural half-width T_2^{-1}. This was demonstrated for sodium particles smaller than 5 μm imbedded in paraffin; the skin depth at 296 K and 300 MHz for Na is 6 μm.

When the thickness of the sample is large compared to δ, the line shape is determined by the ratio T_D/T_2. When this ratio approaches zero, as in the

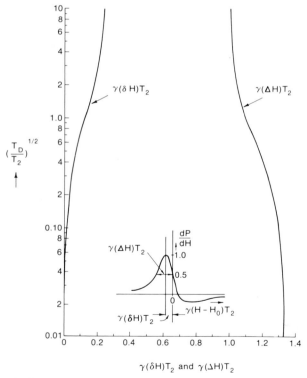

Fig. 9.33 $\gamma(\Delta H)T_2$ and $\gamma(\delta H)T_2$ versus $(T_D/T_2)^{1/2}$ for the derivative of the power absorption of EPR in thick metal plates. [After Feher and Kip (1955).]

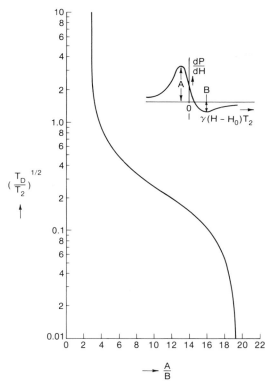

Fig. 9.34 A/B versus $(T_{D/T_2})^{1/2}$ for the derivative of the power absorption of EPR in thick metal plates. [After Feher and Kip (1955).]

case of highly conducting metal at low temperature and narrow EPR line, the asymmetry ratio A/B should approach a value of 19 (Figs. 9.33 and 9.34). In the other limit of broad linewidth, so that $T_D/T_2 \gg 1$, the ratio A/B is small, approaching a value of 2.7. This case applies to slowly diffusing magnetic dipoles or to paramagnetic impurities distributed throughout the volume of the metal.

Some of the significant implications of the parameter A/B are the following. For static paramagnetic impurities located at the surface of the metal, $A/B = 1$. If they are distributed throughout the sample and its line shape is Lorenthzian, $A/B = 2.7$; for a Gaussian line, $A/B = 2.0$. The ratio of 1 can also be due to metal particles with dimensions much greater than the skin depth. The ratio 2.7 could not be applied to polyacetylene because of its narrow EPR linewidth. Slow increase of A/B with y would imply growth of the sizes of metallic domains.

9.4   VISIBLE AND INFRARED ABSORPTIONS

9.4.1   Interband and Midgap Transitions

The strong visible absorption band with edge at 1.4 eV and peak at 1.9 eV for *trans*-polyacetylene and a similar band at slightly higher energies for *cis*-polyacetylene has been attributed to transitions from the density of states in the valence band to that in the conduction band. It corresponds to the $\pi-\pi^*$ transition. Fincher *et al.* (1978) first noted that after doping with iodine, the trans polymer shows a small decrease in the magnitude of the interband transition and a new absorption appeared at lower frequencies. This observations was extended by Fincher *et al.* (1979a), who showed that doping of *cis*-polyacetylene with iodine showed the visible absorption band to be changed progressively to that of the trans isomer, while at the same time absorption intensity at the lower frequency increases with doping. Similar experiments of doping the cis polymer with AsF_5 showed nearly the same changes in absorption at low frequencies, but the interband transition disappeared at the highest doping level and the absorption becomes free carrierlike. Subsequent studies identify the lower frequency absorption as transitions from density of states in the valence band to a state deep in the gap, which will be referred to as the midgap state.

Suzuki *et al.* (1980) recorded spectra of *trans*-polyacetylene with progressive AsF_5 doping (Fig. 9.35) up to $y \approx 5 \times 10^{-3}$. There is a monotonical decrease of the interband intensity, which goes to the transition centered at 0.9 eV. Relative to the valence band edge, the midgap state into which electrons are excited lies at about 0.7 eV, i.e., half of the band gap of 1.4 eV. Compensation with NH_3 is interesting and is shown as Curve 4 of Fig. 9.35. Compared to Curve 3 before compensation, the midgap absorption decreased to nearly the level before doping. However, the main absorption was not affected by compensation; it did not recover to the undoped state. In other words, compensation removes the midgap state but does not restore the depleted density of states of the valence band. Upon closer examination of the low-frequency region a slight tail of absorption can be discerned in the undoped polymer. Lowering of temperatures to 77 K showed a weak but definite midgap peak (inset of Fig. 9.35). Therefore, the midgap transition also occurs in pristine *trans*-polyacetylene consistent with the existence of neutral and charged solitons.

Doping with BF_3 also resulted in the suppression of the interband transition in favor of the midgap transition (Fig. 9.36) (Tanaka *et al.*, 1980). This study of doping up to 7.7 mole % exhibited an isobestic point at 1.5 eV

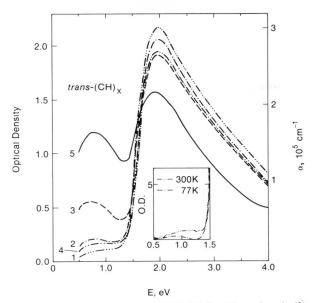

Fig. 9.35 Absorption spectra of *trans*-$[CH(AsF_5)_y]_x$: (1) undoped; (2) $y = 10^{-4}$; (3) $y = 10^{-3}$; (4) compensated with NH_3; (5) $y = 5 \times 10^{-3}$. The inset shows the temperature dependence of undoped *trans*-polyacetylene. [After Suzuki *et al.* (1980).]

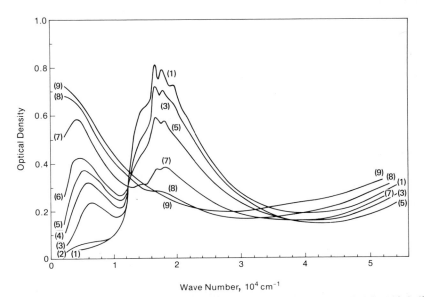

Fig. 9.36 Absorption spectra of *cis*-$[CH(BF_3)_y]_x$ with y values: (1) 0; (2) 4.5×10^{-3}; (3) 9.5×10^{-3}; (4) 1.8×10^{-2}; (5) 2.5×10^{-2}; (6) 2.9×10^{-2}; (7) 3.8×10^{-2}; (8) 5.1×10^{-2}; (9) 7.7×10^{-2}. [After Tanaka *et al.* (1980).]

(12,000 cm^{-1}) showing that only two distinct transitions are involved and that the two absorbing states are in equilibrium. There is a continuous red shift of the midgap transition. At the highest level of doping ($y = 0.077$), the midgap transition appears not to have a maximum and there is no maximum at the wavelength of the interband transition. This suggests the absence of bond alternation, and the absorption is due to free carriers (Curve 9 of Fig. 9.36).

trans-Polyacetylene doped with Na to $y = 0.2$ has the free-carrier-type absorption curve similar to Curve 9 of Fig. 9.36 (Chung *et al.,* 1981). Compensation of this metallic substance by exposure to moist air gave absorption spectra showing broad midgap absorption centered at about 0.9 eV but reduced intensity compared to undoped polymer for the interband transition.

Protonic-acid doping also causes the conversion of the interband oscillator strength to the midgap transition (Tanaka *et al.,* 1981). The results of HClSO$_3$-doped polyacetylene resemble closely those shown in Fig. 9.36 for BF$_3$-doped polymer. The isobestic point occurs at the slightly higher energy of 1.75 eV (14,000 cm^{-1}) than that in BF$_3$ doping. The doping level was not reported. However, the most heavily doped polymer has $\sigma = 250$ (Ω cm)$^{-1}$ and is metallic. In this case there still remains a peak for the midgap transition at ~ 0.5 eV, whereas maxima were not resolved for polyacetylene doped to the metallic state with the other dopants.

The most quantitative and elegant way to follow the effect of doping in optical transition is by optoelectrochemical spectroscopy (Chung *et al.,* 1982; Feldblum *et al.,* 1982a,b). In this technique, *trans*-polyacetylene itself is one of the electrodes and lithium metal is the other. They are situated in two arms of a cell joined at the bottom filled with 1 *M* LiClO$_4$ in propylene carbonate as the electrolyte. The thin polymer film was polymerized onto a conducting glass substrate, and electrical leads were wrapped about the specimen. The entire cell was sealed. The light source was a McPherson EV 700 series monochromator. An Analog Devices MacSym II microcomputer was used for data acquisition and analysis. In these actual experiments, the effective source spectrum and all extraneous absorptions were obtained and stored in the microcomputer with the reference cell. With the polyacetylene specimen in the light beam, the absorption was measured for the initial film. Then with the monochromator set at 0.8 eV, the voltage of the cell was set to a desired value. The optical density was monitored until pseudoequilibrium was reached, as seen by a drop of current level from about 5×10^{-7} to about 2×10^{-8} A. The absorption spectrum was measured and the extraneous absorption was subtracted from it. The process was repeated to give Fig. 9.37.

As doping proceeds, the midgap intensity increases at the expense of the

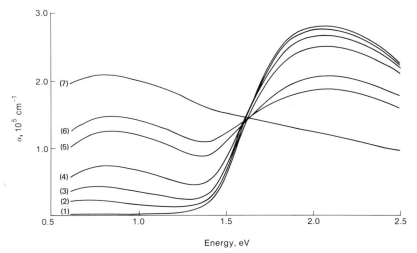

Fig. 9.37   Absorption spectra taken during the doping cycle at different voltages: (1) 2.2 v ($y = 0$); (2) 3.28 v ($y = 7 \times 10^{-4}$); (3) 3.37 v ($y = 5 \times 10^{-3}$); (4) 3.57 v ($y = 1.6 \times 10^{-2}$); (5) 3.64 v ($y = 2.3 \times 10^{-2}$); (6) 3.73 v ($y = 5 \times 10^{-2}$); (7) 3.8 v ($y = 0.1$). [After Feldblum *et al.* (1982a).]

interband oscillator strength. The midgap band is centered at $0.75 - 0.8$ eV, it persists up to $y = 0.078$, but nearly disappears at $y \simeq 0.1$, the spectrum of which is of a free-carrier kind. This metalliclike absorption can be seen to appear at $y = 0.078$ but not for the lighter-doped samples. There is a clean isobestic point at ~ 1.6 eV. The oscillator strength of the interband transition is uniformly suppressed by doping over its entire spectrum. This is illustrated by Fig. 9.38 in which the solid line are the data from the undoped *trans*-polyacetylene film. The broken curie is both the data from $[CH(ClO_4)_{0.01}]_x$ and those generated by multiplying the pristine-polymer results by a constant scale factor of 0.89 in agreement with the prediction of Eq. (9.10).

The above observations show that there is a midgap state that appears at $0.75 - 0.8$ eV in the optical spectrum of polyacetylene upon doping. The absorption intensity is proportional to the concentration of dopant molecules, its extinction coefficient is much greater than that for the interband transition. The results are consistent with the soliton model, which requires that there be a localized electronic state associated with each soliton. The symmetry of *trans*-polyacetylene places the energy of the state midway between the valence band and conduction band. One would think that the Coulombic interaction between the dopant molecule and the charged soliton would raise the energy. However, the large sizes of the charged soliton

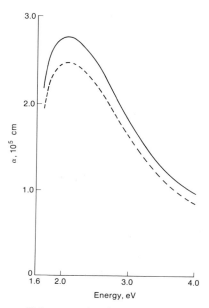

Fig. 9.38   Absorption coefficient over the photon energy range between 1.6 eV and 4 eV. Solid curve from *trans*-polyacetylene film; broken curve is data for doped to $[CH(ClO_4)_{0.01}]_x$ as well as generated by multiplying solid curve by a constant factor of 0.89. [After Feldblum *et al.* (1982a).]

and the counterion imply that the effect will be small. Moreover, this effect is balanced by electron–electron interaction. During the transition, the resulting configuration with a localized spin in the midgap state is lower in energy than that of the charged soliton. The two effects may nearly compensate each other. For the *p*-doped polymer, the transition is from the density of states in the valence band to the soliton band, converting an S^+ to $S\cdot$; for the *n*-doped polymer, an electron is excited from S^- to the conduction band.

In chemical compensation such as in the case of exposing *p*-doped polyacetylene to NH_3, S^+ is converted to a quanternary ammonium ion. The removal of the soliton state suppresses the midgap transition. Yet the density of states in the valence band is not replenished, so the interband transition is weakened. In this process the charged center is compensated, but the π-electron kinks remain on the chain.

The densities of the midgap states are low. Yet the transition involved is remarkably intense. This can be understood by the fact that the effect of a soliton is amplified by the delocalization of the soliton wave function. This can be shown by theoretical considerations (Ikehata *et al.*, 1980; Takayama *et al.*, 1980; Maki and Nakahara, 1981). The effective Hamiltonian for a linear chain in the continuum limit is

$$H = iv_F\sigma_3\, \partial/\partial x + \Delta(X)\sigma_1, \tag{9.5}$$

where $v_F = 2t_0 a$ is the Fermi velocity, σs are the Pauli matrices, and $\Delta(X)$ is the order parameter. The momentum operator is expressed as

$$p_x = 2i\sigma_3 M_x = 2\sigma_3 h \int \phi(x, y, z)\, \frac{\partial\phi(x - a, y, z)}{\partial x}\, dx\, dy\, dz, \tag{9.6}$$

with ϕ being the atomic $\pi(p_z)$ orbital of the carbon atom. For equilibrium Peierls distortion, $\Delta(X) \equiv \Delta_0$, and the interband absorption coefficient for a perfect chain without solitons per carbon atom is α_0 (Ikehata *et al.*, 1980):

$$\alpha_0(\omega) = Af_0(\omega), \tag{9.7}$$

$$f_0(\omega) = \frac{(E_G/\hbar\omega)^2 E_G}{[(\hbar\omega)^2 - E_G^2]^{1/2}} \tag{9.8}$$

$$A = \frac{16\pi he^2|M_x|^2}{m_e^2 nc W E_G}, \tag{9.9}$$

where c is the velocity of light, $E_G = 2\Delta_0$, n is the index of reflection, and W is the full bandwidth ($= 4t_0 \sim 10$–12 eV).

Now let us introduce a soliton and a moving domain wall $\Delta(X) = \Delta_0 \tanh(X/\xi)$. The absorption coefficient α_s for transitions between a soliton level and the band states became

$$\alpha_s(\omega) = A(\pi^2\xi/a)f_s(\omega) \tag{9.10}$$

$$f_s(\omega) = \frac{E_G/2}{[(\hbar\omega)^2 - E_G^2/4]^{1/2}}\, \mathrm{sech}^2\left\{\frac{\pi[(\hbar\omega)^2 - E_G^2/4]^{1/2}}{E_G}\right\} \tag{9.11}$$

The factor of $\pi^2\xi/a$ is the source for the enhancement of the soliton transition because $\xi/a \approx 7$ and $\pi^2\xi/a \approx 70$. The result implies that each π kink removes about 10^2 carbon atoms from the interband transition, meanwhile the total oscillator strength is conserved. The absorption coefficient for the interband transition is $\alpha_i(\omega)$,

$$\alpha_i(\omega) = \alpha_0(\omega)(1 - y2\xi/a). \tag{9.12}$$

The physical basis for the effect of the soliton on the interband-oscillator strength is the following. The two phases **A** and **B** on each side of the soliton domain may be approximated to represent wave functions of the conduction and valence bands $\psi^c(\mathbf{A})$ and $\psi^v(\mathbf{B})$. But the symmetry requires

$$\langle\psi^c(\mathbf{A})|p_x|\psi^v(\mathbf{A})\rangle = -\langle\psi^c(\mathbf{B})|p_x|\psi^v(\mathbf{B})\rangle, \tag{9.13}$$

and the total matrix element $\langle \psi^c | p_x | \psi^v \rangle$ vanishes. However, within the soliton domain, the matrix element is nonzero. So the total absorbance is reduced by the fractional part ξ/L of the contributing chain. Note that L is not the actual molecular chain length, otherwise it would have suppressed the interband completely. Instead, L is the length over which the electronic wave functions are phase coherent, i.e., of the order of the electron mean free path.

Theoretical analyses have also been performed by Gammel and Krumhansl (1981), Lin-Liu (1982), and Kivelson *et al.* 1982).

9.4.2   Infrared-Active Vibration Modes

As either *cis-* or *trans-*polyacetylene is doped, new infrared bands appeared. This was first described by Fincher *et al.* (1979b). To show these absorptions more clearly, the spectra of undoped polymers was subtracted from those of doped polymers. Figure 9.39 shows that upon doping *cis-*polyacetylene with AsF_5 there appeared a narrow and intense band at 1370 cm^{-1} (0.17 eV, half width 50 cm^{-1}) and a broad band centered around 900 cm^{-1} (0.11 eV, half width 400 cm^{-1}). These bands were also observed in iodine-doped *cis-*polyacetylene. In addition there was a very weak band at 1280 cm^{-1} (0.16 eV) (Fig. 9.40). Na-doped *cis-*polyacetylene and AsF_5-doped *trans-*polyacetylene both have the two more prominent new IR bands at 1370 and 900 cm^{-1}. Because these bands were seen in both lightly doped cis and trans polymers, the absorptions cannot be associated with neutral

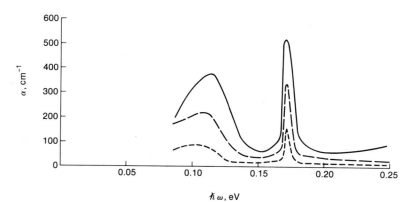

Fig. 9.39   Additional infrared absorption of *cis*-$[CH(AsF_5)_y]_x$ over undoped polymer: (---) $y = 3 \times 10^{-4}$, (– – –) $y = 6 \times 10^{-4}$, (——) $y = 10^{-3}$. [After Fincher *et al.* (1979b).]

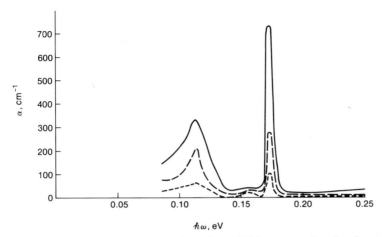

Fig. 9.40 Additional infrared absorption of *cis*-$(CHI_y)_x$ over undoped polymer: (---) $y = 1.5 \times 10^{-4}$, (– – –) $y = 5 \times 10^{-4}$; (——) $y = 10^{-3}$. [After Fincher *et al.* (1979b).]

solitons. The fact that both *p*- and *n*-doping produced the same absorptions suggests that the transition is probably associated with a charged soliton independent of its charge. Doping-induced infrared transitions are apparently a general phenomenon. They were observed in BF_3-doped polyacetylene (Tanaka *et al.*, 1980) as well as in HF-doped polymer (Fig. 9.41; McQuillan *et al.*, 1982).

Spectra obtained with oriented polyacetylene film showed the IR-active vibrational modes to be polarized along the polymer chain; Fig. 9.42 shows the parallel and perpendicular absorption coefficients of polarized light with greater absorption for radiation with the electron vector along the chain.

New infrared bands were also produced when *trans*-$(CD)_x$ was doped with AsF_5 (Etemad *et al.*, 1981a). They differ from those for the corresponding *trans*-$(CH)_x$ in two important aspects. The positions of the prominent bands are shifted to 780 cm^{-1} (0.098 eV) and 1120 cm^{-1} (0.14 eV) in the deuterated polymer; the higher-frequency band is significantly broadened and the lower-frequency band is slightly narrowed so that they have nearly the same half width of ~ 270 cm^{-1} (Fig. 9.43).

There are several theories for the origin of these absorptions. Mele and Rice (1980b) and Etemad *et al.* (1981a) believe the bands arise from the presence of charged solitons. We return to the Hamiltonian [Eq. (6.7)], which includes a set of delocalized force constants due to interaction of atomic displacements with the extended π-electron states. Therefore, the electronic susceptibility affects the phonon frequencies for the backbone stretching vibrations. In turn, the π-electron density fluctuations are cou-

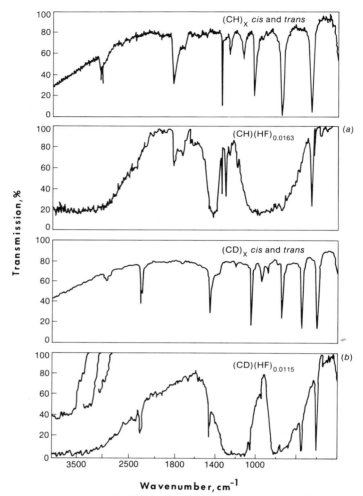

Fig. 9.41   Infrared spectra: (a) $[CH(HF)_{0.0163}]_x$; (b) $[CD(HF)_{0.0115}]_x$. [After McQuillan *et al.* (1982).]

pled through the phonons. This effect is expressed as $\delta L_{\mu\nu}$, which couples displacements μ and ν with the wave vector q:

$$\delta L_{\mu\nu}(q) = 2 \sum_k \frac{\langle n,\, k|\, V_{\mu}^*(q)|n',\, k+q\rangle \langle n',\, k+q|\, V_{\nu}(q)|n,\, k\rangle}{E_{n',k+q} - E_{n,k}} \quad (9.14)$$

where $V_{\mu}(q)$ denotes a modulation of the electronic Hamiltonian linear in $\mu(q)$ and the $|n,\, k\rangle$ ($|n',\, k\rangle$) are the filled (empty) eigenstates in the π manifold

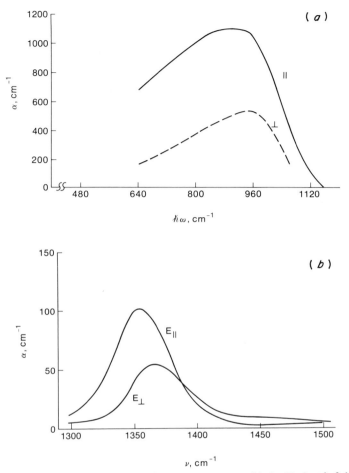

Fig. 9.42 Polarization dependence of absorption spectra with the IR electric field parallel ($E_\parallel$) and perpendicular ($E_\perp$) to the partially aligned sample of *cis*-polyacetylene lightly doped with iodine (<0.1%): (a) for the 900 cm^{-1} band; (b) for the 1350 cm^{-1} band. [After Fincher *et al.* (1979b).]

with eigenvalues $E_{n,k\,(E_{n',k})}$. The V_μ are defined by the geometry of the polymer chain and β. It was stated in Section 6.1.4 that by using $\beta = 8$ eV Å^{-1}, the Raman-active band frequencies for *trans*-(CH)$_x$ were correctly calculated. In similar calculations, without changing any of the parameters, the correct Raman bands for *trans*-(CD)$_x$ were also obtained (Fig. 9.44).

Now let us introduce a charged soliton into the polyacetylene chain. It causes strong effects on the phonon density of states. First, the "bare" force

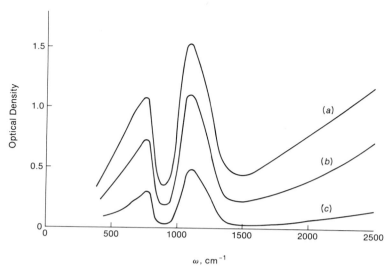

Fig. 9.43 Additional infrared absorption of *trans*-[CD(AsF$_5$)$_y$]$_x$ over undoped polymer doped with successively higher concentrations of AsF$_5$; curve (a) has the highest level of doping of $y \approx 10^{-3}$. [After Etemad *et al.* (1981a).]

constants are modified as the magnitude of bond alternation decreases smoothly through zero and changes sign at the center of the soliton domain. In the vicinity of the π kink, the electronic structure is altered, which changes the electronic contribution to $\delta L_{\mu\nu}$. The two normal modes most affected by the charged soliton are u_1 and u_2, shown in Fig. 9.45, for *trans*-(CH)$_x$. The u_1 mode at 980 cm^{-1} involves an overall contraction of the weaker C—C bonds on one side of the defect, marked by the vertical dotted line, and the expansion of the corresponding bonds on the other side of the defect. The u_2 mode at 1360 cm^{-1} entails similar displacements except that they involve the stronger C—C bonds. These motions drive the charge of the soliton back and forth across the defect center. It is this oscillation of the charge in response to the lattice vibration that gives the high absorbance of these infrared-active bands. It was calculated that the integrated oscillator strength of the u_1 mode is about 40% greater than that of the u_2 mode. It is important to note that neutral defects would not contribute to the infrared-active vibrational modes. There is no phonon-induced contribution to the electronic polarizability for the neutral soliton because, lacking a charge, there is no modulation of the Hamiltonian by a net migration of charge in this case.

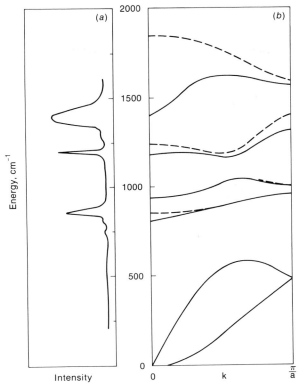

Fig. 9.44 (a) Experimental Raman spectrum of *trans*-(CD)$_x$ (excited at 514.6 nm); (b) Phonon spectra of *trans*-(CD)$_x$: (----) without electron–phonon screening ($\beta = 0$), (——) with screening ($\beta = 8$ eV Å^{-1}). [After Etemad *et al.* (1981a).]

Etemad *et al.* (1981a) had calculated the oscillator strength and frequencies of the charged soliton-induced infrared absorptions for doped *trans*-(CH)$_x$ and *trans*-(CD)$_x$. The theoretical results are in good agreement with the observed spectra shown in Fig. 9.46. The increased width and asymmetry of the higher-frequency mode in doped (CD)$_x$ seems to be accounted for by the presence of a third ungraded mode at ~ 1270 cm^{-1}, which is absent in *trans*-(CH)$_x$. This mode is closely related to the presence of an additional intense Raman line at 1200 cm^{-1} in the deuterated polymer. The shifts in frequency upon deuteration of $\sim 20\%$ are much larger than that expected simply for the $M^{-1/2}$ dependence in the normal modes. Instead, the heavier deuteron mass causes major changes in the normal modes of vibration of polyacetylene. In the case of sodium-doped (CD)$_x$ (Francois *et*

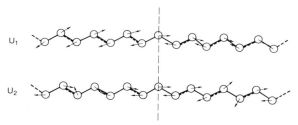

Fig. 9.45   Normal modes for the charged soliton-induced vibrations u_1 and u_2. [After Mele and Rice (1980b).]

al., 1981a), all three infrared-active vibration modes were resolved at 1260, 1070, and 700 cm^{-1} (Fig. 9.47).

The above interpretation of the infrared vibrations induced by the charged soliton for doped polyacetylene is not accepted by some workers. Zannoni and Zerbi (1982) proposed that a centrosymmetric defect, sandwiched between two dopant molecules, may account for the observed spectra. Rabolt *et al.* (1979) interpreted the spectra according to vibronic intensity enhancement of the bands in pristine polyacetylene by doping. In the case of (CH)$_x$, the band at 1385 cm^{-1} was attributed to the Raman active C—C and C=C stretching modes. The Raman mode was said to be weakened by the electron transfer to the dopant molecule, and coupling of the charge oscillation along the polyacetylene chain with the skeletal stretching of the backbone makes it infrared active. This interpretation requires the shifting of the C—C stretching modes from about 1550 cm^{-1}. The large bandwidth was thought to arise from the presence of regions of uninterrupted conjugation length of 4–100 double bonds. The very broad band located at 900 cm^{-1} in *trans*-(CH)$_x$ was proposed to originate from a vibronically activated A_g mode. Because there is no Raman band in this region for pristine polyacetylene, it was suggested that the band was shifted from 1080 cm^{-1} due to a decrease in frequency by doping. A medium Raman band at 1016 cm^{-1} was thought to experience a frequency shift to 900 cm^{-1} and become infrared active if this CH out-of-plane bending mode is strongly coupled with the charge oscillation parallel to the polyacetylene molecular axis.

Etemad *et al.* (1981a) objected to the interpretation of Rabolt *et al.* (1979). They argued that the frequency shifts upon doping with respect to undoped polymer would be expected to be proportional to the doping level. Furthermore, if these A_g modes became activated in the infrared because of the interaction with collective charge oscillation parallel to the chain axis, the absorption intensity should be saturated at low levels of doping. These expectations are contrary to the observations that at low levels of doping the

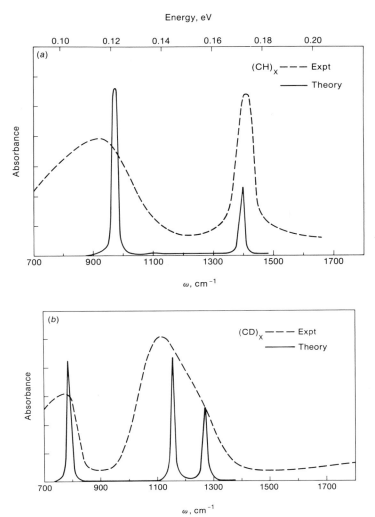

Fig. 9.46 Comparison of observed infrared absorption in (a) doped *trans*-(CH)$_x$ and (b) doped *trans*-(CD)$_x$ shown in broken curve with theoretical curve in solid curve based on a charged soliton model. [After Etemad *et al.* (1981a).]

infrared intensities are proportional to the dopant concentration, whereas the frequencies are concentration independent.

Rabolt *et al.* (1982) countered with a new Fourier-transform infrared study of AsF$_5$-doped polyacetylene. They observed that the two bands of interest do in fact shift with dopant concentration. Thus the high-frequency

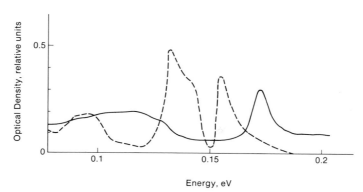

Fig. 9.47   Additional infrared absorption for (----) sodium-doped $(CD)_x$, (——) sodium-doped $(CH)_x$. [After Francois *et al.* (1981a).]

bands were found at 1403, 1401, and 1385 cm^{-1} for y values of 3×10^{-4}, 10^{-3}, and 0.1, respectively. There appears to be a linear decrease of frequency with increase of log y. The low-frequency bands show similar behavior; the band positions are at 901, 898, and 832 cm^{-1} for AsF$_5$ concentrations of 3×10^{-4}, 10^{-3}, and 0.1, respectively.

Horovitz (1982) offered rigorous arguments for the interpretation that the intense IR bands in doped polyacetylene arise from Raman bands rendered active upon the introduction of charges and that the behavior is a universal result of the translation degree of freedom of added charge independent of its configuration. He considered N phonon bands in polyacetylene with uniform bond length whose frequencies in the absence of charges are ω_n° at a zone boundary $q = \pi/a$ with dimensionless electron–phonon coupling constants λ_n. Each CH unit in *trans*-polyacetylene has four degrees of freedom for vibration within the polymer plane, which may couple to the electronic charge; therefore, $N = 4$ and $n = 1, 2, 3, 4$. When the molecule undergoes Peierls distortion to bond alternation, the zone boundary phonons have frequencies ω_n^R, which are Raman active. These modes result from oscillation of the amplitude of the disorder parameter, which are IR inactive. When a unit charge is introduced, a defect of domain width ξ results and all translation modes become IR active with frequencies ω_n^{IR}.

The phonon equation of motion was solved by inducing a pinning potential due to the charge distribution. The important conclusions are the following. The coulomb interaction of the charge distribution in doped polyacetylene with the dopant ion endowed a finite frequency ω_1 to the zero frequency translation mode. The frequencies of ω_1 for both IR and Raman should be less than the corresponding base-phonon frequency ω_n°. There are a pair of "product rules" of the frequencies:

$$\prod_{n=1}^{4} \left(\frac{\omega_n^{IR}}{\omega_n^0}\right)^2 = \beta \tag{9.15}$$

and

$$\prod_{n=1}^{4} \left(\frac{\omega_n^{R}}{\omega_n^0}\right)^2 = 2\lambda, \tag{9.16}$$

where β is the electron–photon coupling constant and $\lambda = \sum_n \lambda_n$. The ratio of Eqs. (9.15) and (9.16) gives

$$\prod_{n=1}^{4} \left(\frac{\omega_n^{IR}}{\omega_n^{R}}\right)^2 = \frac{\beta}{2\lambda}. \tag{9.17}$$

Because both β and λ involve only electronic properties independent of the isotope, the "product rule" is also isotope independent. Third, there should be the same number of Raman and IR modes. In resonance Raman scattering, the intensity of phonons that are coupled to the π-electron system is enhanced. In $(CH)_x$ there are two coupled modes and three such modes in $(CD)_x$. This implies that the coupling of a CD bending mode with the electrons is stronger than that of the corresponding CH mode. In Table 9.4 the vibrational parameters are summarized. The isotopic independence of the product rule [Eq. (9.17)] is demonstrated by the value of the left-hand side of the equation, which is 0.61 for $(CH)_x$ and 0.55 for $(CD)_x$.

Mele and Rice (1980b) and Etmad *et al.* (1981a) assigned the two IR modes in doped $(CH)_x$ to u_1 and u_2 modes and predicted the pinning mode to lie at a much lower frequency. Horovitz (1982) disagreed with their analysis because unless there is a third phonon mode there should not be three vibrational modes in *trans*-$(CH)_x$. Because both ω_1^{IR} and ω_1^{R} must be smaller than ω_1^0, he assigned the pinning mode to be at 900 cm^{-1} in doped

TABLE 9.4

Vibrational Parameters of Doped Polyacetylene[a]

Polymer	ω_n^{IR} (cm^{-1})	ω_n^{R} (cm^{-1})	ω_n^0 (cm^{-1})	λ_n/λ
$(CH)_x$	900	1075	1210	0.08
	1370	1470	2110	0.92
$(CD)_x$	700	850	890	0.04
	1070	1200	1220	0.007
	1260	1340	2040	0.953

[a] After Horovitz (1982).

trans-(CH)$_x$ and at 700 cm^{-1} in doped *trans*-(CD)$_x$. Although the pinning force is isotope independent, the pinning frequency is isotope dependent inasmuch as the latter involves the balance of all the masses in the system.

9.4.3   Raman Spectra

The Raman results on iodine-doped polyacetylene were described in Section 8.3.2 in connection with the identification of the I$_3^-$ and I$_5^-$ species. In both the iodine- and AsF$_5$-doped *cis*-polyacetylene, there is significant intensity reduction for the resonance-enhanced C—C and C=C stretching modes, whereas the weak second C—C stretching mode at 1292 cm^{-1} was not much affected by doping (Kuzmany, 1980; Lefrant *et al.*, 1979). Bands characteristic for *trans*-polyacetylene appeared indicative of cis–trans iso-merization. The resonance-enhanced C=C stretching at 1470–1520 cm^{-1} in undoped *cis*-polyacetylene is broadened upon doping; the maximum shifted to 1630 cm^{-1} in samples heavily doped with AsF$_5$ (Fig. 9.48, compare Curves c and a). Kuzmany (1980) interpreted this to mean a great shorten-ing of conjugation length to less than four monomer units. However, because the corresponding shift was not observed in iodine-doped polymer, the effect was attributed to distortion caused by intercalated AsF$_6^-$ ions.

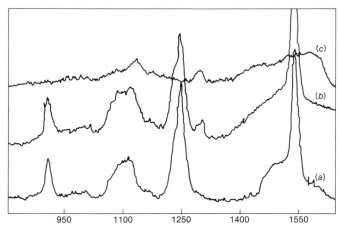

Fig. 9.48   Raman spectra of (a) *cis*-polyacetylene; (b) lightly doped with AsF$_5$; (c) heavily doped with AsF$_5$, $y \approx 0.12$. [After Kuzmany (1980).]

9.5 NUCLEAR MAGNETIC RESONANCE

NMR had been used to investigate various aspects of doped polyacetylene, although some studies were of a very cursory nature. The low-temperature narrowing of ^{1}H NMR of *trans*-polyacetylene (Section 6.8) was proposed to be due to large-amplitude librational or translational motion. The temperature for this narrowing decreases with AsF$_5$ doping. Thus this temperature is 155, 110, and 95 K for *trans*-[CH(AsF$_5$)$_y$]$_x$ with y values of 0, 0.026, and 0.053, respectively (Ikehata *et al.*, 1981). If the line narrowing is attributable to full rotation of the polymer segment as a kink goes by during soliton diffusion, then the results imply that these processes are facilitated by AsF$_5$ doping. The increased number of solitons and greater interchain separation of the polymer upon doping may be contributing factors. However, this transition, which is said also to occur in undoped *trans*-polyacetylene, could not be reproduced.

The second moment of ^{1}H NMR had been measured as a function of iodine doping for cis- and trans-rich polyacetylene by Mihàly *et al.* (1979a) (Fig. 6.29). The second moments in both cases decreased upon doping. The value for the saturation-doped material was definitely smaller than the value found for undoped *trans*-polyacetylene. This was attributed to an increase in the separation of the polyacetylene chain, at least in the vicinity of the dopant, in accordance with the intracalation model. The greater decrease of the second moment for doping of *cis*-polyacetylene at low y was attributed to isomerization. According to this interpretation, major cis–trans isomerization occurs at about $y = 0.005$ because of the large observed change in the second moment. This is a much smaller level of doping than those found necessary for isomerization by other techniques. This ^{1}H NMR study of iodine-doped polyacetylene was marred by a very narrow component at $y = 6 \times 10^{-3}$. The authors suggested mobile methyl or ethyl end groups to be the origin for the narrow linewidth component. This seems unlikely in view of the high molecular weight of polyacetylene and the fact that this feature was not observed with the undoped polymer.

The more recent investigation on the effect of iodine doping on the second moment of *cis*-polyacetylene by Meuer *et al.* (1982) gave results different from those of Mihàly *et al.* (1979a). The new study found that it takes 5% of iodine doping to reduce the second moment of the cis polymer to that of the trans isomer. Also there was no mention of the narrow component reported previously.

The presence of unpaired-electron spin can produce a strong local magnetic field due to hyperfine interactions and resulting in a shift that is proportional to the spin susceptibility and the unpaired-electron spin density at the nucleus. This NMR shift, which was first observed in metals where the paramagnetism comes from conduction electrons, is called the Knight shift. However, paramagnetic ions will cause the same shift. For instance, the various protons in *n*-propanol are shifted by Co^{2+} ions up to 100 ppm. Peo *et al.* (1979a) reported ^{13}C NMR spectra of $[CH(AsF)_{5y}]_x$ for $y = 0, 0.01$, 0.056, 0.088, and 0.124, which are basically similar to those described above (Clarke *et al.*, 1983, Figs. 8.10–8.12). Namely, above about 6–7 mole % dopant, the NMR line broadened and shifted about 20–40 ppm. These authors attributed this to be the Knight shift and concluded it to be consistent with the onset of Pauli paramagnetism. Clarke *et al.* (1983) did not agree with this interpretation. The observed downfield shift was considered not to be unusual when compared to the shifts observed for the carbon atoms of delocalized carboncation where no conduction electrons are present. For instance, the ^{13}C NMR spectrum of benzene has a chemical shift of 128.7 ppm, whereas the signal of the tropylium ion, which contains one positive charge equally distributed over seven carbon atoms, is shifted to 155.3 ppm. The current uncertain state of magnetic susceptibility of polyacetylene does not warrant definitive attribution of the ^{13}C NMR shift. For instance, the Schumacher–Slichter and Faraday measurements on the intermediately AsF_5-doped polymer show no Pauli susceptibility, whereas it was said to be about ten times greater based on EPR studies. Also Peo *et al.* (1979b) reported that the Curie susceptibility is almost one-half of Pauli susceptibility for $[CH(AsF_5)_{0.07}]_x$, which can also contribute toward the chemical shift. In fact, the downfield shift may be the superposition of any Knight shift from conduction electrons, and unpaired spins and chemical shifts arising from the removal of electrons from the π-system creating positive solitons. Finally, the Knight shift is strongly temperature dependent. It would seem that the ^{13}C NMR spectra of doped polyacetylenes should be measured over a wide temperature range.

The proton nuclear relaxation data for two samples of metallic $[CH(AsF_5)_{0.1}]_x$ are shown in Fig. 5.32. The spin diffusion is expected to be dominated by the carriers. There is a small positive intercept on the T_1^{-1} axis in the figure, indicating that ω_N is smaller than the three-dimensional crossover frequency and $f(\omega_N) \simeq$ const, while $f(\omega_e)$ remain as in Eq. (5.36). Nechtschein *et al.* (1980) used the Pauli susceptibility determined by Weinberger *et al.* (1979) and the slope of the plot in Fig. 5.32 to obtain $\tilde{D}_\| = 1.7 \times 10^{17}$ rad sec^{-1}. Based on the absence of saturation of the relaxation rate at the lowest pumping frequency (see above), $\tilde{D}_\perp$ is estimated to be less than 4×10^{10} rad sec^{-1}.

If the spin diffusion reflects the motion of the charge carriers, then one can presumably deduce from $\tilde{D}_\|$ and $\tilde{D}_\perp$ the mean free path along the polymer chain, L, and the anisotropy for electrical conductivity. For a one-dimensional metal, Nechtschein *et al.* (1980) find

$$L = 2\pi\hbar\chi D_\| a, \tag{9.18}$$

and $L \simeq 35a$. According to Belinsky (1976) the conductivity for the one-dimensional metal is

$$\sigma_\| = 2e^2 L(\pi\hbar\Sigma) \simeq 5 \times 10^4 \ (\Omega \ \text{cm})^{-1}, \tag{9.19}$$

where Σ is the cross section per polyacetylene chain. The anisotropy is given by

$$\sigma_\|/\sigma_\perp = D_\| a^2/D_\perp a_\perp^2 \geqslant 4 \times 10^5. \tag{9.20}$$

Whereas the estimate for $\sigma_\|$ is within an order of magnitude of experimental value, the estimated anisotropy is at least four orders of magnitude greater than the currently observed anisotropy.

9.6 SPECIFIC HEATS

The specific heat for undoped polyacetylene is discussed in Section 6.4. Figure 6.24 shows the results for $[\text{CH}(\text{AsF}_5)_{0.12}]_x$. Unlike the pristine polymers, the heavily doped metallic samples have the traditional form for heat capacity

$$C(T) = \gamma T + \beta T^3 \tag{9.21}$$

The relation is obeyed up to 18 K, $\gamma = 0.86 \ \text{mJ mole}^{-1} \ \text{K}^{-2}$ and $\beta = 1.89 \ \text{mJ mole}^{-1} \ \text{K}^{-4}$. The results were identical for *cis*- and *trans*-polyacetylene doped to the same concentration of AsF_5.

The linear term in undoped *trans*-polyacetylene is very small; $\Delta C = \gamma T$, where $\gamma = 0.29 \ \text{mJ mole}^{-1} \ \text{K}^{-2}$. The value of γ is larger for a doped polymer, which may arise from a combination of increased disorder on doping and the formation of a degenerate electron gas in the metallic state. If we assume the increase in γ to be due to the latter, then

$$\Delta C = (2\pi^2/3)N(0)k_\text{B}^2 T, \tag{9.22}$$

where $N(0)$ is the density of states at the Fermi energy for one sign of spin. From the value of γ, an estimate of 0.12 states $(\text{eV C atom})^{-1}$ was obtained for $N(0)$.

Doping causes large increases in heat capacity and the T^3 dependence. This probably arises from the incorporation of the high-mass dopant ions and the resulting change in lattice dynamics. The dopant may weaken the interchain coupling, leading to new degrees of freedom due to the dopant ion vibrations about equilibrium. Interactions between the dopant ions and between the dopant ion and the polymer chains can give rise to a Debye-like spectrum and the T^3 dependence for heat capacity.

9.7   PHOTOELECTRON SPECTRA

Photoelectron spectroscopy has been used to investigate the core and valence-level energies in doped polyacetylene. In these experiments monochromatic X-ray or ultraviolet photons were used to irradiate the sample kept at $< 10^{-8}$ torr. The electrons emitted from the specimen were collected at or near normal direction. The results were found to be independent of the takeoff angle of the electron because of the fibrous nature of polyacetylene.

One important characteristic of the XPS technique is that the efficiency of detecting the electron depends on its kinetic energy. In other words, the escape depth λ of the photoelectron is energy dependent. The peak area detected for photoelectrons from a certain ion-state energy i can be written

$$I_i = C \int_0^\infty S_i N_i(x) \exp\left(\frac{-x}{\lambda_i}\right) dx, \tag{9.23}$$

where $x = 0$ at the sample surface, C is an experimental constant, and $N_i(x)$ is the number of atoms per unit volume of the species from which the ith electron is ejected. $S_i = T(E_i) \cdot \sigma_i$, where σ_i is the photoionization cross section for level i, and $T(E_i)$ is the transmission function of the spectrometer for electrons of kinetic energy E_i.

It should be appreciated that XPS technique is surface sensitive. The typical electron elastic mean free path encountered in XPS studies lies in the region of $1-3$ nm. Thus for $\lambda \approx 2$ nm, only 10% of the volume of a typical 20-nm diameter polyacetylene fibril is probed by XPS. Typically, for 1.4 keV electrons about 63% of the XPS signal originates within the top 2.1 nm of the sample, 82% emanates from about 4.2 nm, etc. Operationally, one might say that the top 2λ of the sample surface is interrogated by XPS. Consequently, the relative signal intensities of the ith and jth levels of intensity will depend on their respective photon energies. In the gaseous phase there is no need to include the effect of λ. Then the relative peak areas of the two energy levels are

$$R_{ij}(\text{gas}) = I_i/I_j = S_i/S_j \tag{9.24}$$

from the same atoms of the same species. If the atoms are found in a solid specimen, then

$$R_{ij}(\text{solid}) = S_i\lambda_i/S_j\lambda_j. \tag{9.25}$$

For instance, the F($1s$) electron ($E = 560$ eV) and F($2s$) electron ($E = 1210$ eV) in the solid AsF$_5$ or AsF$_3$ give $\lambda(\text{F}2s)/\lambda(\text{F}(1s) \sim 1.6$; and $R_{ij}(\text{solid}) = 1.6R_{ij}(\text{gas})$. If the two photoelectrons originate from levels i and j of two different atoms a and b, then Eq. (9.25) is generalized to

$$R_{ia,jb} = (SN\lambda)_{ia}/(SN\lambda)_{jb}. \tag{9.26}$$

If the species being investigated has some volatility, then the XPS intensity will further depend on the temperature and the duration that the specimen is kept under the very high vacuum of the sample chamber of the spectrometer.

9.7.1   Iodine-Doped Polyacetylene

Several groups have studied iodine-doped polyacetylene with XPS (Hsu *et al.*, 1978; Ikemoto *et al.*, 1979; Salaneck *et al.*, 1980; Thomas *et al.*, 1980). The $3d$ core level of iodine has a high cross section and a reasonably narrow linewidth. In the case of model compounds CsI$_3$ and cobalt phthalocyanane I$_3^-$, there are only sharp lines at 620.6 and 619.0 eV, respectively. In (CHI$_{0.22}$)$_x$ the two lines of I($3d_{3/2}$) and I($3d_{5/2}$) are well separated, each broadened, and the latter can be deconvoluted into two components: one at higher binding energy (620.6 eV) and another of lower binding energy (619.0 eV), as shown in Fig. 9.49. In comparison with model compounds, the lower binding energy component is assignable to I$_3^-$. The higher binding-energy component was attributed to a less negatively charged iodine species by Hsu *et al.* (1978). It decreased in intensity relative to the peak with lower binding energy upon thermal annealing or the application of mechanical stress, which is analogous to the behavior of the line at 150–160 cm^{-1} in Raman spectroscopy. Hence the 620.6-eV XPS line is consistent with photoemission from I$_5^-$. Salaneck *et al.* (1980) observed similar results. The initial *trans*-(CHI$_{0.28}$)$_x$ samples have a 2:1 ratio for the 620.5 eV:619 eV peaks. The two peaks have comparable heights after evacuation. These authors postulated that an excess of I$_2$ was present; removal of this excess led the remainder to form I$_2$ + I$_3^-$ $\longrightarrow$ I$_5^-$. This interpretation would suggest I$_2$ and I$_5^-$ to have the same $3d_{5/2}$ core energy. Because most workers routinely evacuated samples after doping, it would seem that occluded I$_2$ would have been removed. This process should continue when the doped polyacetylene

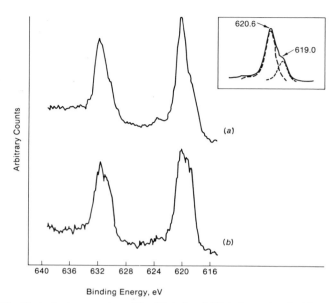

Fig. 9.49   Photoelectron spectra of the $I(3d_{3/2})$ and $I(3d_{5/2})$ core levels: (a) *trans*-$(CHI_y)_x$; (b) *cis*-$(CHI_y)_x$; insert is deconvoluted of the $I(3d_{5/2})$ level of (a). [After Hsu *et al.* (1978).]

is placed in the XPS instrument and pumped to less than 10^{-8} torr, which is a much higher vacuum than those usually used in sample preparation. An explanation, more in line with the Mössbauer results, is that in *trans*-$(CHI_{0.28})_x$ there was about twice as much I_5^- as I_3^-. In the XPS experiments I_5^- is lost by $I_5^- \rightarrow I_3^- + I_2$. The I_3^- ion is much more stable in this regard because its elimination would require the abstraction of a proton from polyacetylene to give I_2 + HI.

The C(1s) core level photoelectrons have a binding energy of 284.2 ± 0.2 eV. Hsu *et al.* (1978) found that the 1.6-eV-wide line was not affected by iodine doping. Salaneck, *et al.* (1980) observed a chemical shift of 0.2 ± 0.2 eV (i.e., on the order of the experimental sample-to-sample reproducibility). If this shift is real, then, based on the literature value of the C(1s) chemical shift slope of 5.6 (± 0.2) eV/electron, there is a charge transfer of no more than 0.03 electron per carbon atom from polyacetylene to the dopant. This corresponds to a distribution of charge over 30 carbon atoms for each I_3^- or I_5^-.

The C(2p) and I(5p) electrons give rise to the XPS valence-band spectra. In the case of heavily doped *trans*-$(CHI_y)_x$, the position of the band maximum is 3.2 eV below E_F; more importantly, the onset of the valence-band structure appears 0.5–0.7 eV below E_F. On the other hand, a similarly doped cis polymer showed a finite density of states at the Fermi level (Fig. 9.50). The latter result is consistent with the metallic conductivity of the material. The difference was suggested to be due to surface depletion of iodine from the heavily doped *trans*-$(CHI_y)_x$ in the spectrometer vacuum. Whereas this is certainly a possibility, the greater ease of removal of surface iodine from the trans polymer than that from the cis isomer is difficult to

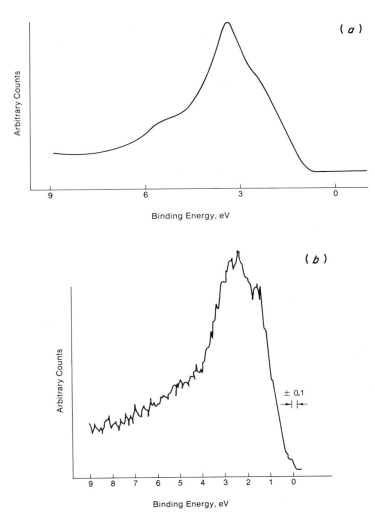

Fig. 9.50   Valence-band XPS spectra: (a) *trans*-$(CHI_{0.22})_x$; (b) *cis*-$(CHI_{0.22})_x$. [After Hsu *et al.* (1978).]

explain because the latter is believed to have been isomerized to the trans structure upon saturation doping.

The kinetic energy of photoelectrons from the $I(3p_{3/2})$ and $I(3p_{3/2})$ levels, as measured by $MgK_{\alpha1,2}$ radiation, differ by about a factor of two. For iodine molecules in the gas phase, the $I(3d):I(3p)$ intensity ratio is found to be 3.0 ± 0.1; it is about 4.8 ± 0.1 for solid I_2 (Salaneck *et al.*, 1980). For all *trans*-$(CHI_y)_x$ samples initially inserted into the spectrometer, this intensity ratio is about 4.0 ± 0.2. This indicates an approximate uniform distribution in the near-surface region. After pumping overnight, the $I(3d):I(3p)$ ratio is lowered to 3.5 ± 0.2, suggesting less uniform distribution of the dopant. Ultraviolet photoelectron spectroscopy (UPS) was also obtained by Salaneck *et al.* (1980), using 21.2- and 40.8-eV photons. The UPS spectrum of *trans*-$(CHI_{0.28})_x$ after pumping resembles the UPS spectrum of undoped polymer. Therefore, there is surface depletion and no surface localization of the iodine-dopant species.

9.7.2   Bromine-Doped Polyacetylene

The $3d$ XPS line of Br decreases in intensity during measurement, indicating loss of dopant. However, unlike the case of iodine-doped poly-acetylene, the $3d_{3/2}$ line changed shape by pumping (Section 9.7.1), and the $3d$ line of bromine-doped polymer retains its symmetric line shape and line width of $\sim 2.3-2.6$ eV. Therefore, the dopant exists almost exclusively as Br_3^-, in agreement with Raman spectra. This is consistent with the fact that bromine is lost as both Br_3 and HBr. Furthermore, the binding energy of $Br(3d)$ is the same for all levels of doping.

The $C(1s)$ peak of *trans*-$(CHBr_y)_x$ is sensitive to the doping level (Fig. 9.51) and consists of two components separated by $1.7-2.0$ eV. If we denote the one with the higher binding energy as $C^*(1s)$ and the component of lower binding energy as $C^\circ(1s)$, then the intensity of the former increases whereas that of the latter decreases with increasing bromine content. $C^*(1s)$ cannot be attributed to the carbon atoms forming $C-Br$ bonds because its intensity decreases in parallel to the decrease of bromine content and the chemical shift difference of $C(1s)$ between $C-Br$ and $C-H$ carbon atoms is expected to be less than 1 eV. Ikemoto *et al.* (1982) proposed that $C^*(1s)$ is associated with positively charged carbon atoms that are interacting with Br_3^- atoms. Elimination of bromine will convert these atoms to others, such as $C-Br$ or $C\equiv C$, with a corresponding decrease in $C^*(1s)$ intensity. Based on the chemical-shift difference between the two components of the XPS spectra, the charge transfer is about 0.3 electron per carbon atom. This

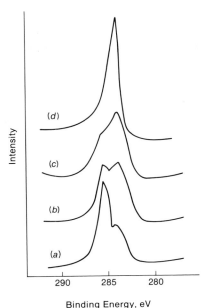

Intensity

(d)

(c)

(b)

(a)

290   285   280

Binding Energy, eV

Fig. 9.51 C($1s$) core XPS spectra of *trans*-(CHBr$_y$)$_x$ with y values: (a) 8×10^{-3}; (b) 0.127; (c) 0.368; (d) 0.696. [After Ikemoto *et al.* (1982).]

tenfold larger electron density as compared to iodine-doped polyacetylene may be due to the fact that the Br$_3^-$ ion is much smaller than the I$_3^-$ ion.

As pointed out above, XPS is a surface probe. Therefore, Ikemoto *et al.* (1982) compared the bromine to carbon atomic ratio for the bulk polymer by elemental analysis and the ratio at the surface by the integrated intensity ratio of Br($3d$) and C($1s$) peaks. The results are summarized in Table 9.5. The values of y obtained by the two methods are nearly equal for $y > 0.127$; the values obtained by XPS are larger than those from elemental analysis for $y < 0.127$. This shows that bromine content is more concentrated in the surface region than in the bulk at low doping, but is more homogeneously distributed in heavily doped specimens.

Comparison of y values obtained before and after XPS measurements showed that there is a decrease of only a few percent for samples up to 30% in bromine by elemental analysis. On the other hand according to XPS (Columns 5 and 6 of Table 9.5) the decrease of the Br/C ratio is as large as 3.8 for most heavily doped polymers. This decrease became progressively smaller with the lowering of doping level. In all the specimens investigated, most of the bromine, as determined by XPS, was lost in the first eight hours in vacuo, after which there was little change. These observations imply

TABLE 9.5

XPS Results of Bromine-Doped Polyacetylene[a]

Sample	Electrical conductivity $(\Omega\ cm)^{-1}$	Elemental analysis		$\dfrac{Br(3d)^c}{C(1s)}$		$\dfrac{C*(1s)^b}{C(1s)}$
		Initial[d]	Final[e]	Initial	Final	
$trans\text{-}(CHBr_y)_x$	$<10^{-8}$	0.697	0.665	0.76	0.20	0.74
	1.2×10^{-8}	0.661	0.647	0.70	0.34	0.75
	1.8×10^{-2}	0.633	0.576	0.63	0.19	0.69
	3.5	0.368	0.262	0.39	0.18	0.43
	1.3×10^1	0.127	0.112	0.31	0.12	0.33
	3.0	0.042	0.035	0.19	0.09	0.28
	1.5×10^{-1}	0.008	0.002	0.03	0.02	—
	2.9×10^{-3}	~ 0	~ 0	~ 0	~ 0	—
$cis\text{-}(CHBr_y)_x$	2.0×10^1	0.194	0.188	0.27	0.14	0.39
	5.6	0.264	0.246	0.39	0.16	0.53

[a] From Ikemoto *et al.* (1982).
[b] Relative area ratio of the $C*(1s)$ component to the total $C(1s)$ peak.
[c] Relative area ratio of the $Br(3d)$ to the $C(1s)$ peaks.
[d] Sample first placed in instrument.
[e] Sample after 15–20 hr of repetition of XPS measurements.

preferential depletion of surface dopant, as was concluded also in iodine-doped polyacetylene.

Comparison of Columns 5 and 7 of Table 9.5 showed that the $C*(1s)/C(1s)$ ratios are nearly the same as the $Br(3d)/C(1s)$ ratios. Therefore, the ratio of positively charged $C*$ atoms to Br is always about 1:1 and the two independent estimates for charge on carbon atoms are in agreement (Table 9.6).

TABLE 9.6

Charge per Carbon Atom in Bromine-Doped Polyacetylene

Sample	Estimated for $C*/Br$ ratio by $C*(1s)/Br(3d)$ of XPS		Estimate from the chemical shift of $C*(1s)$ and $C°(1s)$	
	$C*/Br_3^-$	Charge/C atom	ΔE (eV)	Charge/C atom
$trans\text{-}(CHBr_{0.127})_x$	3.2	0.31	1.7	0.30
$trans\text{-}(CHBr_{0.042})_x$	4.5	0.22	1.5	0.26
$cis\text{-}(CHBr_{0.194})_x$	4.4	0.23	2.0	0.26

9.7.3 AsF$_5$-Doped Polyacetylene

Photoelectron spectra have been obtained with XPS, UPS, and XAES (X-ray-induced Auger electron spectroscopy) (Salaneck *et al.,* 1979a,b; Thomas *et al.,* 1980) for *trans*-[CH(AsF$_5$)$_y$]$_x$ and model compounds AsF$_3$, AsF$_5$, and NaAsF$_6$. For the latter the binding energies for F(1s) and F(2s) are 0 and 654.6 $\pm$ 0.1 eV, respectively, and all kinetic energies are relative to the F(1s) level. The As(3d) levels, referenced to the F(2s) level, are at 639.5 $\pm$ 0.2, 638.5 $\pm$ 0.2, and 637.9 $\pm$ 0.2 eV, respectively, for AsF$_3$, AsF$_5$, and NaAsF$_6$. The As(3d)/F(2s) ratio was found to be 10.8/F atom. Four samples of *trans*-[CH(AsF$_5$)$_y$]$_x$ were studied with y values (for samples): 0.11 (A), 0.013 (B), 0.002 (C), and 0.001 (D).

From the As(3d)/F(2s) peak-area intensity ratios, the As:F ratio was found to be 1:4.9 for the most heavily doped sample A. However, the ratio is only 1:3 for the other three samples. Therefore, the surface dopants are deficient in fluorine atoms in lightly doped samples.

Even though the escape depth of the F(2s) photoelectron is expected to be $\sim$ 1.6 times greater than that of the F(1s) core level (see above) in the condensed state, the F(1s)/F(2s) peak ratio for sample A is near that for gaseous AsF$_5$. This was interpreted to mean that there is a tendency for the dopant to be concentrated in a region near the surface of the fibrils. On the other hand, the dopants in the other three specimens are not so obviously localized. However, the observation that the F(1s)/F(2s) intensity ratios in these samples are actually greater than those in free gaseous AsF$_5$ molecules is perplexing and has no easy explanation.

The C(1s) level spectra of heavily doped polyacetylene differ from those of undoped polymer in two aspects (Fig. 9.52). The main peak is moved to higher binding energy by about 0.6 eV. From Siegbahn's experimental correlation, this chemical shift corresponds to 0.11 $\pm$ 0.01 electrons per carbon atom, which is equivalent to about one electron per AsF$_5$ moiety. In addition, the C(1s) spectra can be deconvoluted into two components separated by 1.5 eV. Salaneck *et al.* (1979b) thought that a nonuniform distribution of dopant ion may somehow contribute to the phenomenon. However, in view of the similar higher binding-energy component seen in (CHBr$_y$)$_x$, (Section 9.7.2), the component here may also be attributed to positively charged carbon atoms. The smaller energy difference of 1.5 eV separating the two C(1s) components, as compared to up to 2.0 eV in bromine-doped polyacetylene, is consistent with a lower positive charge on the carbon atom (0.1/C), as compared to higher values (Table 9.6) in the AsF$_5$- and Br$_2$-doped polymers, respectively.

The valence-level XPS spectra of AsF$_5$-doped polyacetylene is shown

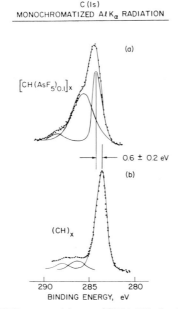

C (1s)
MONOCHROMATIZED AℓKₐ RADIATION

$\left[\text{CH}\left(\text{AsF}_5\right)_{0.1}\right]_x$

(a)

0.6 ± 0.2 eV

(b)

(CH)ₓ

290   285   280
BINDING ENERGY, eV

Fig. 9.52   C($1s$) core XPS spectra: (a) *trans*-[CH(AsF$_5$)$_{0.1}$]$_x$; (b) undoped *trans*-polyacety-lene. [After Salaneck *et al.* (1979b).]

along with gaseous AsF$_5$, condensed AsF$_3$, and undoped polyacetylene in Fig. 9.53. The energy scale is that of gaseous AsF$_5$, and all other spectra are referenced to it. The spectra of AsF$_5$ and [CH(AsF$_5$)$_{0.11}$]$_x$ are very similar, and they are very different from that of AsF$_3$. Therefore, AsF$_3$ is not an important dopant. This is supported by the As($3d$)–F($2s$) core level energy spacing of 638.3 ± 0.2 eV for the doped polyacetylene, which is very close to the value of 638.5 ± 0.2 eV for AsF$_5$.

The 40.8-eV photon USP spectra (Fig. 9.54) show that the onset of photoemission, referenced to the cutoff of the secondary energy spectrum, occurs at 4.5 ± 0.1 eV and 5.7 ± 0.1 eV for *trans*-(CH)$_x$ and *trans*-[CH(AsF$_5$)$_{0.11}$]$_x$, respectively. Therefore, the binding energy (work function) of the polymer is raised about 1.2 eV upon doping, as expected if electronic charge is withdrawn from low binding-energy states of the polymer. There is no indication from the spectra of a well-defined step at the Fermi energy level. As the density of states is of the order of 0.1 states eV^{-1} per carbon atom (both signs of spin), it would be difficult to detect these states by means of photoemission.

There is in the valence-electron photoemission spectrum of *trans*-

Valence Band

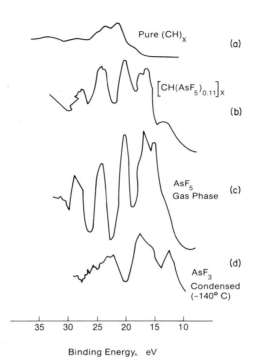

Fig. 9.53 Valence-band XPS spectra: (a) pristine *trans*-polyacetylene; (b) *trans*-$[CH(AsF_5)_{0.11}]_x$; (c) AsF_5 gaseous; (d) AsF_3 condensed at 133 K. [After Slaneck *et al.* (1979b).]

$[CH(AsF_5)_{0.11}]_x$ a new hole-state level that is absent in either AsF_5 or undoped polymer and occurs at a lower binding energy relative to the latter. This new level is marked with the arrow in Fig. 9.54b. Also in Fig. 9.54a there is no intensity in this energy region of *trans*-polyacetylene corresponding to the new level in 9.54b of the doped polymer. Further evidence for this additional energy level is provided by a comparison of the highest kinetic-energy positions of the fluorine KLL Auger spectra of AsF_5-doped polyacetylene and AsF_5 vapor (Fig. 9.55). The new hole-state level is seen at about 4.4 eV above the highest energy peak in AsF_5. These XAES results show that there is an additional occupied energy level in *trans*-$[CH(AsF_5)_{0.11}]_x$ that is lower in binding energy than the highest occupied orbital in the AsF_5 molecule and has a finite amplitude on the F atom. The most probable explanation is that the dopant ion is $As_2F_{10}^{2-}$, although $HAsF_4$, AsF_6^-, and more extended configurations of $(AsF_5)_{2n}^{2n-}$ are other possibilities.

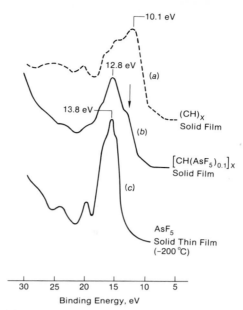

Fig. 9.54 The 40.8-eV photon UPS spectra: (a) pristine *trans*-polyacetylene; (b) *trans*-[CH(AsF$_5$)$_{0.11}$]$_x$; (c) AsF$_5$ condensed at 73 K. [After Salaneck *et al.* (1979b).]

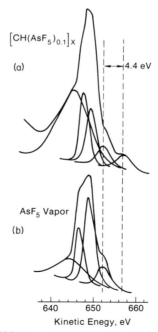

Fig. 9.55 The fluorine KLL Auger electron spectra: (a) *trans*-[CH(AsF$_5$)$_{0.11}$]$_x$; (b) AsF$_5$ gaseous. [After Salaneck *et al.*(1979b).]

9.8   STABILITY

One of the frequently posed questions concerns the stability of conductive polyacetylenes. Under pyrolysis conditions, polyacetylene doped with I_2 and AsF_5 are similar in stability to undoped polyacetylene (Section 7.1.1.3). In most respects, doped polyacetylenes are surprisingly stable, considering that the *p*-doped polyacetylenes are essentially macrocarbocations. In some instances the stability of polyacetylene increases upon doping. The rapid autoxidations of polyacetylene and poly(methylacetylene) were treated in Section 7.5. The doped polyacetylenes are by comparison relatively stable toward autoxidation. The chemical reactivity of a dopant can be greatly reduced upon charge-transfer reaction with polyacetylene. The most noted example is AsF_5, which reacts instantly with water vapor when exposed to laboratory air, yet as a dopant it is rather unreactive to water vapor. An exception is the unusual thermal instability of $[CH(ClO_4)_y]_x$.

9.8.1   Iodine- and Other Halogen-Doped Polyacetylenes

During the doping of polyacetylenes by exposure to iodine, there exist in the polymer I_3^- and I_5^-, and maybe some that has not undergone charge transfer. The ratio of the two polyiodide ions is dependent on the doping level. Free iodine usually is present in small amounts; it is readily removable and does not contribute to the transport properties. For example, when a *cis*-polyacetylene sample is mounted on a four-probe and doped to $y = 0.25$, it has a resistance of $1.5\ \Omega$. Pumping at room temperature reduced y to 0.18 without much change in resistivity. After 24 hr under vacuum, y is lowered to ~ 0.16 and the resistance is changed to $4.5\ \Omega$. This slow change continued so that after 200 hr y is about 0.14, whereas the resistivity remains relatively constant. The above behavior may be interpreted as initial rapid removal of free iodine, whereas the much slower change in composition may be attributed to the elimination process: $I_5^- \longrightarrow I_3^- + I_2$. During the above treatment the film changes from a silvery luster to a shiny black. The loss of the metallic luster may be due to depletion of dopants in the surface regions of the fibrils. The silver appearance can be restored by redoping, indicating the above processes to be reversible.

Iodine-doped polyacetylene is relatively stable to both oxygen and moisture. We have doped polyacetylene foam (density $0.166\ \mathrm{g\ cm^{-3}}$) to $y = 0.25$ and a conductivity of $113\ (\Omega\ \mathrm{cm})^{-1}$. The sample was successively exposed to oxygen, pumped, and exposed to water vapor. The results for duplicate

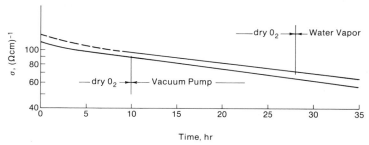

Time, hr

Fig. 9.56   Change in conductivity of $(CHI_{0.25})_x$ foam with time at 298 K subjected to indicated treatments.

specimens (Fig. 9.56) showed that the gradual decrease in conductivity is neither accelerated nor abated by the treatments. Kiess *et al.* (1980) have followed the change of current of polyacetylene saturation-doped with iodine as a function of time of exposure to oxygen. The rise in resistivity was exponential both for undoped and doped polymers. In the presence of oxygen at room temperature the half-life is about one day for undoped polymer and two to three days for the iodine-doped polyacetylene.

The enhanced oxidative stability of heavily p-doped $(CHI_y)_x$ may be attributed to reduced availability of sites for further p-doping by oxygen, which is prerequisite for further chemical reactions, leading to loss of conductivity. It seems reasonable to propose that there is a competition between the dopants:

$$(CHI_y)_x + O_2 \rightleftharpoons [CHI_{y'}O_{2y''}]_x \rightleftharpoons [CHO_{2y}]_x \tag{9.27}$$

where y', $y'' < y$. The formation of $(CHO_{2y})_x$ eventually leads to a nonconducting material, such as through reactions proposed in Section 7.5.

Pron *et al.* (1981) have studied the thermal properties of $(CHI_y)_x$. Using TGA and DSC, they found exotherms and weight loss at 145 and 267°C, as shown in Fig. 9.57. The main products are HI and I_2 because the mass spectra showed peaks only for I_2^+, HI^+. I^+, and I^{2+}. The analysis for the initial material corresponds to $(CHI_{0.14})_x$, which becomes $(CHI_{0.009})_x$ after thermolysis.

The kinetics for weight loss are given in Fig. 9.58 at several temperatures showing increasing amounts of decomposition with increasing temperature. At the same time, there is corresponding loss of conductivity (Fig. 9.59). These losses in iodine and conductivity occurring at elevated temperatures are not reversible on redoping with iodine. In these reactions, the I_5^- (162 cm^{-1}) Raman band decreased immediately as the p-doped polyacetylene is heated to 100°C, whereas the I_3^- band intensity at 107 cm^{-1} increased. On further heating, both bands continuously decrease in intensity. Figure 9.60 shows this change where the bands at 214 and 320 cm^{-1} are the first two

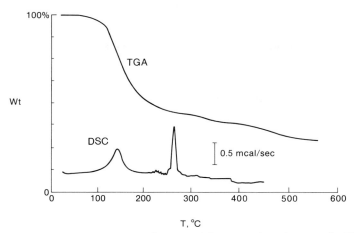

Fig. 9.57 Differential scanning calorimetry and thermogravimetric curves for $(CHI_{0.14})_x$. [After Pron *et al.* (1981).]

overtones of the 107-cm^{-1} line. The initial reactions are probably

$$[CH(I_5)_{y'}]_x \xrightarrow{\Delta} I_2 + [CH(I_3)_{y''}]_x \xrightarrow{\Delta} I_2 + (CHI_{y'''})_x \qquad (9.28)$$

These reversible reactions are followed at elevated temperatures of iodination and dehydroiodination processes analogous to those described for bromination and dehydrobromination in Section 7.3, though there are probably differences in details. That almost all iodine can be removed from the doped polyacetylene, that HI was produced, and that organic iodides were separated and identified in the pyrolysis of iodine-doped polyacetylene (Section 7.1.1.3) support the above mechanism.

The principal difference between the stabilities of iodine- and bromine-

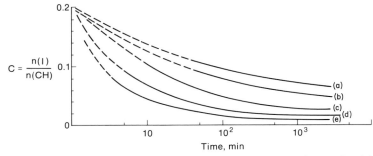

Fig. 9.58 Rate of desorption of iodine from $(CHI_{0.14})_x$ under dynamic pumping: (a) 353 K; (b) 373 K; (c) 393 K; (d) 423 K; (e) 453 K. [After Pron *et al.* (1981).]

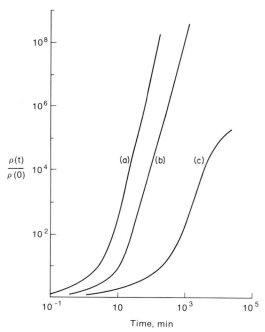

Fig. 9.59   Variation of resistivity of $(CHI_{0.14})_x$ under dynamic pumping: (a) 413 K; (b) 393 K; (c) 353 K. [After Pron *et al.* (1981).]

doped polyacetylenes is the product distribution upon heating. In the former case, heating produced more I_2 than HI; aged samples ($\sim$ 2 months) yield I_2 and $\sim$ 40% HI. In the case of $(CHBr_y)_x$, HBr is the dominant thermolysis product with small amounts of Br_2 corresponding to $\sim$ 5% of HBr. This trend is extended to $(CHCl_y)_x$, which upon heating liberates only HCl. This can be interpreted as indicating that either dehydrochlorination is the only reaction during thermolysis, or that any Cl_2 produced reacts immediately to chlorinate the double bonds (Allen *et al.*, 1979). By contrast with the other halogens, the reaction of fluorine with polyacetylene is so vigorous that the polymer film is destroyed. Large amounts of white, Teflon-like flakes were all that remained after treatment with F_2.

9.8.2   AsF$_5$- and SbF$_5$-Doped Polyacetylenes

Polyacetylenes heavily doped with AsF$_5$ responded to dynamic pumping, like those doped with I_2. For instance, a sample of $[CH(AsF_5)_{0.147}]_x$

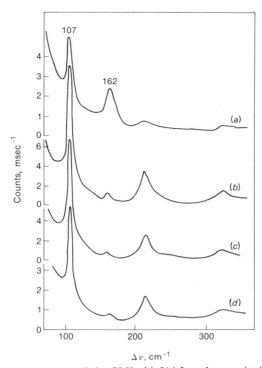

Fig. 9.60 Raman spectra recorded at 78 K with 514.5-nm laser excitation of iodine-doped *trans*-polyacetylene: (a) initial sample and after evacuation at 383 K for (b) 1.75 hr; (c) 3.75 hr; (d) 7.75 hr. Note changes in coordinate scales. [After Rolland *et al.* (1981a).]

subjected to room-temperature evacuation loses 14% of the dopant very rapidly. This was followed by much slower decrease of AsF_5 content reaching a $y \approx 0.12$; the resistivity change during this period was only from 1 to 1.5 Ω after 30 hr. Further pumping up to 120 hr caused no measurable reduction of dopant. There was no marked change in the appearance of the polymer film. The results can be explained by removal of some free AsF_5 or dissociation of other dopant species in heavily doped material.

AsF_5-doped polyacetylene is relatively insensitive to oxygen, like other p-doped polymers. What is remarkable is the apparent stability of the dopant toward moisture. We have subject $[CH(AsF_5)_y]_x$ to repeated cycles of exposure to oxygen, water vapor, and evacuation. The results given in Figs. 9.61 and 9.62 show that the reduction in conductivity was always less than one order of magnitude and that some of the effects of oxygen and water are detrimental but reversible and at other times even slightly beneficial.

Weber *et al.* (1981) have prepared many $[CH(AsF_5)_y]_x$ samples with y

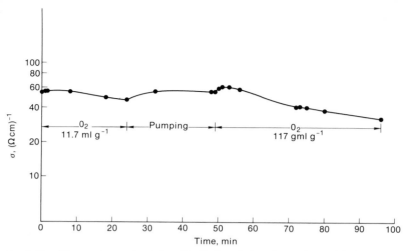

Fig. 9.61 Changes in conductivity of AsF$_5$-doped *cis*-polyacetylene with time at 298 K subjected to indicated treatments.

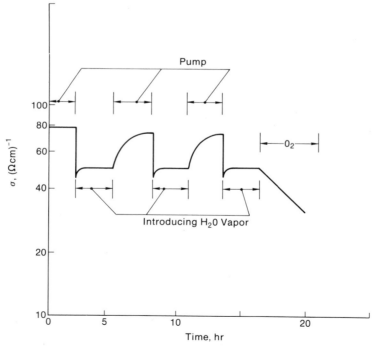

Fig. 9.62 Changes in conductivity of AsF$_5$-doped *trans*-polyacetylene with time at 298 K subjected to indicated treatments.

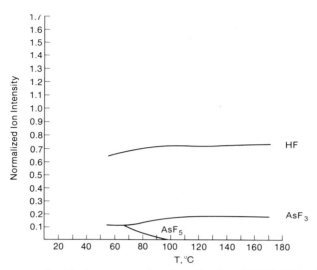

Fig. 9.63   Normalized ion intensity as a function of heating of $[CH(AsF_5)_y]_x$. [After Weber *et al.* (1981).]

ranging from 0.026 to 0.108. Elemental analysis showed that the F:As ratio varied from 5.02 to 6.0. Therefore, the dopant exists as different amounts of $As_2F_{10}^{2-}$ and AsF_6^-. Upon heating to 328 K, evolution of AsF_5, AsF_3, and HF commences (Fig. 9.63). The formation of AsF_5 decreases with increase of temperature and it is not detectable above 373 K. On the other hand, the amounts of HF and AsF_3 are in a constant ratio of about 4:1 up to 453 K. Based on these results, the main decomposition process of the dopant appears to be

$$[CH(AsF_6)_y]_x \xrightarrow{\Delta} xyAsF_3 + 3xyHF + (CH)_x \qquad (9.29)$$

form the AsF_6^- dopant ion. For the AsF_5 dopant species, it can first dispro-portionate:

$$[CH(AsF_5)_y]_x \rightarrow [CH(AsF_6)_{5y/6}]_x + (xy/6)AsF_3 \qquad (9.30)$$

followed by the reaction of Eq. (9.29).

SbF$_5$ is one of the more stable *p*-type dopants. It is dissociated from doped polyacetylene slowly at 353 K and occurs at appreciable rate only at elevated temperatures (Fig. 9.64). The increase in resistivity with heating is shown in Fig. 9.65. There is virtually no change in conductivity after heating at 353 K for 160 hr.

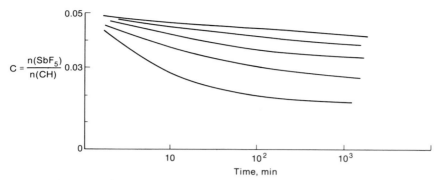

Fig. 9.64   Loss of dopant from $[CH(SbF_5)_{0.05}]$ under dynamic pumping: from top to bottom 353 K, 373 K, 393 K, 423 K, 453 K. [After Rolland *et al.* (1981a).]

9.8.3   Perchlorate-Doped Polyacetylenes

Perchloric acid-doped polyacetylenes appear to be very insensitive to O_2 and H_2O. We have doped polyacetylene foam (density 0.09 g cm^{-3}) with $HClO_4$ to $y = 0.09$ and conductivity of 65 $(\Omega\,cm)^{-1}$. Note that low conductivity is associated with low polymer density. Figure 9.66 shows the effect of O_2 and pumping on its conductivity. Initial exposure to 760 torr of

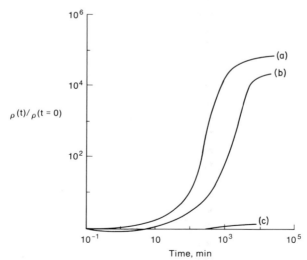

Fig. 9.65   Variation of resistivity of $[CH(SbF_5)_{0.05}]_x$ under dynamic pumping: (a) 413 K; (b) 393 K; (c) 353 K. [After Rolland *et al.* (1981a).]

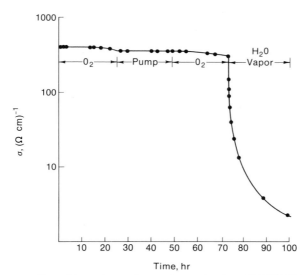

Fig. 9.66   Effect of O_2 and pumping on the conduction of $[CH(HClO_4)_{0.09}]_x$.

O_2 resulted at first in a slight increase in σ_{RT}, suggesting a slight doping effect and then a decrease. During this treatment 11.7 ml g^{-1} of oxygen was absorbed by the polymer, its composition is approximately $[CH(HClO_4)_{0.09}O_{0.013}]_x$. Pumping restored the sample to its original conductivity. Finally, the specimen was exposed to oxygen, upon which it absorbed 117 ml g^{-1} of oxygen, resulting in a net loss of σ_{RT} of only 40%. This is remarkable because the composition of the polymer is about $[CH(HClO_4)_{0.09}O_{0.13}]_x$. This shows that oxygenation is relatively ineffective in disrupting the transport properties of $HClO_4$-doped polyacetylene.

The above polymer was also insensitive to H_2O, except for reversible adsorption and desorption processes, as shown in Fig. 9.67. Exposure to water vapor caused a 40% drop in conductivity, which is virtually all recovered by pumping. The process can be repeated with no net loss in transport property. Apparently the water vapor increases the solvation of the ClO_4-dopant. The H_2O molecules of solvation are weakly coordinated and are readily eliminated by evacuation.

It is interesting to compare polyacetylene doped with ClO_4^- by electrochemical or chemical methods with the $HClO_4$-doped material. Sample A was electrochemically doped at 3.8 V in the presence of $LiClO_4$ (10 wt % in propylene carbonate). The initial current was 1.37 mA and doping time was 130 min. The product $[CH(ClO_4)_{0.057}]_x$ has $\sigma_{RT} = 403$ (Ω cm)$^{-1}$. Figure 9.68 shows O_2 to have virtually no effect on the transport property. However,

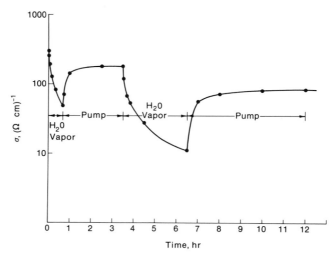

Fig. 9.67 Effect of adsorption and desorption of H_2O on conductivity of $[CH(HClO_4)_{0.09}]_x$.

exposure to water vapor reduced σ_{RT} drastically by more than two orders of magnitude to ~ 2 $(\Omega\ cm)^{-1}$. This effect of H_2O was partially reversible, as shown in the same figure. After two complete cycles of exposure to water vapor and pumping, the sample lost only 20% of its conductivity even though during the entire experiment σ_{RT} has been lowered by 40-fold.

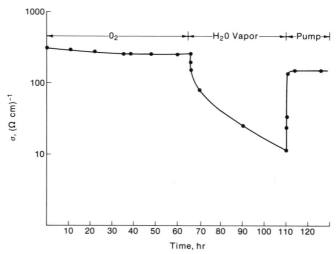

Fig. 9.68   Effect of O_2, pumping, and H_2O on the conductivity of electrochemically doped $[CH(ClO_4)_{0.057}]_x$.

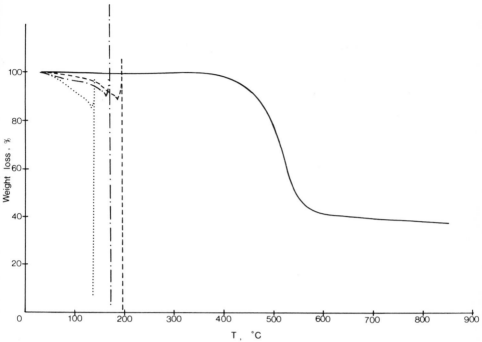

Fig. 9.69 TGA curves: (———) undoped polyacetylene; ($\cdots$) $Fe(ClO_4)_3$-doped $[CH(ClO_4)_{0.065}]_x$; (– – – –) perchloric acid-doped $[CH(HClO_4)_{0.11}]_x$; (– · – · · – · –) electrochemically doped $[CH(ClO_4)_{0.057}]_x$.

We have shown that polyacetylene can be doped with an organic solution of $Fe(ClO_4)_3$ (Section 8.2.3). This process is the redox equivalent of the electrochemical process. The polymer $[CH(ClO_4)_{0.065}]_x$ obtained by the new method responds virtually the same to water vapor as did the electrochemically doped polymer.

The main difference between $[CH(HClO_4)_y]_x$ lies probably with the fact that protonic acid-doped material contains a hydrated anion whereas the anion is unsolvated in the latter materials. Exposure to water vapor in the latter solvates the anions to alter their coulombic interactions with the positive charge on the polymer chain. Water vapor has much less of an effect on the already solvated anion in protonic acid-doped polyacetylene.

Polyacetylenes containing ClO_4^- ions showed extreme thermal instabilities. Figure 9.69 gives the TGA curves. The perchloric acid-doped $[CH(HClO_4)_{0.11}]_x$ began to lose weight at 303 K; the sample exploded at 403 K at 15% weight loss. The $Fe(ClO_4)_3$-doped polymer $[CH(ClO_4)_{0.005}]_x$ exploded at 453 K after 10% weight loss. The electrochemically doped specimen is the most stable of the three. The specimens that by analysis were found to have the composition $[CH(ClO_4)_{0.05}]_x$ lost 10.5% by weight between 313 and 456 K, at which point explosion occurred. The sensitivity to explosion is probably not due to different doping methods but rather to a function of the ClO_4^- content. The sensitivity appears to be related to increasing $[ClO_4^-]$ in the polymer. Some of the lightly doped sample did not explode even when heated to 1273 K. The mechanism of weight loss beginning at about room temperature is being studied.

Chapter 10

Theoretical Models

10.1 BAND STRUCTURE

The principles for band theory were given in Section 1.2, and the merging of molecular orbitals into energy bands, such as with the increase of the length of the $(CH)_x$ chain, was discussed in Section 1.7. In the Hückel molecular-orbital (MO) model for cyclic conjugation molecules of $2N$ atoms in quantum chemistry, the eigenvalues are

$$E_n = \alpha + 2\beta \cos(n\pi/N), \qquad n = -N, -N+1, \ldots, +N. \quad (10.1)$$

The one-electron parameter α is the Coulomb integral, and β is the resonance integral. As the chain length, i.e., $2N$, increases, the density of eigenstates becomes continuous and the density of states is being counted in the momentum space; k, the energy, is

$$E(k) = \alpha + 2t_0 \cos ka, \qquad -\pi/a \leqslant k \leqslant \pi/a. \quad (10.2)$$

In solid-state physics t_0 is referred to as the transfer or hopping integral. The relation between Eqs. (10.1) and (10.2) is obviously

$$k \equiv n\pi/Na, \quad (10.3)$$

and β and t_0 are the same integrals.

The energy of the lowest excitation in short polyenes scales with inverse chain length according to MO theory by Lennard-Jones (1937) and Coulson

(1938), and the bond length tends to a constant value of 1.38 Å. Whereas MO theory predicts a vanishing gap between valence and conduction bands, spectroscopic data showed an asymptotic energy gap of ~ 2 eV for very long polyenes. Kuhn (1948) proposed that bond alternation should persist in long polymers. Longuet-Higgins and Salem (1959) proved that according to linear-combination-of-atomic-orbitals (LCAO) theory an infinite polyene with uniform bond length is unstable and is stabilized by bond alternation. The uniform bond-length molecule is said to undergo "dimerization" to give two degenerate bond alternation molecules of lower energy. (Note to polymer chemists: this dimerization should not be mistaken as the reactions of two monomers to a dimer in a step-growth polymerization.)

Simple band theory also predicts distortion of all linear chains of atoms (Peierls, 1955). An electron is in a periodic field of force, with a potential energy $V(r)$,

$$V(r + a) = V(r), \tag{10.4}$$

where $V(r)$ has the translational symmetry of the lattice. If all atoms are equidistant, separated by a, then all multiples of a are also lattice vectors. The basic cell in reciprocal space is the interval

$$-\pi/a < k < \pi/a. \tag{10.5}$$

If the chain is distorted by displacing each atom a little, the displacement repeats every n atom. This immediately reduces the translational symmetry. The unit cell now contains n atoms and has $-\pi/(na) < k < \pi/(na)$. Only multiples of na are lattice vectors. The energy curve $E(k)$ for an electron will be modified by the distortion; this curve became n different bands with the reduced reciprocal cell. The effect of distortion is to separate any two energy values that are close. The mean of the two energies remains unchanged except for second-order terms arising from the interaction with more distant states. The gain of energy is greatest when n is a small number. The most favorable case is for $k = \pi/2a$, or one electron per atom. In this case $n = 2$, or every other atom is displaced. Therefore, a one-dimensional model could never have metallic properties.

This principle can be readily applied to $(CH)_x$. Let us consider a hypothetical *trans*-polyacetylene molecule with uniform C—C bond length, i.e., ideally conjugated system. In Fig. 10.1 each dot represents a CH unit, and a is the unit cell vector. Since there are two electrons in the unit cell, the polymer should be an insulator. However, there is a screw axis of symmetry, that is, through a half-cell translation $(x \rightarrow x - a/2)$ followed by inversion through the yz plane $(y \rightarrow -y)$, which is the screw-axis operation which carries one CH in the unit cell to the other one, i.e., $V(x, y) = V(x - a/2, -y)$.

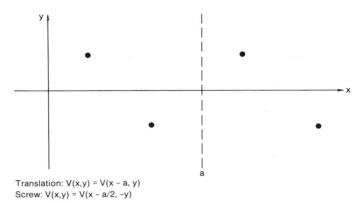

Translation: $V(x,y) = V(x - a, y)$
Screw: $V(x,y) = V(x - a/2, -y)$

Fig. 10.1 Schematic representation of CH units in a *trans*-polyacetylene unit cell with screw symmetry. [After Grant and Batra (1979a).]

The effect of this screw-axis symmetry on the conducting property of polyacetylene is shown in Fig. 10.2. In the absence of this symmetry, the reduced plane wave representation of the systems showed a band gap at $k = \pi/a$ (Fig. 10.2b), the first Brillouin zone is filled, and the system is an insulator. However, with the screw-axis symmetry, the potential coefficient at the reciprocal lattice point $k = \pi/a$ is eliminated and the gap is removed (Fig. 10.2a), and the first Brillouin zone is half-filled. Consequently, *trans*-polyacetylene with uniform bond length as well as screw-axis symmetry should be an intrinsic metal. This is contrary to all experimental evidence showing a band gap of 1.4 eV. Peierls distortion of index two results in the alternation in C—C bond lengths, which restores the gap at odd integral multiples of π/a. The slightest degree of bond alternation will mix states just above and just below the Fermi level, crossing at $k = \pi/a$, producing a gap to lower the ground-state energy of the π system at the expense of a smaller energy increase due to σ compression. The process may be thought of as a solid-state manifestation of the Jahn–Teller effect familiar to chemists for inorganic complexes endowed with electronic degeneracy in the ground state. The reason such distortion does not occur in low molecular weight aromatic compounds like benzene is that the ground and excited states are separated by more than 4 eV so the mixing is very small. In the case of higher molecular weight annulenes, the energy separation is much smaller and a tendency toward bond-alternation distortion has been noted.

The effect of bond alternation on the energy bands of one-dimensional *trans*-polyacetylene has been calculated by Grant and Batra (1979a,b) using the LCAO extended tight-binding approximation (Fig. 10.3) with the Slater local exchange parameters. Figure 10.3a showed that for uniform bond

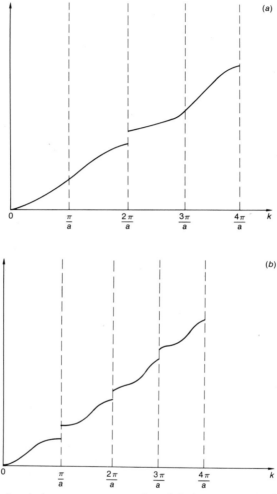

Fig. 10.2   Reduced plane wave representation of the band structure of *trans*-polyacety-lene: (a) with screw-axis symmetry; (b) without screw-axis symmetry.

length, there is no band gap at the Fermi level, and the band structure is that for a metal. For the case of complete bond alternation (Fig. 10.3c), i.e., the double and single bond lengths correspond to those in ethylene and ethane, respectively, and $E_G \approx 2.3$ eV. One weakly alternating structure is shown in Fig. 10.3b; from the experimentally estimated bond alternation (Section

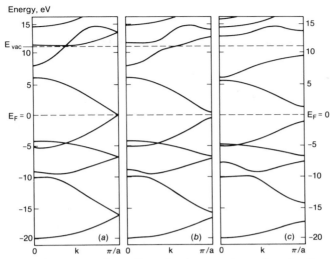

Fig. 10.3 Band structure of one-dimensional *trans*-polyacetylene for three choices of bond lengths: (a) uniform (1.39 Å); (b) weakly alternating (C=C: 1.36 Å; C—C: 1.43 Å); (c) strongly alternating (C=C: 1.34 Å; C—C: 1.54 Å). [After Grant and Batra (1979).]

3.7) the band structure is intermediate between those shown in Figs. 10.3a and b.

The unit cell of *cis*-polyacetylene is twice that of the *trans*-polyacetylene unit cell. An internal screw-axis symmetry is present regardless of whether bond alternation exists or not. There is always an even number of electrons in the unit cell because of the screw-axis operation, and the valence band is full. There is no Peierls-type distortion. Instead, bond alternation must be of some other origin such as nonbonded interaction of the hydrogen atoms. The two cis structures, cis-transoid and trans-cisoid, have different energies, as was pointed out in Section 3.2.

10.2 MICROSCOPIC DISCRETE MODEL FOR A NEUTRAL SOLITON

The concept of π-phase kinks were first introduced by Rice (1979), Rice and Mele (1980), and Su *et al.* (1979) for the defects in *trans*-polyacetylene. This model is now discussed in this section.

10.2.1   SSH Hamiltonian

Su *et al.* (1979, 1980) introduced a simple Hückel-type Born–Oppenheimer Hamiltonian that was able to describe most of the properties of a neutral soliton. Let us denote by **U** the structure of *trans*-(CH)$_x$ with uniform bond length; then **A** and **B** are the dimerized structures

$$(10.6)$$

This Peierls distortion is expressed in terms of displacement in the configuration coordinate u_n for the *n*th CH group along the symmetry axis x of the chain. Therefore, to form structure **A** from **U**, the distortions are $u_n < 0$ and $u_{n+1}, u_{n-1} > 0$, which shortens the bond between carbon atoms $n - 1$ and n, and lengthens the bond between carbon atoms n and $n + 1$. The opposite distortions lead to structure **B**. The idealized Hamiltonian for these linear chains is

$$H = -\sum_{ns} t_{n+1,n} \left(C^{\dagger}_{n+1,s} C_{ns} + C^{\dagger}_{ns} C_{n+1,s} \right)$$

$$+ \frac{1}{2} \sum_n K(u_{n+1} - u_n)^2 + \frac{1}{2} \sum_n M\dot{u}_n^2 - \sum \Gamma(u_{n+1} - u_n). \quad (10.7)$$

The first term is the tight-binding Hamiltonian written in the second quantitization formalism using the creation and annihilation operators $C^{\dagger}_{n,s}$ and $C_{n,s}$, respectively, for π electrons in the atomic orbital n and with spin projection $s = \pm 1$. The important electronic operator is the bond order

$$P_{n,n+1} = \frac{1}{2} \sum_s \left(C^{\dagger}_{n,s} C_{n+1,s} + C^{\dagger}_{n+1,s} C_{n,s} \right). \quad (10.8)$$

The mean value of P_{nn+1} for the incommensurate state is $\pi/2$, and

$$t_{n+1,n} = t_0 - \alpha(u_{n+1} - u_n), \quad (10.9)$$

where α is the electron–phonon coupling constant and not the Coulomb integral in Eqs. (10.1) and (10.2). The second term of Eq. (10.7) approximates the change in σ-bond energy as a Hookean spring with K as the effective spring constant. The third term is the kinetic energy of the nuclear motion, the Born–Oppenheimer approximation being assumed. The fourth term is introduced, which is a repulsive energy and linear in the atomic displacements, in order to stabilize a finite chain against a tendency to contract uniformly. This term can be neglected as long as one is not concerned with specific chain-end effects (Vanderbilt and Mele, 1980).

10.2.2 Bond Alternation

If we assume simple bond alternation due to Peierls distortion so that each odd- (even-)numbered carbon atom undergoes a negative (positive) displacement u as shown in Eq. (10.6), then

$$u_n = (-1)^n u, \tag{10.10}$$

where the magnitude of displacement u is constant. Then one can calculate E_0, the ground-state energy, as a function of u by the variation principle. This is done by taking the derivative of Hamiltonian (10.7) with respect to u and equating it to zero. Because of the symmetry in Eq. (10.10), if a minimum value of E_0 is found for a value of u, it is also true for $-u$. Therefore, the distortion in Eq. (10.10) leads to a doubly degenerate ground state, as required by the Peierls theorem. Let us perform this calculation for that portion of Hamiltonian (10.7) containing terms linear only in u, i.e.,

$$H(u) = -\sum_{ns} [t_0 + (-1)^n 2\alpha u] \, (C_{n+1,s}^\dagger C_{n,s} + C_{n,s}^\dagger C_{n+1,s}) + 2NKu^2, \tag{10.11}$$

where N is the total number of carbon atoms or π electrons. Diagonalization of (10.11) gives the π-orbital energies to be

$$E_k = (\epsilon_k^2 + \Delta_k^2)^{1/2}. \tag{10.12}$$

The ϵ_k are the energy for the zero-order ($u = 0$) valence E_k^{0v} and conduction E_k^{0c} bands, respectively,

$$-\epsilon_k = E_k^{0v} = -2t_0 \cos ka,$$
$$\epsilon_k = E_k^{0c} = +2t_0 \cos ka. \tag{10.13}$$

The distortion energy is contained in Δ_k, which is termed the gap parameter and is given by

$$\Delta_k = 4\alpha u \sin ka. \tag{10.14}$$

The ground-state energy of polyacetylene is obtained by summing k over the first Brillouin zone from $-\pi/2a$ to $+\pi/2a$, which is

$$E_0(u) = 2 \sum_k E_k + 2NK_u^2, \tag{10.15}$$

where the first term corresponds to the π energy and the second term to the σ energy. We can replace the sum by an integral

$$E_0(u) = -\frac{2L}{\pi} \int_0^{\pi/2a} [(2t_0 \cos ka)^2 + (4\alpha u \sin ka)^2]^{1/2} \, dk + 2NKu^2$$

$$\approx -\frac{4Nt_0}{\pi} + \frac{NK\, t_0^2\, z^2}{2\alpha^2} \tag{10.16}$$

where

$$z = t_1/t_0 = 2\alpha u/t_0, \tag{10.17}$$

$$L = Na. \tag{10.18}$$

The approximation of the integrated results of Eq. (10.16) is good when $z \ll 1$, which is valid when the contribution to the hopping integral from displacement is small compared to the transfer integral.

The density of states per spin is

$$\rho_0(E) = \frac{L}{2\pi|dE_k/dk|} = \begin{cases} \dfrac{N}{\pi} \dfrac{|E|}{[(4t_0^2 - E^2)(E^2 - \Delta^2)]^{1/2}}, & \Delta \leqslant |E| \leqslant 2t_0 \\ 0 & \text{otherwise.} \end{cases} \tag{10.19}$$

Δ is the gap parameter, defined as

$$\Delta \equiv \Delta_{\pi/2a} = 4\alpha u_0 = 2t_1. \tag{10.20}$$

The wave functions for the valence and conduction electrons are

$$\psi_k^v(n) = [\alpha_k + i\beta_k\, (-1)^n] \exp\,(ikan)/\sqrt{N},$$
$$\psi_k^c(n) = [i\alpha_k(-1)^n - \beta_k] \exp\,(ikan)/\sqrt{N}, \tag{10.21}$$

with the transformation coefficients given by

$$\alpha_k = [\tfrac{1}{2}(1 + \epsilon_k/E_k)]^{1/2}, \qquad \beta_k = [\tfrac{1}{2}(1 - \epsilon_k/E_k)]^{1/2}. \tag{10.22}$$

In the treatment above, several assumptions and approximations were made, which will now be discussed along with the parametrization. The Born–Oppenheimer Hamiltonian is for individual $(CH)_x$ chains; interchain effects are neglected as they usually are in the quantum-mechanical treatment of organic molecules. The σ and π electrons were treated separately as is standard practice for Hückel-type calculations. With regard to the σ electrons, they are treated with adiabatic approximation justifiably because the gap between the π bonding and antibonding state is large (~ 10 eV) compared to the soliton and photon energies, which are less than 0.5 eV. The σ-bonding energy is assumed to vary with the distortion to the second order [second term on the right-hand side of Eq. (10.7)]. Su *et al.* (1980)

assumed K to have the value of the elastic constant of ethane, which is 21 eV Å^{-2}. The π-electron energy is determined by the hopping integral $t_{n,\,n+1}$, which depends on the bond length and is expanded in powers of u. This should be carried out to the quadratic term, which would, of course, be of the same order as the elastic energy. This term is absorbed by the spring constant K in Eq. (10.7).

Semiempirical results are obtained with the following parametrization: t_0 is directly related to the π-band widths (Fig. 10.4), which is calculated to be 10 eV (Grant and Batra, 1979a); thus t_0 is taken to be 2.5 eV. From the definition [Eq. (10.14)] for the gap parameter, the band gap is equal to $4t_1$, as shown in Fig. 10.4. E_G is experimentally found to be 1.4 eV; therefore $t_1 = 0.35$ eV. In this model E_G depends on the u_0 and a. The electron–phonon coupling constant may be estimated from Eq. (10.12) by requiring that the value of u_0 that minimizes the ground-state energy also fit the correct value for t_1. In this manner $\alpha t_0^{-1} \simeq 1.65\ \text{Å}^{-1}$ and $\alpha \simeq 4.1$ eV Å^{-1}. The ground-state energy is plotted as a function of u in Fig. 10.5. The two energy minima correspond to the **A** $(+u_0)$ and **B** $(-u_0)$ phases in Eq. (10.6) as the result of Peierls distortion. Using the above parameters of t_0, t_1, α, and K, Su *et al.* (1980) estimated $u_0 \simeq 0.04$ Å and the corresponding bond length change to be ± 0.07 Å. This is much larger than the direct experimental estimate of bond-length alternation obtained from electron and X-ray diffraction data (Section 3.7). There is some doubt whether it is appropriate to take K from the elastic constant of ethane because properties of small molecules are not expected to be simply transferable to polymers. Mele and Rice (1981) estimated from Raman data of polyacetylene a much larger value of K and consequently much smaller bond-length alternation of ± 0.04 Å.

The energy barrier between the two "dimerized" minima is

$$-E_c/N = 1/N[E_0(u_0) - E_0(0)] = -0.015 \quad \text{eV.} \qquad (10.23)$$

The one-electron density of states for *trans*-polyacetylene is shown in Fig. 10.6.

10.2.3   Neutral Soliton

So far we have considered a *trans*-polyacetylene chain to be perfectly dimerized, i.e., having either structure **A** or **B**. If a chain contains both **A** and **B** phases, then there exists a topological defect:

$$(10.24)$$

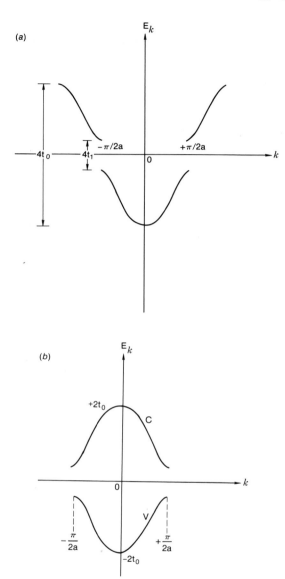

Fig. 10.4 π-band structure of "dimerized" *trans*-polyacetylene: (a) extended representation; (b) reduced representation. [After Su *et al.* (1980)].

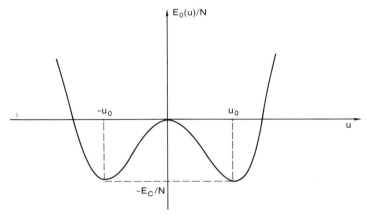

Fig. 10.5 Ground-state energy of "dimerized" *trans*-polyacetylene. [After Su *et al.* (1980).]

The π-wave functions change phases and their amplitudes vanish at the defect. Such a localized topological defect has been referred to as a π kink or a neutral soliton S $\cdot$. This one-particle state is without charge, has spin one half, and can be observed with EPR. Since the neutral soliton is not localized, a more descriptive representation for it than that of Eq. (10.24)

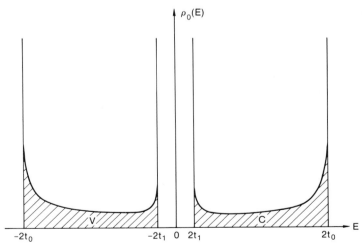

Fig. 10.6 One-electron density of states for "dimerized" *trans*-polyacetylene. [After Su *et al.* (1980).]

is

$$\text{(10.25)}$$

and S$\cdot$ is a domain wall between the **A** and **B** phases. The neutral soliton represents a midgap state, i.e., nonbonding MO.

If a *trans*-polyacetylene chain is initially perfect in structure, i.e., either **A** or **B**, then there are problems in creating a single S$\cdot$. For any finite chain length, there will be another unpaired spin at the chain end for conservation of spin multiplicity. Also, creation of a single S$\cdot$ will leave a hole in the Fermi sea. To keep the valence band full and spin paired, we require the creation of a soliton–antisoliton pair S$\cdot\bar{\text{S}}\cdot$. In this way the former condition is satisfied; both S$\cdot$ and $\bar{\text{S}}\cdot$ are without charge but have spin $\frac{1}{2}$; and Kramer's theorem, which requires that the spin of a system with an even number of electrons be an integer, is not violated. This process of soliton "excitation" leads to

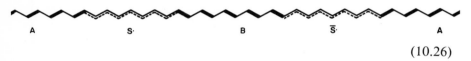

$$\text{(10.26)}$$

The problems to be solved are the width, energy, mass, spin, and other properties of the soliton. The soliton is placed at rest at the origin $n = 0$ symmetrically about the **A** and **B** phases. A staggered order parameter ψ_n is defined by

$$\psi_n = (-1)^n u_n, \qquad u_n = (-1)^n \psi_n. \qquad \text{(10.27)}$$

For the two ground states $\psi_{0n} = -u_0 (+u_0)$ for the **A** (**B**) phase and $\psi_{00} = 0$ by symmetry. For simplicity, S$\cdot$ and $\bar{\text{S}}\cdot$ are assumed to be widely separated so that they do not interact and the energy change for the formation of a S$\cdot\bar{\text{S}}\cdot$ pair is just twice the soliton creation energy E_s. Let the S segment be $2v + 1$ in length. The three relevant segments are **A** $(n \geqslant v)$, **B** $(n \leqslant -v)$, and S $(-v < n < v)$. The hopping integral is defined to be zero between CH groups in the S segment; in the other segments **A** (**B**) the hopping integrals are $t_0 - (-1)^{n-v} t_1$ $(t_0 + (-1)^{n-v} t_1)$ for $n \leqslant -v$ $(n \geqslant v)$. The soliton is treated as a perturbation so that the full Born–Oppenheimer Hamiltonian consists of

$$H = H_0 + \hat{V}, \qquad \text{(10.28)}$$

with H_0 the zero-order Hamiltonian of Eq. (10.7). The perturbation $\hat{V}$ is the missing hopping integral within the **S** segment,

$$-\hat{V}_{n, n+1} = -\hat{V}_{n+1, n} = t_0 + (-1)^n \alpha(\psi_{n+1} + \psi_n). \qquad (10.29)$$

The ψ_n are varied to minimize the energy. The shift in the ground-state energy is expressed as

$$\Delta E = \frac{2}{\pi} \int_{-\infty}^u \text{Im ln det}[1 - G^0(\omega)]\hat{V} \, d\omega, \qquad (10.30)$$

where $G^0(\omega)$ is the Green's function as described by Su *et al.* (1980), who also proposed the trial function for the staggered order parameter to be

$$\psi_n \begin{cases} u_0, & n \leqslant -v \\ -u_0 \tanh(n/\xi), & -v < n < v \\ -u_0, & n \geqslant v. \end{cases} \qquad (10.31)$$

Using this trial function, ΔE was calculated for three values of E_G: 1.0, 1.4, and 2.0 eV. For $E_G = 1.4$ eV, the soliton creation energy has a minimum value at $E_s \simeq 0.42$ eV and the width parameter $\xi \simeq 7$ (Fig. 10.7). Therefore E_s is only about 30% of $E_G = 2\Delta$. This width parameter is in agreement with the soliton domain width estimated from the result of EPR linewidth analysis (Section 5.4.2).

The change in the density of states that is due to the presence of the soliton is shown in Fig. 10.8. The gap center state, $\phi_0(n)$, gives a δ function corresponding to one electron. This is compensated for by densities taken away from both the valence and conduction bands, each integrating to a one-half state. As was pointed out above, to satisfy the Kramer's theorem an antisoliton was also created with the function

$$\bar{\psi}_n = +u_0 \tanh(n/\xi). \qquad (10.32)$$

This decreases the density of states in the valence and conduction band by another one-half state, each leaving them both paired before and after $S \cdot \bar{S} \cdot$ creation. The antisoliton and soliton wavefunctions differ only by a sign change.

The soliton behaves like a WKB state, tunnels through a barrier of height Δ, modulated by the zone-edge wave function for $k = \pm \pi/2a$,

$$\phi_0(n) \simeq \frac{1}{\xi} \text{sech}\left(\frac{n}{\xi}\right) \cos\left(\frac{\pi n}{2}\right) \qquad (10.33)$$

The probability distribution of the soliton state localized at the gap center is illustrated in Fig. 10.9 for $\xi = 7$.

From the kinetic energy of the soliton, its effective mass can be esti-

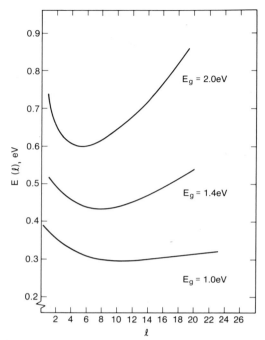

Fig. 10.7 Soliton energy versus half-width l (the same as ξ in the text) for three values of E_G. [After Su *et al.* (1980).]

mated. This is done by adding time dependency to the trial function ψ_n of Eq. (10.31) and writing

$$\psi_n(t) = u_0 \tanh[(na - v_s t)/\xi a], \tag{10.34}$$

where v_s is the velocity of soliton. Within the adiabatic approximation

$$\tfrac{1}{2} M_s v_s^2 = \tfrac{1}{2} M_{CH} \sum_n \dot{\psi}_n^2, \tag{10.35}$$

where M_s and M_{CH} are the mass of the soliton and the CH unit, respectively. The sum is

$$\sum \dot{\psi}_n^2 = \left(\frac{u_0 v_s}{\xi a}\right)^2 \sum_n \operatorname{sech}^4\left(\frac{n}{\xi}\right). \tag{10.36}$$

Therefore, using the various parameters already given above, one obtains from

$$M_s = (4/3\xi)(u_0/a)^2 M_{CH}, \tag{10.37}$$

a value of $M_s \approx 5 m_e$ for $u_0 = 0.04$ Å estimated from theory, or $M_s \approx m_e$ in

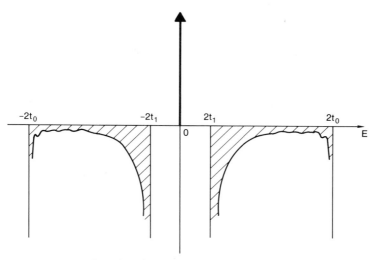

Fig. 10.8   Changes of density of states due to the presence of a soliton. [After Su *et al.* (1980).]

the case of $u_0 = 0.014$ Å obtained from electron diffraction. In other words, the mass of the soliton varies with the square of the distortion length. The smallness of M_s requires that solitons be treated as quantum particles. Furthermore, it must be very mobile. In fact, it will be shown below that this is indeed true.

In addition to the neutral soliton there are low-energy-charged soliton states S^+ (S^-) corresponding to the removal (addition) of an electron from the soliton state ϕ_0. Their relations were shown in Fig. 8.23.

cis-$(CH)_x$ can be treated in a manner similar to that above but with some important differences. Bond alternation distortion also takes place in the cis polymer. The distortion coordinate u_n is perpendicular to the C—H bond

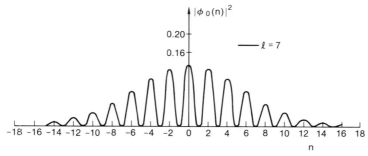

Fig. 10.9   Soliton probability density plotted for $l(\xi) = 7$ centered at $n = 0$. [After Su *et al.* (1980).]

lying in the plane of the molecule:

(10.38)

(10.39)

In the cis-transoid structure **C**, u_n and u_{n+2} are positive and u_{n-1} and u_{n+1} are negative; their signs are reversed for the trans-cisoid **D**. The two states differ only about 8 meV in favor of the cis-transoid; but the two cis structures are not degenerate. It is recalled that the condensation energy in *trans*-polyacetylene is 15 meV/CH·unit.

Because of the small energy difference, if one synthesizes a perfect cis-transoid chain, it readily isomerizes. Neutral defects were produced during this isomerization (Section 5.6); the neutral defect in the cis-rich polymer will be referred to as S·′. Electron paramagnetic resonance spectroscopy showed that S·′ has a delocalized spin like the soliton S· in *trans*-(CH)$_x$. They differ in an important respect: S· is highly mobile, whereas S·′ is not. The movement of the soliton domain wall in *trans*-(CH)$_x$ involves degenerate phases **A** and **B**, and preliminary calculations show that E_s varies by approximately 2 meV as the center of the soliton moves between lattice sites. This indicates that relatively free translational motion of soliton would take place in an otherwise perfect lattice to as low as 20–40 K. The same is not true for S·′ in the cis polymer. It was also shown earlier (Sections 5.2.1 and 5.6) that the trans-cisoid segment is unstable toward further isomerization to the trans-transoid structure.

10.3   CONTINUUM MODEL

The SSH Hamiltonian [Eq. (10.7)] can be rewritten for the continuum limit (Takayama *et al.*, 1980)

$$H = \frac{MK}{4\alpha^2 a} \int dx\, \Delta^2(x)$$
$$+ \int dx\, \psi^+(x)\,[-iv_F\,\sigma 3\frac{\partial}{\partial x} + \Delta(x)\sigma_1]\psi(x). \quad (10.40)$$

Here the second term is the electronic part of the mean field Hamiltonian containing the Pauli matrices σ_i, and $\psi(x)$ is the spinor representation of the electronic field. Approximation was made in the above deviation for the weak-coupling limit

$$-2t_0 \cos[(k \pm k_F)a] = \pm 2t_0 \sin ka \simeq \pm 2t_0ak \equiv \pm v_Fk. \quad (10.41)$$

From the Hamiltonian one obtains a pair of Bogoliubov–de Gennes equations, the solution of which is the total mean field energy

$$\pm E_k = \pm[(v_Fk)^2 + \Delta_0^2]^{1/2}, \quad (10.42)$$

where v_F is the Fermi velocity, k the wave vector of the solution, and Δ_0 the order parameter for uniform displacement.

The creation energy of a soliton is the difference between the mean field energy in the presence of a soliton and that in its absence. This is found to be

$$E_s = 2\Delta_0/\pi, \quad (10.43)$$

with $\Delta_0 = 0.7$ eV. E_s is 0.46 eV, in good agreement the value of 0.42 eV obtained by Su *et al.* (1980) for the discrete model. Another theoretical conclusion from the model is that solitons can form a regular lattice that is stable at least to a soliton concentration of 20% as long as the continuum limit is a good approximation (Nakahra and Maki, 1981).

10.4 COULOMBIC INTERACTION AND CORRELATION EFFECT

In the above discrete lattice model and its continuum limit approximation, neither the electron–electron correlation nor the Coulombic interaction was explicitly included. They are thought to be partially included by using screened values of transfer integral and the electron–lattice displacement coupling constant. But Ovchinnikov *et al.* (1978) argued that the band gap actually results from electronic correlation and not from bond alternation.

Horsch (1981) examined the effect of correlation on bond alternation. The π electron system is represented by a tight-binding model of the Pariser–Parr–Pople type. The Hamiltonian used is

$$H = H_0 + H'_{ra} + H'_{er}, \quad (10.44)$$

where H_0 is given by Eq. (10.11). The intra-atomic interaction Hamiltonian H'_{ra}, consists of a Hubbard term U and a term that gives a correct ionization potential I in the atomic limit. Thus

$$H'_{ra} = U \sum_n C^\dagger_{n\uparrow}C_{n\downarrow} - I \sum_{n\sigma} C^\dagger_{n\sigma}C_{n\sigma}, \tag{10.45}$$

where $n_{n\sigma} = c^\dagger_{n\sigma}c_{n\sigma}$ is the π-electron number operator for atomic $2p_z$ orbitals at site n and spin σ. The interatomic interaction Hamiltonian H'_{er}, consists of three terms:

$$H'_{er} = \frac{1}{2} \sum_{m,n;\sigma\sigma'}{}' V_{mn}C^\dagger_{n\sigma}C_{n\sigma}C^\dagger_{m\sigma}C_{m\sigma'} - \sum_{m,n;\sigma}{}' V_{mn}C^\dagger_{n_\sigma}C_{n\sigma}$$

$$+ \frac{1}{2} \sum_{m,n}{}' V_{mn}, \tag{10.46}$$

where the first term is for electron–electron repulsion, the second term for electron–ion attraction, and the third term arises from ion–ion repulsion. V_{mn} is taken to be 14.397 $((14.397/11.13) + r_{mn})^{-1}$, and r_{mn} is the interatomic distance in angstroms.

The kinetic energy for small distortion has an asymptotic form:

$$E_{\text{kin}} = \frac{4}{\pi} t_0 \left\{ 1 + \frac{Y^2}{4} \left[2 \ln\left(\frac{4}{Y}\right) - 1 \right] \right\}, \tag{10.47}$$

with $Y = t_1/t_0 = 2\alpha u_0/t_0$. This kinetic energy contribution of $Y^2 \ln Y$ is the driving force for Peierls instability since it lowers the energy of H_0 more than the increase due to the σ-bond elastic energy term.

To calculate the contribution of the correlation effect, the correlated ground state $|\psi\rangle$ is constructed from the independent particle ground state $|\phi_0\rangle$ by means of a set of a local two-particle operator κ, i.e.,

$$|\psi\rangle = \exp \kappa |\phi_0\rangle. \tag{10.48}$$

κ comprises two parts, one to reduce double occupancy of atomic $2p_z$ orbitals

$$\kappa_1 = \sum_n \eta_0 C^\dagger_{n\uparrow}C_{n\uparrow}C^\dagger_{n\downarrow}C_{n\downarrow}, \tag{10.49}$$

and the second to take into consideration long-range correlations

$$\kappa_2 = \sum_{n+m;\sigma,\sigma'} \eta_{nm}C^\dagger_{n\sigma}C_{n\sigma}C^\dagger_{n\sigma'}C_{n\sigma'}. \tag{10.50}$$

The parameters η are obtained by the variational method of minimizing

$$E(\eta) = \langle \exp \kappa(\eta) |H| \exp \kappa(\eta) \rangle. \qquad (10.51)$$

The correlation energy E_c for Hubbard interaction is

$$E_c = -\frac{\langle \kappa_1 H' \rangle_c^2}{\langle \kappa_1 H_0 \kappa_1 \rangle_c}. \qquad (10.52)$$

It shows a nearly linear increase with Y in the region $Y = t_1/t_0 = 0$–0.2. Inclusion of long-range correlation with κ_2 gives less than 4% additional gain in correlation energy. The total energies for the Hubbard interaction model are strongly dependent on U. With $U = 0$, the energy is that of H_0 alone. There is a weak minimum for finite distortion at $Y = 0.06$ Å and a lowering of energy of only 0.002 eV. Here, Y is related to the change in bond length d by the Coulson–Golebiewski rule,

$$Y = d/0.3727 \text{ Å}. \qquad (10.53)$$

For $u_0 = 0.024$ Å, $Y = 0.064$ Å, which is in good agreement with experimental results. However, the condensation energy is only 13% of that estimated by Su *et al.* (1980). With the inclusion of U, the correlation energy increases significantly. For $U = 8$ eV, the condensation energy is about 0.08 eV, which is an increase of 40-fold over that for $U = 0$. Furthermore, Y becomes about 0.18, which implies a very strong bond alternation value of ~ 0.067 Å. This would give a value of $E_G = 8\alpha u_0 = 2.5$ eV, which is too large. The implication is that U has a value much less than 8 eV.

The effect of electron correlation is to favor bond alternation. However, whether it is the main cause of bond alternation instead of the kinetic driving force of Peierls instability is uncertain. Introduction of Coulombic interaction tends to stabilize the incommensurate state of uniform bond length. However, if this interaction is allowed to be weakly screened by a static dielectric constant, which is assumed to arise from the polarizability of the σ electrons, then the effect on stabilization of bond alternation is restored. A value of as small as 2.0 for the dielectric constant is sufficient to overcome the Coulombic interaction.

Another way to include electron correlation is to use alternate molecular orbitals within an "unrestricted Hartree–Fock" scheme (Dugay and Rouston, 1983). The electron–electron interactions originate the formation of domain walls and distribution of spin densities of the neutral soliton comparable to that given by Su *et al.* (1980). Nakano and Fukuyama (1980) included strong correlation for a one-dimensional π system with Coulombic interactions between two electrons on the same site and on different sites. The Hamiltonian is similar to that of Eq. (10.45); two transformations converted the system to a quantum sine–Gordon system. The soliton solution with appropriate choice of parameters gives the following estimates:

$$u_0 = 0.018a, \quad E_c = -0.018 \text{ eV}, \quad \text{and} \quad E_s = 0.34 \text{ eV}. \quad (10.54)$$

These values are nearly the same as those obtained by Su *et al.* (1980).

10.5   PROPERTIES OF SOLITONS

10.5.1   Photogeneration

Absorption of photon of energy $\geqslant E_G$ by polyacetylene creates an electron–hole pair. Su and Schreiffer (1980) had carried out real-time integration of the classical equations of motion for the Hamiltonian (10.7). The result demonstrated that a photogenerated e–h pair would transform into a neutral soliton–antisoliton pair within a time of order 10^{-13} sec or one molecular vibration. However, Flood *et al.* (1982) were unable to detect any photogenerated unpaired spin. Therefore, the quantum efficiency for photoproduction of neutral solitons is small ($<2 \times 10^{-7}$). Blanchet, *et al.* (1983) had monitored the infrared spectra during the irradiation of a thin film of *trans*-polyacetylene with a CW-argon ion laser ($h\nu = 2.4$ eV). The spectra have three main features. The strong absorption at 3870 cm^{-1} (0.48 eV) has an asymmetric line shape similar to that predicted for the transition between the midgap state and the valence band on the conduction band. This is about 0.2 eV lower in energy than the midgap absorption for doped polyacetylenes. This may be due to the effect of the Coulomb interaction on the soliton electronic energies shifting the energy of the doubly occupied gap state off the gap center. There are sharp absorptions at 1370 and 1250 cm^{-1}, both of which were also seen at the same frequencies in doped sample. Finally, a well-defined absorption was observed at about 500 cm^{-1}. In doped (CH)$_x$ the mode at 900 cm^{-1} was attributed to the binding of the charged soliton to the donor or acceptor impurity. It appears that the photogenerated charged solitons are also pinned by intrinsic pinning potential.

The preference of photogeneration of the charge soliton–antisoliton pair ($S^{\mp}S^{\pm}$) over the neutral soliton–antisoliton pair ($S \cdot \bar{S} \cdot$) is attributable to suppression of the latter by two separate symmetry effects (Ball *et al.*, 1982). Following the photocreation of an electron–hole pair in *trans*-polyacetylene, an $S \cdot \bar{S} \cdot$ pair begins to evolve, resulting in the splitting off of two states ψ_+ and ψ_- from the conduction and valence band edges having even and odd parity, respectively. The excited electron and hole rapidly decay to occupy ψ_- and ψ_+. The conduction band remains completely empty and the valence band filled. The parity of the excited state wave function with ψ_+ and ψ_- singly occupied is odd while both the conduction and valence bands have even parity. The Pauli principle requires that the spatial part of the wave function be even because the spin part is odd. Thus the electron in the ψ_- state drops into the ψ_+ state to form a $S^{\mp}\overline{S}^{\pm}$ pair.

Second, the charge conjugation or particle–hole symmetry of the half-filled π band in *trans*-$(CH)_x$, the matrix element for the electron–phonon coupling Hamiltonian vanishes. Therefore, even though phonons can have mixed parity in general, the photoproduction of a $S \cdot \overline{S} \cdot$ pair remains suppressed under charge conjugation symmetry.

10.5.2   Effect of Pinning Potential

The properties of free and pinned charged solitons were compared with those of neutral solitons by Rice and Mele (1980). They used the general Hamiltonian

$$H = \tfrac{1}{2} M_i(\xi)\dot{\xi}^2 + V_i(\xi) + \lambda V(x,\xi) + \tfrac{1}{2} M_s(\xi)\dot{x}^2. \tag{10.55}$$

The first term is the internal kinetic energy of the soliton in which $\dot{\xi}$ is $d\xi/dt$, and $M_i(\xi)$ the internal inertial mass of soliton. The second term denotes the internal potential with the form

$$V_i(\xi) = (A + \sigma U_0/\epsilon)/\xi + B\xi, \tag{10.56}$$

where A/ξ describes the deformation energy of the soliton due to the spatial gradient in the local amplitude of u_n, $B\xi$ represents the deformation energy arising from the modulation of the magnitude of u_n, and $U_0/\xi\epsilon$ is the Coulomb repulsion term that is present for $S^{\pm}$.

The $\lambda V(x, \xi)$ term describes the Coulombic attraction between $S^{\pm}$ and the dopant ion, where x is the distance between the soliton domain and the ion. λ is zero for $S \cdot$ and $S^{\pm}$; it is one for $S_p^{\pm}$. Assuming A^- to be a point charge and the charge of soliton distributes uniformly over 2ξ, the potential is given by

$$V(x, \xi) = -\frac{e^2}{\xi\epsilon}\log\left(\frac{x + \xi/2 + \sqrt{(x + \xi/2)^2 + d^2}}{x - \xi/2 + \sqrt{(x - \xi/2)^2 + d^2}}\right), \tag{10.57}$$

where d is the perpendicular distance from A^- to the chain. It is usual to use $\epsilon = 10$ for Eqs. (10.56) and (10.57). The actual dielectric constant may be

TABLE 10.1

Properties for Three Forms of Solitons[a]

	S ·	S$^+$	S$_p^{\pm}$
Internal kinetic energy	$\frac{1}{2}M_i(\xi)\dot{\xi}^2$	$\frac{1}{2}M_i(\xi)\dot{\xi}^2$	$\frac{1}{2}M_i(\xi)\dot{\xi}^2$
Internal potential energy	$A/\xi + B\xi$	$A/\xi + U_0/\epsilon\xi + B\xi$	$A/\xi + U/\epsilon\xi + B\xi$
Coulombic attraction	0	0	$V(x,\xi)$
Translational kinetic energy	$\frac{1}{2}M_s(\xi)\dot{x}^2$	$\frac{1}{2}M_s(\xi)\dot{x}^2$	$\frac{1}{2}M_s(\xi)\dot{x}^2$
ξ_s/a	10	16.6	14
$E_s(\xi_s)$ (eV)	0.6	1.0	0.68
$\omega_i(\xi_s)$ (cm^{-1})	1284	1284	1380
$M_s(\xi_s)$, m_e	15.1	9.1	10.8
$\omega_s(\xi_s)$ (cm^{-1})	0	0	315
$U(\xi_s)$ (eV)	0	0.32	0.38

[a] After Rice and Mele (1980).

much smaller. The last term of the Hamiltonian of Eq. (10.55) is the translational kinetic energy of the soliton, and $M_s(\xi)$ is translational inertial mass.

Evaluation of the Hamiltonian gives

$$M_i(\xi) = \epsilon_c/9\omega_A^2\xi a,$$

$$M_s(\xi) = 64\epsilon_c/3\omega_A^2\xi a,$$

$$A = 32\epsilon_c c_0^2/3\omega_A^2 a, \tag{10.58}$$

$$B = 2\epsilon_c/3a,$$

where $a = 1.22$ Å is the lattice parameter, ω_A is the frequency of optical phonon of polyacetylene, c_0 is the velocity characterizing the phonon dispersion $\omega_A^2(q) = \omega_A^2 + c_0^2q^2$ ($|q| \ll \pi/2a$), $M_i(\xi) = M_s(\xi)/12$, and ϵ_c is the stabilization energy per carbon atom of the "dimerized" chain referenced to the incommensurate structure.

To estimate the magnitude of various quantities, Rice and Mele (1980) used some of the old parameters. For instance, ξ_s^o/a was taken to be 10 instead of 7, E_s^o was taken to be 0.6 eV instead of 0.42 eV, λ_0 was taken to be 2.5 instead of 1.75, ϵ_c to be 0.045 eV instead of 0.11 eV, and $U(\xi_s^o) = 0.53$ eV instead of 0.37 eV. However, in view of the other approximations mentioned above, it does not seem necessary at this time to make corrections for these parameters and we give just the results of Rice and Mele (1980) in Table 10.1 and Fig. 10.10.

The energy of formation of S$_p^+$ is only slightly greater than that for S · by ~ 0.08 eV. The extension of domain wall is smallest for S · and greatest for S$^+$, the latter because of electron correlation effect. In the case of S$_p^+$, this

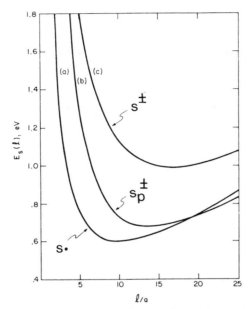

Fig. 10.10   Potential energy curves for $E_s(l)$ versus l/a: (a) S $\cdot$; (b0 S$_p^\pm$; (c) S$^\pm$. [After Rice and Mele (1980).]

effect was partially compensated by the attractive Coulomb interaction with the dopant ion. Since the soliton mass is inversely related to the soliton width, the results for M_s given in Table 10.1 can be understood; however, all the values are probably too high. There is no Hubbard U term for the neutral soliton. The Coulomb repulsion energy is greater for S$_p^+$ than S$^+$ because of the greater confinement of the former soliton. Finally, only S$_p^+$ will possess a pinning frequency $\omega_s(\xi_s)$.

Since the ξ_s for S$_p^+$ is large, one inquires about its intrinsic stability. Because in the absence of distortion, S$_p^+$ would at 0 K enter a conventional donor state position at an energy E_B below the conduction band. The criterion here is therefore $E_s(S_p^+) < \Delta - E_B$. The value of E_B is estimated to be ~ 0.02 eV. In the case of free charged soliton, the Coulomb repulsion was not compensated by attraction as in the pinned species. Therefore, its energy of creation is much higher. The stability criterion for S$^+$ is then that $U(\xi_s^o)$ must be smaller than a certain value, so that $E_s < \Delta$. In other words,

$$U(\xi) < (E_s^o/2)[(\Delta/E_s^o)^2 - 1] < 0.47\Delta, \qquad (10.59)$$

where the superscript o denotes neutral soliton. If this condition is not satisfied, the free S$^+$ would become a conventional electron or hole state. Analogous arguments apply to negative solitons S^- and S_p^-.

10.6   POLARON

   Campbell and Bishop (1981) showed that the Bogoliubov–de Gennes equations of the Hamiltonian contains another solution. They found the solution by comparison with the relativistic-field-theory model by Gross and Neveu. The new solitons have a single electron spectrum with the two states symmetrically placed in the gap at $\epsilon_+ \equiv \omega_0$. The states exist only when there is a single unpaired electron occupying the ϵ_+ state, which is the localized "electron" state, or when there is a single unpaired electron occupying the ϵ state which is the localized "hole" state. The electron (hole) is trapped in the lattice and coupled to phonons, this excitation is polaron-like (Fig. 8.24).

   The form of the polaron solution is shown in Fig. 10.11 and is compared with those of the neutral solution. The gap parameter $\Delta(x)$ for the polaron is of the same sign on both sides of the center (Fig. 10.11a) whereas $\Delta(x)$ changes sign for the soliton, which is a π kink. The value of $\Delta(x)$ at the

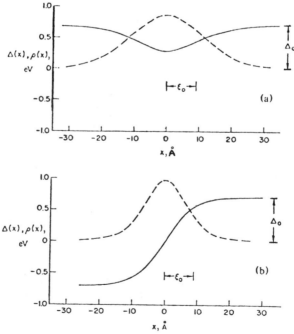

Fig. 10.11   A comparison of (a) polaron excitation and (b) soliton excitation: (——) $\Delta(x)$ and (----) $\rho(x)$. [After Campbell and Bishop (1981).]

minimum is $\Delta_0(\sqrt{2} - 1) = 0.41\Delta_0$. The polaron has a characteristic width

$$2x_0 = 4t_0\, a[(\Delta_0\sqrt{2})\ln(\sqrt{2} + 1)]^{-1} \approx 1.24\,\xi_0. \qquad (10.60)$$

Consequently, the electron density of the polaron shown by dashed lines is more extended than that for soliton. The energy of the polaron is $E = (2\sqrt{2}\Delta_0)/\pi \approx 0.90\Delta_0$. So the polaron has a binding energy of $0.10\Delta_0 \approx 0.07$ eV. Therefore, for *trans*-polyacetylene, the polaron is the lowest energy state available to a single electron. Other estimates of the polaron binding energy are 0.3 eV by Su and Schrieffer (1980) and ~ 0.05 eV by Bryant and Glick (1982) and Brédas (1982). The estimates for polaron half-width vary from 10 to 30 lattice parameters.

Brédas *et al.* (1982) had considered the interaction between two polarons. The probability of this interaction increases with the level of doping. The two polarons would have their charges located at positions fixed by the dopant ion. The spins are pulled away from charged defects; it takes ~ 0.25 eV for complete separation (Fig. 8.25). Two neutral defects will then recombine, leaving behind two pinned positive solitons. The process begins with two polarons ($E \approx 2 \times 0.65 = 1.3$ eV), leading to four separated defects ($E \approx 4 \times 0.45 = 1.8$ eV) and ends up with two charged solitons ($E \approx 2 \times 0.42 = 0.9$ eV). Figure 8.26 shows that there is a barrier to recombination. With increased doping, which is assumed to be homogeneous, the energy barrier is lowered because full separation into four noninteracting defects before annihilation of $S\cdot$ is not required. At $y \approx 0.02$ the charges are on the average 50 sites apart; there is a barrier of 0.1 eV. At $y \approx 0.03$ (charges about 35 sites apart) the barrier disappears. At high-hopping level, a newly created polaron transforms itself and a preexisting one into S_p^+, leading to a lattice of charged solitons.

There is another kind of interaction that affects the band structure, as illustrated in Fig. 10.12. At infinite separation, $S\cdot$ and S^+ are noninteracting, both occupying midgap states (Fig. 10.12a). Two neutral solitons would have both states singly occupied in this figure. As the separation decreases, the electronic states interact, leading to a bonding and an antibonding state within the gap. Brédas *et al.* (1982) had calculated the level positions; the results are given in Table 10.2.

The midgap absorption upon doping had been previously attributed to transition to soliton states (Section 9.4). Figure 10.12 indicates that there could be three other transitions when polarons are present. However, this may be difficult to discern spectroscopically for several reasons. At light doping both charged solitons and polarons are created. Also, polarons on the same chain are prone to be converted to charged solitons. For the polaron, there is a distribution of defect separation, and consequently, a distribution of the three transition energies. Finally, the average absorption energy for

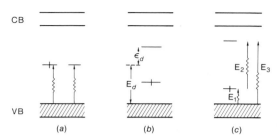

Fig. 10.12   Effect of separation between S · and S⁺ on the band structure of polyacetylene. Vertical bar denotes electron occupancy; arrow indicates optical transition. Separation Δn: (a) ∞; (b) 7; (c) 1. [After Brédas *et al.* (1983).]

the polaron model is also about midgap. Therefore, the presence of polaron may cause a broadening of the midgap absorption.

The increase of more than 13 orders of magnitude in conductivity that occurred with the doping of polyacetylene must be very complicated. Significant approximations and assumptions are usually made in theoretical analysis (Rice and Timonen, 1979; Rice and Mele, 1981; Horovitz, 1981; Brédas *et al.*, 1982; Mele and Rice, 1981). The results of the last work are summarized below.

Mele and Rice (1981) started with the Hamiltonian (10.7) selected for the parameters $t_0 = 3$ eV, $K = 68.6$ eV Å⁻², and $\alpha = 8$ eV Å⁻¹, and carried out numerical calculations for the static system with a chain length of 256 atoms. Changing of the electronic states resulted in lattice distortion and was treated as a spatial modulation of bond alternation. For band filling of the half-filled band by a single carrier, the modulation has the form of Eq. (10.32). However, for greater distortion, the modulation is sinusoidal. A continuous function which bridges the tanh and sine limits is the elliptical

TABLE 10.2

Energies and Level Positions for Interacting Neutral and Charged Defects[a]

Separation Δn	E_d (eV)	ϵ_d (eV)	Transition energy (eV)		
			E_1	E_2	E_3
1	0.652	±0.434	0.20	0.62	0.81
3	0.656	±0.395	0.23	0.56	0.79
5	0.667	±0.357	0.25	0.50	0.75
7	0.680	±0.318	0.28	0.45	0.73

[a] Notations are given in Fig. 10.17. [After Brédas *et al.* (1982).]

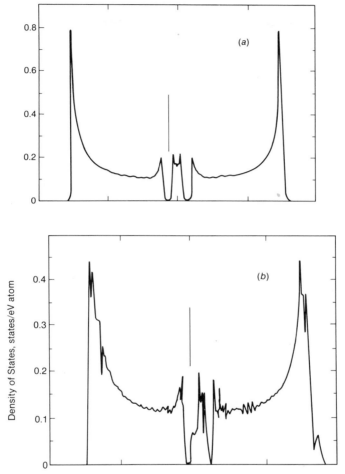

Fig. 10.13   Densities of states for π electrons in polyacetylene with 5.5% acceptor doping: (a) ideal incommensurate structure; (b) dopant potential treated in the coulomb coupling model; and (c) with three-dimensional interaction. The vertical bar gives the Fermi energy. [After Mele and Rice (1981).]

sine (sn),

$$u(x) = u_0 \, \text{sn}(\zeta x + \zeta; m), \tag{10.61}$$

where the wave period $\zeta = 4K(m)$ and K is the complete elliptical integral of the first kind.

One effect of doping is the introduction of dopant ions, which tends to alter the periodic nature of the above-mentioned modulation of bond

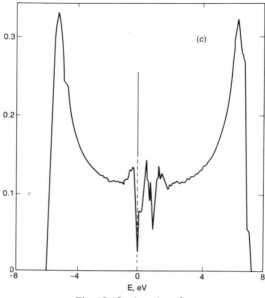

Fig. 10.13   (*continued*)

alternation because of the injection of carriers. This disordering was modeled by either a screened Coulomb interaction or a local modification of π-electron basis orbitals adjacent to the dopant.

The calculation was first performed for the polymers doped to various concentrations for ideally incommensurate structure without considering disorders introduced by the dopant ions. In the center of the gap the excess charges due to doping induced holes are paired with an equal number of "hole states" from the conduction band. The Fermi energy lies midway between the hole state and valence band. As y increases, the midgap "band" grows accordingly, and the Fermi energy lowers as required by Peierls theorem. The ground state is thus always diamagnetic and the midgap band represents a "condensate band" into which frequency regime is added and intrinsic charges condense and separate from the valence and conduction band. The π-electron charge density is localized at low values of y. As y increases, the charge begins to localize on the molecule and takes on sinusoidal oscillation form as the charge is uniformly distributed on the chain. The order parameter u_0 decreases at $y \approx 0.06$, reaching a minimum value of ~ 0.01 Å just below $y \approx 0.10$. However, at $y = 0.14$ there is no indication of closing of the band gap.

In order to overcome the Peierls instability, the effect of ionized dopants

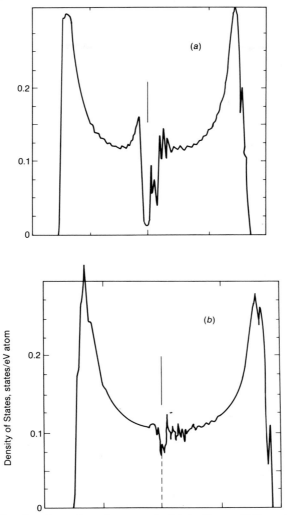

Fig. 10.14   Density of states for π electrons calculated with a multidimensional Coulomb-coupling model. The Fermi energy is given by the vertical bar: (a) $y = 0.031$; (b) $y = 0.078$; (c) $y = 0.109$. [After Mele and Rice (1981).]

has to be included. This was done by Mele and Rice (1981) in two different ways. The first is the covalent bond model, which assumed charge transfer to form a bond between the dopant and a π orbital of the polymer. This model lacks physical reality. For instance, the pyrolysates of AsF_5-doped polyacetylene are no different from the undoped polymer except for the product

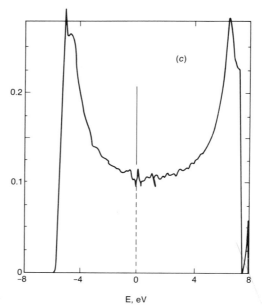

Fig. 10.14 (*continued*)

distribution (Section 7.1.1.3). Also, for iodine-doped polyacetylene, the dopant exists as discrete ions I_3^- and I_5^-, which are in equilibrium. The second approach is simply to introduce a pinning potential on the chain due to the screened Coulomb interaction between the dopant ion and the π electrons on the polyacetylene. The potential is given as

$$\epsilon_n = \sum_i v \left[1 + \left(\frac{x_n - x_i}{d} \right)^2 \right]^{1/2}, \qquad (10.62)$$

where $v = 0.6$ eV, $d = 2$ Å, x_n is the location of the dopant ion at the nth carbon atom, and the sum is over sites in the vicinity of the dopant. This is referred to as the Coulomb-coupling model. Furthermore, since the system is no longer one dimensional when heavily doped, interchain interaction is introduced. Each site in the lattice is coupled strongly to two nearest-neighbor sites on the same chain and weakly to z nearest neighbors on adjacent chains ($z = 6$). The interchain bandwidth is taken to be 0.2–0.3 eV near the Peierls edges. The results are shown in Fig. 10.13 for $y = 0.055$. In the ideally incommensurate structure midgap states are separated from the valence and conduction bands. The introduction of Coulomb coupling serves only to broaden the midgap band, which is still prominent. However, when the three-dimensional interaction is introduced, the gaps are closed.

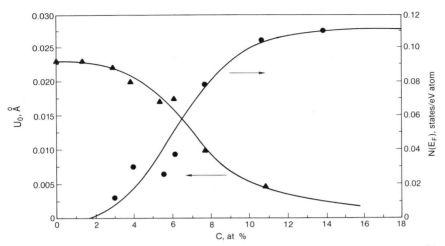

Fig. 10.15 Variation of Fermi-level density of states $N(E_F)$ and order parameter u_0 with dopant concentration obtained with the multidimensional disordered models. [After Mele and Rice (1981).]

Nevertheless, the appearance of the midgap state band remains. Therefore, it is necessary to include the interaction of random dopant ion potential in multidimension.

The changes in the electronic structure of polyacetylene with increasing doping level are shown in Figs. 10.14 and 10.15. At $y = 0.031$, there is a deep pseudogap, the Fermi density of states $N(E_F)$ begins to appear, and the order parameter decreases. A smeared soliton band can still be discerned above the Fermi level. For $y = 0.078$, the order parameter is only about half of the value for ideally incommensurate structure, and $N(E_F) \approx 0.07$ states (eV atom)$^{-1}$. There remains a shallow gap and indication of remnants of the soliton band. Finally, the gap is closed at $y = 0.109$ and $N(E_F)$ approaches the asymptotic value of 0.1 states (eV atom)$^{-1}$. The polymer strongly prefers a structure that is commensurate with the dopant ion potential. This random potential introduces off-diagonal disorder on the polymer as well. The bond alternation is greatly reduced. The calculation has two important points. The charged soliton states persist well beyond the doping level referred to as the semiconductor–metal transition. Perhaps the rapid increases in conductivity is better referred to as the "soliton glass transition." This will permit differences in the sharpness of transition with doping species, doping procedures, and past history of the polyacetylene sample. Second, even at $y = 0.14$, the bond alternation did not completely vanish. This may be due to strong polaronic effect even for the metallic state of heavily doped polyacetylene.

Chapter 11

Electrical Properties

In the early study of the effect of halogen and AsF_5 doping on *trans*-polyacetylene (Chiang *et al.*, 1978b; Park *et al.*, 1980), the conductivities were measured from few tenths of a mole percent and up. Based on these results the authors concluded that the semiconductor-to-metal transition occurred at dopant concentrations of 1–3%. Subsequent experimental measurements of Pauli susceptibility (Section 9.3.2) and theoretical calculations on the effect of heavy doping on the midgap states showed that it requires more than 5% of dopant for polyacetylene to attain metallic characteristics. On the other hand, Chien *et al.* (1983h) found that the thermopower changes from a p-type semiconductor to a metallic value at $y \approx 10^{-3}$ within a twofold to threefold increase in dopant concentration. In other words there is a transition for intrinsic transport property occurring at a dopant level about two orders of magnitude lower than that for conversion to a metallike substance predicted by theory (Section 10.7). The results presented in this chapter will show that there are at least three regimes for the transport properties of polyacetylene as a function of dopant concentration over an increase in conductivity of 13 orders of magnitude.

The transport behaviors for metals and crystalline intrinsic and extrinsic semiconductors have been described in Section 1.3. Undoped polyacetylene is not like an intrinsic semiconductor. Lightly doped polyacetylenes do not behave as extrinsic semiconductors. Heavily doped polyacetylene cannot be considered metallic in the classic sense. In the past decade the study of physical properties of amorphous semiconductors has been an active field in solid-state physics. The most intensively investigated materials are multi-component chalogenide glasses. Semiconducting amorphous states generally contain group-III, -IV, -V, and -VI elements. These elements possess highly directional interatomic binding forces and, as a consequence, show strong local order in their amorphous compounds. In fact, the nearest-neighbor configuration in many amorphous materials is very similar to that of the crystalline state. However, small deviations in bond distances and angles lead to a loss of translation order after a few coordination spheres. Polyacetylene also possesses strong local order and has a high degree of crystallinity. On the other hand, this order does not extend to very long range, as shown by the large disorder in the Ruland analysis of X-ray diffraction data (Section 3.4). In the case of amorphous semiconductors, the basic features of the electronic structure of the crystal are believed to be preserved because of the short-range order. To account for the translational disorder, modifications have been proposed for the band structure of the amorphous solids. This is achieved by introducing the basic idea of the presence of localized states at the band extremities. Transport properties have been formulated based on various modified band models. Attempts have been made to analyze the transport data of polyacetylene according to these models.

In addition, conduction by polaron and by intersoliton hopping had been proposed. Finally, there is the view held by some that doping created small metallic domains in polyacetylene and conduction occurs by virtue of percolation. It seems expedient to discuss the various mechanisms for conduction first so that the observed transport and spectroscopic properties can be compared with the predictions according to various mechanisms.

11.1 MECHANISMS FOR TRANSPORT

11.1.1 Modified Band Models

The main features of the energy distribution of the density of electronic states of crystalline semiconductors are the sharp structures in the valence and conduction bands. The density of states terminates abruptly at the valence band maximum, the conduction band minimum forming a well-defined forbidden energy gap. Within the band the states are said to be extended, meaning that the wave function occupies the entire volume. In amorphous solids the configurational disorder causes fluctuations in the potential, leading to the formation of localized states. The states are termed localized in the sense that an electron placed in a region will not diffuse at 0 K to other regions with the same potential fluctuations. Various models differ in the nature of localized states.

Figure 11.1a shows two new features. The first is the tails of localized states, which are narrow and extend a few tenths of an electron volt into the

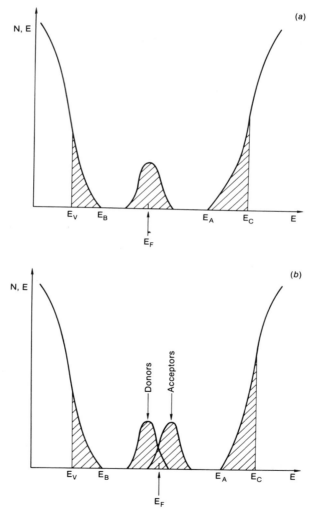

Fig. 11.1   Schematic drawing of density of states: (a) band with tail or localized states and a band of compensated levels near the middle of the gap; (b) the center band split into acceptor and donor bands in the pseudogap; (c) "real" disordered semiconductor with defect states.

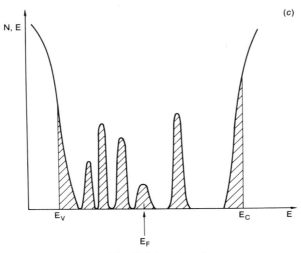

Fig. 11.1 (*continued*)

forbidden gap. There is a sharp boundary between the energy ranges of the localized states (shaded) and extended states (unshaded) with energies E_c and E_v. Second, there exists a band of compensated levels near the middle of the gap, originating from defects such as dangling bonds and vacancies. The reader will find close resemblance between this figure and Fig. 10.13, which describes the band structure of polyacetylene with midgap soliton states. In going from extended to localized states, the mobility decreases by several orders of magnitude, producing a mobility edge. The concept of localized states is that the mobility is zero at 0 K. The region between E_c and E_v is now called the pseudogap or the mobility gap.

A refinement of the above model is the splitting off from the tail states various localized states located at well-defined energies in the gap (Fig. 11.1c). This is proposed to explain luminescence, photoconductivity, and drift mobility results.

If a charge carrier causes distortion of its surrounding atomic lattice by polarization, it is said to be in a self-trapped polaron state. The presence of disorder in an amorphous solid helps to slow down a carrier, and if the carrier stays at an atomic site sufficiently long to permit atomic rearrangement to take place, a polaron is formed.

The theories of transport in amorphous semiconductors have been advanced by Davis and Mott (1970), Cohen *et al.* (1969), Emin (1973), Mott (1970), Anderson (1958), Fritzsche (1971), Nagels *et al.* (1970), and Nagels (1976). The discussions of these subjects are based on these papers.

11.1.2   Conduction by Thermal Activation
to Extended States

The conductivity for any semiconductor can be expressed as

$$\sigma = -e \int N(E)\mu(E)kT \frac{\partial f(E)}{\partial E} dE. \tag{11.1}$$

The Fermi level is situated near the midgap, which is sufficiently far from E_c that Boltzmann statistics can be used instead of the Fermi–Dirac distribution function to describe the occupancy of states,

$$f(E) = \exp[-(E - E_F)/k_B T]. \tag{11.2}$$

For the nondegenerate case and with the assumption of a constant density of states and mobility, the conductivity due to electrons excited beyond the mobility edge into the extended states is obtained from Eqs. (11.1) and (11.2) as

$$\sigma = eN(E_c)k_B T \mu_c \exp[-(E_c - E_F)/k_B T]. \tag{11.3}$$

It can be shown readily that the mobility varies inversely with temperature. Conductivity in this regime would be more appropriately considered as Brownian motion, and the mobility can be obtained from Einstein's relation:

$$\mu = eD/k_B T. \tag{11.4}$$

The diffusion coefficient D can be written

$$D = \tfrac{1}{6} v a^2, \tag{11.5}$$

where a is the interatomic distance and v the jump frequency. Combining Eqs. (11.4) and (11.5) gives

$$\mu_c = \tfrac{1}{6} (ea^2/k_B T)v, \tag{11.6}$$

and the mobility varies inversely with temperature. Thus at 300 K, the mobility is of the order 6 cm^2 (V sec)$^{-1}$, which corresponds to a mean free path of the order of a.

Combination of Eqs. (11.3) and (11.6) finds the dc conductivity to assume the form

$$\sigma_{dc} = \sigma_0 \exp[-(E_c - E_F)/k_B T]. \tag{11.7}$$

If the excitation is into localized states at band edges with impurity band at E_A

$$\sigma_{dc} = \sigma_0 \exp\left(\frac{E_A - E_F + W}{k_B T}\right), \tag{11.8}$$

where W is the activation energy for hopping.

For ac transport we expect $\sigma(\omega)$ to be given by a formula of the Drude type

$$\sigma(\omega) = \frac{\sigma(0)}{1 + \omega^2 \tau^2}. \tag{11.9}$$

Since at most ac frequencies $\omega^2 \tau^2 \ll 1$, we find $\sigma_{ac} = \sigma_{dc}$.

A general expression for the thermopower has been given by Fritzsche (1971):

$$S = \frac{k_B}{e} \frac{\int \mu(E) N(E) [(E - E_F)/k_B T] \exp[-(E - E_F)/k_B T] \, dE}{\int \mu(E) N(E) \exp[-(E - E_F)/k_B T] \, dE}. \tag{11.10}$$

Integration under the same assumption used to obtain Eq. (11.3) gives

$$S = \pm \frac{k_B}{|e|}\left(\frac{E_c - E_F}{k_B T} + A\right), \tag{11.11}$$

where A is a constant depending on the scattering mechanism.

The lowest value of the electrical conductivity before the start of an activation process is called by Mott (1970) the "minimum metallic conductivity" given by

$$\sigma_{min} = \text{const } e^2/\hbar a. \tag{11.12}$$

The constant has a value of about 0.026, and σ_{min} is in the range of 200–300 $(\Omega \text{ cm})^{-1}$. The mobility at room temperature is about the same as that estimated from Eq. (11.6).

11.1.3   Conduction by Activated Hopping in Band Tails

If the wave functions are localized, conduction can occur by hopping between states in the band tails. An electron moves from one localized state to another with energy provided by exchanging with a phonon. The mobility will have an Arrhenius form,

$$\mu = \mu_0 \exp[- W(E)/k_B T] \tag{11.13}$$

and

$$\mu_0 = \tfrac{1}{6} v_{ph} e R^2 / k_B T, \tag{11.14}$$

where v_{ph} is the phonon frequency and R is the distance of hopping. For $W \approx k_B T$ and $v_{ph} \approx 10^{13}$ sec^{-1}, the mobility is of the order 10^{-2} cm^2 (V sec)$^{-1}$, which is at least a factor of 100 slower than the mobility for extended state conduction given by Eq. (11.6).

The conductivity is obtained by integrating over all localized states. However, the density of localized states is not known. If it is assumed that $N(E)$ varies with some power m of E,

$$N(E) = N(E_c)\left(\frac{E - E_A}{E_c - E_A}\right)^m. \tag{11.15}$$

Substitution of Eqs. (11.14)–(11.16) into Eq. (11.1) with the assumption that $m = 1$ gives

$$\sigma = \frac{v_{ph}}{6} \frac{e^2 R^2 N(E_c) k_B T}{E_c - E_A}\left\{1 - \exp\left(-\frac{E_c - E_A}{k_B T}\right)\right\}\left\{1 + \frac{E_c - E_A}{k_B T}\right\}$$
$$\times \exp\left[-(E_A - E_F + W)/k_B T\right]. \tag{11.16}$$

For ac conductivity Austin and Mott (1969) have derived an expression

$$\sigma(\omega) = \tfrac{1}{3}\pi e^2 k_B T[N(E_F)]^2 \alpha^{-5} \omega [\ln(v_{ph}/\omega)]^4, \tag{11.17}$$

where α is defined as the rate at which the atomic wave function falls off with distance.

$$\Psi \approx \exp(-\alpha r). \tag{11.18}$$

The assumptions involved are that hopping is between independent pairs of centers, that multiple hopping can be neglected, and that there is no correlation between the hop energy and hop distance.

Equation (11.17) can be approximated as

$$\sigma(\omega) \simeq \text{const } \omega^s, \tag{11.19}$$

where

$$s = \frac{d \ln \sigma}{d \ln \omega} = 1 - \frac{4}{\ln(v_{ph}/\omega)}. \tag{11.20}$$

The value of s is 0.84 for $\omega = 10^2$ sec^{-1} and 0.65 for $\omega = 10^8$ sec^{-1}. It is often taken to be about 0.8. In addition there is the contribution of an activation energy to create charge carriers into the band tails so that the ac conductivity has the functional dependence

$$\sigma_{ac}(\omega) \propto \omega^{0.8} k_B T \exp[-(E_A - E_F)/k_B T]. \qquad (11.21)$$

To obtain the thermopower, one combines Eqs. (11.11) and (11.15), assuming $m = 1$, to find

$$S = \pm \frac{k_B}{|e|} \left[\frac{E_A - E_F}{k_B T} + C \right], \qquad (11.22)$$

where

$$C = \frac{2 - \{\exp(-\Delta E/k_B T)[2 + 2(\Delta E/k_B T) + \Delta E/kT)^2]\}}{1 - \{\exp(-\Delta E/k_B T)[1 + (\Delta E/k_B T)]\}} \qquad (11.23)$$

with $\Delta E = E_c - E_A$.

11.1.4 Variable-Range Hopping Conduction

Consider a semiconductor with strong interaction between the carrier and the atom. Each wave function is confined to a small region of space falling off exponentially with distance as $\exp(-\alpha R)$. So states at E_F are localized, and the material has been called "Fermi glass." When the Fermi energy lies in the range of energies where states are localized, conduction by variable-range hopping is possible. This is illustrated in Fig. 11.2, where an electron from state **A** below the Fermi energy hops to state **B**. The probability per unit time that hopping occurs is the product of the following: the Boltzmann factor $\exp(-W/k_B T)$, where W is the difference between the energies of the two states **A** and **B**; a factor ν_{ph} is the phonon-assisted attempt frequency, which depends on the phonon spectrum; and a factor depending on the overlap of the wave functions. For the case of strong localization,

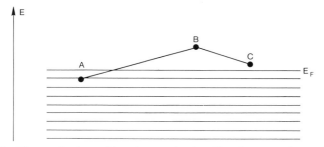

Fig. 11.2 The mechanism of hopping conduction. Two hops are shown, from **A** (an occupied state) to **B** and from **B** to **C**.

hopping will be only between nearest neighbors. In the presence of a weak external field

$$\sigma = 2e^2 R^2 v_{ph} N(E_F) \exp(-2\alpha R - W/k_B T). \tag{11.24}$$

The hopping distance R increases with decreasing temperature. At a given low temperature T the electron will normally hop to a site at a distance smaller than a value R. This implies that it will have available $4/3\pi (R/a)^3$ sites. Furthermore, it will normally jump to a site for which the energy W is as low as possible:

$$W = 3/4\pi R^3 N(E_F). \tag{11.25}$$

The average hopping distance is

$$\overline{R} = \left(\int_0^R r^3 \, dr\right)\left(\int_0^R r^2 \, dr\right) = \frac{3R}{4}. \tag{11.26}$$

The probability of a hop per unit time is

$$v_{ph} \exp[-2\alpha R - W/k_B T]. \tag{11.27}$$

If one assumes v_{ph} to be relatively independent of both $\overline{R}$ and T, then the maximum hopping probability occurs when

$$3\alpha/2 = 9[4\pi \tilde{R}^4 N(E_F) k_B T]^{-1}, \tag{11.28}$$

where $\tilde{R}$ is the optimum hopping distance,

$$\tilde{R} = \{3/[2\pi \alpha N(E_F) k_B T]\}^{1/4}. \tag{11.29}$$

The actual hopping distance is $\frac{3}{4}R$. The hopping therefore becomes

$$v_{ph} \exp[-B/T^{1/4}], \tag{11.30}$$

where

$$B = B_0 \{\alpha^3/k_B N(E_F)\}^{1/4} \tag{11.31}$$

and

$$B_0 = 2(3\pi/2)^{1/4} = 1.66. \tag{11.32}$$

Consequently the conductivity is

$$\sigma_{dc} = e^2 N(E_F) \overline{R}^2 v_{ph} \exp[-B/T^{1/4}] = \sigma_0(T) \exp[-B/T^{1/4}]. \tag{11.33}$$

In principle, the two parameters α and $N(E_F)$ can be evaluated from the slope of a plot of $\ln[\sigma(T)T^{1/2}]$ versus $T^{-1/4}$ and from the intercept at $T^{-1/4} = 0$. Alternatively, we can get an idea of $\overline{R}$ by taking $\alpha^{-1} = 10$ Å, $N(E_F) = 10^{19}$ cm^{-3} eV^{-1}, to find $\overline{R} = 250T^{-1/4}$ Å, which yields 80 Å at 100 K.

If one uses the optical phonon frequency for ν_{ph}, which is about 3.6×10^{13} sec^{-1}, the value of σ calculated from Eq. (11.33) can be two to three orders of magnitude too low. This is found for the calculation of conductivity for amorphous Si by the variable-range hopping model. Colson and Nagels (1980) have suggested that $\nu_{ph}e^{2\alpha R}$ should be used instead.

The frequency dependence of conductivity is that given in Eq. (11.17). Two important features are (1) $\sigma(\omega)$ varies linearly with temperature, and (2) the slope of a plot ln $\sigma(\omega)$ versus ln ω is not constant, but decreases slightly with increasing frequency of the applied field with the approximate functional form

$$\sigma(\omega) \propto T\omega^{0.8}. \tag{11.34}$$

The thermopower has been suggested to be

$$S = \frac{\pi^2 k_B^2 T}{3e}\left[\frac{d \ln \sigma(E)}{dE}\right]_{E_F}. \tag{11.35}$$

A linear (log σ)–$T^{-1/4}$ dependence has been widely used as evidence for transport by variable range hopping. However, one has to exercise caution, as indicated by the following observations. Redfield (1975) made a thorough experimental study on transport properties in energy band tails. The system was crystalline n-GaAs heavily doped with impurities that have large Bohr orbits and thus small binding energies so that easily attainable concentrations can cause strong interactions among the impurities with resulting energy levels overlapping the crystal band. The carriers are free electrons in states above the band tail, which are "drained off" to bring the Fermi energy into the range of energies covered by the tail states. Therefore, the shallow donors are closely compensated by almost equal numbers of acceptors. A result of heavy doping and close compensations is that the number of mobile carriers that can screen the impurity charges is significantly reduced, the strength of potential fluctuation correspondingly increases, and the random potential distribution becomes symmetrical. The density of state in the tail is close to half Gaussian shape:

$$N(E) = N_c \exp\left[-\left(\frac{E - E_c}{E_0}\right)^2\right], \tag{11.36}$$

where E_0 is the tail-width parameter. The starting n-GaAs has $N_D = 2 - 3 \times 10^{18}$ cm^{-3}, and the concentration of free electron $n = N_D$ at room temperature. It was compensated in two ways. One was doping with Cu to a ratio of $N_A/N_D \simeq 0.97$, which has Fermi level close to the top of the band tail. It was also compensated by irradiation with 1-MeV electrons to produce deep acceptors. This second method allows very fine adjustment of the compensation ratio.

Alternating-current conductivity was measured for the above materials at low temperatures to ensure that the electrons remain in the lowest available states. The frequency dependence of $\omega^{0.8-1.0}$ [Eq. (11.34)] is obeyed for measurements at 4.2 K and 77 K from 10^2 to 5×10^5 Hz and 1 V cm^{-1}. The possibility of transport in an impurity band was eliminated by Hall coefficient measurements. However, the dc conductivity data for different levels of compensation, i.e., different Fermi levels, from $1.5 \leqslant T \leqslant 300$ K fit $\log \sigma \approx T^{-1/2}$ better than $\log \sigma \approx T^{-1/4}$. These results of Redfield (1975) are cited here to caution readers that for closely compensated heavily doped *n*-GaAs specially prepared to possess band tails having states that have varying depth of localization, the experimental results are not in agreement with theory. Simple explanations are not available. One possibility is that the conduction is by percolation of finely dispersed metal particles in insulating matrix (Section 11.1.7). Another explanation is that the material is pseudo-one-dimensional. The $\log \sigma \propto T^{-1/4}$ relation was derived for three-dimensional systems. Conduction by variable-range hopping in two dimensions has $\log \sigma \propto T^{-1/3}$ (Kurkijarvi, 1973), and it is expected to be $\log \sigma \propto T^{-1/2}$ for pseudo-one-dimensional conduction (Shante and Varma, 1973; Brenig *et al.*, 1973).

11.1.5   Conduction by Polaron

In the bound state, the polaron cannot move without changing the positions of the neighboring atoms. The polaron has a lower energy than free electrons but has a larger effective mass. The reduction in energy of the polaron relative to that of the electron in the undistorted lattice corresponds to its binding energy E_b. There are two opposite contributions to this energy: a lowering of energy due to the displacements of the neighboring atoms, and an increase (half that of the decrease) from the strain energy of the induced lattice distortion.

A polaron can move by hopping from one lattice position to an equivalent one, which can only be created by a similar distortion of the lattice surrounding that site. The deformation energy comes from phonons. Therefore, the motion is phonon assisted tunneling between nearby sites. Analogous to Eqs. (11.4) and (11.5), we write

$$\sigma = ne\mu = ne^2 D/k_B T = (ne^2 a^2/k_B T)P, \tag{11.37}$$

where P is the total hopping probability. P contains two probabilities: the probability of a site suitable for jumping P_1 and the probability for charge transfer P_2. The former is

$$P_1 = v_{ph} \exp(-W/k_B T), \qquad (11.38)$$

and W is the minimum energy required to create two sites having equivalent distortion. W is related to the polaron binding energy by $W = E_b/2$. The expression for P_2 is

$$P_2 = \frac{1}{\hbar v_{ph}} \left(\frac{\pi}{W k_B T}\right)^{1/2} t_0^2. \qquad (11.39)$$

The conductivity due to polaron hopping is given by

$$\sigma = \frac{ne^2 a^2 t_0^2}{k_B Th} \left(\frac{\pi}{W k_B T}\right)^{1/2} \exp(-W/k_B T). \qquad (11.40)$$

At low temperatures, conductivity will increase exponentially over a wide temperature range. In the range $k_B T \approx W$, the preexponential term becomes predominent and $\sigma \propto T^{-3/2}$. If the probability for charge-transfer following a hop is high, then $P_2 = 1$ and Eq. (11.40) is simplified to

$$\sigma = \frac{ne^2 a^2 v_{ph}}{k_B T} \exp\left(\frac{-W}{k_B T}\right). \qquad (11.41)$$

In other words, at the temperature range of $k_B T \approx W$, the conductivity varies with T^{-1}.

The ac conductivity has the functional form

$$\sigma(\omega) = \sigma_0 \omega^s \exp[-W(1-s)/k_B T]. \qquad (11.42)$$

The thermopower for this conduction has the classic form

$$S = \pm \frac{k_B}{|e|}\left[\frac{E}{k_B T} + A\right], \qquad (11.43)$$

where E is the energy associated with the thermal generation of carriers and is greater than the minimum energy W. The kinetic term A is very small if there is no transfer of vibrational energy involved with a polaron hop.

11.1.6 Intersoliton Hopping Conduction

The above mechanisms of transport in semiconductors are basically three dimensional whether the material is crystalline or amorphous. However, the soliton model proposed for polyacetylene is noted for its extremely one-dimensional character. In the ideal description solitons are topologically confined to one polymer chain. Thus, one might expect no transverse conduction from one chain to another in such a system. Furthermore, solitons are very susceptible to effects of disorder, and it is impossible to avoid defects in synthetic polymers (i.e., impurities, chain ends, cross-links, etc.). Any disorder will localize the solitons. We are thus forced to conclude that soliton transport is a very inefficient process.

The neutral solitons are not responsible for the transport properties. Thus the conductivity of p-doped *trans*-$(CH)_x$ is lowered many orders of magnitude by NH_3 compensation without at the same time decreasing the EPR signal intensity of the neutral soliton. Also, the sign and magnitude of the thermopower of pristine polyacetylene requires that the carrier be positively charged. On the other hand, doping with an electron donor or acceptor is expected to produce localized states that are either doubly occupied or unoccupied, respectively. Such a system would be quite insulating. When there are present both neutral and charged solitons, they can bind to form a polaron. The polaron is not topologically confined to a single polyacetylene chain and the three-dimensional large polaron could contribute toward electronic conduction. However, polaron hopping requires a large activation energy, making it unlikely to contribute significantly to conductivity in lightly doped polyacetylene.

Kivelson (1981a,b, 1982) postulated an intersoliton hopping (ISH) mechanism, which can proceed multidimensionally. Even though transport is anisotropic, it is not so sensitive to defects on the chain as discussed above. Also, the activation energy will be quite small and conductivity can be moderately large even for very low carrier concentration and low temperatures. The theory is predicated on the presence of both a charge soliton (taken to be S^+) and a neutral soliton ($S \cdot$), which is situated near a dopant ion D^-, referred to as an impurity site. This requirement can be readily seen from the following considerations. If there is only an isolated charged soliton bound to D^-, little transport can be associated with its motion if $k_B \ll E_b$. Since $E_b \geqslant 0.3$ eV, the translation of S^+ cannot be effective in charge transport. The presence of a neutral soliton in a nearby chain alters this situation somewhat. Since there will in general be a nonzero overlap between the electronic states associated with the two differently charged solitons, it is possible for the electron to make a phonon assisted hop from one state to the other as follows,

$$(11.44)$$

However, as S^+ is strongly bound to D^- and the binding energy for $S \cdot$ is zero to first order in the impurity potential, this situation is energetically not more favorable than the hopping of isolated S^+. If on the other hand there is a D^- ion situated in the vicinity of $S \cdot$, then electron hopping between the two states would require little or no activation energy, as illustrated by

$$(11.45)$$

The main energy involved in the process would be the difference in the Coulombic energies for the initial and final S^+ sites, depending on the relative positions of D^- with respect to S^+.

To obtain the rate of intersoliton hopping, let us first introduce the parameters involved. The two sites i and j are separated by a vector $\mathbf{R} = \mathbf{R}_i - \mathbf{R}_j$ with components $\mathbf{R}_\parallel$ and $\mathbf{R}_\perp$. The average separation between D^- is

$$R_0 = (4\pi[D^-]/3)^{-1/3}. \tag{11.46}$$

The three-dimensional average electronic decay length is

$$\xi = (\xi_\parallel \xi_\perp^2)^{1/3}. \tag{11.47}$$

Let $N(\gamma)$ be the mean number of sites to which the hopping rate from a given site at the origin is greater than γ:

$$N(\gamma) = R_0^{-3}(\xi_\parallel/2)(\xi_\perp/2)^2 \, [\ln(\gamma_0/\gamma)]^3. \tag{11.48}$$

Conductivity is determined by the critical hopping rate γ_c, which defines the percolation fraction $\mathbf{P} = N(\gamma_c)$ and $\mathbf{P} = 2.71$ in three dimensions. Thus,

$$\gamma_c = \gamma_0 \exp[-2BR_0/\xi], \tag{11.49}$$

where $B = \mathbf{P}^{1/3} = (2.71)^{1/3} = 1.39$.

The factors that determine the hopping rate between a pair of solitons are the following: when soliton i undergoes a transition from an initial charge state **1** to the neutral state **2**, it will give up (or gain) an energy ε. This energy is furnished by electron–phonon coupling; the complicated thermal average of the electron–phonon coupling function is $\gamma(T)$. The probability that $S \cdot$ is found near a $\overline{D}$ is N^{-1}; $\gamma(T)/N$ is then proportional to the fraction of time a pair of solitons are so situated that the initial and final soliton states are within $k_B T$ of each other. The transition rate is proportional to the square of electronic overlap, which dominates the position dependence of the sites, and is assumed to depend exponentially on $\overline{R}$ or $\exp[-\{(R_\parallel/\xi_\parallel)^2 + (R_\perp/\xi_\perp)^2\}^{1/2}]$. Finally, the rate of the process is dependent on the fraction of occupied S $\cdot$ (S$^+$) sites represented by y^0 (y^+). Kivelson (1981, 1982) derived the following expressions for transport by intersoliton hopping:

$$\sigma_{dc} = \frac{Ae^2\gamma(T)}{k_B T}\left(\frac{\xi}{R_0^2}\right)y^0\frac{y^0 y^+}{(y^0 + y^+)^2}\exp\left(\frac{-2BR_0}{\xi}\right), \qquad (11.50)$$

$$\sigma_{ac}(\omega) = \frac{e^2\xi^3[D^-]^2\omega}{384 k_B T}\frac{y^0 y^+}{(y^0 + y^+)^2}\left\{\ln\left[\frac{2\omega(y^0 + y^+)^2}{y^0\gamma(T)y^0 y^+}\right]\right\}^4, \qquad (11.51)$$

and

$$S = \pm\frac{k_B}{|e|}\left\{\frac{\overline{\epsilon}(T)}{k_B T} + \ln\left(\frac{y_0}{y^+}\right) + \sigma[\ln(T)]\right\}, \qquad (11.52)$$

where $\overline{\epsilon}(T)$ is the average energy transported per hop and $A = 0.45$ and $B = 1.39$ are dimensionless. The implied assumptions are that all the impurity sites are equivalent, that the temperature dependence of $\gamma(T)$ arises from ϵ, that the factor $y^0 y^+(y^0 + y^+)^{-2}$ turns a quantum-mechanical transition rate into a net transition rate, and that the discrete nature of the polyacetylene chain has been neglected. The last assumption requires a simple three-dimensional approximation of the microscopic one-dimensional model of Su *et al.* (1979). The value of ξ_1 is estimated to be $b/\ln(\Delta_0/t_1)$, which is about 2.3 Å if $\Delta_0 \gg t_1$. The electron–phonon coupling function is approximated as

$$h\gamma(T) \simeq 500 \text{ eV}\left[\frac{T}{300 \text{ K}}\right]^{10}. \qquad (11.53)$$

The main predictions of the intersoliton hopping model are the following. (1) Dopant concentration does not affect the temperature dependence of conductivity. (2) Thermopower is approximately independent of dopant concentration. (3) Conductivity varies with temperature as $\sigma\alpha(T)^{x+1}$, which is slower than $\exp(-E_b/k_B T)$. (4) Conductivity varies rapidly with dopant concentration, $\ln\sigma \approx [D^-]^{-1/3}$. (5) There is strong dependence of conductivity on frequency, $\sigma_{ac}(\omega) - \sigma_{dc} \approx \omega(\ln\omega)^4$. (6) Thermopower is very weakly temperature dependent, $S = (k_B/e)\{(x + 2)/2 + \ln(y^0 y^+) + \ln(k_B T/h\omega_s)\}$, where $h\omega_s$ is the soliton vibrational energy, ~ 0.06 eV. (7) Non-Ohmic effects are expected to become apparent for electric fields greater than $\mathcal{C}_0$, $\mathcal{C}_0 = k_B T/(eR^*)$, where R^* is the characteristic hopping distance on the percolation path, i.e., $R^* = (\xi_\parallel/\xi)BR_0$.

There are two interesting aspects of the intersoliton hopping mechanism. The first is that in the regime of light doping for which the theory is valid, the interchain process is more efficient than intrachain hopping and the anisotropy is small. The frequency for a three-dimensional hop is given by

$$\ln \nu_{3D} \approx \ln \gamma - 2R_0/\xi; \qquad (11.54)$$

for the one-dimensional process it is

$$\ln \nu_{1D} \approx \ln \gamma - 2R_1/\xi_\parallel, \qquad (11.55)$$

where R_1 is the typical separation between randomly placed dopant ions on a given polyacetylene chain, which is $R_1 \approx (b^2[D^-])^{-1}$. The interchain separation b is 4.39 Å. Taking $\xi_\parallel \approx 8.5$ Å and $\xi_\perp \approx 2.3$ Å, Kivlson finds with Eqs. (11.54) and (11.55) that interchain hopping always dominates intrachain hopping at low doping levels. This is because of the linear dependence of R_1 on $[D^-]^{-1}$, whereas R_0 depends on $[D^-]^{1/3}$. For instance, at $y = 10^{-3}$, the value for $2R_0/\xi \simeq 13$ which is much smaller than $2R_1/\xi_\parallel \simeq 660$. Under these conditions conduction is purely three dimensional. However, the conduction becomes increasingly one dimensional as the dopant concentration is increased. Intrachain hopping becomes more probable when $[D^-] > \frac{1}{3}\pi(\xi_\perp/\xi_\parallel)(1/b)^3$, which corresponds to $y > 0.18$. However, this dopant concentration lies in the metallic regime and different conduction mechanisms would apply at this or even much lower doping levels.

The second point of interest concerns the spin diffusion constant. The transverse component is expected to be related to the conductivity in those directions by an Einstein relation

$$D_\perp = k_B T\sigma_\perp/e^2[D^-]. \qquad (11.56)$$

For an intrinsic impurity concentration of $y \approx 4 \times 10^{-4}$ and a typical room-temperature conductivity of 10^{-5} $(\Omega\text{ cm})^{-1}$, $D_\perp/b^2 \approx 5 \times 10^7$ sec^{-1} or $D_\perp \approx 9.5 \times 10^{-8}$ cm^2 sec^{-1} for an interchain separation of 4.35 Å. The experimental value for $D_\perp$ estimated from the effect of dopant on EPR saturation is 1.6×10^{-7} cm^2 sec^{-1} (Table 5.3). The agreement between theory and experiment is very good. On the other hand, the in-chain diffusion constant should be much greater than $D_\parallel$ because of the high in-chain diffusivity of the neutral solitons:

$$D_\parallel \gg k_B T\sigma_\parallel/e^2[D^-]. \qquad (11.57)$$

The anisotropy for dc conductivity is estimated to be $\sigma_\parallel\sigma_\perp \approx 50$. In contrast, experimental estimates for $D_\parallel/D_\perp$ range from 10^3 to 10^5 (Section 5.5.1).

Finally, the conductivity of pristine *trans*-polyacetylene of $\sim 3 \times 10^{-6}$ $(\Omega\text{ cm})^{-1}$ can be obtained from Eq. (11.50) if y^0 is taken to be 4×10^{-4} and

y^+ is about 2×10^{-4}. The former corresponds well with the presence of about one-half $S \cdot$ per polyacetylene chain, or 3×10^{-4}. On the other hand, the origins of the charged soliton in undoped polyacetylene and the nature of the impurity that is required to be found in the vicinity of $S \cdot$ in order to enable the intersoliton hopping to take place are unknown. A possible explanation may be found as the natural consequence of the chemistry and physics of the system. It has been explained in Section 2.6 that the polyacetylene chain is initiated by a Ti^{3+} species, which remains bound to the polymer. Equations (5.56)–(5.58) showed how positive solitons may be formed by autoxidation. According to this hypothesis each polyacetylene chain that has a bound Ti^{3+} ion can potentially form a S^+ pinned to an O_2^{2-}. Kinetic results (Section 2.7.3) showed that in a typical polymerization about one-third to one-fifth of the polyacetylene chains are thus endowed. This corresponds to $y^+ \simeq 1.2 \times 10^{-4}$ to 2×10^{-4} in agreement with various estimates. Oxygen can also act as the impurity for the neutral soliton:

$$
\begin{array}{c}
O_2^{\delta^-} \\
\text{\Large /\!\backslash\!/\!\backslash\!/\!\backslash\!/\!\backslash} \\
\delta^+
\end{array}
\qquad (11.58)
$$

The polarized neutral soliton will have energy closer to the pinned charged soliton than a free $S \cdot$.

Finally, the presence of charged solitons in undoped polyacetylene is supported by two observations. First, exposure of *trans*-polyacetylene to NH_3 leads to compensation of a large fraction of the charged carriers resulting in a five-order-of-magnitude decrease in σ_{RT} (Chiang *et al.*, 1978a). Similarly, NH_3 compensation also lowers the conductivity of pristine *cis*-polyacetylene. Second, depletion experiments on heterojunctions (Ozaki *et al.*, 1980) showed about 1.8×10^{18} cm^{-3} or $y^+ = 0.97 \times 10^{-5}$ of charged carriers in undoped *trans*-polyacetylene.

11.1.7   Fluctuation-Induced Tunneling Conduction

The conduction mechanisms discussed above share one common assumption, which is that the material is homogeneous in the sense that the localized states, the polarons, or the charged and neutral solitons are randomly distributed. There is, however, a conduction mechanism, which stems from the study of composite materials of conducting particles imbedded in an insulating matrix, which seems to be particularly applicable to doped polyacetylene. First, polyacetylene, by virtue of its fibrillar morphology, is not a continuous material; the fibrils are finite in dimension. Second, there is ample evidence that doping can be grossly inhomogeneous, thus creating tiny metallic domains separated by insulating regions. Even if doping is uniform, the carriers are delocalized and free to move over distances very large compared to the atomic dimension. For such systems the electrical conduction may be dominated by carrier transfer between conducting segments rather than by hopping between localized states. The premise is that the carriers tend to tunnel between conducting regions at points of their closest approach; the relevant tunnel junctions are small in size and are subject to thermally activated voltage fluctuations across the junction. The voltage fluctuation modulates the potential barrier and introduces a temperature dependence to the tunneling probability. This fluctuation-induced tunneling mechanism has been developed by Sheng and co-workers (Sheng *et al.*, 1978; Sichel *et al.*, 1978; Sheng, 1980).

The conducting domains are irregular in shape, and tunneling is dominated by the region of closest approach. The voltage fluctuation across this region is due to transient excess or deficit charges. The tunnel junction may be simulated by a parallel-plate capacitor of capacitance C. The average square of the thermal fluctuation voltage across the junction $\langle V_{\mathrm{T}}^2 \rangle$ is

$$\langle V_{\mathrm{T}}^2 \rangle = k_{\mathrm{B}} T / C. \tag{11.59}$$

The mechanism gives a field and fluctuation-induced conductivity

$$\sigma_i = \sigma_{i0} \exp[-(T_1/T)(\epsilon^*)^2 - (T_1/T_0)\phi(\epsilon^*)], \tag{11.60}$$

where the parameters

$$T_1 = a\epsilon_0^2/k_{\mathrm{B}}, \tag{11.61}$$

$$T_0 = T_1/2\chi w \xi(0), \tag{11.62}$$

represent, respectively, the temperature below which the conductivity becomes temperature independent (corresponding to a pure elastic tunneling mechanism) and the temperature above which the conduction mechanism across the junctions is purely thermally activated. The function $\phi(\epsilon^*)$, where ϵ^* is a reduced electric-field- and temperature-dependent parameter, contains another parameter λ that governs the shape of the image-force-corrected rectangular potential barrier,

$$\lambda = 0.795e^2/4wKV_0. \tag{11.63}$$

In Eqs. (11.61) and (11.62), ε_0 is the electric field for maximum potential equal to zero.

$$\chi = (2mV_0/\hbar^2)^{1/2} \tag{11.64}$$

is the tunneling constant, m is the effective mass of the carrier, V_0 is the height of the potential barrier, and w in Eq. (11.63) is the parameter to express the separation of junction in reduced variable.

If the intrinsic conductivity of the conducting regions is attributed to the electrical transport in a three-dimensional disordered material where

$$\sigma_j = \sigma_{j0} \exp[-B/T^{1/4}]. \tag{11.65}$$

The total resistivity $R(T)$ can be written as the sum of the resistance of the polyacetylene, and for the tunneling,

$$R(T) = R_i(T) + R_j(T). \tag{11.66}$$

The thermopower is expressed by Eq. (11.35). Audenaert *et al.* (1981) combined the models of variable-range hopping and fluctuation-induced tunneling to find the following relation for the thermopower:

$$S = \pm \frac{T}{A'f(T) + 1}[B' - C'T^{-1/4}], \tag{11.67}$$

where

$$f(T) = \frac{\exp[(T_1/T)\epsilon^{*2} + (T_1/T_0)\phi(\epsilon^*)]}{\exp[B/T^{1/4}]}. \tag{11.68}$$

B is defined by Eq. (11.32), and A', B', and C' are given by

$$\begin{aligned} A' &= R_{i0}/R_{j0}, \\ B' &\approx \frac{1}{R_{i0}} \frac{dR_{i0}}{dE}\bigg|_{E = E_F}, \\ C' &\approx 0.5\left(\frac{\alpha^3}{k_B N(E)}\right)^{1/4} \frac{1}{N(E)} \frac{dN(E)}{dE}\bigg|_{E = E_F}. \end{aligned} \tag{11.69}$$

Equation (11.60) may be expressed in simpler functional form. At sufficiently low fields and at concentrations of metallic domains below the percolation threshold, the conductivity is dominated by excitation to produce a pair of oppositely charged grains from a pair of neutral grains and the probability of carrier tunneling between these grains. The temperature dependence of conductivity has the form

$$\sigma \approx \exp[-AT^{-1/2}]. \tag{11.70}$$

At high electric fields, non-Ohmic behavior is expected for the model.

For concentrations of metallic domains above the percolation threshold, if the domains are highly disordered, the conductivity has a temperature dependence like that of variable-range hopping in three dimensions:

$$\sigma \approx \exp[-A'T^{-1/4}]. \tag{11.71}$$

Even above the percolation threshold there may still be potential barriers between the metallic domains, and conduction requires tunneling. Thermal fluctuations can modulate the potential barrier to facilitate transport resulting in a temperature dependence of the form

$$\sigma \approx \exp[-T_1/(T + T_0)]. \tag{11.72}$$

However, the percolation threshold is quite dependent on the sizes and shapes of the metallic grains. Therefore, no simple relation between dopant concentration and conductivity can be given.

In this model the conductivity would show field dependence resulting from the concentration of voltage drop across the highly resistive barrier of the insulating matrix. This is observed both in metal–dielectric composites and in anisotropic conductors with random barriers. There will also be a frequency dependence for σ resulting from the capacitive coupling between the highly conducting regions, typically of the form $\sigma \approx \omega^k$. If one assumes $\sigma \approx 10^2$ $(\Omega$ cm$)^{-1}$ for the conducting and $\sigma \approx 10^{-3}$ $(\Omega$ cm$)^{-1}$ for the nonconducting regions, with the volume fraction of the conducting regions about one-third of the total volume, the effective medium theory of Springett (1973) predicts the onset of frequency dependence at $\sim 10^2$ Hz. The dependence of σ on electric field should follow Eq. (11.60).

11.2 THERMOPOWER

With the fibrous morphology, polyacetylene is not a continuous material. Its dc conductivity is expected to be limited by interfibril contacts; this expectation has been demonstrated. Under such circumstances the intrinsic metallic conductivity may be considerably higher than the measured dc values. This difficulty is not present in thermoelectric power (TEP) measurements, because there is no flow of current. The interfibril contacts should play a minimal role allowing an evaluation of the intrinsic transport properties of the material.

The thermopower coefficient (Seebeck coefficient) is measured by mounting a rectangular polyacetylene film specimen between two copper blocks using pressure contacts. One of the copper blocks is heated to give a temperature difference of 2–5 K measured with a copper constant differential thermocouple. The voltage generated was measured with a microvoltmeter.

11.2.1   Sign of S

At low dopant concentration TEP should be large because of a few carriers with many possible states per carrier. TEP can be viewed as a measure of entropy per carrier; it is positive or negative depending on the sign of the charge carrier. As-prepared *trans*-polyacetylene was first reported by Park *et al.* (1979b, 1980) to have values of S ranging from $+800$ to $+1000$ μV K^{-1}. The results suggest the presence of positively charged carriers in undoped polymer. Higher values of S of 1450 ± 150 μV K^{-1} are now obtained consistently in our laboratory for rigorously purified polymer, isomerized by optimum procedure, and measured with higher external impedance. Such specimens have apparently lower concentration of intrinsic positive carriers. Measurements on undoped *cis*-polyacetylene have not been reported; the high resistivity of the material makes such measurement difficult.

Moses *et al.* (1981) obtained TEP of n-type material. *trans*-Polyacetylene was electrochemically doped with tetrabutylammonium to form $[CH(Bu_4N)_y]_x$ with $y \simeq 0.03$. The sample was mounted in a sample holder designed for measurements of S versus R. It was compensated by exposure to air over a three-day period through a slow, controlled leak. The resistance increased from 10^2 Ω with $S = -45.3$ μV K^{-1} to 10^8 Ω with $S = -850 \pm 130$ μV K^{-1} near the compensation point. These results established the dependence of sign of S on the charge of the carriers.

11.2.2   Effect of Temperature

The temperature dependence of TEP for *trans*-$(CHI_y)_x$ is given in Fig. 11.3. The undoped polymer and *trans*-$(CHI_{0.017})_x$ have temperature-independent TEP. This was initially explained by Park *et al.* (1979b, 1980) as indicating carrier hopping among a set of localized states. TEP is given by the Heikes formula

$$S = +\frac{k_B}{|e|}\ln\left(\frac{1-\rho}{\rho}\right),\tag{11.73}$$

where $\rho = n/N$ is the ratio of number of holes n to the number of available sites N. The expression applies to spinless Fermi ions. The experimental value of S would give $\rho \approx 10^{-4}$ or a carrier concentration of 2×10^{18} cm^{-3} in undoped polyacetylene.

Subsequently, Moses *et al.* (1981) analyzed the temperature independence of TEP in terms of the intersoliton hopping transport. Within the

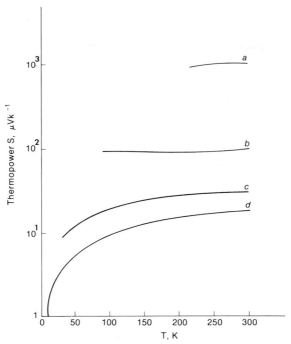

Fig. 11.3 Temperature dependence of the thermopower for *trans*-(CHI⁻)ₓ with y values: (a) 0; (b) 0.017; (c) 0.03; (d) 0.22. [After Park *et al.* (1979b).]

temperature range of 230–280 K, Kivelson's equation (11.52) can be recast into

$$S = \pm \frac{k_B}{|e|} \left\{ \frac{x+2}{2} + \ln\left(\frac{y^0}{y^+}\right) \right\}. \tag{11.74}$$

Substitution of $S \simeq 850~\mu V~K^{-1}$ and $x = 13$ gives $\ln(y^0/y^+) \simeq 25$. EPR results give $y^0 \approx 3 \times 10^{-4}$ ($1.5 \times 10^{19}~cm^{-3}$), therefore $y^+ \simeq 3 \times 10^{-5}$ ($1.8 \times 10^{18}~cm^{-3}$), which is the same as estimated for the carrier concentration by Eq. (11.73). The sample of *trans*-(CHI₀.₀₁₇)ₓ has temperature-independent $S \approx 100~\mu V~K^{-1}$. If we assume the dopant ion to be I_3^-, i.e., $y(I_3^-) = 5.7 \times 10^{-3}$, then according to Eq. (11.73), S should have a value of $440~\mu V~K^{-1}$. That the observed value is lower than expected suggests that the intersoliton hopping model may be failing at this doping level.

Kwak *et al.* (1979a) had measured TEP of AsF₅-doped *trans*-polyacetylene. At a nominal dopant concentration of about 0.34%, TEP was found to increase with the decrease of temperature. The data in the 240–300 K range have the functional form of the variable-range hopping mechanisms [Eqs. (11.12), (11.23), and (11.43)]. The authors used the equation

$$S = \frac{k_B}{|e|} \left\{ \frac{\alpha + 1 + n}{2} + \frac{\zeta}{k_B T} \right\}, \tag{11.75}$$

where $\zeta = 0.4$ eV is the hole activation energy, n is the dimensionality, and α is ~ 1 in three dimensions, but ~ 2 in one dimension.

TEP of polyacetylene heavily doped with AsF_5 decreases linearly with temperature, and $S \rightarrow 0$ as $T \rightarrow 0$ K (Fig. 11.4). Park *et al.* (1979b) found

$$S = +\frac{k_B}{|e|}[3.6 \times 10^{-4} \, T]. \tag{11.76}$$

Kwak *et al.* (1979b) reported virtually identical results for both *trans*- and *cis*-$[CH(AsF_5)_{0.1}]_x$ with a slope of 3.8×10^{-4}. Therefore, this material is said to be a degenerate electron gas. The data implies the absence of any anomalous electron scattering mechanism that would contribute a nonlin-

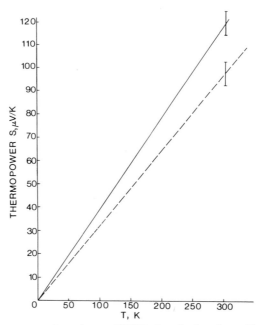

Fig. 11.4   Temperature dependence of TEP in heavily doped metallic $[CH(AsF_5)_y]_x$. (---) $[CH(AsF_5)_{0.147}]_x$ non oriented; (——) $[CH(AsF_5)_{0.095}]_x$ oriented, $l/l_0 = 3.2$. [After Park *et al.* (1980).]

ear term to the thermopower. Moreover, this linearity is inconsistent with a narrow impurity band.

In the metallic state, doped polyacetylene has nearly filled bands. The thermopower can be written

$$S = +\frac{\pi^2}{3}\frac{k_B}{|e|} k_B T \left[\frac{d \ln \sigma(E)}{dE}\right]_{E_F},   (11.77)$$

where $\sigma(E) = n(E)|e|\mu(E)$ and $n(E)$ is the number of carriers contributing to $\sigma(E)$, $dn(E)/dE$ is the density of states (both signs of spin), and $\mu(E)$ is the energy-dependent mobility. Assuming energy-independent scattering, i.e., $\mu(E)$ independent of E,

$$S = +\frac{k}{|e|}\frac{\pi^2}{3}kT\eta(E_F),   (11.78)$$

where $\eta(E_F)$ is the density of states per carrier, $dn(E)/N\ dE$. Park *et al.* (1979b) considered the experimental results to be in good agreement with the equation, and suggested that $\eta(E_F)$ is 1.36 states per electron volt per carrier. Since there are 0.15 carriers per carbon atom in $[CH(AsF_5)_{0.15}]_x$, the thermopower data gives 0.2 states per electron volt per CH. The same density of states was obtained by Goldberg *et al.* (1979) and Weinberger *et al.* (1979) from spin resonance and magnetic susceptibility studies.

Figure 11.4 also showed that the TEP results for the metallic AsF$_5$-doped polyacetylenes are nearly the same for both unoriented *cis*-$[CH(AsF_5)_{0.147}]_x$ and partially oriented *trans*-$[CH(AsF_5)_{0.095}]_x$ with $l/l_0 = 3.2$. Therefore, in contrast to the anisotropic electrical conductivity (Section 11.2.9), the TEP is isotropic. This is consistent with expectation since TEP is a zero-current transport coefficient and is not affected by interfibril contact as is the case for conductivity. On the other hand, the polymers heavily doped with iodine show a curvature in their S-versus-T curve (Figs. 11.3 and 11.5) and higher values of S than the AsF$_5$-doped material. This behavior can be interpreted as due to the presence of energy-dependent scattering. Audenaert *et al.* (1981) had reanalyzed these data of $(CHI_{0.22})_x$ thermopower data of Park *et al.* (1980) according to the mechanism of fluctuation-induced tunneling [Eqs. (11.67)–(11.69)] with the parameters $B = 14K^{1/4}$, $T_0 = 13.6K$, and $T_1 = 150K$ and found excellent fitting (shown in Fig. 11.6).

11.2.3   Concentration of Dopant

The variation of TEP with dopant concentration was investigated by Park *et al.* (1979b, 1980) for $(CHI_y)_x$. The results showed that S is un-

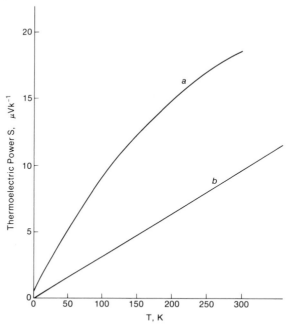

Fig. 11.5   Temperature dependence of TEP for heavily doped metallic polyacetylenes: (a) *trans*-$(CHI_{0.22})_x$, $\sigma_{RT} = 40$ $(\Omega\ cm)^{-1}$; (b) *cis*-$[CH(AsF_5)_y]_x$, $\sigma_{RT} = 640$ $(\Omega\ cm)^{-1}$. [After Park *et al.* (1979b).]

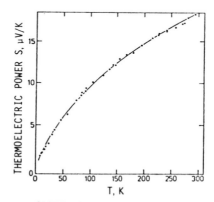

Fig. 11.6   Thermopower of $(CHI_{0.22})_x$ versus T: data points from Park *et al.* (1979b); curve according to fluctuation-induced tunneling theory. [After Audenaert *et al.* (1981).]

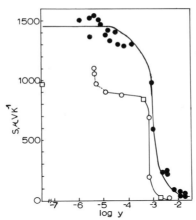

Fig. 11.7 Variation of TEP with dopant concentration: (O) *trans*-[CH(AsF$_5$)$_y$]$_x$ doped with "cyclic" procedure (Warakomski *et al.*, 1983); (●) *trans*-[CH(I$_3$)$_y$]$_x$ (Warakomski *et al.*, 1983); (□) *trans*-[CH(AsF$_5$)$_y$]$_x$ of Moses *et al.* (1982).

changed by doping up to $y \approx 3 \times 10^{-3}$ and then falls steeply for $0.003 < y < 0.03$. For $y > 0.03$, $S = +18.5 \ \mu$V K^{-1} in the heavily doped metallic limit and is nearly independent of y. The authors considered the results to imply a critical concentration of $n_c \simeq 1$ mole % I$_3^-$.

We have recently measured room-temperature TEP of *trans*-polyacetylene doped over a wide range of concentrations with iodine. Our undoped samples have higher values of S of $1450 \pm 150 \ \mu$V K^{-1}. Very low levels of iodine doping causes a gradual decrease of S (Fig. 11.7). Precipitous decrease of S occurs between 4.5×10^{-4} and 10^{-3} for y(I$_3^-$). In the case of AsF$_5$ doping in *trans*-polyacetylene, even at very low levels of $y = 4 \times 10^{-6}$, the thermopower is $1050 \pm 50 \ \mu$V K^{-1}. Figure 11.7 suggests that there is a drop of S at ppm AsF$_5$ doping, which may be due to preferential doping of the surface of the polymer fibrils. There is a region of relatively constant S of $\sim 900 \ \mu$V K^{-1} for $10^5 \leqslant y \leqslant 3 \times 10^{-4}$. As in the iodine case, thermopower drops very sharply for y between 4.4×10^{-4} and 9×10^{-4}. Very heavily iodine-doped polymer has slightly larger S values than heavily AsF$_5$-doped *trans*-polyacetylene. These results showed transformation of individual fibrils from semiconducting to metallic occurs abruptly spanning only about twofold changes in dopant concentration. The transitions are comparably sharp for iodine- and AsF$_5$-doped *trans*-polyacetylenes found both at $y \approx 7 \times 10^{-4}$. The midpoint of the transition is significantly lower than previously reported values based on either thermopower or conductivity data.

11.3   ELECTRICAL CONDUCTIVITY

11.3.1   Concentration of Dopant

11.3.1.1   *Iodine-Doped Polyacetylene*

The early results on conductivity of iodine-doped polyacetylene by Chiang *et al.* (1978b) have been cited above. We have carried out extensive conductivity measurements on polyacetylenes doped to the intermediate and especially very low (ppm) concentrations. The results on *trans*-$(CHI_y)_x$ are shown in Fig. 11.8. Also included are the data of Epstein *et al.* (1981b), Chiang *et al.* (1978b), and Sichel *et al.* (1982). There are several important features. First, σ_{RT} remained constant between $y = 10^{-6}$, which has the same value as undoped polymer, and $y = 10^{-4}$. Second, σ_{RT} begins to

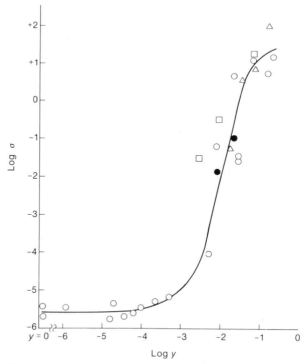

Fig. 11.8   Variation of conductivity with y for *trans*-$(CHI_y)_x$: (O) Warakomski *et al.* (1983); (□) Epstein *et al.* (1981b); (●) Sichel *et al.* (1982); (△) Chiang *et al.* (1978b).

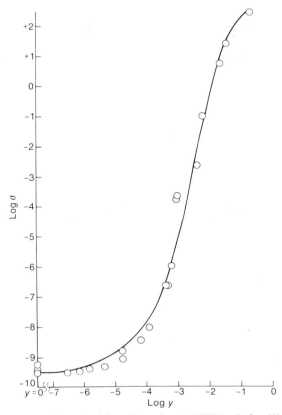

Fig. 11.9   Variation of conductivity with y for *cis*-(CHI$_y$)$_x$. [After Warakomski *et al.* (1983).]

increase gradually at $y > 10^{-4}$. From $y = 10^{-3}$ upward, the conductivity increases rapidly from 10^{-5} (Ω cm)$^{-1}$ to 10^2 (Ω cm)$^{-1}$ at $y = 0.32$. The linear portion in Fig. 11.8 represents the steepest part of log σ versus log y dependence where σ increases by four orders of magnitude between $y = 4 \times 10^{-3}$ and 2.9×10^{-2}.

The increase of conductivity with iodine doping of *cis*-polyacetylene is shown in Fig. 11.9. The room-temperature value for $y < 3 \times 10^{-6}$ is 3×10^{-10} (Ω cm)$^{-1}$, which is the same as pristine *cis*-polyacetylene. But conductivity begins to increase when y exceeds 3×10^{-6}. Beginning at $y = 1.8 \times 10^{-5}$ conductivity increases from 2×10^{-9} (Ω cm)$^{-1}$ rapidly until it reaches 300 (Ω cm)$^{-1}$ at $y = 0.2$. The linear portion of Fig. 11.9 has log σ increasing from -5 to -1 over a mere 4.5-fold increase in dopant concen-

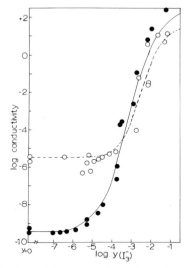

Fig. 11.10   Replot of the data in Figs. 11.8 and 11.9 in the form of σ_{RT} versus $y(I_3^-)$: ○ trans polymer; ● cis polymer.

tration, corresponding to a mobility increase of about 2200 times. The midpoint of the transition occurs at $y(I_3^-) = 7 \times 10^{-4}$.

Various spectroscopic evidences were cited in Section 8.3.2 that the iodine-doped polyacetylene contains mainly I_3^- ions with some I_5^- ions at very high dopant concentrations. It is a common practice to assume the dopant to be largely I_3^-. Therefore we recast and combine the results of Figs. 11.8 and 11.9 into Fig. 11.10 in the form of log σ versus log $y(I_3^-)$ for two purposes. One is for better comparison with data of AsF$_5$-doped polyacetylene. The other is to show how remarkably the same are the conductivities of *cis*- and *trans*-polyacetylenes in the dopant concentration ranges of $1.8 \times 10^{-4} \leq y(I_3^-) \leq 10^{-2}$. This agreement implies the same increase in carrier mobilities with doping for the two isomers.

11.3.1.2   AsF$_5$-Doped Polyacetylene

We present first the results obtained with the "slow" doping procedure. Figure 11.11 gives the data for the *trans*-[CH(AsF$_5$)$_y$]$_x$, including those of Park *et al.* (1980). The increase of σ(RT) with y is monotonic beginning at the lowest dopant concentration, which is less than 1 ppm. Below $y < 10^{-4}$, we found $\sigma \propto y^{1.25}$; above $y > 10^{-3}$ the conductivity increases more rapidly

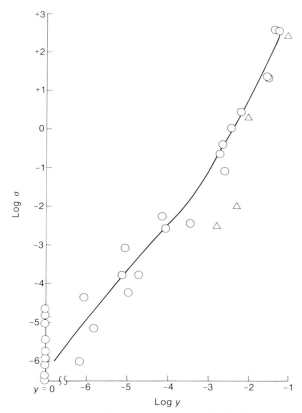

Fig. 11.11 Variation of conductivity with y for *trans*-$[CH(AsF_5)_y]_x$ doped with the "slow" procedure: ($\bigcirc$) Warakomski *et al.* (1983); ($\triangle$) Park *et al.* (1980).

with dopant to the 1.9th power. There is no clear transition from semiconducting to metallic conductivity. The results for *cis*-$[CH(AsF_5)_y]_x$ are even more surprising (Fig. 11.12). The (log σ)-versus-(log y) plot is nearly linear for $10^{-6} \leqslant y \leqslant 0.14$, with a slope of 2.3. These results imply that the "slow" AsF_5 doping created metallic domains even at the lowest dopant concentration. Increasing doping produces more and larger metallic islands. Probably, the surface of the fibrils near the surface of the film is first doped to metallic conductivity. Doping progresses toward the cores of the fibrils and the interior of the specimens. In this figure the data points of usually high conductivity were obtained for specimens that were purposely doped rapidly and thus nonuniformly.

The conclusion of nonuniform AsF_5 doping by the "slow" method is

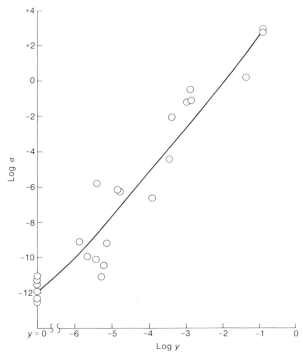

Fig. 11.12   Variation of conductivity y for *cis*-[CH(AsF$_5$)$_y$]$_x$ doped with the "slow" procedure. [After Warakomski *et al.* (1983).]

supported by the different $\sigma-y$ behaviors obtained with "cyclic" doping in which gaseous doping products were removed by evacuation (Section 8.2.1). The results in Fig. 11.13 show very similar $\sigma-y$ dependences for cyclic doped *trans*-[CH(AsF$_5$)$_y$]$_x$ and *trans*-[CH(I$_3$)$_y$]$_x$, the latter shown as the broken curves. The remarkable agreements between the conductivities of the iodine- and AsF$_5$-doped materials impart confidence that they are both homogeneously doped at least insofar as the variation of conductivity with dopant concentration is concerned. This is further supported by our finding that the same doping curve was obtained for polyacetylene electrochemically doped with ClO$_4^-$ ion. In this procedure the approach to equilibrium doping was followed precisely with coulometry.

We might add that in the early study of iodine doping of *cis*-polyacetylene by Mihàly *et al.* (1979b) they reported linear dependence in log–log plots of σ_{RT} with y from $y \approx 5 \times 10^{-4}$ to $y \approx 0.2$ for a 10-order-of-magnitude change in conductivity. Apparently, iodine doping can be also grossly inhomogeneous if not done sufficiently slowly (Epstein *et al.*, 1981d).

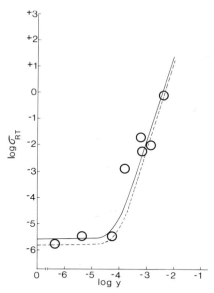

Fig. 11.13 Variation of conductivity with y for *trans*-$[CH(AsF_5)_y]_x$ doped with the "cyclic" procedure. Broken curve is that of *trans*-$[CH(I_3)_y]_x$ of Fig. 11.10. [After Chien *et al.* (1983h).]

11.3.2 Effect of Temperature

The temperature dependence of conduction of pristine *trans*-polyacetylene has been extensively investigated. Epstein *et al.* (1981c,d) had analyzed the data according to various models. Figure 11.14 is an Arrhenius plot; the line is drawn for an activation energy of 0.31 eV. The poor fit ruled out the models of hopping conduction involving extended states, band tails, and polarons. The data plotted according to the model of metallic islands (Fig. 11.15) also showed deviation from theory at low temperatures. The line was drawn for $\sigma(T) = \sigma_0 \exp[-(1.8 \times 10^5/T)^{0.5}]$. Figure 11.16 is a plot of the data according to the variable-range hopping model. The fit is excellent. However, the line drawn is for $\sigma = \sigma_0 \exp[-(1.9 \times 10^9/T)^{0.25}]$. The coefficient B [cf. Eq. (11.33)] is two orders of magnitude larger than that found typically in amorphous semiconductors. Also, the ac conductivity data (Section 11.3.3) and the pressure dependence of dc conductivity (Section 11.3.5) cannot be rationalized with this model. Finally, the data were plotted according to the power-law dependence for the intersoliton hopping model (Fig. 11.17). The slope gives $\sigma \propto T^{13.7}$. Moses *et al.* (1981) also reported a log–log plot with an exponent of ~ 13.

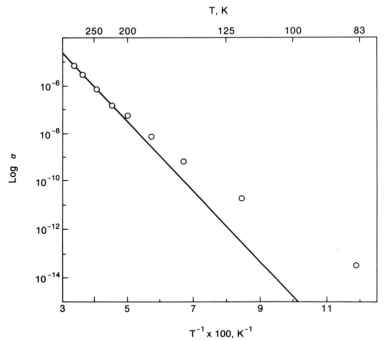

Fig. 11.14   Arrhenius plot for conductivity of *trans*-polyacetylene. [After Epstein *et al.* (1981d).]

The amount of data available on the temperature dependence of conductivity for pristine *cis*-polyacetylene is not large. Epstein *et al.* (1980d) had reported measurements on 70% *trans*-polyacetylene compensated with NH_3, which is said to be like *cis*-polyacetylene. We have plotted their data in three ways. The power-law plot gave a straight line for $\sigma(T) \propto (T)^{17.9}$ (Fig. 11.18a); the Arrhenius plot is linear with an activation energy of 1.11 eV (Fig. 11.18b); and a plot for $T^{-1/4}$ dependence gave $\sigma(T) \propto \exp[-(2.5 \times 10^{11}/T)^{1/4}]$ (Fig. 11.18c). Thus, the analysis of temperature dependence of conductivity for this polymer did not shed any light on the probable mechanism of conduction.

Epstein *et al.* (1981d) had measured the temperature dependence of conductivity for a lightly doped *trans*-polyacetylene with iodine. The specimen was not determined for the iodine concentration. However, from the tenfold increase in conductivity, it may be estimated that $y \approx 1.9 \times 10^{-3}$. The samples were found to have the same T^{14} dependence for log σ_{dc} as the undoped polymer (Fig. 11.19). The authors concluded that this result fulfills

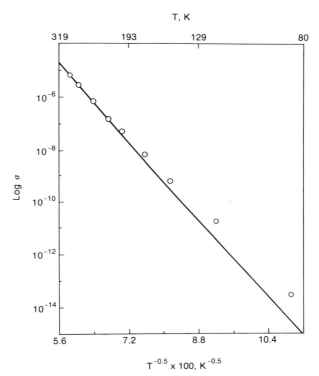

Fig. 11.15 Plot of conductivity of *trans*-polyacetylene versus $T^{-1/2}$ according to the model of metallic islands. [After Epstein *et al.* (1981d).]

the predication of the intersoliton electron hopping mechanism. Equation (11.50) can be factored into temperature-dependent and dopant-dependent components, and therefore the temperature dependence of σ_{dc} should be independent of dopant concentration. If the dopant is present as I_3^-, then the samples have $y^+ \simeq 4.7 \times 10^{-4}$ for unit charge transfer from the chain to the dopant. The neutral soliton concentration was unchanged from pristine *trans*-polyacetylene according to Fig. 9.8. We can use now the parameters for the intersoliton hopping model and Eq. (11.50) to obtain a theoretical increase of conductivity of 15-fold in good agreement with experiment. This suggests that the model may be valid at least to near the onset of rapid increase in conductivity with further doping.

At high levels of doping, Chiang *et al.* (1978b) obtained for $(CHI_y)_x$ linear plots of $\log[\sigma(T)/\sigma(RT)]$ versus $T^{-1/4}$ (Fig. 11.20). The same was observed for bromine-doped polyacetylenes (Fig. 11.21). Epstein *et al.* (1980) confirmed this for $(CHI_{0.12})_x$ from about 4 K to 218 K. This may be

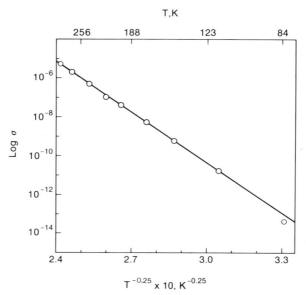

Fig. 11.16 Plot of conductivity of *trans*-polyacetylene according to the model of variable-range hopping. [After Epstein *et al.* (1981d).]

taken as an indication of conduction by variable-range hopping for the halogen-doped polyacetylene. The decreasing slope of the plots with increasing y may suggest the increase in $N(E_F)$ and α^{-1} with doping according to Eqs. (11.24) and (11.28).

Audenaert *et al.* (1981) prepared several samples of *trans*-$(CHI_y)_x$ (isomerized at 453 K for 1 hr) and found them to have different dependences of conductivity on temperature. Two samples with $y = 0.11$ and 0.19 act like simple three-dimensional disordered semiconductors with log σ varying linearly with $T^{1/4}$. Two other samples with $y = 0.03$ and 0.29 have conductivity dependency on temperature consistent with the fluctuation-induced tunneling model. The authors thought that the samples may have different textures; they may also be improperly doped.

The temperature dependence of the normalized conductivity of stretched-aligned *trans*-$[CH(AsF_5)_y]_x$ films is shown in Fig. 11.22. At low concentrations, the curves are concave and appear to be thermally activated, and $\sigma \to 0$ as $T \to 0$. Above some critical doping level, the curve becomes convex and the conductivity remains finite as $T \to 0$. The temperatue dependence of conductivity is apparently also consistent with the model of metallic islands for these early samples, which may not be doped uniformly. Tomkiewicz *et al.* (1981) showed (Fig. 11.23) that the $\sigma(y, T)$ data can be

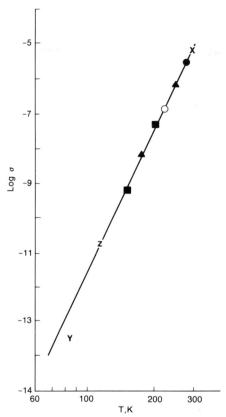

Fig. 11.17 Plot of conductivity of *trans*-polyacetylene according to the intersoliton hopping model. [After Epstein *et al.* (1981d).]

fitted with Eq. (11.70) for *cis*-[CH(AsF$_5$)$_{0.0095}$]$_x$ and with Eq. (11.71) for *cis*-[CH(AsF$_5$)$_{0.167}$]$_x$. A fit with the same quality for the latter sample can also be achieved with Eq. (11.72) (Sheng, 1980).

Even for samples that have nonexponential temperature dependence for conductivity, it is a common practice to obtain the "activation energy" from the slope of Arrhenius plot near ambient temperature. By this method, the "activation energy" was found by Park *et al.* (1980) to be 0.3 and 0.5 eV for undoped *trans*- and *cis*-polyacetylene, respectively. With increasing dopant concentration, there is a sharp decrease between $10^{-3} \leqslant y \leqslant 10^{-2}$ to a dopant-concentration-independent activation energy of a few milli-electron-volts (Fig. 11.24).

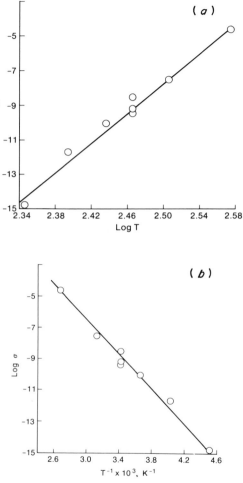

Fig. 11.18   Variation of conductivity with temperature data of Epstein *et al.* (1981d) for 70% *trans*-polyacetylene compensated with NH_3: (a) power-law plot; (b) Arrhenius plot; (c) (log σ)-versus-$T^{-1/4}$ plot.

There are qualitative differences between polyacetylene doped with AsF_5 and I_2 in the metallic regime. Figure 11.25 showed the behaviors of stretch-oriented materials ($l/l_0 = 3.1$) with σ_{RT} of 2450 $(\Omega \, cm)^{-1}$ for $[CH(AsF_5)_{0.14}]_x$ and 1500 $(\Omega \, cm)^{-1}$ for $(CHI_{0.25})_x$. The former first increases in conductivity on cooling below room temperature, goes through a maximum at 220 K, then decreases and becomes constant below 5 K, approaching $\sigma(0 \, K)/\sigma(300 \, K) = 0.66$. The conductivity maximum varies from sample to sam-

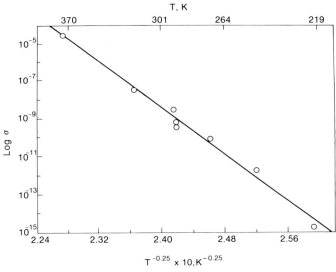

Fig. 11.18c

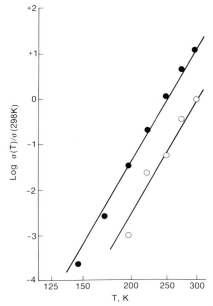

Fig. 11.19   log σ_{ac} versus log T for undopd *trans*-polyacetylene (●), and the same sample doped with iodine (○). [After Epstein *et al.* (1981d).]

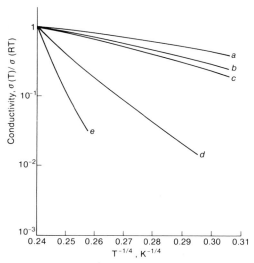

Fig. 11.20   ln σ versus $T^{-1/4}$ for $(CHI_y)_x$ for y: (a) 0.19; (b) 3.7×10^{-2}; (c) 2.9×10^{-2}; (d) 1.3×10^{-2}; (e) 0. [After Chiang *et al.* (1978b).]

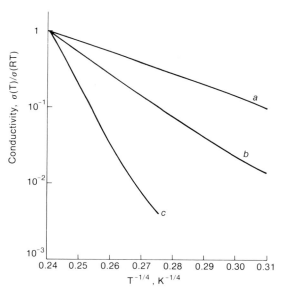

Fig. 11.21   ln σ versus $T^{-1/4}$ for $(CHBr_y)_x$ for y: (a) 0.195; (b) 0.014; (c) 5.7×10^{-3}, 2.9×10^{-3}, 0. [After Chiang *et al.* (1978b).]

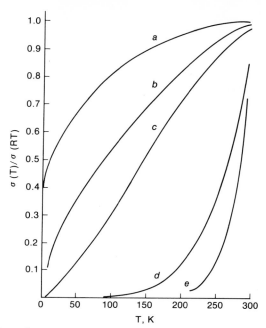

Fig. 11.22 Normalized temperature dependence of *trans*-[CH(AsF$_5$)$_y$]$_x$ for $l/l_0 = 3.0$ and y values of (a) 0.1; (b) 0.03; (c) 0.008; (d) 0.004; (e) 0. [After Park *et al.* (1980)].

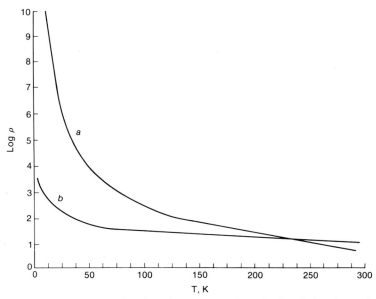

Fig. 11.23 Resistivity as a function of temperature for AsF$_5$-doped *cis*-polyacetylene: (a) $y = 9.5 \times 10^{-3}$; (b) $y = 0.167$. [After Tomkiewicz *et al.* (1981).]

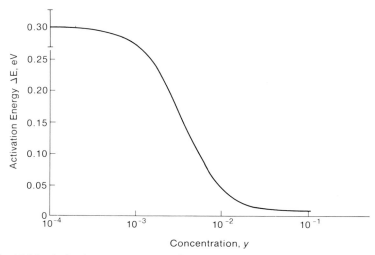

Fig. 11.24   Activation energy versus y for *trans*-$[CH(AsF_5)_y]_x$ and *trans*-$(CHI_y)_x$. [After Park *et al.* (1980).]

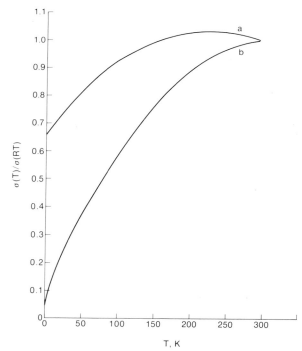

Fig. 11.25   Temperature dependence of normalized conductivity of heavily doped metallic polyacetylenes: (a) *cis*-$[CH(AsF_5)_y]_x$; (b) *cis*-$(CHI_y)_x$. $l/l_0 = 3.1$. [After Park *et al.* (1980).]

ple, but in general *cis*-polyacetylene starting materials lead to a larger maximum than *trans*-polyacetylene. In the case of $(CHI_{0.25})_x$, the conductivity decreases monotonically and appears to be going to zero as $T \to 0$.

Temperature dependence can apparently take different forms for various dopants. In the case of SbF_5, Rolland *et al.* (1980b) observed exponential variation of σ with T for $y \geqslant 1.5 \times 10^{-2}$ but deviation from Arrhenius equation for low-temperature data with $y \leqslant 6.5 \times 10^{-3}$. The undoped *trans*-polyacetylene has log $\sigma \propto T^{-1}$ instead of the $T^{-1/4}$ or power-law dependences found by other investigators. Figure 11.26 shows the Arrhenius plot of resistivity of *trans*-$[CH(SbF_5)_y]_x$. Based on the slope of the linear region, the "activation energies" were obtained and given in Fig. 11.27. The intercept indicates a transition at $y \approx 0.008$.

The variation of σ with y and T for CF_3SO_3H-doped polyacetylene is like that of SbF_5 doping (Rolland *et al.*, 1980c). The Arrhenius relation was valid for $y > 0.03$, and deviations were noted for low dopant concentrations. The activation energies decrease with y and showed a break at $y \approx 0.02$. The same can be said about other dopants. Deviation from Arrhenius dependence of σ on temperature was observed for BF_3 doping at $y \leqslant 6.5 \times 10^{-2}$, for ClO_3H doping at $y \leqslant 5 \times 10^{-3}$, and for any concentration of H_2SO_4 doping from y values of 10^{-3} to 9.3×10^{-2}. The conduction transition, as judged by the variation of activation energy with y, occurred at $y = 3 \times 10^{-2}$ for $[CH(BF_3)_y(H_2O)_w]_x$, at $y = 2.6 \times 10^{-3}$ for $[CH(H_2SO_4)_y(H_2O)_w]_x$, and at $y = 1.4 \times 10^{-2}$ for $[CH(ClSO_3H)_y(H_2O)_w]_x$.

Epstein *et al.* (1980) found two distinct behaviors for highly doped samples of $(CHI_y)_x$ cut from the same doped film. One has a logarithmic dependence on temperature,

$$R = R(T_0)[1 - S_T \ln(T/T_0)], \tag{11.79}$$

as illustrated in Fig. 11.28, and a $T^{-1/4}$ dependence,

$$R = R(T_A) \exp[(T_A/T)^{1/4}], \tag{11.80}$$

such as shown in Fig. 11.29.

The results cited in this section suggest that the nature of the dopant, the doping procedure, and doping level, the initial isomeric content, and other unknown factors influencing the sample quality have strong influences on the functional form for the temperature dependence of conductivity. Consequently, one should exercise caution in drawing conclusions concerning possible conduction mechanism based on this type of data and analysis.

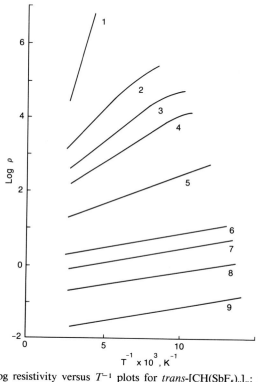

Fig. 11.26 Log resistivity versus T^{-1} plots for *trans*-[CH(SbF$_5$)$_y$]$_x$: y values (1) 0; (2) 1.2 × 10^{-3}; (3) 3 × 10^{-3}; (4) 4 × 10^{-3}; (5) 6.5 × 10^{-3}; (6) 1.5 × 10^{-2}; (7) 2.4 × 10^{-2}; (8) 3.7 × 10^{-2}; (9) 6 × 10^{-2}. [After Rolland *et al.* (1980b).]

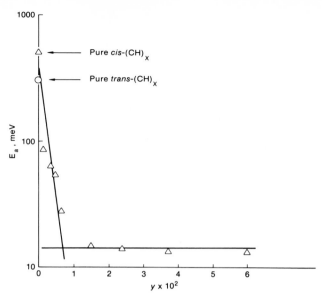

Fig. 11.27 Activation energy for conductivity of *trans*-[CH(SbF$_5$)$_y$]$_x$ as a function of y. [After Rolland *et al.* (1980b).]

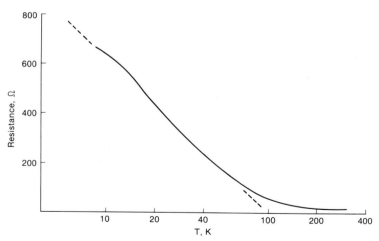

Fig. 11.28 Variation of resistivity with log T for a sample of (CHI$_{0.17}$)$_x$ with $\sigma(295 \text{ K}) = 5$ $(\Omega \text{ cm})^{-1}$. [After Epstein *et al.* (1980).]

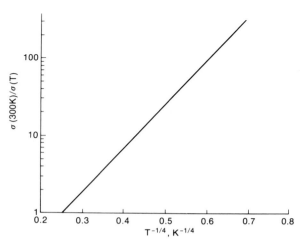

Fig. 11.29   Variation of $\sigma(300 \text{ K})/\sigma(T)$ versus $T^{-1/4}$ for a sample of $(CHI_{0.12})_x$. [After Epstein *et al.* (1980).]

11.3.3   Frequency Dependence

Conductivities of *cis*-polyacetylene and trans-rich polyacetylene compensated with NH_3 were measured from 10^1 to 10^5 Hz by Epstein *et al.* (1981d). The results of $\sigma(\omega)$ for the compensated polymer show marked frequency dependence even at 100 Hz with $\sigma(\omega) \propto \omega^{0.8}$ (Fig. 11.30). Whereas the σ_{dc} decreases markedly with decreasing temperature, there is only weak temperature dependence for σ_{ac}. In fact, at 10^5 Hz the conductivity of the polymer was nearly the same between 183 K and 373 K. This behavior for $\sigma(\omega, T)$ is very similar to a broad class of disordered semiconductors and several physical mechanisms have been proposed for these materials including the presence of surface barriers, variable range hopping, and the presence of ionic dipoles. The behavior of *cis*-polyacetylene is like that of NH_3-compensated polyacetylene with certain differences. Figure 11.31 shows that σ_{ac} is temperature dependent at all frequencies and that it depends on frequency to the first power. The $\sigma(\omega, T)$ have the same exponential temperature dependence at all frequencies, and $\sigma(\omega, T) \propto \exp[-1600K/T]$ (Fig. 11.32).

The conductivity of pristine *trans*-polyacetylene was measured at constant temperature from 10 to 10^6 Hz. Figure 11.33 presents the total conductivity as a function of temperature and frequency. The ac conductivity was obtained from $\sigma_{ac} = \sigma_{total} - \sigma_{dc}$ and given in Fig. 11.34. At 294 and 274 K, $\sigma_{ac} = \sigma_{dc}$. As temperature is lowered, the frequency dependence becomes observable. All the lines in Fig. 11.32 have the same slope, i.e., $\sigma_{ac} \propto \omega^{0.6}$. It was recalled that the dc conductivity data for *trans*-polyacetylene can be fitted nearly equally well with either the intersoliton hopping or the variable-range hopping models (Section 11.3.2). The present $\sigma(\omega, T)$ data afforded a differentiation of the two models. For the latter model, $\sigma(\omega) \propto T$. However, the observed dependence on temperature is much stronger; σ_{ac} changes by 10^3-fold for 3-fold changes in temperature. On the other hand, the ac conductivity data can be described by the intersoliton model. This is shown by recasting Eq. (11.51) into

$$\sigma_{ac}(\omega) = K'(f/T)\,[\ln(K''f/T^x)]^4 = K'Z. \qquad (11.81)$$

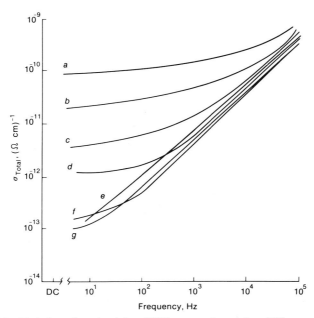

Fig. 11.30 Variation of conductivity of 70% *trans*-polyacetylene NH_3 compensated with frequency at (a) 373 K; (b) 350 K; (c) 329 K; (d) 312 K; (e) 294 K; (f) 239 K; (g) 183 K. [After Epstein *et al.* (1981d).]

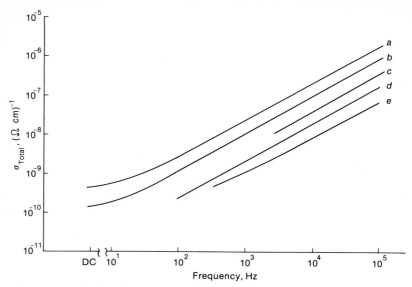

Fig. 11.31 Variation of total conductivity versus frequency for *cis*-polyacetylene at (a) 295 K; (b) 273 K; (c) 247 K; (d) 221 K; (e) 197 K; (f) 173 K. [After Epstein (1982).]

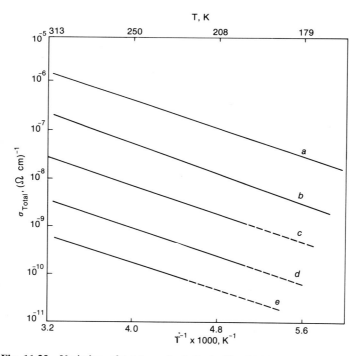

Fig. 11.32 Variation of total conductivity in Fig. 11.31 versus T^{-1} at (a) 10^5 Hz; (b) 10^4 Hz; (c) 10^3 Hz; (d) 10 Hz; (e) dc. [After Epstein (1982).]

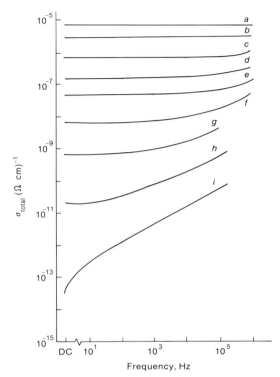

Fig. 11.33 Variation of total conductivity versus frequency for *trans*-polyacetylene at (a) 294 K; (b) 274 K; (c) 245 K; (d) 220 K; (e) 201 K; (f) 174 K; (g) 149 K; (h) 118 K; (i) 84 K. [After Epstein *et al.* (1981c).]

The experimental values of $\sigma_{ac}(\omega, T)$ are plotted accordingly using $x = 14.7$ and the parameter K'' varied for best linear fit (Fig. 11.35). The line was drawn in the figure for $K'' = 1 \times 10^{24}$ sec $K^{14.7}$ and $K' = 3 \times 10^{-15}$ $(\Omega\text{ cm})^{-1}$. According to Eq. (11.51) and parameters given in Section 11.1.6, the calculated values are $K'' = 4 \times 10^{23}$ sec $K^{14.7}$ and $K' = 6.3 \times 10^{-16}$ $(\Omega\text{ cm})^{-1}$.

We return to Fig. 11.8, containing data of the σ_{dc} for lightly doped *trans*-$(CHI_y)_x$. Epstein *et al.* (1981d) argued that doping increased y^+ from 2.35×10^{-4} to 3.0×10^{-4}. Comparison of Eqs. (11.50) and (11.51) suggests that there should be a $\sigma_{ac}(\omega, T)$ increase of 3.5-fold upon doping. The measured increase of a factor of 2.5 in $\sigma_{ac}(\omega, T)$ is in reasonable agreement with the intersoliton electron hopping model.

For heavily iodine-doped polyacetylene, $y > 9 \times 10^{-3}$ or 3×10^{-3} I_3^- per CH unit, the conductivity is frequency independent, as shown in Fig. 11.36.

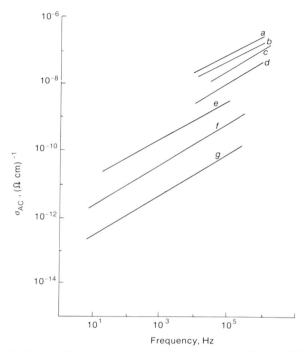

Fig. 11.34   Variation of σ_{ac} versus frequency for *trans*-polyacetylene at different temperatures: (a) 274 K, 245 K; (b) 220 K; (c) 201 K; (d) 174 K; (e) 149 K; (f) 118 K; (g) 84 K. [After Epstein *et al.* (1981c).]

Thus the intersoliton model may be valid for y up to 1.4×10^{-3} but not for values of $y > 9 \times 10^{-3}$ in iodine-doped polyacetylene. Heavy doping will cause the charged soliton to become less localized. Grant and Krounbi (1980) noted also that the conductivity of AsF_5-doped polyacetylene is independent of frequency.

Very different results were reported by Mihàly *et al.* (1979b). They measured the microwave conductivity at 9 GHz and found large frequency dispersion. The ac conductivity for undoped *cis*-polyacetylene is 10^3 times greater than σ_{ac}. The undoped polymer has $\sigma_{RT} \approx 10^{-8} (\Omega \, cm)^{-1}$, which is a little higher than current values. It will be recalled that they observed that $\sigma_{dc} \propto y\alpha$ for iodine doping, and there was no conduction transition up to $y > 0.1$. The frequency dispersion becomes progressively smaller until $\sigma_{ac} \approx \sigma_{dc}$ for $y > 0.03$ (Fig. 11.37). They interpret the results as due to hopping between localized states. As was pointed out earlier, their sample may be grossly nonuniformly doped.

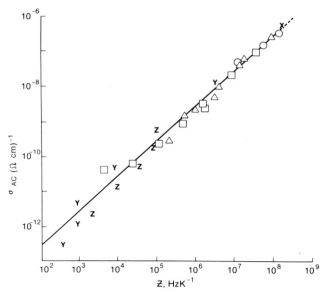

Fig. 11.35 Plot of log σ_{ac} versus log Z, with $Z = (f/T)[\ln(1 \times 10^{24} f T^{14.7})]^4$: (x) 294 K; (●) 274 K; (△) 245 K; (○) 220 K; (□) 201 K; (△) 174 K; (□) 149 K; (Z) 118 K; (Y) 84 K. [After Epstein *et al.* (1981d).]

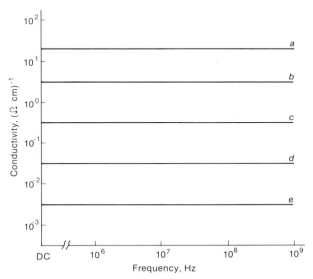

Fig. 11.36 Plot of σ_{RT} versus frequency for $(CHI_y)_x$ for y values: (a) 0.2; (b) 0.12; (c) 0.025; (d) 0.010; (e) <0.010. [After Epstein (1981b).]

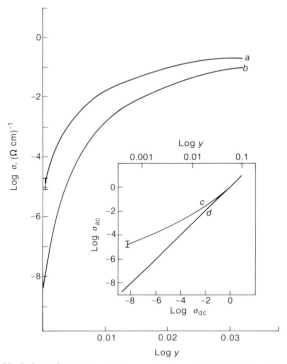

Fig. 11.37   Variation of (a) σ_{ac} and (b) σ_{dc} with y for *cis*-$(CHI_y)_x$. Inset is σ_{ac} versus σ_{dc}; (c) experimental; (d) for $\sigma_{dc} = \sigma_{ac}$. All plots are log–log. [After Mihaly *et al.* (1979b).]

11.3.4   Field Dependence

Epstein *et al.* (1980, 1981b) reported that lightly doped semiconducting $(CHI_y)_x$ displayed a linear current–voltage relation for fields up to 10^4 V cm^{-1} and $T > 125$ K. These materials are Ohmic. Non-Ohmic behavior was observed for heavily doped "metallic" $(CHI_y)_x$. The field dependence is weak, as shown in Fig. 11.38. The slow decrease of resistance with field follows the relation

$$R = R(E_0,T)[1 - S_v \ln(E/E_0)], \qquad (11.82)$$

with E_0 arbitrary and S_v the logarithmic slope. Because of the nonexponential dependence, the origin of non-Ohmic behavior was not believed to be due to metallic islands. Instead, the authors thought there was disorder localization of the wave functions leading to extensive band tailing. Because

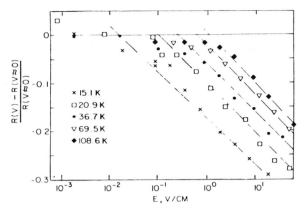

Fig. 11.38 Variation of resistance with electric field for a "metallic" $(CHI_y)_x$ sample $[\sigma_{RT} \approx 3\,(\Omega\,cm)^{-1}]$ at (a) 108.6 K; (b) 69.5 K; (c) 36.7 K; (d) 20.9 K; (e) 15.1 K. [After Epstein *et al.* (1980).]

of the $\ln T$ dependence [Eq. (11.79)] observed for these samples, it was proposed that the doped polymer resembles a two-dimensional disordered metal. The electric field dependence may be due to electron heating, which would reflect the temperature dependence of the resistance. The two-dimensional behavior was attributed to a fibril core, which remains undoped, producing two-dimensional conductivity along an annulus.

Mortenson *et al.* (1980) reported different non-Ohmic behavior for semiconductivity AsF_5-doped *cis*-polyacetylene. It was shown in Fig. 11.39 that their samples, typically *cis*-$[CH(AsF_5)_{0.005}]_x$, have a $R \propto T^{-1/2}$ dependence. The effect of field was measured with a pulse technique, with typical pulse width of 300 nsec and a repetition rate of 5–50 Hz. It was verified that there was neither a heating effect during the pulse nor an average heating effect. Fields up to 6×10^4 V cm^{-1} were achieved. At 4.2 K the resistance has an exponential dependence on E^{-1} over at least three orders of magnitude at the high field region. On the other hand, the high field limit cannot be reached at 77 K. The data obtained at several temperatures below 10 K (Fig. 11.40) showed a temperature-independent value for E_0 of 2×10^5 V cm^{-1}, where all the curves converge. The results were said by the authors to be consistent with isolated metallic particles dispersed in a continuous dielectric matrix. In such material an electron has to be removed from a neutral metallic grain and placed on another neutral grain in order to generate a charge carrier. Therefore, if the field is too low, i.e., when the voltage drop between adjacent grains is smaller than $k_B T$, then the carriers are thermally activated. At high fields, the number of carriers thus generated is much

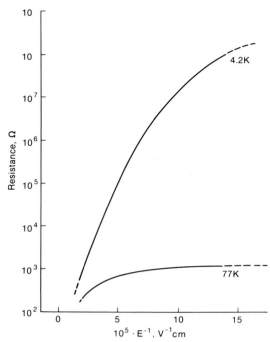

Fig. 11.39 Resistance as a function of E^{-1} of $[CH(AsF_5)_{0.005}]_x$ at 4.2 and 77 K. [After Mortenson *et al.* (1980).]

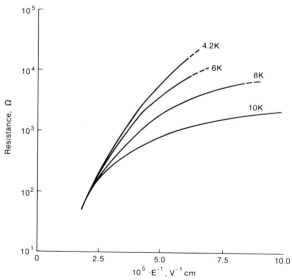

Fig. 11.40 Resistance as a function of E^{-1} of $[CH(AsF_5)_{0.05}]_x$ at several temperatures below 10 K. [After Mortenson *et al.* (1980).]

greater than the thermal equilibrium, and the resistance is given by

$$R(E) = R(0) \exp(E_0/E). \tag{11.83}$$

This was observed at 4.2 K. At 77 K the number of the thermally generated carriers far exceeds that obtainable from the field at the field strength available.

Two methods are used to estimate the average distance r between metallic grains. The first is through the relation between r and E_0 and T_0 by

$$r \approx (k_B/4e)T_0/E_0. \tag{11.84}$$

The second is from the crossover field E_c between Ohmic and non-Ohmic behavior occurring at

$$eE_c r = k_B T. \tag{11.85}$$

From these two independent methods, r was estimated to be ~ 60 Å. The above results for dopant concentration below the percolation threshold indicate the existence of metallic domains with high tunneling barriers. Above the threshold, measurements showed a much weaker field dependence suggesting that although barriers still do exist, they are much smaller and the temperature and field dependences should be interpreted by fluctuation-induced tunneling.

Non-Ohmic behavior of heavily iodine-doped polyacetylene was reported by Phillipp *et al.* (1982). They studied two samples with an iodine concentration of 14% and 30% with room-temperature conductivities of 68 and 320 $(\Omega \, cm)^{-1}$, respectively. Below the liquid-nitrogen temperature there was large nonlinear field-dependent deviation of conductivity from the zero-field value. The deviation is greater and is more field dependent the lower the temperature. For $(CHI_{0.14})_x$, the non-Ohmic behavior disappeared above 155 K; it is ~ 200 K for $(CHI_{0.3})_x$. The data could be fitted by a modified form of Eq. (11.60).

11.3.5 Effect of Pressure

Moses *et al.* (1981) measured the transport properties of undoped *trans*-polyacetylene over a range of pressure up to 8.74 kbar. This was done in a Teflon cell using fluorinert as the pressure medium. Since fluorinert wets the polymer, the effect of pressure is on the fibrils rather than to collapse the film to increase the gross density. There is about a twofold decrease in resistivity at 8.74 kbar (Fig. 11.41 inset). The variations of resistance with temperature at two applied pressures are shown in this figure. The linear

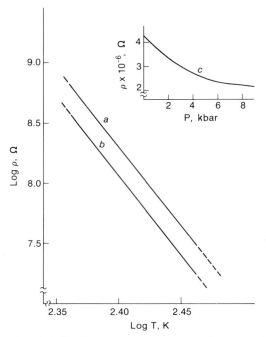

Fig. 11.41 Log–log plot of resistance versus temperature for *trans*-polyacetylene at two pressures: (a) 0.219 kbar; (b) 8.74 kbar. Inset shows variation of resistance with pressure. [After Moses *et al.* (1981).]

log–log plots with identical slopes indicate that the effect of pressure and temperature can be factored.

Among the models described in Section 11.1 only the intersoliton hopping and the variable-range hopping models contain pressure-sensitive terms. For the former model, Eq. (11.50) can be rewritten

$$\ln \sigma = \ln a - 2 \ln R_0 - 2BR_0/\xi, \qquad (11.86)$$

where a contains all the constants and temperature-dependent terms. The derivative with respect to pressure of $\ln R_0$ is small. Therefore, one obtains

$$\frac{\partial(\ln \sigma)}{\partial P} \simeq \frac{-2BR_0\kappa}{\xi}, \qquad (11.87)$$

where κ is the compressibility. From the slope of Fig. 11.41, $\partial(\ln \sigma)/\delta P \simeq 7 \times 10^{-11}$ cm^2 dyn^{-1}. Since $R_0/\xi \simeq 6$ from the analysis of the temperature dependence of conductivity according to Eq. (11.50), we find $\kappa \approx 4.2 \times 10^{-12}$ cm^2 dyn^{-1} for *trans*-polyacetylene. This is in agreement with the compressibility measured by Ferraris *et al.* (1980) of 4×10^{-12} cm^2 dyn^{-1} for *cis*-polyacetylene. The factorization of σ into temperature and pressure dependences requires that the ratio of conductivities at two pressures to be independent of temperature. This is realized in Fig. 11.42.

For the variable-range hopping model, it is the density of states near the Fermi level that can be affected by pressure. So Eq. (11.33) can be rewritten

$$\ln \sigma = \ln a' - \tfrac{1}{2} \ln N(E_F) + \tfrac{1}{4}a'' TN(E_F).$$ (11.88)

With increasing pressure the density of localized states in the gap increases,

$$\frac{\partial N(E_F)}{N(E_F)} \simeq \left[\frac{\partial E_G}{E_G} + \frac{\partial V}{V}\right],$$ (11.89)

and Moses *et al.* (1981) showed that

$$\sigma = \sigma_0 \exp\left\{-\left(\frac{B}{T^{1/4}}\right)[1 - \tfrac{1}{4}(\varepsilon + \kappa)P]\right\},$$ (11.90)

where $\delta E_G/E_G \equiv \varepsilon P$ is the fractional change in energy gap with pressure, which is about 10% at 8.74 kbar. Therefore, conductivity cannot be factored into pressure and temperature terms and the ratio of conductivities at two pressures would vary with temperature. This variation is shown in Fig. 11.42, which is inconsistent with experimental data.

High-pressure conductivity measurements at room temperature have been made by Ferraris *et al.* (1980) with a tetrahedral anvil press that had been previously calibrated against several fixed-point resistive transitions. The *cis*-polyacetylene has a constant resistivity of 1.7×10^6 Ω cm from 0 to over 60 kbar. In these experiments, unlike those of Moses *et al.* (1981), the polymer film is being compressed to a density of 1.135 g cm^{-3}. Since the samples were precompressed, the lack of effect of pressure can be understood. The pressurization is to decrease the void volume of the polymer to 95% of the calculated density for void-free polyacetylene but does not influence either the intrachain or interchain transport processes. Measurements on iodine-doped *cis*-polyacetylene showed a decrease of resistivity by 22-fold for $(CHI_{0.0094})_x$. The decreases are smaller for samples in the metallic regime; 3.5 times for $(CHI_{0.03})_x$ and 2 times for $(CHI_{0.219})_x$. The minimum resistivities were obtained for these samples at about 20 kbar. This decrease has been attributed by the authors to either reduced interfibril resistance or increased density of conducting fibrils per unit volume.

At pressures exceeding 20 kbar the resistivity of iodine-doped polyacetylene increases. The increase is greater for the more heavily doped samples. Ferraris *et al.* (1980) thought that either iodine is being pushed out of the polymer or the pressure induces iodination of polyacetylene. Cross-linking and scission under pressure in the presence of dopant were also suggested.

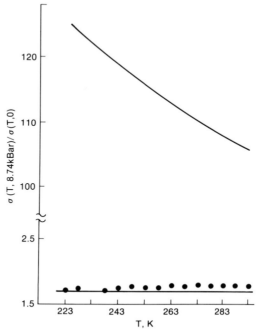

Fig. 11.42   Comparison of $\sigma(T, 8.74\ \text{kbar})/\sigma(T, 0)$ (solid dots) with intersoliton hopping model (lower line) and variable-range hopping model (upper line). [After Moses *et al.* (1981).]

These speculations should have been followed by characterization of the compressed polyacetylenes, e.g., elemental analysis and Raman spectroscopy.

11.3.6   Effect of Dopant Ions

Sichel *et al.* (1982) proposed that the size and shape of dopant ions can influence the electrical properties of polyacetylene. They distinguish the linear dopants Br_3^- with a diameter of 3 Å and I_3^- with a diameter of 3.37 Å from the more spherical ones, i.e, the octahedron AsF_6^- ($a = 4.41$ Å, $b = 6.24$ Å), and $IrCl_6^{-2}$ ($a = 6.08$ Å, $b = 8.6$ Å). They doped polyacetylene with $H_2IrCl_6 \cdot 6H_2O$ dissolved in dry niromethane to obtain $[CH(IrCl_{5.55})_y(H_2O)_w]_x$. The resistivity of these samples has an $\exp(T^{-1/2})$ dependence (Fig. 11.43). The results were interpreted in terms of formation

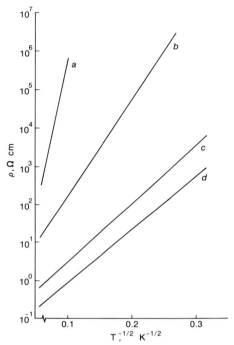

Fig. 11.43   Log resistivity versus $T^{-1/2}$ for polyacetylene doped with $IrCl_6$ from solution. The weight percent of dopant is (a) 37.4%; (b) 45.9%; (c) 82.9%; (d) unknown. [After Sichel *et al.* (1982).]

of conducting islands. It was suggested that the first large dopant molecules to reach the fibril surface opens up channels, thus making it energetically favorable for the next molecules to follow established pathways resulting in inhomogeneous doping.

11.3.7   Ultralow Temperature

Resistance of metallic $[CH(AsF_5)]_x$ stretch oriented to $l/l_0 = 2.8$ has been measured down to 0.4 K by Gould *et al.* (1981). The entire range of temperature of measurement is shown in the inset of Fig. 11.44. The resistance increases with decrease of temperature until it reaches a plateau at 1 K. Between 30 and about 7 mK, it varies logarithmically in T quite accurately. The electric field used was small (<200 nV/cm), so there is no

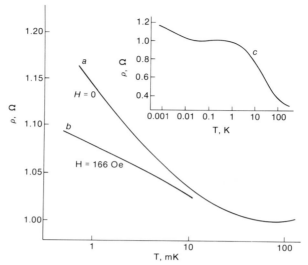

Fig. 11.44 Electrical resistance of stretch-aligned polyacetylene heavily doped with AsF$_5$ as a function of temperature from 0.4 to 100 mK: (a) $H = 0$; (b) $H = 166$ G. Inset is from 0.4 mK to 100 K. [After Gould *et al.* (1981).]

field effect. The results at ultralow temperature can be described by

$$R(T, H_0) = \text{const} - 0.0501 \ln T \qquad (11.91)$$

at $k_B T \gg g\mu_B H_0$.

There is an interesting effect of a small applied magnetic field. A field of only 166 G decreases the resistance at temperatures below 10 mK (cf. Fig. 11.44). The effect of field strength at various temperatures is shown in Fig. 11.45. At low temperatures the resistance is much greater than the plateau resistance above 30 mK, and the effect of the magnetic field in suppressing this excess resistance is pronounced. At higher temperatures, i.e., 9.8 mK, the magnetoresistance is low and so is the excess resistance above the plateau resistance. Since the curves are not straight, the results cannot be described as logarithmic in temperature with a field prefactor. The results of Fig. 11.45 can be fitted with

$$R(T, H) = R_0(T)\{1 - \alpha \ln[(k_B T)^2 + \beta(g\mu_B H)^2]\}, \qquad (11.92)$$

where $g = 2$, $\alpha = 0.013$, $\beta = 0.149$, and $R_0(T)$ is a temperature-dependent factor and is $\sim 1.145 - 0.02 \ln T$ (mK). The asymptotic field dependence can be approximated by

$$R(T_0, H) = \text{const} - S_H \ln H \qquad \text{for} \quad kT \ll g\mu_B H_0, \qquad (11.93)$$

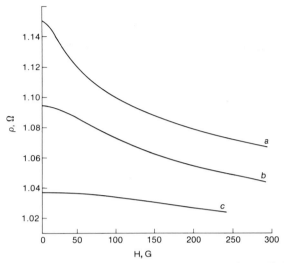

Fig. 11.45   Effect of small magnetic field on the resistance of metallic $[CH(AsF_5)_y]_x$ at ultralow temperatures: (a) 0.96 mK; (b) 2.7 mK; (c) 9.8 mK. [After Gould *et al.* (1981).]

and the asymptotic temperature dependence [Eq. (11.91)] can be recast into

$$R(T, H_0) = \text{const} - S_T \ln T \qquad \text{for} \quad kT \gg g\mu_B H_0. \qquad (11.94)$$

There are several noteworthy points. The ratio S_T/S_H is slightly temperature dependent, but is always near 2. The value of S_T is 0.0501 Ω for the sample used in this study. The resistance in the presence of a magnetic field is never greater than the plateau resistance above 30 mK. Finally, above 40 mK no magnetoresistance was observed.

The logarithmic increase in resistance at ultralow temperature is expected from current theories, which maintain that electron transport in one- and two-dimensional materials are not metallic in the classic sense. The plateau resistance can be interpreted as dominated by the interfibril contacts in the polymer material and transport occurs by virtue of fluctuation-induced tunneling. It is possible that the tunneling characteristics of the interfibril contacts are modified at ultralow temperature. The extended temperature-independent plateau below 1 K is a direct result of this tunneling mechanism.

Generally speaking, the logarithmic increase of resistance below 1 mK must be due to some kind of localization. Impurities and defects are expected to lead to a localized wave function, so that such systems do not remain conductive down to absolute zero. However, localization can occur at several levels, such as along a single chain, within a single fibril but across

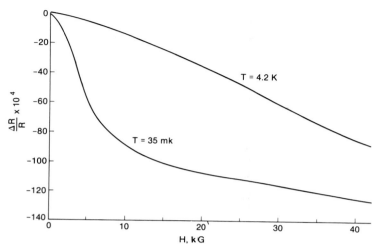

Fig. 11.46   Magnetoresistance of polyacetylene heavily doped with AsF_5. [After Kwak *et al.* (1979a).]

many chains, across several interconnected fibrils, or even across the entire specimen. Two important points of the above results are that the resistance increase below 30 mK is distinct from the plateau resistance above 30 mK, which is temperature independent over nearly two decades, and that the excess resistance is suppressed by a small magnetic field. This negative magnetoresistance is expected in the localization model.

Kwak *et al.* (1979a) has also studied conductivity of polyacetylene heavily doped with AsF_5; their results are in line with those described above. They also measured transverse magnetoresistance from 35 mK to 4.2 K (Fig. 11.46). It is small and negative, the resistance decreasing by 0.5% to 1% at 50 kG. the field dependence at 4 K is approximately quadratic. As the temperature is lowered, the magnetoresistance tends to saturate at high field. At low temperatures most of the resistive shift occurred at low fields. The negative sign of the magnetoresistance is not normally found in metals. A negative effect was observed in $[SN]_x$ crystals, which has been attributed to a high degree of disorder. Similar effects occur in doped covalent semiconductors, which were attributed to localized spins. Both these considerations are aplicable to doped polyacetylenes.

11.3.8   Effect of Density

Polyacetylene can be synthesized in the form of low-density (0.02 – 0.04 g cm^{-3}) foams, as discussed in Section 2.2.7. Scanning electron micrographs of this foamlike material show fibril morphology characteristic of $(CH)_x$ shown in Fig. 11.47. The typical fibril diameter appears to be $\sim 60-80$ nm, i.e., much larger than the ~ 20-nm fibril diameter of the as-grown film. The wet polymer foam was pressed to higher densities of 0.1 and 0.4 g cm^{-3}; the latter density is the same as the as-polymerized film. The transport properties were compared and summarized in Table 11.1.

Since the thermopower is a zero-current transport coefficient, the inter-fibril contacts should be unimportant, allowing evaluation of the intrinsic

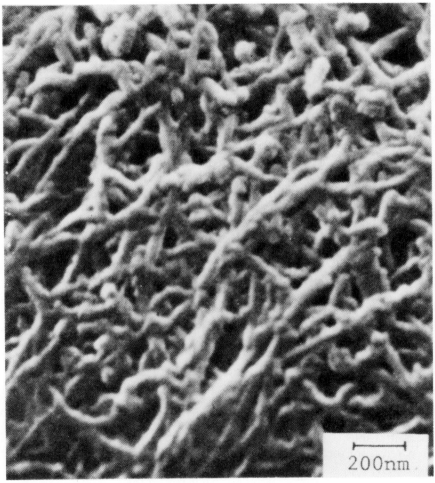

Fig. 11.47 Scanning electron micrograph of the low-density foamlike $(CH)_x$ prepared from the gel (see text). The fibrillar structure is evident with fibril diameters in the range 600–800 Å. [After Wnek *et al.* (1979a).]

TABLE 11.1

Comparison of Transport Properties of Polyacetylene of Varying Density

Sample	Foam			Pressed film		
	Density[a] (g cm^{-3})	σ(RT) (Ω cm)$^{-1}$	S(RT) (μV K^{-1})	Density[a] (g cm^{-3})	σ(RT) (Ω cm)$^{-1}$	S(RT) μV K^{-1}
trans-$(CH)_x$[b]	0.02–0.04	1.1 × 10^{-6}	+900	0.4	7.9 × 10^{-5}	+920
cis-$(CHI_{0.06})_x$[c]	0.02–0.04	8.1	+18.7	0.4	350	+18.4
cis-$(CHI_{0.06})_x$				0.1	11.2	
cis-$[CH(AsF_5)_{0.06}]_x$	0.02–0.04	81.3	+8.0	0.1	176	+8.9d

[a]Density of sample before isomerizing and doping.
[b]Isomerized by heating at 453 K for 2 hr in vacuo.
[c]IR showed 80–85% cis before doping.

properties. Moreover, since S can be viewed as a measure of the entropy per carrier, the results depend only on the properties of the conducting fibrils and not on the number of fibrils per unit volume. As indicated in Table 11.1, the thermopower of the undoped polymer is insensitive to the density; the foamlike material and the pressed film yield the same thermopower values of approximately $+900$ μV/K. Any variations are comparable with the typical variations observed in as-grown film samples from different synthetic preparations. Similarly, after heavy doping with iodine and AsF$_5$ the results are insensitive to the density with values for the foamlike material, pressed film, and as-grown films in good agreement for each dopant. Comparison with the variation in S as a function of dopant concentration has been reported by Park et al. (1979b).

The electrical conductivity data are consistent with this conclusion. The conductivity of the undoped foamlike material is nearly two orders of magnitude below that of the high-density pressed film. Although there may be some increase in the interfibril contact resistance, the reduction in volume-filling fraction is of major importance. This conclusion is strengthened by the observation that the conductivity activation energy (obtained from the temperature variation of the conductivity near room temperature) is 0.25 eV for both samples. This value is comparable to the 0.3 eV value typically obtained from as-grown films (Chiang et al., 1978b; Shirakawa et al., 1978).

Similar results are obtained with the heavily doped polymers. As indicated in Table 11.1, the conductivity (176 (Ω cm)$^{-1}$) of $[CH(AsF_5)_{0.06}]_x$, prepared from the low-density pressed cis-$(CH)_x$ films ($\rho = 0.1$ g/cm^3), is correspondingly lower than that of the conductivity 560 (Ω cm)$^{-1}$ and 1200 (Ω cm)$^{-1}$ for $[CH(AsF_5)_{0.14}]_x$ and $[CH(AsF_5)_{0.10}]_x$, respectively, prepared from as-grown cis-$(CH)_x$ film ($\rho = 0.4$ g/cm^3) (Chiang et al., 1978b). The

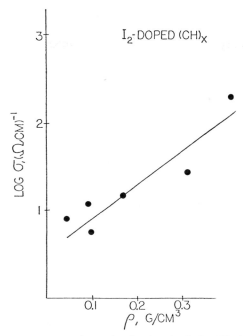

Fig. 11.48   Conductivity versus density for $(CHI_y)_x$. [After Wnek *et al.* (1979a).]

room-temperature conductivity of $(CHI_{0.06})_x$ increases by a factor of 40 in going from the low-density foamlike material to the high-density pressed film. Figure 11.48 shows the variation of conductivity with the density of the polymer. Note that the increased fibril size in the foamlike material does not appear to alter significantly the resulting transport properties of the doped polymers.

The temperature dependences of the conductivities for several samples (both doped and undoped) are shown in Fig. 11.49. In each case the normalized temperature depends primarily on the dopant and the dopant concentration, whereas the absolute value depends critically on the density (Table 11.1). The foamlike and pressed-film samples typically behave as observed earlier in as-grown film. For metallic samples heavily doped with iodine, σ decreases slowly with decreasing temperature. For metallic samples heavily doped with AsF_5, σ remains nearly constant upon cooling, increasing slightly between room temperature and 250 K, then decreasing slowly as the temperature is lowered. In this case, the doping is probably nonuniform as the behavior is consistent with existence of metallic domains.

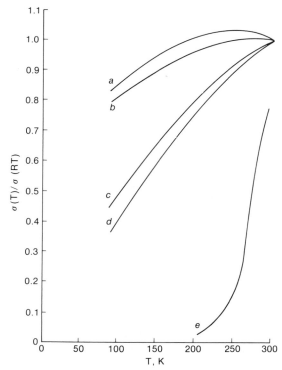

Fig. 11.49 Normalized temperature dependence of the conductivity for foamlike (CH)$_x$ and pressed film (CH)$_x$: (a) pressed film (CH)$_x$: [CH(AsF$_5$)$_y$]$_x$, heavily doped; (b) foamlike (CH)$_x$: [CH(AsF$_5$)$_y$]$_x$, heavily doped; (c) pressed film (CH)$_x$: (CHI$_y$)$_x$, heavily doped; (d) foamlike (CH)$_x$: (CHI$_y$)$_x$, heavily doped; (e) pressed film *trans*-(CH)$_x$, undoped; and foamlike (CH)$_x$: *trans*-(CH)$_x$, undoped. [After Wnek *et al.* (1979a).]

In conclusion, the various forms of the doped and undoped polymers are essentially microscopically identical. The lower-density materials simply consist of fibrils at a smaller filling fraction. The resulting doped and undoped low-density polyacetylene can be viewed as effective media in which the dc electrical transport is determined by the volume-filling fraction of conducting fibrils.

11.3.9   Effect of Orientation

As a pseudo-one-dimensional conductor, doped polyacetylene should show anisotropic transport properties. This has been demonstrated by

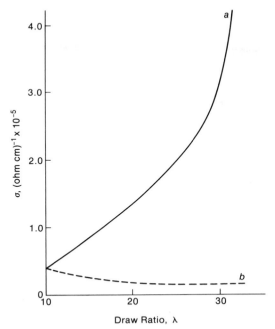

Fig. 11.50   Electrical conductivity of heavily doped *trans*-polyacetylene as a function of draw ratio: (a) $\sigma_{\parallel}$; (b) $\sigma_{\perp}$. [After Shirakawa and Ikeda (1979).]

Shirakawa and Ikeda (1979) and by Park *et al.* (1979). Figure 11.50 showed that the $\sigma_{\parallel}$ increases rapidly with elongation, especially toward the high draw ratios. Values of $\sigma_{\perp}$ decrease slightly; the maximum observed anisotropy at $l/l_0 = 3.0$ is $\sigma_{\parallel}/\sigma_{\perp} \approx 27$. According to the intersoliton hopping model the expected anisotropy for lightly doped polyacetylene is

$$\sigma_{\parallel}/\sigma_{\perp} \approx (\xi_{\parallel}/\xi_{\perp})^2 \approx 10. \tag{11.95}$$

The observed anisotropy is much greater. It will be recalled that the model anticipated increase in anisotropy with increased doping as transport reduces in dimensionality.

   The anisotropy in electrical conductivity is apparently not the same for different dopants. Figure 11.51 showed that the AsF_5-doped samples have greater anisotropies than iodine-doped materials at comparable elongation and dopant concentration. Electron micrographs of the stretched polyacetylene revealed only modest alignment of the fibrils. The theoretical anisotropy could be greater than that observed so far.

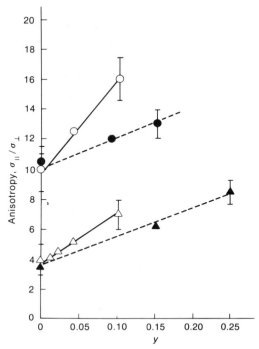

Fig. 11.51   Electrical anisotropy $\sigma_\parallel/\sigma_\perp$ as a function of dopant concentration for partially aligned polyacetylene films at different elongations: (▲) $[CHI_y]_x$ at $l/l_0 = 2.11$; (△) $[CH(AsF_5)_y]_x$ at $l/l_0 = 2.11$; (●) $(CHI_y)_x$ at $l/l_0 = 2.91$; (○) $[CH(AsF_5)_y]_x$ at $l/l_0 = 2.92$. [After Park *et al.* (1979a).]

11.3.10   Acetylene–Methylacetylene Copolymers

Poly(methylacetylene) is orange colored; the electronic spectra (Fig. 7.26) showed it to have a band gap of ∼ 2.7 eV. The undoped polymer is devoid of any EPR signal, and none is observable after it has been heated to 373 K. Poly(methylacetylene) is nearly an insulator; its resistivity is not affected by AsF_5. On the other hand, the polymer can be doped by iodine to 17% and a conductivity by $10^{-3}\,(\Omega\,cm)^{-1}$ (Table 11.2). But the iodine-doped poly(methylacetylene) still gives no EPR signal. The material is devoid of neutral defects.

Acetylene–methylacetylene copolymers of four compositions were synthesized; the molar ratios of the comonomer feed are given by the two numbers following AMA. Copolymer AMA-11 displayed a dull greenish-gold luster on both sides of the film; its morphology (Fig. 11.52a) has a

TABLE 11.2

Conductivities and Compositions of Doped Homo- and Copolymers
of Acetylene and Methylacetylene

Sample[a]	Mol % C_2H_2 in Polymer	Composition-doped polymer	σ [$(\Omega\ cm)^{-1}$]
$(CH)_x$	100	$[CH(AsF_5)_{0.12}]_x$	400
AMA-31	~ 55	$(CH_{1.18}I_{0.24})_x$	36
		$(CH_{1.18}I_{0.16})_x$ (after pumping above sample overnight)	18
		$[CH_{1.18}(AsF_5)_{0.08}]_x$	45
		$(CH_{1.18}Br_{0.17})_x$	4×10^{-2}
		Na doped; composition not determined	2.5×10^{-1}
AMA-11	~ 33	$(CH_{1.24}I_{0.16})_x$	1.5
		$[CH_{1.24}(AsF_5)_{0.1}]_x$	1.0
AMA-13	~ 15	$(CH_{1.31}I_{0.11})_x$	2×10^{-2}
		$[CH_{1.31}(AsF_5)_{0.05}]_x$	2×10^{-3}
$(C_3H_4)_x$		$(CH_{1.33}I_{0.17})_x$	10^{-3}

[a] Acetylene methylacetylene mole ratio in feed indicated by digits following AMA.

spongy texture. The luster was lost and the film became black upon wetting with a solvent (e.g., pentane, toluene), suggesting solvent swelling of the polymer occurred. The luster was recovered upon evaporation of the solvent. Wetting also caused the film to become very elastic. In fact, wetted films could be stretched to extension ratios as high as ~ 7 and, upon solvent evaporation, a fixed elongation could be achieved. Wetting of such stretched, dried film resulted in considerable retraction.

The AMA-13 film was less lustrous and more elastic when wetted with solvent than the AMA-11 film. It has an irregular "clumplike" morphology (Fig. 11.47b). Our attempts to prepare AMA-14 copolymer failed to produce a tractable film. Thus the methylacetylene content of AMA-13 (Table 7.2) was nearly the upper limit for the production of free-standing film.

Copolymer AMA-31 film has a shiny green front side of relatively smooth appearance, while the back side displayed a dull gray–black coloration. Electron micrographs in Fig. 11.53 show no evidence of a fibrillar texture but ill-defined "clumps." Copolymer sample AMA-61 exhibited a lustrous and rippled surface similar to those of the polyacetylene film prepared under similar conditions (Fig. 11.54a). The material was very brittle and disintegrated into several small pieces during washing, revealing fibrillar morphology along the crack in film surface (Fig. 11.54b).

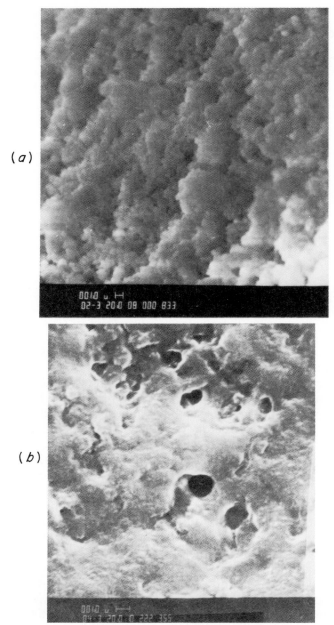

Fig. 11.52 Electron micrographs of copolymer films: (a) AMA-11 at ×1500 magnification; (b) AMA-13 at ×3000 magnification. [After Chien *et al.* (1981).]

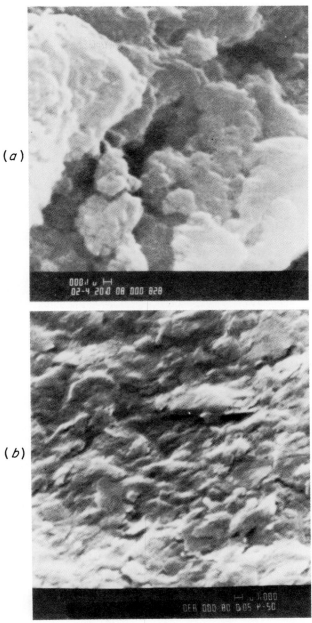

Fig. 11.53   Electron micrographs of AMA-31 copolymer film: (a) shiny side at ×37,500 magnification; (b) dull side at ×15,000 magnification. [After Chien *et al.* (1981).]

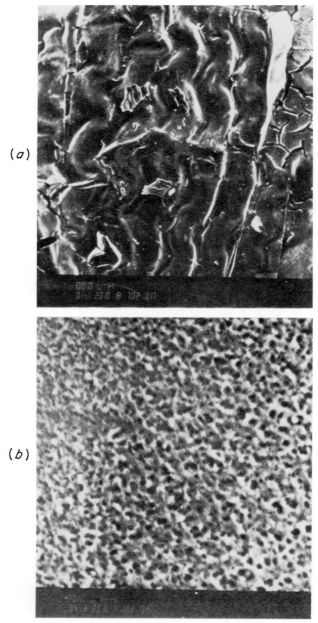

Fig. 11.54   Electron micrographs of AMA-61 film: (a) film surface at ×7.5 magnification; (b) crack in film surface at ×3000 magnification. [After Chien *et al.* (1981).]

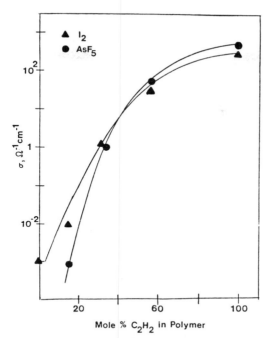

Fig. 11.55   Variation of conductivity with acetylene–methylacetylene copolymer composition for films doped with iodine and AsF$_5$. [After Chien *et al.* (1981).]

All the copolymers can be doped by both AsF$_5$ and iodine (Table 11.2 and Fig. 11.55). Since AsF$_5$ is not a dopant for poly(methylacetylene), it is a poorer dopant than iodine for copolymers rich in this monomer. However, for copolymers rich in acetylene, AsF$_5$ becomes a better dopant than iodine. In the case of AsF$_5$-doped copolymers the conductivity increases rapidly by more than four orders of magnitude, going from 15 to 55 mole % acetylene. The difference in conductivity between the last sample and $[CH(AsF_5)_{0.12}]_x$ is only tenfold.

The EPR results of poly(acetylene-*co*-methylacetylene) are shown in Fig. 11.56. All copolymer samples displayed Lorentzian resonance with g values in the range 2.0023–2.0026. The linewidth is ~ 10 G for copolymers rich in methylacetylene. As the acetylene content is increased the linewidth approaches 6 G, which is about the same as that of pristine *cis*-polyacetylene. Therefore, the neutral defects are immobile in the copolymer, and the spin delocalization is less in methylacetylene-rich copolymer. Very interesting is the pronounced effect of line narrowing when the copolymers were heated to 473 K for 30 min then cooled to room temperature for EPR observation. This may be due partly to increase in backbone conjugation. It will be

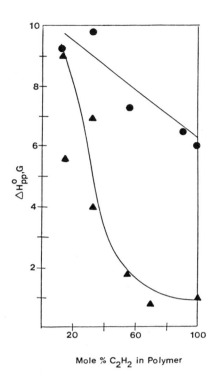

Mole % C_2H_2 in Polymer

Fig. 11.56   Variation of EPR linewidth with acetylene content in copolymer films: (●) room temperature; (▲) after heating to 473 K for 30 min and then cooled to room temperature for EPR observation. [After Chien *et al.* (1981).]

recalled that pyrolysis of poly(methylacetylene) at relatively low tempera- tures of 500 K (Section 7.1.2) resulted in the formation of proton-enriched products of very low molecular weight. This was interpreted as occurrence of facile electron–proton and electron–methyl exchanges. Such processes would tend to increase the backbone conjugation. However, it still seems unlikely that a *trans*-$(CH)_x$-like backbone resulted allowing the formation of highly diffusive solitons. The observations on the copolymers indicate that spin diffusion is not significantly hindered by strong bond alternation in the trans isomer of the methylacetylene units.

The qualitative results concerning morphology, solubility, and "solvent swelling" as described above suggest that true copolymers were synthesized rather than a mixture of homopolymers. The IR spectra (Chien *et al.*, 1981) supports this conclusion. Definitive evidence is, however, available. First, pyrolysis–GC–mass spectrometry showed that the most abundant aro- matic pyrolysis product of polyacetylene is benzene followed by toluene and xylenes, but no mesitylene (Section 7.1.1). The reverse order was obtained upon pyrolyzing poly(methylacetylene) (Section 7.1.2). For the copolymers, depending on their composition, it is toluene or xylene that is the dominant pyrolysate. Furthermore, polyacetylene of variable density upon doping showed the conductivity to bear a semilog relationship to the density (Section 11.3.8). In other words, conductivity depends on the volume-filling factor. Since poly(methylacetylene) is not doped by AsF_5, then, if the copolymers are merely a blend of homopolymers, a semilog plot of conductivity with mole percent acetylene should be linear. Figure 11.55 shows this not to be the case. Furthermore, above 30 mol % acetylene the conductivity of the copolymers doped with AsF_5 and iodine is the same.

The conductivity of polyacetylene is only ten times greater than that of copolymers containing 40 mol % methylacetylene. If the copolymers are random in structure, each acetylene unit is likely to be followed by a methylacetylene unit. Consecutive acetylene sequences are few and short. Since it is unlikely that doping would be able to render all the bonds including those of the methyl acetylene units uniform in length, the results imply that interchain carrier transport is very efficient in heavily doped copolymers.

11.4 SPECTROSCOPIC CONDUCTIVITY

The infrared spectra of heavily doped polyacetylene such as the one shown in Fig. 9.13 are characteristic of free-carrier absorption. One can obtain an estimate of the magnitude of electrical conductivity from each spectrum. The infrared absorption coefficient can be expressed as

$$\alpha = \sqrt{2}(\omega/c)/[(\varepsilon_1^2 + \varepsilon_2^2)^{1/2} - \varepsilon_1]^{1/2}, \tag{11.96}$$

where ε_1 and ε_2 are the real and imaginary parts of the dielectric function, respectively. Using the free-carrier Drude approximation, one can write

$$\varepsilon_1 = \varepsilon_c - \frac{4\pi\sigma\tau}{1 + \omega^2\tau^2}, \tag{11.97}$$

$$\varepsilon_2 = \frac{4\pi\sigma}{\omega}\frac{1}{1 + \omega^2\tau^2}, \tag{11.98}$$

where ε_c is the high-frequency core dielectric constant and τ the carrier scattering time, which is short, of order 10^{-15} sec. Therefore $\varepsilon_2/|\varepsilon_1| \gg 1$, and in the long-wavelength limit of $\omega\tau < 1$, the absorption coefficient is given by the classic skin-depth expression

$$\alpha = (2/c)[2\pi\sigma\omega]^{1/2}. \tag{11.99}$$

Using the experimental results of $\alpha \simeq 3 \times 10^5$ cm^{-1} at 0.65 eV, we obtain a value of $\sigma \simeq 3 \times 10^3$ (Ω cm)$^{-1}$. This value certainly has been attained in stretch-aligned samples in dc conductivity measurements.

Reflectance spectra have been obtained from oriented polyacetylene film doped with AsF$_5$ by Fincher *et al.* (1979a). At the highest doping level are spectra that are clearly metallic in characteristics. The reflectance increases with decreasing energy, reaching 90% reflectance at the longest wavelengths. For light doping ($y < 0.01$) the reflectance remains small even at long wavelengths, consistent with semiconducting behavior. The sharp increase in reflectance below 0.2 eV in the $y = 0.034$ sample is indicative of the presence of free carriers. The broad reflectance peak centered near 0.5 eV has corresponding absorption maxima in iodine- and AsF$_5$-doped polyacetylene. This has been attributed to the soliton state above. The spectra suggest the formation of small metallic domains that gradually grow and coalesce into an extended metallic substance at the highest dopant concentration.

The effective number of electrons per molcule participating in the free-carrier optical transitions for energies lower than the interband transition is given by the oscillator strength

$$8 \int_0^{\omega_c} \sigma(\omega) \, d\omega = \frac{4\pi N e^2}{m^*} \eta_{\text{eff}}(\omega_c), \qquad (11.100)$$

where $\eta_{\text{eff}}(\omega_c)$ is the fractional number of carriers contributing to the metallic conductivity. Based on $v_c = 11,000$ cm^{-1}, the oscillator strength from Eq. (11.100) is $\sim 4 \times 10^{31}$, which is approximately equal to the total π-electron oscillator strength in the polyacetylene. This result indicates the removal of bond alternation upon heavy doping with all π electrons contributing to the transpot and $\eta_{\text{eff}} \sim 10$. However, Raman spectra suggests the presence of remnants of bond-alternated semiconducting regions coexisting with undistorted metallic domains.

11.5   MICROWAVE CONDUCTIVITY

To measure the microwave conductivity and the dielectric constant, the waveguide voltage-standing-wave-ratio (VSWR) and cavity-perturbation techniques were used at 33 and 10 GHz, respectively. For each doping level, three samples of polyacetylene foam were cut: a piece to fit into a Q-band waveguide, a sample to be mounted for four-probe dc conductivity measurements to allow in situ monitoring of the doping, and a third small reference sample. The two unmounted samples were weighed and then placed in a platinum cage attached to the back side of the four-probe apparatus. The entire system was made of glass with platinum wiring; electrical contact to the monitor sample was made with Electrodag.

For VSWR measurements, the sample was placed into a section of waveguide terminated with a movable short. The shorted section was placed on the output side of a slotted line, with the input side connected to a sweep oscillator, isolator, wavemeter, and a precision calibrated attenuator. The VSWR was determined by using the attenuator to maintain a constant voltage at the detector on the slotted line; the position of the VSWR minimum was read directly from the slotted line. Depending on conductivity and the length of the samples, two possible conditions can result: either the microwave field could penetrate through the sample and be reflected from the short at the back surface, or the sample could be sufficiently strongly absorbing that reflection from the back surface was insignificant. In the first case, the single impedance method described by Roberts and von Hippel (1946) was used. This method requires the solution of a complex transcendental equation and thus has an infinite number of discrete solutions for σ and ε. One must either carry out measurements as a function of sample length or have independent information on the magnitudes to determine which branch to use. Typically, the values of ε change greatly from branch to branch, whereas σ is less sensitive. In the second case, the sample is treated as a semi-infinite slab and the single reflection problem is solved in terms of the complex dielectric function. Since the value obtained for ε is very sensitive to the precise shift in position of the wave pattern when terminated with the sample as compared with when terminated by the short, the front reference plane of the sample must be very exactly determined (an uncertainly of 0.5 mm leads to an order-of-magnitude error in ε). Thus the roughness in the cut sample surface limits the accuracy in ε. Again, the value of σ is relatively insensitive to such effects and thus is determined with higher accuracy.

For the cavity-perturbation measurements, a small needle-shaped sample was subsequently cut from the initial waveguide sample. The method used was that described by Buravov and Shchegolev (1971). A small ellipsoidal sample is placed in the center of a transmission caivity, and the cavity resonance frequency shift and change in half-width are used to determine σ and ε. The sample volume and depolarization factor η are determined from the sample dimensions ($\eta = (bc/a^2)\{\ln[4a/(b + c)] - 1\}$, where a, b, and c are the three orthogonal semiaxes).

In Fig. 11.57 the results of the different methods for determining the conductivity of the sample are plotted on a semilogarithmic scale. The results are given in Ω^{-1} cm^{-1}. The three independent measurements on the

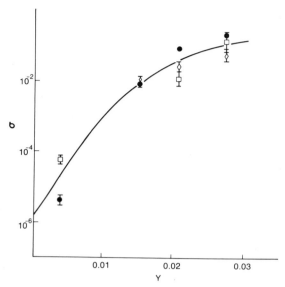

Fig. 11.57 Microwave conductivity of foamlike $(CHI_y)_x$ doped with iodine, $(CHI_y)_x$: (●) dc four-probe; (□) cavity perturbation (10 GHz); (W) VSWR (30 GHz). The solid curve represents $\sigma = 0.1\sigma_f$, where σ_f corresponds to the values obtained from $(CH)_x$ films and the factor of 0.1 comes from the ratio of filling factors. [After Feldblum *et al.* (1981).]

foamlike samples, four-probe dc, VSWR, and cavity perturbation, were in generally good agreement. Variations are probably the result of residual variations in doping level and uncertainties in the precise sample dimensions (particularly in the determination of η). The cavity measurements for the $y = 0.027$ sample at room temperature could not be directly evaluated because of the strong absorption and subsequent reduction in Q. We therefore cooled this sample to 100 K and estimated the room-temperature result by multiplying the 100 K value by the factor $\sigma(RT)/\sigma(100\ K)$ for iodine-doped polyacetylene films at the same dopant concentration. In addition, we have plotted in Fig. 11.57 the values of σ appropriate to the foamlike material using the values obtained from $(CH)_x$ films (σ_f) and scaling the results according to the simple effective medium equation $\sigma \simeq \sigma_f/10$, where the factor of $\frac{1}{10}$ comes from the ratio of filling factors.

The dielectric constant can be determined directly from analysis of the cavity perturbation data (Fig. 11.58). The VSWR results were independent, but used the cavity-perturbation results to determine the correct branch. Reliable dielectric constant data could not be obtained from the semi-infinite slab method because of front-surface irregularities as described above. The microwave conductivity and dielectric constant scale approximately

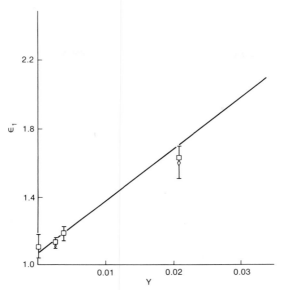

Fig. 11.58 Microwave dielectric constants of iodine-doped low-density polyacetylene: (□) cavity perturbation at 10 G Hz; (◇) VSWR at 30 G Hz. The solid line represents $\epsilon = 1 + (\epsilon_f - 1)/10$. [After Feldblum *et al.* (1981).]

with the density (or filling factor). The agreement between the VSWR results and the cavity-perturbation results independently implies that the low-density material can be treated as an effective medium. The VSWR experiment measures the properties of the macroscopic medium that terminates the line, whereas the cavity measurement explicitly integrates over the volume of the sample and over the depolarization field induced by the presence of the dielectric. The volume used was the total volume, which is about 30 times the actual volume of polymer fibrils. Moreover, the depolarization factor was calculated assuming a single three-axis ellipsoid with dimensions equal to those of the piece of foamlike polyacetylene, implying no significant charge buildup other than at the macroscopic surface of the sample.

The skin depth δ at 30 GHz was calculated for the foamlike polymer as a function of dopant concentration using the full expression

$$\delta = \frac{c\sigma}{\omega\sqrt{\varepsilon}}\left(\frac{2}{[1 + (4\pi\sigma/\omega\varepsilon)^2]^{1/2} - 1}\right)^{1/2}. \tag{11.101}$$

The results are plotted in Fig. 11.59. An experimental estimate of the skin depth was obtained using the VSWR apparatus. Starting with the sample length *l* in the infinite slab limit ($l \gg \delta$), the sample was successively sliced

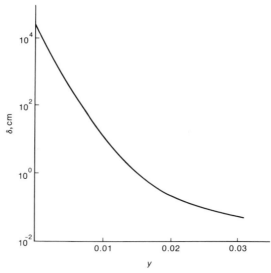

Fig. 11.59   Calculated skin depth δ versus y for low-density $(CHI_y)_x$ using Eq. (11.101). [After Feldblum *et al.* (1981).]

away until the effect of the moving short could be quantitatively detected. In this manner we determined, for example, that the skin depth for $y = 0.021$ was approximately 2 mm.

In order to obtain a quantitative description of the electromagnetic properties, we model the variable-density polymer as a composite effective medium, consisting of the $(CH)_x$ fibrils in a matrix. The composite medium problem was solved for isotropic spheres. Tanner *et al.* (1976) considered the case of randomly oriented ellipsoids with highly anisotropic properties $(\varepsilon_\| \gg \varepsilon_\perp)$. The complex effective medium dielectric function for such a system is written

$$\langle \varepsilon \rangle = \varepsilon_m + \frac{\frac{1}{3}f\varepsilon_m(\varepsilon^i - \varepsilon_m)}{\varepsilon_m + g(1 - \frac{1}{3}f)(\varepsilon^i - \varepsilon_m)}, \qquad (11.102)$$

where ε_m is the dielectric constant of the matrix (for air, $\varepsilon_m = 1$), ε^i the intrinsic complex dielectric constant of the particles of interest, f the fractional volume-filling factor, and g the depolarization factor of the eillipsoidal particles. In our case, we identify the ellipsoid particles with the polyactylene fibrils. The factor of $g(1 - \frac{1}{3}f)(\varepsilon^i - \varepsilon_m)$ in the denominator of Eq. (11.102) represents the reduction in internal field due to charge buildup at the boundaries. Since the fibrils are interconnected, there is no significant charge buildup except at the macroscopic boundaries of the sample. This was verified in the cavity-perturbation experiments described above. Thus we take the $g \rightarrow 0$ limit and obtain

$$\text{Re}\langle\varepsilon\rangle \equiv \langle\varepsilon_1\rangle = \varepsilon_m + \tfrac{1}{3}f(\varepsilon_1^i - \varepsilon_m),$$
$$\text{Im}\langle\varepsilon\rangle \equiv \langle\varepsilon_2\rangle = \tfrac{1}{3}f\varepsilon_2^i, \qquad (11.103)$$

where we have assumed ε_m to be real. For $\varepsilon_m = 1$, Eq. (11.102) is in agreement with experimental results.

A potentially useful application of this variable-density conducting polymer is as a microwave absorbing material. Such material should have $\langle\varepsilon_1\rangle$ close to unity so that reflection is minimized and determined by the conductivity of the medium. As shown above, these conditions are fulfilled for the foamlike polyacetylene.

The general expression for the reflectance (normal incidence) of microwaves from a surface in free space is given by

$$R = [(n - 1)^2 + k^2]/[(n + 1)^2 + k^2], \qquad (11.104)$$

where $n + ik$ is the complex index of refraction and $\langle\varepsilon_1\rangle + i\langle\varepsilon_2\rangle = (n + ik)^2$. Minimizing R with n as the free parameter, we obtain optimum conditions $n^2 = 1 + k^2$, so that $\varepsilon_1 = n^2 - k^2 = 1$ and $\varepsilon_2 = 2nk = 2k(1 + k^2)^{1/2}$. Under these conditions one has minimum reflectance

$$R_{min} = [1 + k^2 - (1 + k^2)^{1/2}]/[1 + k^2 + (1 + k^2)^{1/2}], \qquad (11.105)$$

and R_{min} is a function of k only. The power-absorption coefficient is given by $\alpha = 2(\omega/c)k$, so that the skin depth is $\delta = c/\omega k$. Thus under optimum conditions ($\varepsilon_1 = 1$), the reflected power and skin depth are directly related

$$\delta = \frac{\lambda_0/2\pi}{\{[(1 + R_{min})/(1 - R_{min})]^2 - 1\}^{1/2}}, \qquad (11.106)$$

$$R_{min} = \frac{1 + (\lambda_0/2\pi\delta)^2 - [1 + (\lambda_0/2\pi\delta)^2]^{1/2}}{1 - (\lambda_0/2\pi\delta)^2 + [1 + (\lambda_0/2\pi\delta)^2]^{1/2}} \qquad (11.107)$$

where λ_0 is the free-space wavelength.

For microwave-absorber applications, it would be desirable to have a relatively thin coating capable of absorbing, for example, 90% of the incident microwave power. For $\lambda_0 = 1$ cm (35 GHz), $R_{min} = 10\%$ corresponds to $\delta = 2$ mm, using Eqs. (11.106) and (11.107).

For doped polyacetylene, we can use the values of σ and ε to calculate δ [Eq. (11.101)] and the free-space (normal incidence) reflection coefficient R. The results are shown in Figs. 11.59 and 11.60. As can be seen from the figures, foamlike polyacetylene doped in the range $y \simeq 0.02$ is close to the optimum conditions as a microwave absorber. Moreover, numerical esti-

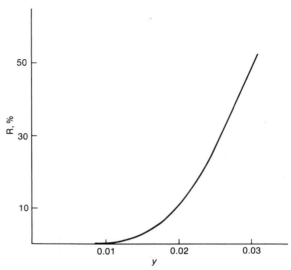

Fig. 11.60   Calculated reflection coefficient versus concentration y for low-density $[CHI_y]_x$ using Eq. (11.104). [After Feldblum *et al.* (1981).]

mates indicate that using a dielectric filler with $\epsilon_m < 2$ could lead to improved mechanical properties and a good adhesion without seriously degrading the electromagnetic properties of such a coating.

We note finally that since the quantity λ_0/δ is proportional to $\omega^{-1/2}$, the reflectance decreases as ω increases, but the absorption coefficient increases. The resulting coating would be broadband with better performance at higher frequencies.

11.6   HOMOGENEITY OF DOPED POLYACETYLENE

The homogeneity of doping is probably the predominant cause of discordance in results reported by various researchers. This is likely to be also true for other organic conductive polymers. Since there are no stoichiometric complexes formed between any dopant and polyacetylene, polyacetylene is said to be homogeneously doped when the distribution of dopant ions in the polymer is random and statistically uniform.

That polyacetylene is structurally heterogeneous renders uniform doping a difficult task. For instance, the initial cis–trans content, the polymer molecular weight, the degree of crystallinity, the morphology of the polymer, and the history of the sample prior to doping all may contribute toward the uniformity of doping. It is known that the trans isomer dopes more readily than the cis. Certainly, the amorphous phase would be doped first before the crystalline regions. The influence of molecular weight in this respect has not yet been investigated. It is likely that fibrils of small diameters would be doped more uniformly than those with larger diameters. We have found that polyacetylene having high density dopes less readily than low-density specimens. Finally, for *trans*-polyacetylene, the doping process may be affected by how the polymer was isomerized. It is likely that materials obtained from thermal isomerization of *cis*-polyacetylene using optimum procedures may dope differently than samples incompletely isomerized or excessively heated.

We consider next the homogeneity of heavily doped polyacetylenes. The dopant probably does not form stoichiometric compounds or complexes with the polymer. There is for each dopant a maximum dopant concentration determined by the size of the dopant ion, effectiveness of charge transfer, and the magnitude of Coulomb interaction. The dopant ions are probably not present in polyacetylene with any order in one or more dimensions. Their placements are likely to be statistical. Since intercalation of large ions will distort the interchain relationship, there is also a fluctuation of interchain separation in heavily doped polyacetylene. There is some theoretical reason to believe that these fluctuations may in fact contribute toward the incommensurate–commensurate transition (Section 10.8).

As long as heavily doped polyacetylene cannot become a classic metal because of the limitation in dimensionality, the material should have discontinuous domains. According to Horovitz (1981) heavily doped polyacetylenes tend to phase-separate into conducting and nonconducting domains.

The question of heterogeneity is of great concern for doped polyacetylene, particularly at low and intermediate dopant levels. Only the I_2 and AsF_3 systems have been studied with sufficient thoroughness about which this point can be commented on with some confidence. The most sensitive techniques are magnetic resonance and conductivity. A corollary to this is that nonuniform doping would affect these measurements markedly. The two techniques are complementary; magnetic resonance measures the concentration of unpaired spins and its relaxation processes by various EPR and NMR methods, while electrical conductivity and TEP measures the transport properties of the spinless charged carriers. Based on the mechanisms of doping and conduction proposed above, we suggest the following criteria for homogeneous doping.

Nonuniform doping of *cis*-polyacetylene can be readily seen by the apearance of two EPR signals having different linewidths. The narrow soliton signal indicates the formation of domains of *trans*-polyacetylene at low doping levels. The observation of Dysonian EPR signal or temperature-independent Pauli susceptibility at intermediate dopant concentrations ($y < 10^{-2}$) indicates the formation of conducting domains resulting from nonuniform doping.

Other EPR characteristics seem to be associated with samples doped with the best-known procedures. The EPR intensity and linewidth of the neutral soliton or defect are not significantly affected by doping up to $y \approx 10^{-3}$. We found that any significant changes in EPR intensity or linewidth in this regime are correlated with heterogeneous doping according to transport properties as described below.

If all the individual chains are doped simultaneously to a conductive state, then one expects an abrupt decrease of thermopower as a function of dopant concentration. Uniformly doped *trans*-polyacetylene has this transition complete within a twofold to threefold change in y, with the midpoint occurring at $\sim 7 \times 10^{-4}$.

The electrical conductivity of *trans*-polyacetylene is unaffected by doping up to the amount of intrinsic carrier, which is $\sim 10^{-4}$. We found that rapidly doped samples have high conductivities at low dopant concentration, indicating the presence of heavily doped domains. This statement may be altered if purer polyacetylenes are synthesized containing fewer intrinsic carriers. In the vicinity of $4 \times 10^{-4} \leq y \leq 10^{-2}$ the conductivity increases rapidly with doping. Very sharp transition is associated with uniform doping; nonuniformly doped materials may not exhibit such transition.

To achieve uniform doping, high-quality polyacetylene should be used as the starting material. This means either freshly prepared cis polymer or this cis polymer immediately isomerized at 453 K for 5–15 min. According to our experience uniform doping with iodine can be obtained by exposing the polymer to very low vapor pressure of iodine. In the case of AsF_5 even low-vapor-pressure slow doping apparently gives nonuniform products. We have tried to decrease this rate by keeping the polymer at low temperatures but found no significant differences in conductivity from polymers doped to the same concentration at room temperature. The preferred method of AsF_5 doping is by the "cyclic" method.

Although the preceding criteria were largely based on results for iodine- and AsF_5-doped polyacetylenes, they are probably also applicable to other dopants. For instance, we found that electrochemical *p* doping with perchlorate ion to equilibrium produced polymers that meet all the preceding criteria.

11.7 PROBABLE TRANSPORT MECHANISMS

The very large increase of conductivity upon doping of polyacetylene requires that different mechanisms may be necessary to account for the transport properties at certain doping levels. Various mechanisms have been used to interpret experimental data that are sample dependent. We discuss those probable transport mechanisms in order to stimulate further research. It is understood that the discussion is tentative and speculative. Most proposed mechanisms seem to be based on the unique properties of Peierls distorted *trans*-polyacetylene having broken symmetry ground states. There is a need to stress the similarity between doped cis and trans polymers in their transport properties. Rationalization of these similarities may be a key to the understanding of transport phenomenon in doped polyacetylenes.

11.7.1 Lightly Doped Polyacetylenes

That pristine *trans*-polyacetylene has conductivity as high as 10^{-5} (Ω cm)$^{-1}$ is unexpected. The elementary excitations in conventional semiconductors are electrons, holes, and phonons. In the case of *trans*-polyacetylene, because of the twofold degenerate ground state, there is an additional excitation of the topological soliton. The energy required to form the soliton is 0.42 eV $\simeq 2\Delta/\pi$. Therefore, instead of the usual mechanism of adding electrons (or holes) upon doping, which requires an energy 2Δ, it is energetically more favorable to form an S^+S^+ (or S^-S^-) pair by about 0.25 eV per charge.

Within the soliton model, two charge transport mechanism can be envisioned for lightly doped polyacetylenes. The first is probably applicable to low doping concentration and high temperature. The conduction is attributable to activated transport of charged solitons, which are normally bound. The process is characterized by an Arrhenius dependence, $\sigma\alpha$ $\exp(-E_b/k_B T)$ and $S \approx (e/k_B)(E_b/k_B T + \text{const})$, where E_b reflects the thermal population of unbound $S^{\pm}$. There is only a weak dependence on the concentration of impurities. Because of the large E_b, free $S^{\pm}$ conduction is very inefficient at low temperatures. This mechanism may be applicable to pristine *cis*-polyacetylene.

An interesting mechanism has been proposed by Kivelson, invoking electron hopping between soliton midgap states. In this ISH model an electron in a charged soliton bound state makes a phonon-assisted transition to a nearby neutral soliton. In order for the process to be energetically favorable, the neutral soliton is required to be in the vicinity of another charged impurity so that charge is transferred from impurity to impurity without the intermediacy of a free neutral soliton. Even though experimental support for the ISH model is strong, we note those characteristics of lightly doped polyacetylenes, which may be inconsistent with the model thus requiring further refinement of the mechanism. The most definitive data is probably the thermopower coefficient, which is not affected by interfibril transport processes intrinsic to lightly doped polyacetylenes. According to Eq. (11.74), the thermopower is only weakly dependent on the concentrations of neutral and charged solitons. However, the values of thermopower of $\sim$ 900 and 1450 μV K^{-1}, obtained by Park *et al.* (1980) and Chien *et al.* (1983f), respectively, would imply a 500-fold difference in y^+. The lower y^+ or higher thermopower may be suggestive of unintentional compensation. However, that the thermopower is not noticeably affected by iodine doping to y (I$_3^-$) $\approx$ 10^{-5} seems to rule out the compensation argument.

The validity of the ISH model for undoped *trans*-polyacetylene is predicted on the intrinsic concentration of charged and neutral solitons and the presence of impurity near the neutral soliton. Therefore, the mechanism is dependent on the intricate complicated impurity chemistry. We have suggested a source for the impurity and an indirect way to scrutinize the ISH model by the transport behaviors of polyacetylenes of different molecular weights.

For very lightly doped *trans*-polyacetylene, σ is independent of doping. This is consistent with the fact that the neutral soliton concentration is unchanged according to EPR and the dopant concentration is less than that of the intrinsic carriers.

The conductivity of undoped *cis*-polyacetylene is four to five orders of magnitude lower than that of the trans isomer; the EPR spin concentration is about one tenth. Doping commences to affect conductivity at $y \approx$ 10^{-6}, suggesting that any intrinsic carrier concentration would be of that order. This low concentration would be of that order. This low concentration may be attributed to the low mobility of the neutral defects in the cis polymer as judged by its broad temperature-independent EPR linewidth and the lack of effect of dopant on EPR T_1 below $y =$ 10^{-3} concentration. Consequently, reactions (5.56)–(5.58) have low probabilities. Together with the few num-

ber of neutral defects, the low intrinsic carrier concentration may be rationalized. Above $y > 10^{-6}$, σ_{RT} increases with doping as a result of carrier
injection. Conduction is probably by variable-range hopping.

11.7.2 Intermediately Doped Polyacetylenes

There are three important observations in this regime of $10^{-4} \leqslant y \leqslant$
10^{-2}. The first is that the TEP transition at $y \approx 7 \times 10^{-4}$ showed that for the
intrinsic transport property the conversion of polyacetylene chains from a
semiconducting to a conducting state occurred within about twofold increase of dopant concentration. If we assume that the polyacetylene has a
molecular weight of 11,000, then this transition occurs when there is about
one charged soliton per polyacetylene chain. Incidentally, there is also about
0.5 neutral soliton per chain. The second observation is the increase of dc
conductivity with y, which occurs over a broader range of dopant concentration. Nevertheless, the midpoint of this change is also $y \approx 7 \times 10^{-4}$, as in
TEP. The third and perhaps the most significant observation is that for
$10^{-4} \leqslant y \leqslant 10^{-2}$, *trans-* and *cis*-polyacetylenes have identical (log σ)-versus-
(log y) dependences. Assuming each dopant introduces one charge carrier,
one finds the carrier mobility to increase 2200 times over a 4.5-fold increase
in y.

Figure 11.61 is a replot of the variation of log σ with log y (I_3^-) for
iodine-doped *trans*-polyacetylene, assuming I_3^- to be the major dopant
species. The broken curve in Fig. 11.62 is the theoretical one according to
the ISH model with the assumption that one positive charge is added to

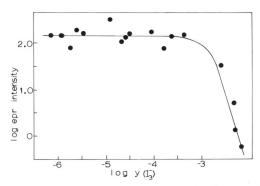

Fig. 11.61 Variation of EPR intensity versus doping for *trans*-$[CH(I_3)_y]_x$.

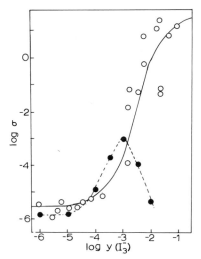

Fig. 11.62   Comparison of the data of Fig. 11.10 for *trans*-[CH(I$_3$)$_y$]$_x$: (O) with conductivity calculated with Eq. (11.50) of the intersoliton hopping conduction model (●).

polyacetylene by each I$_3^-$ ion and that there is an intrinsic carrier concentration of $\sim 2-3 \times 10^{-4}$. There is a rapid decrease of conductivity when y exceeds 10^{-3}.

A possible conduction mechanism is one involving electron hopping between a polaron and a charged soliton. It was discussed in Section 8.8 that doping creates both these states. As in the case of ISH model, this polaron soliton hopping model (PSH) has the final states after electron hopping very similar to the initial states.

That the polyacetylene fibrils become intrinsically conducting at $y \approx 7 \times 10^{-4}$ implies that the defects are not strongly pinned by the impurity potential. If this is the case, then the large decrease in thermopower may be regarded as a soliton melting transition, from a soliton glass to a soliton liquid state.

The conductivity versus y in this regime is nearly the same for iodine-doped *cis*- and *trans*-polyacetylenes. Present data show there is little isomerization at these low doping levels. There is more than one possible explanation. The pair of charged and neutral defects of a polaron does not cause π-amplitude kink and may be able to diffuse in a predominantly cis chain. Alternatively, doping results in the isomerization of a segment of the chain sufficiently long so that PSH can occur between states on similar segments of adjacent polyacetylene molecules.

11.7.3   Heavily Doped Polyacetylenes

As y exceeds 10^{-2}, the EPR signals disappear in uniformly doped polyacetylene. The charged carriers probably form a band. Then conduction may occur by variable-range hopping between states in this band. Epstein *et al.* (1983) have found for $[CH(I_3)_y]_x$ with y from 0.017 to 0.048 that

$$\sigma(T) = 0.39[N(E_F)/\alpha k_B T]^{1/2}\, v_0 e^2 \exp[-(T_0/T)^{1/4}], \qquad (11.108)$$

where $v_0 = v_{ph} e^{2\alpha R}$ has been corrected for the decay of wave function with R and $v_{ph} \approx 3.6 \times 10^{13}\ \sec^{-1}$. This mechanism seems to be able to account for the variation of TEP with temperature. Therefore, the expression

$$S^{vrh}(T) = \tfrac{1}{2}\left(\frac{k_B}{e}\right) k_B \left(\frac{B}{T}\right)^{1/4} N^{-1}\left(\frac{dN}{dE}\right), \qquad (11.109)$$

where N is the energy-dependent density of states. The equation gave a good fit with the experimental data for $y = 0.017$ and 0.033 with $N^{-1}\, dN/dE \approx 0.4\ eV^{-1}$. At higher levels of doping, the TEP results are intermediate between that for metallic substances.

$$S^m(T) = \left(\frac{2\pi^2}{3}\right)\left(\frac{k_B}{e v k_B T N^{-1}}\right)\frac{dN}{dE}. \qquad (11.110)$$

for $y \approx 0.04 - 0.05$, Eq. (11.109) gave $N^{-1}\, dN/dE = 0.6\ eV^{-1}$, whereas the value is $1.9\ Ev^{-1}$ using Eq. (11.110). Experimental data suggest that $N^{-1}\, dN/dE \approx 0.4\ eV^{-1}$ gives $N^{-1}\, dN/dE$ values of 0.6 and 1.9 eV^{-1}, respectively.

At still higher level of doping, polyacetylene became metallic.

Chapter 12

Technology

The discovery of new and unique materials should undoubtedly lead to new technological innovations. The lag times between these events are, however, difficult to predict, depending on the economics, the market demand, and the social needs. The case of conducting polymers is no exception. Their applications for electrical conductors, semiconducting devices, solar cells, and storage batteries have been explored and the last is being developed commercially. The purpose of this chapter is to document them. It will, however, depend on the technologists to improve and render each device economically attractive and practically useful.

Even more exciting is the possibility that there are technological innovations that are not yet conceived. Close cooperation between scientists doing basic research and others interested in application may generate ideas for innovation.

12.1 RECHARGEABLE BATTERIES

Current storage batteries are heavy in weight and have certain limitations in their charging and discharging characteristics. MacDiarmid and co-workers (Nigrey *et al.*, 1979, 1981a, b; MacInnes *et al.*, 1981; Kaneto *et al.*, 1982) have fully developed polyacetylene-based, lightweight, recharge-

able batteries and showed their superior charging and discharging characteristics, energy density, and power density.

12.1.1 Battery Construction

A number of configurations have been described by MacDiarmid and
co-workers. The latest one uses two collectors: a piece of 52-mesh platinum
gauze of about 2 cm² with a length of platinum wire spot welded to it and a
piece of 109-mesh nickel grid with a length of nickel wire spot welded to it.
The nickel grid collector is folded tightly around a piece of lithium ribbon
about half the size of the grid. This is placed on top of the separator, which is
in turn placed on top of a polyacetylene film, and finally on top of the
platinum gauze, all three elements having about the same size. The whole
assembly is then tightly folded in half and inserted into a rectangular glass
tube (3 × 10 mm) containing the electrolyte and some basic alumina. The
tube is finally sealed under vacuum. In its final form, from the center to the
exterior are the lithium electrodes, nickel collector, separator, polyacetylene
electrode, and finally, the platinum collector. The two electrodes are separated by about 0.5 mm. The Ni and Pt wires protrude from the sealed
tubing. Polypropylene film is used as the separator to measure cell potential
and energy density; glass filter paper dried at 873 K is used in power-density
studies. In the following description of battery performance the electrolyte
solution is 1 M LiClO$_4$ in propylene carbonate.

12.1.2 Charging Characteristics

The negative terminal of a galvanostat is connected to the Li electrode
and the positive terminal to the polyacetylene electrode. The battery can be
charged simply by maintaining the current at 1 mA for about 30 min. The
polyacetylene is doped to $[CH(ClO_4)_{0.06}]_x$ in this process. The anode charge
reaction is

$$(CH)_x + 0.06x ClO_4^- \rightarrow [(CH)^{+0.06}(ClO_4^-)_{0.06}]_x + 0.06x e^- \tag{12.1}$$

and the cathode charge reaction is

$$0.06x Li^+ + 0.06x e^- \rightarrow 0.06x Li^0 \tag{12.2}$$

The charging characteristics were studied by a stepwise procedure measuring the open circuit voltage $V_{oc}(0)$ periodically. Here (0) denotes the
measurement of open circuit voltage within a few seconds of the interrup-

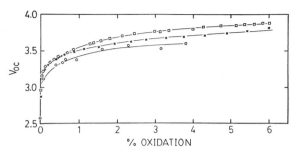

Fig. 12.1 Relation between the open circuit voltage and degree of charging for a $(CH)_x|LiClO_4|Li$ cell: $(\square)$ $V_{oc}(0)$; (X) $V_{oc}(0.08)$; (O) $V_{oc}(24)$. [After Kaneto *et al.* (1982).]

tion of the charge process; (24) would then refer to measurements made after the battery has rested for 24 hr. At a constant applied voltage the initial current would be high, ~ 0.1 mA mg^{-1} of polyacetylene; when the current dropped by about 90%, the applied voltage was increased slightly. The procedure was repeated until the desired final V_{oc} was reached. The results for $V_{oc}(0)$, $V_{oc}(0.08)$, and $V_{oc}(24)$ are shown in Fig. 12.1. The net charging reaction is the sum of Eqs. (12.1) and (12.2),

$$(CH)_x + xyLi^+ClO_4^- \rightarrow [(CH)^{y+}(ClO_4)_y^-]_x + yxLi^0 \tag{12.3}$$

In Fig. 12.1 V_{oc} was measured without noticeable discharging of the cell; the percent oxidation z was obtained from the coulombs passed as measured with a coulometer and the known weight of the polyacetylene electrode. It is seen from the figure that at a given z the values of $V_{oc}(0)$ were greater than $V_{oc}(24)$ whereas those of $V_{oc}(0.08)$ lie between. Or for a given V_{oc}, the extent of charging is higher if the V_{oc} is measured after the battery has rested for 24 hr. The curves in Fig. 12.1 were drawn according to

$$V_{oc}(24) = 3.43 + 0.14 \ln z \tag{12.4}$$

The differences in the variation of $V_{oc}(t)$ with z are attributable to the diffusion of the counterion into the polyacetylene fibrils, which is a slow process, whereas V_{oc} is determined by the state of doping of the surface of the fibrils in contact with the electrolyte. Therefore, when the charging is interrupted, $V_{oc}(0)$ measures the degree of oxidation on the exterior of the fibrils. During the time the battery is allowed to stand, the excess ClO_4^- ions at the surface diffuse toward the interior of the fibrils and $V_{oc}(t)$ decreases with time. Thus $V_{oc}(0) > V_{oc}(0.08) > V_{oc}(24)$.

In fact, the change of V_{oc} with t can be used to study the diffusion process. For a 20-nm polyacetylene fibril, the diffusion time constant τ is about 2 days. This is the time it takes to reach diffusion equilibrium. For instance, if the fibril is considered to be a uniform solid cylinder immersed in an electrolyte solution of concentration C_0 and the average concentration of dopant ion in the cylinder at any given time is $\overline{C}$, then the approach to equilibrium expressed as $\overline{C}/C_0$ is 0.74 at $t = \tau$, 0.91 at 2τ, 0.97 at 3τ, and 0.99 at 4τ. Hence, diffusion equilibrium is virtually reached after $t = 4\tau$. This description is a crude one. As we have shown, the polyacetylene fibril comprises microfibrils, and the fibrils are not always cylindrical but quite often ribbon-shaped. For the typical 20-nm fibrils, it takes about eight days to establish near-diffusion equilibrium for ClO_4^- ions. Of course, τ may vary for different ionic species. Returning to Fig. 12.1, it is the $V_{oc}(24)$ curve that more accurately describes the V_{oc} versus z relation than the other two. Hereafter, it is understood that the extent of oxidation, percent doping, and y are all "apparent" values because the system was seldomly ever at diffusion equilibrium.

12.1.3 Discharge Characteristics

The short-circuit discharge characteristics are as follows. A $[CH(ClO_4)_{0.06}]_x$ film was prepared by passing 1.7 C, the V_{oc} was 3.7 V and the short-circuit current I_{sc} was 25 mA for a 0.5 cm^2 ($\sim$ 3 mg) sample of the material (I_{sc} can vary from $\sim$ 15 to 35 mA from sample to sample). I_{sc} falls rapidly as the battery discharges. For instance, after $\sim$ 30 sec of short-circuit discharge, I_{sc} falls to 17 mA whereas V_{oc} is unchanged. During this period more than half of the coulombs put in while charging the battery are released. After 1 min V_{oc} begins to decrease to 3.2 V; I_{sc} is 3.2 mA. At the end of 3 min, the fall of V_{oc} becomes rapid, I_{sc} is only 0.1 mA and a total of 1.0 C is released. The polyacetylene electrode now has the composition $[CH(ClO_4)_{0.024}]_x$ and its conductivity begins to decrease rapidly.

The anode discharge reaction of the above process is

$$0.036x Li^0 \longrightarrow 0.036x Li^+ + 0.036x e^- \tag{12.5}$$

the cathode discharge reaction is

$$[(CH)^{0.06+}(ClO_4^-)_{0.06}]_x + 0.036x e^- \longrightarrow [(CH)^{0.024+}(ClO_4^-)_{0.024}]_x + 0.036x ClO_4^- \tag{12.6}$$

and the net discharge reaction is

$$[(CH)^{0.06+}(ClO_4^-)_{0.06}]_x + 0.036x Li^0 \longrightarrow [(CH)^{0.024+}(ClO_4^-)_{0.024}]_x + 0.036x LiClO_4 \tag{12.7}$$

The discharge process can also be carried out in steps. Figure 12.2 shows the battery charged to various values of z by a series of constant, applied

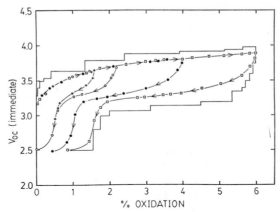

Fig. 12.2 Relation between $V_{oc}(0)$ and apparent percent of oxidation z for a $(CH)_x|LiClO_4|Li$ cell charged to z and then discharged: $(+)$ $z = 1.54\%$, $(\bigcirc)$ $z = 2.17\%$, $(\bullet)$ $z = 4\%$, $(\square)$ $z = 6\%$. [After Kaneto *et al.* (1982).]

potentials (the upper steps). It is then discharged at a series of lower constant, applied potentials (the lower step). $V_{oc}(0)$ is determined periodically, and z is calculated from $Q_{in(tot)} - Q_{out}$, where $Q_{in(tot)}$ is the total number of coulombs charged, and $Q_{out(tot)}$ gives the coulombs liberated at a given V_{oc}.

The coulomb efficiency can be obtained by discharging the battery at 2.5 V for about 16 hr and recording the total coulombs released, $Q_{out(tot)}$. The coulomb efficiency in percent is given by $100(Q_{out}/Q_{in})_{tot}$ in Table 12.1, Column 2. The corresponding energy efficiency in percent is given by $100(E_{out}/E_{in})_{tot}$ in Table 12.1, Column 3. The total energy (coulombs × volts) expended in a charge cycle $E_{in(tot)}$, is obtained from the area under the applied stepped-up voltage-charging curve; $E_{out(tot)}$ is given by the

TABLE 12.1

Coulombic and Energy Efficiencies of a $(CH)_x|LiClO_4|Li$ Cell[a]

Apparent oxidation (%)	Coulombic efficiency (%)	Energy efficiency (%)
1.54	100.0	88.7
2.01	99.2	89.5
2.17	100.1	89.2
2.51	95.8	87.7
4.0	89.5	78.8
6.0	85.7	74.8

[a] After Kaneto *et al.* (1982).

area under the applied stepped-down voltage-discharge curve. The energy efficiency obtained under different charge–discharge conditions will be different.

Figure 12.2 shows that there is hysteresis in the charge–discharge curves. In general, slower rates of charging and discharging will yield higher energy efficiencies because of reduction in I^2R loss. Also the effect of diffusion, described above for the charging characteristics, applies here as well but in an opposite way. During discharge the dopant ion at the fibril surface is first removed and $V_{oc}(0)$ is lower than that if measured after the battery has rested and the dopant ion in the interior diffuses toward the exterior. Consequently, at higher discharge rates the discharge potential will be lower because of this depletion of charge on the exterior of the fibrils.

There are several sources of incomplete coulomb recovery such as chemical reaction of the electrolyte, reaction of impurities with the doped polyacetylene, and the absence of a diffusion–equilibrium system. A trace of air can cause the voltage of a cell to fall sharply on standing. However, the stability may be enhanced by several charge–discharge cycles. Apparently, impurities are scavenged during these electrochemical processes.

12.1.4 Rechargeability

Nigrey *et al.* (1981a) have charged and discharged the $(CH)_x|LiClO_4|Li$ battery at the same constant currents and have recorded the voltage change continuously. An example is shown in Fig. 12.3 for three complete charge–discharge cycles. During a discharge, the voltage dropped immediately from 4.1 to 3.7 V and then decreased gradually to 3.2 V. For the charging process, the voltage rose immediately from 3.2 to 3.9 V and then reached 4.1 V at the

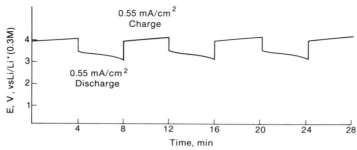

Fig. 12.3 Three complete charge–discharge cycles of a $(CH)_x|LiClO_4|Li$ battery at a constant current of 0.55 mA. [After Nigrey *et al.* (1981a).]

end of the cycle. The figure shows that the three cycles behaved identically. Each cycle involved 120 mC and took 4 min. The composition of the polyacetylene at the beginning of each of the charge cycles was $[CH(ClO_4)_{0.045}]_x$, which became $[CH(ClO_4)_{0.05}]_x$. The battery did not lose charge spontaneously. Thus a charged cell had the same V_{oc} and discharge current after standing for 48 hr.

To demonstrate rechargeability, a 0.5-cm² piece of polyacetylene film was coated with a 60-nm layer of gold on one side to act as contact and doped to $[CH(ClO_4)_{0.097}]_x$ electrochemically with the passage of 2.73 C. The battery constructed from it was discharged and recharged for 326 cycles, each cycle consisting of a 100-sec discharge at 2 mA (0.2 C, 7.3% discharge) and a 100-sec charge at the same current. The battery initially had $V_{oc} = 4.13$ V; after 320 cycles it was 4.16 V.

12.1.5 Energy Density

The ideal storage battery should show little or no change in voltage until it is almost completely discharged; then the voltage drops sharply. Figure 12.4 shows the polyacetylene battery to possess this characteristic. The cell was constructed of 3.3 mg of polyacetylene, doped to 7% and discharged to 2.5 V immediately after charging. There was a small initial drop in voltage followed by a plateau region of nearly constant voltage, which was longer for a smaller discharge current. Of course, greater amounts of polyacetylene will also increase the time for discharge at a given current. The total energy released upon discharge can be obtained from the area under each curve in Fig. 12.4. The energy density is calculated with the weight of $[CH(ClO_4)_{0.07}]_x$ and the weight of Li consumed in the process and summarized in Table 12.2 together with the discharge rate. The lead–acid automobile battery has an energy density of ~ 31 W hr kg^{-1}. If one takes into account a factor-of-7 increase in weight for a completely packaged battery, then the polyacetylene system will have an energy density comparable to that of the lead–acid battery.

The Coulomb efficiencies are given in Column 4 of Table 12.2. If the battery is allowed to rest for 20 min, an increase in V_{oc} results. An additional constant potential discharge at 2.9 V for 16 hr causes further release of coulombs. The increased coulombic efficiencies are given in parentheses of Column 4 of the table. This additional coulomb release is due to diffusion of ClO_4^- ion from the interior to the exterior of a fibril during the 20-min rest period.

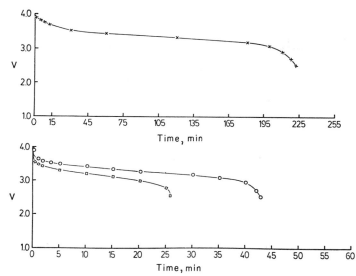

Fig. 12.4   Change of voltage with discharge time of a $(CH)_x|LiClO_4|Li$ battery at a constant current: (x) 0.1 mA; (O) 0.55 mA; (□) 1.0 mA. [After Kaneto *et al.* (1982).]

12.1.6   Power Density

The maximum power P_{max} may be measured in several ways:

$$P_{max} = V_m I_m, \tag{12.8}$$

$$P_{max} = V_m^2/R_l, \tag{12.9}$$

$$P_{max} = I_m^2/R_l, \tag{12.10}$$

TABLE 12.2

Energy Densities of a $(CH)_x|LiClO_4|Li$ Cell[a]

Discharge current (mA)	Discharge rate (A kg^{-1})	Energy density (W hr kg^{-1})	Coulomb efficiency (%)
0.1	19.5	258	74.0 (80.3)
0.55	107	255	79.3 (84.4)
1.0	195	254	86.7 (87.2)

[a] 3.3 mg of $(CH)_x$ doped to 7%, discharged to 2.5 V. [After Kaneto *et al.* (1982).]

TABLE 12.3

Typical Maximum Power Densities of $(CH)_x|LiCl]_4|Li$ Cells[a]

Weight of $(CH)_x$ (mg)	Total weight of electro-active materials (mg)	y (%)	$V_{oc}(15)$ (V)	R_i (Ω)	R_l (Ω)	V_m (V)	I_m (mA)	P_{max} kW kg^{-1}
4.2	4.9	2.1	3.52	22.4	22.5	1.80	—	28 ± 5
4.2	4.9	2.1	3.54	21.3	22.0	—	86	33 ± 5
7.0	8.2	2.0	3.52	13.0	11.0	1.83	—	37 ± 7
7.0	8.4	2.5	3.55	11.0	10.5	1.80	—	36 ± 7
12.0	17.4	5.5	3.66	8.0	10.0	2.08	208	25 ± 1

[a] After Kaneto *et al.* (1982).

where the internal resistance of the battery, R_i, is matched to the resistance of the external load, R_l, and V_m and I_m are the voltage and current, respectively, measured at the very beginning of a discharge cycle. R_i may be obtained in the following manner. The battery is charged and allowed to stand for 5 hr. The $V_{oc}(5)$ and discharging current I are measured as well as the resistance R_{ac} due to the ammeter and coulometer. Then

$$R_i = [R_{ac}I - V_{oc}(5)]/I. \tag{12.11}$$

To obtain P_{max}, the cell is recharged with the same number of coulombs released during the above measurement. The cell is allowed to rest for 15 hr after which $V_{oc}(15)$ is recorded. The cell is discharged through a circuit with matched R_l and R_i, and the values of the voltage and current measured after 2 sec are V_m and I_m, respectively. Typical results are given in Table 12.3.

The values of P_{max} were found not to vary greatly with y or the weight of the polyacetylene used for $y \leqslant 2\%$. P_{max} is about 30 kW kg^{-1} for the $(CH)_x|LiClO_4|Li$ battery. Therefore, even after including a sevenfold weight increase for packaging, P_{max} for the polyacetylene system is still an order of magnitude greater than the lead–acid counterpart.

A number of factors contributed to the extremely high P_{max} values. The effective surface area of a 4-mg polyacetylene is very high, which is $\sim 2.4 \times 10^3$ cm^2, using a surface area of 60 m^2 g^{-1}. Because the polyacetylene fibrils are only ~ 20 nm in diameter, any portion of the polyacetylene is less than ~ 10 nm from the electrolyte. The good accessibility of the polyacetylene to the dopant ion results in very high current density. The high conductivity of the doped polyacetylene permits rapid removal or injection of charges, the rate being limited by the diffusion of the counterion. Furthermore, the former process will increase the electrostatic repulsion between the remaining dopant ions and thus accelerate their diffusion to the fibril surface.

12.1.7   Types of Polyacetylene Batteries

A unique property of polyacetylene is that the polymer can be transformed into metallic conductors by either *p*-type or *n*-type doping. The electrochemistry involved has been discussed in Section 5.3.3. This makes it possible to fabricate batteries of three different types: (1) those in which a polyacetylene derivative acts as the anode; (2) those in which a polyacetylene derivative serves as the cathode; and (3) those in which polyacetylene derivatives constitute both the anode and the cathode. They can also be classified according to the relation between the polyacetylene doping process and the charge–discharge reactions. If doping corresponds to a discharge process, it is referred to as a class A battery; a class B battery is undoped during a discharge cycle.

12.1.7.1   Polyacetylene Anodes — Type 1

When a polyacetylene film is immersed in a THF solution of alkali metal nephthalide, *n*-type doping occurs spontaneously,

$$(CH)_x + xy Li^+(Naph)^- \rightarrow [(CH)^{y-}Li_y^+]_x + Naph \qquad (12.12)$$

The electrochemical analog of the process occurs with the production of electric current. For instance, when a strip of polyacetylene and a strip of lithium metal are connected through an ammeter and placed in a THF solution of $LiClO_4$, current flow is observed. This is an example of a class A battery, where the doping reaction is the discharge reaction and can act as a battery during this process. With the application of an appropriate potential, the reaction is reversed and the battery is said to be recharged. In Table 12.4, this battery **I** is listed as a class A type 1.

In battery **II** and also in **VI, VII, VIII, XI,** and **XII,** the cation is the organic *n*-Bu$_4$N$^+$. In battery **II** a glass frit is necessary to prevent mixing of *n*-Bu$_4$N$^+$ and Li$^+$ ions. The doping of polyacetylene by *n*-Bu$_4$N$^+$ occurs spontaneously and is therefore the discharge reaction.

12.1.7.2   Polyacetylene Cathode — Type 2

To use polyacetylene as the cathode, it is *p*-doped in the charging process. This can be done, for instance, by immersing a polyacetylene film in

TABLE 12.4

Polyacetylene Battery Systems

Number	Class	Type		Reactions	Electrolyte	Voltage–current characteristic[a]
I	A	1	Cathode	$(CH)_x + 0.06xLi^+ + 0.06xe^- \rightarrow [Li_{0.06}(CH)^{0.06-}]_x$	0.3 M LiClO$_4$ in THF	Charge $V_{oc,m} = 1.5$ V $I_{sc,m} = 0.5$ mA
			Anode	$0.06xLi \rightarrow 0.06xLi^+ + 0.06xe^-$		
			Net	$(CH)_x + 0.06xLi \rightarrow Li_{0.06}^+(CH)^{0.06-}]_x$		
II	A	1	Cathode	$(CH)_x + 0.02x(n\text{-}Bu_4N)^+ + 0.02xe^- \rightarrow [(n\text{-}Bu_4N)_{0.02}^+(CH)^{0.02-}]_x$	1.0 M $(nBu_4N)ClO_4$ in THF	Charge $V_{oc,m} = 1.3$ V $I_{sc,m} = 0.2$ mA
			Anode	$0.02xLi \rightarrow 0.02x Li^+ + 0.02xe^-$		
			Net	$(CH)_x + 0.02x(nBu_4N)^+ + 0.02xLi \rightarrow [(n\text{-}Bu_4N)_{0.02}^+(CH)^{0.02-}]_x$		
III	B	2[e]	Cathode	$[(CH)^{0.06+}(ClO_4^-)_{0.06}]_x + 0.06xe^- \rightarrow (CH)_x + 0.06xClO_4^-$	0.3 M LiClO$_4$ in PC[b]	Discharge[d] $V_{oc,i} = 3.7$ V $I_{sc,i} = 25$ mA
			Anode	$0.06xLi \rightarrow 0.06xLi^+ + 0.06xe^-$		
			Net	$[(CH)^{0.06+}(ClO_4^-)_{0.06}]_x + 0.06xLi \rightarrow (CH)_x + 0.06xLi^+ClO_4^-$		
IV	B	2	Cathode	$[(CH)^{0.001+}(I_3^-)_{0.001}]_x + 0.001xe^- \rightarrow (CH)_x + 0.001xI_3^-$	0.1 M NaI in solid PEO[c]	Discharge $V_{oc,i} = 1.3$ V $I_{sc,i} = 2.1$ mA
			Anode	$0.001xNa \rightarrow 0.001xNa^+ + 0.001xe^-$		
			Net	$[(CH)^{0.001+}(I_3^-)_{0.001}]_x + 0.001xNa \rightarrow (CH)_x + 0.001xNa^+I_3^-$		
V	B	3α[f]	Cathode	$[(CH)^{0.05+}(ClO_4^-)_{0.05}]_x + 0.05xe^- \rightarrow (CH)_x + 0.05xClO_4^-$	0.3 M LiClO$_4$ in THF	Discharge $V_{oc,i} = 3.1$ V $I_{sc,i} = 1.9$ mA
			Anode	$[Li_{0.05}^+(CH)^{0.05-}]_x \rightarrow (CH)_x + 0.05xLi^+ + 0.05xe^-$		
			Net	$[(CH)^{0.05+}(ClO_4^-)_{0.05}]_x + [Li_{0.05}^+(CH)^{0.05-}]_x \rightarrow 2(CH)_x + 0.05xLi^+ClO_4^-$		
VI	B	3α	Cathode	$[(CH)^{0.05+}(ClO_4^-)_{0.05}]_x + 0.05xe^- \rightarrow (CH)_x + 0.05xClO_4^-$	0.3 M $(n\text{-}Bu_4N)^+ClO_4^-$ in THF	Discharge $V_{oc,i} = 2.8$ V $I_{sc,i} = 3.5$ mA
			Anode	$[n\text{-}Bu_4N_{0.05}^+(CH)^{0.05-}]_x \rightarrow (CH)_x + 0.05x(n\text{-}Bu_4N)^+ + 0.05xe^-$		
			Net	$[(CH)^{0.05+}(ClO_4^-)_{0.05}]_x + [n\text{-}Bu_4N_{0.05}^+(CH)^{0.05-}]_x \rightarrow 2(CH)_x + 0.05 (n\text{-}Bu_4N)^+ClO_4^-$		
VII	B	3α	Cathode	$[(CH)^{0.024+}(ClO_4^-)_{0.024}]_x + 0.024xe^- \rightarrow (CH)_x + 0.024xClO_4^-$	0.5 M $(n\text{-}Bu_4N)^+ClO_4^-$ in PC	Discharge $V_{oc,i} = 2.5$ V $I_{sc,i} = 11.1$ mA
			Anode	$[n\text{-}Bu_4N_{0.024}^+(CH)^{0.024-}]_x \rightarrow (CH)_x + 0.024x(n\text{-}Bu_4N)^+ + 0.024xe^-$		
			Net	$[(CH)^{0.024+}(ClO_4^-)_{0.024}]_x + [(n\text{-}Bu_4N)_{0.024}^{+0.024}(CH)^{0.024-}]_x \rightarrow 2(CH)_x + 0.024x(n\text{-}Bu_4N)^+ClO_4^-$		
VIII	B	3α	Cathode	$[(CH)^{0.06+}(PF_6^-)_{0.06}]_x + 0.06xe^- \rightarrow (CH)_x + 0.06xPF_6^-$	0.3 M $(n\text{-}Bu_4N)^+PF_6^-$ in THF	Discharge $V_{oc,i} = 2.5$ V $I_{sc,i} = 4.1$ mA
			Anode	$[n\text{-}Bu_4N_{0.06}^+(CH)^{0.06-}]_x \rightarrow (CH)_x + 0.05x(n\text{-}Bu_4N)^+ + 0.06xe^-$		
			Net	$[(CH)^{0.06+}(PF_6^-)_{0.06}]_x + [(n\text{-}Bu_4N)_{0.06}^+(CH)^{0.06-}]_x \rightarrow 2(CH)_x + 0.05x(n\text{-}Bu_4N)^+PF_6^-$		

						Discharge
IX	B	3β	Cathode	$[(CH)^{0.05+}(ClO_4^-)_{0.05}]_x + 0.021xe^- \rightarrow [(CH)^{0.029+}(ClO_4^-)_{0.029}]_x + 0.021xClO_4^-$	1 M LiClO$_4$ in PC	$V_{oc,i}$ = 0.25 V $I_{sc,i}$ = 0.01 mA
			Anode	$[(CH)^{0.008+}(ClO_4^-)_{0.008}]_x + 0.021xClO_4^- \rightarrow [(CH)^{0.029+}(ClO_4^-)_{0.029}]_x + 0.021xe^-$		
			Net	$[(CH)^{0.05+}(ClO_4^-)_{0.05}]_x + [(CH)^{0.008+}(ClO_4^-)_{0.008}]_x \rightarrow 2[(CH)^{0.029+}(ClO_4^-)_{0.029}]_x$		
X	B	3β	Cathode	$[(CH)^{0.05+}(ClO_4^-)_{0.05}]_x + 0.015xe^- \rightarrow [(CH)^{0.035+}(ClO_4^-)_{0.035}]_x + 0.015xClO_4^-$	0.5 M $(n\text{-}Bu_4N)^+ClO_4^-$ in PC	$V_{oc,i}$ = 0.5 V $I_{sc,i}$ = 0.3 mA
			Anode	$[(CH)^{0.02+}(ClO_4^-)_{0.002}]_x + 0.015xClO_4^- \rightarrow [(CH)^{0.035+}(ClO_4^-)_{0.035}]_x + 0.015xe^-$		
			Net	$[(CH)^{0.05+}(ClO_4^-)_{0.05}]_x + [(CH)^{0.02+}(ClO_4^-)_{0.02}]_x \rightarrow 2[(CH)^{0.035+}(ClO_4^-)_{0.035}]_x$		
XI	B	3α	Cathode	$[(CH)^{0.09+}(ClO_4^-)_{0.09}]_x + 0.07xe^- \rightarrow [(CH)^{0.02+}(ClO_4^-)_{0.02}]_x + 0.07xClO_4^-$	0.5 M $(n\text{-}Bu_4N)^+ClO_4^-$ in PC	$V_{oc,i}$ = 2.5 V $I_{sc,i}$ = 16 mA
			Anode	$[(n\text{-}Bu_4N)_{0.07}^+(CH)^{0.07-}]_x \rightarrow (CH)_x + 0.07x(n\text{-}Bu_4N)^+ + 0.07xe^-$		
			Net	$[(CH)^{0.09+}(ClO_4^-)_{0.09}]_x + [(n\text{-}Bu_4N)_{0.07}^+(CH)^{0.07-}]_x \rightarrow [(CH)^{0.02+}(ClO_4^-)_{0.02}]_x + (CH)_x + 0.07x(n\text{-}Bu_4N)^+ClO_4^-$		
XII	B	3γ	Cathode	$[(n\text{-}Bu_4N)_{0.02}^+(CH)^{0.02-}]_x + 0.025x(n\text{-}Bu_4N)^+ + 0.025xe^- \rightarrow [(n\text{-}Bu_4N)_{0.045}^+(CH)^{0.045}]_x$	0.5 M $(n\text{-}Bu_4N)^+ClO_4^-$ in PC	$V_{oc,i}$ = 0.66 V $I_{sc,i}$ = 1.9 mA
			Anode	$[(n\text{-}Bu_4N)_{0.07}^+(CH)^{0.07-}]_x \rightarrow 0.025x(n\text{-}Bu_4N)^+ + 0.025xe^- + [(n\text{-}Bu_4N)_{0.045}^+(CH)^{0.045}]_x$		
			Net	$[(n\text{-}Bu_4N)_{0.07}^+(CH)^{0.07-}]_x + [(n\text{-}Bu_4N)_{0.02}^+(CH)^{0.02-}]_x \rightarrow 2[(n\text{-}Bu_4N)_{0.045}^+(CH)^{0.045}]_x$		

[a] $V_{oc,m}$ = maximum open circuit voltage, $I_{sc,m}$ = maximum short circuit current, $V_{oc,i}$ = initial open circuit voltage, $I_{sc,i}$ = initial short circuit current.

[b] PC is propylene carbonate.

[c] PEO is polyethylene oxide.

[d] Charging was usually carried out by one of the following methods: (1) a potentiostat was used to apply a constant potential of 1.0 V (versus Ag/Ag$^+$ electrode), (2) a galvanostat was used to supply a constant current of 1 mA, (3) a dc power supply was used to apply a potential of 4 V.

[e] The metal electrode can be Li, Al, or Pt. The Li metal formed on the electrode during a charging reaction is consumed in the discharge reaction.

[f] Stainless steel electrodes were used. A piece of glass filter paper impregnated with the solid electrolyte was used in a separator between the two pieces of polyacetylene film.

a THF or propylene carbonate solution of salts with Li^+ or n-Bu_4N^+ cation and ClO_4^- or PF_6^- anions and by attaching it to the positive terminal of a dc power supply. The battery reaction is the undoping of the p-doped polyacetylene. The $(CH)_x|LiClO_4|Li$ or Al battery **III** have been most extensively studied, and the results are discussed in the above sections. Battery **IV** uses I_3^- as the dopant ion; it has much lower V_{oc} and I_{sc} than **III**.

12.1.7.3   Polyacetylene Anode and Cathode — Type 3

In these systems polyacetylene derivatives are used for both electrodes. Three possibilities have been demonstrated by MacInnes *et al.* (1981). In type 3α the cathode is p-doped polyacetylene and the anode is n-doped polyacetylene, both having the same doping level; examples are **V, VI, VII, VIII, XI**. Batteries **VII** and **XI** appear to have good V_{oc} and I_{sc} values.

In type 3β both electrodes are p-type polyacetylene but doped to different levels. Examples are **IX** and **X**. The more heavily doped material served as the cathode and the other as the anode. During the discharge cycle, electrons flow from the less oxidized to the more oxidized state. The battery is completely discharged when both electrodes have the same oxidation state and composition. In this and type 3γ, the polyacetylene must be predoped before it can be used for the initial charging.

Finally, type 3γ cells have n-doped polyacetylenes as electrodes. Opposite to type 3β, here the more heavily doped polyacetylene is used as the anode and the less heavily doped material serves as the cathode.

12.1.8   All-Polymeric Solid-State Battery

Battery **IV** of Table 12.4 is a solid-state battery because a solid electrolyte is used. It is a complex of NaI with polyethylene oxide (PEO) of molecular weight 5×10^6. The electrical conductivity of a PEO–NaI complex with Na/O = 0.2 was 10^{-4} $(\Omega \text{ cm})^{-1}$ at 358 K and has reasonable chemical stability. The material can be processed into a free standing film 0.1–0.2 mm thick. The battery can be constructed by sandwiching the PEO–NaI film with a polyacetylene film onto one side and a platinum or aluminum reservoir for sodium metal onto the other side. Graphite paste and platinum plates with leads provide electrical connections. The system $(CH)_x|PEO$–NaI$|Na$ is a class A battery, and the cell is in a fully charged state as assembled. In order to render the polyacetylene electrode conducting, the battery is partially discharged to dope the polymer. This electrochemical doping converts the cell to $(CHNa_y)_x|PEO$–NaI$|Na$.

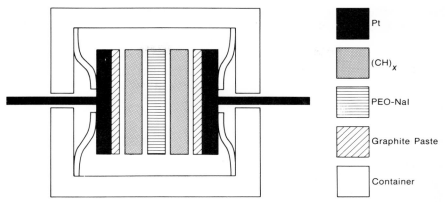

Fig. 12.5   Schematic structure of an all-polymeric solid-state battery. [After Chiang (1981).]

If both electrodes are derived from polyacetylene, then the system is an all-polymeric solid-state battery. A schematic structure of a Type B 3α all-polymeric solid-state battery $[(CH)^{y-}Na_y^+]_x|PEO-NaI|[(CH)^{y+}I_y^-]_x$ is shown in Fig. 12.5.

A typical V_{oc} for such batteries ranged from 2.8 to 3.5 V and I_{sc} from 1 to 12 mA cm^{-2}. The charge–discharge behaviors are shown in Fig. 12.6. During the charging cycle a constant charging current of 0.1 mA was

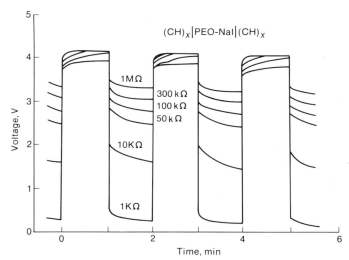

Fig. 12.6   Charge–discharge curves of a $(CH)_x|PEO-NaI|(CH)_x$ battery for a constant charging current of 0.1 mA. The disccharge loads are, from the top curve to the bottom curve: 1 MΩ, 300 kΩ, 100 kΩ, 50 kΩ, 10 kΩ, 1 kΩ. [After Chiang (1981).]

applied. The battery was discharged through a load resistance in series with a multimeter. Each cycle lasts 1 min. There was no significant decay in the battery's characteristics after 20 complete cycles.

The maximum power density for the $(CH)_x|PEO-NaI|(CH)_x$ battery is about $10-20$ W kg^{-1}. For the $(CH)_x|PEO-NaI|Na$ battery the estimated P_{max} is ~ 250 W kg^{-1} and an energy density of 20 W hr kg^{-1}.

12.2   SEMICONDUCTOR DEVICES AND PROPERTIES

12.2.1   Introduction

Whereas the rechargeable batteries make use of the metallic-level conductivity of polyacetylene, other practical applications would lie in the regime with the polymer at semiconducting doping levels. Current investigations appear to be focused on two objectives. Polyacetylene may be the first organic compound to share several electronic properties in common with inorganic covalent semiconductors. One universal manifestation of a semiconductor has been the ability to form Schottky barriers to another substance with appropriate work functions. Junctions have been made, using either undoped *trans*-polyacetylene as the *p*-type semiconductor or heavily doped polyacetylene as the metal. The interest is in how closely polyacetylene resembles conventional semiconductors and whether a Schottky junction made with polyacetylene can be described by the standard thermionic emission model. Knowledge about the nature of the junction profile and about the oxidant-induced acceptor levels are of importance for further technological development.

To pursue the development of a polyacetylene solar cell is the other area of interest. Polyacetylene is an attractive candidate for use as the absorbing layer in a solar cell. Its optical band gap is 1.5 eV, and it is highly absorbing with an absorption coefficient in excess of 10^5 cm^{-1} over the photon energy range $1.5-3.0$ eV, providing an excellent match to the solar spectrum. Of crucial importance, polyacetylene is potentially inexpensive, requiring virtually no energy for its preparation in marked contrast to the energy-intensive manufacturing of inorganic semiconductors. It has been estimated, though somewhat optimistically, that polyacetylene could be manufactured for a few cents per square foot. At 1% solar conversion efficiency, a polyacetylene-based solar cell would provide electrical energy at a cost of about $0.1/peak watt. The present Department of Energy goal for 1986 for photovoltaic collectors is $0.7/peak watt. However, as we will see, there are technical problems that must be overcome.

There are three general categories of junction devices: (1) Schottky diodes using heavily doped polyacetylene as the metallic electrode on a variety of *n*-type semiconductors; (2) Schottky diodes using undoped *trans*-polyacetylene as a *p*-type semiconductor on a variety of metallic contacts; and (3) *p–n* heterojunction diodes formed by the interface of undoped *p*-type *trans*-polyacetylene with an *n*-type semiconductor.

For the first two types of junction devices, the theory of Schottky barrier formation predicts that a rectifying barrier will be formed at the interface between a simple *p*-type semiconductor and a metal if the metallic work function is smaller than that of the semiconductor. If this condition is satisfied, majority carriers (holes) will be depleted from the region of the semiconductor near the junction, creating a high resistance layer to support the barrier electric field caused by the charge separation. On the other hand, if the work function of the metal is larger than that of the semiconductor, although motion of charge will occur, no high resistance depletion layer will be created, and the contact will be ohmic.

The above simple picture is frequently invalidated when a semiconductor has a high density of surface states due to lattice termination, grain boundaries, etc. These states tend to pin the Fermi level at the surface. A consequence is that the nature of the junction formed is independent of the work function of the metal used at the interface. Whether or not polyacetylene is complicated by high density of surface states will be discussed below. The relation between the work functions of metals and undoped and doped polyacetylenes are shown in Fig. 12.7.

There are three general methods for preparing the devices. One method is to polymerize acetylene directly onto metal or semiconductor substrates using the standard $Ti(OBu)_4$–$4AlEt_3$ catalyst. The thickness of the thin polymer film can be varied from ~ 0.1 to 1 μm, which can be subsequently doped if necessary. The second method is to begin with free-standing polyacetylene film. The Schottky junctions are formed by thermal evaporation of a thin layer of metal onto the surface of the polymer film. The use of the relatively thick polyacetylene films (~ 100 μm) reduces the chance of pinhole shorts. Alternatively, junctions and ohmic contacts were formed between polyacetylene and substrates by pressure contacts. There have been no thorough comparisons of the properties of devices made by the different methods.

12.2.2 Schottky Diode Using Metallic (CH)$_x$

Ozaki *et al.* (1979) polymerized acetylene directly onto the *n*-Si and subsequently doped the polymer to metallic levels by exposure to AsF_5. The

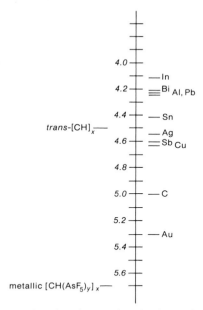

Fig. 12.7   Work functions for metals and polyacetylenes in volts.

current versus voltage $(I-V)$ curve shows the Schottky junction to be rectifying (Fig. 12.8) and fit well by the exponential form

$$I = I_s \exp(qV/nk_BT), \qquad (12.13)$$

where q is the electronic charge, n is the diode quality factor, and I_s is the saturation current

$$I_s = A^*T^2 \exp(-q\phi_b/k_BT). \qquad (12.14)$$

Using an effective Richardson's constant A^* of 264 A cm^{-2} K^2 for Si resulted in an estimate of 0.72 V for the barrier height ϕ_b and $n = 2.7$.

If p-Si is used instead of n-Si, the p-Si: metallic $[CH(AsF_5)_y]_x$ junction is ohmic. Therefore, $[CH(AsF_5)_y]_x$ is a relatively high-work-function metal. Ultraviolet photoemission studies (Salaneck *et al.*, 1979b) showed the work function to be about 5.7 eV.

Metallic polyacetylene also has been shown to form Schottky junctions with n-GaAs by direct polymerization of acetylene on the semiconductor surface (Ozaki *et al.*, 1979). The $I-V$ (Fig. 12.9) and $C-V$ (Fig. 12.10) characteristics showed that the device had a barrier height in the range 0.8–1.0 V.

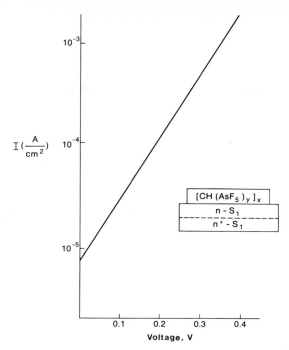

Fig. 12.8 $I-V$ characteristics of n-Si:metallic $[CH(AsF_5)_y]_x$ Schottky junction. [After Ozaki *et al.* (1979).]

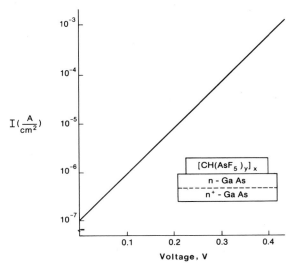

Fig. 12.9 $I-V$ characterization of n-GaAs:metallic $[CH(AsF_5)_y]_x$ Schottky junction: $n = 1.8$, $V_{bi} = 0.81$ V, $A^* = 144$ A cm^{-2} K^{-2}. [After Ozaki *et al.* (1979).]

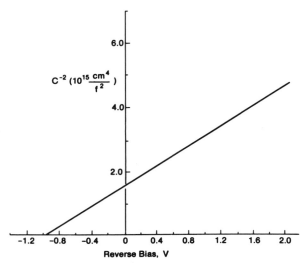

Fig. 12.10 $C-V$ characteristic of n-GaAs:metallic $[CH(AsF_5)_y]_x$. [After Ozaki *et al.* (1979).]

12.2.3   Schottky Diode Using Semiconducting (CH)$_x$

Undoped *trans*-polyacetylene or lightly doped semiconducting polyacetylene has a work function of ~ 4.5 eV, as indicated by photoemission studies (Salaneck *et al.*, 1979b). Therefore, it should form rectifying junctions with metals with smaller work functions, such as Sn, Bi, Al, Pb, and In (Fig. 12.7). This has been demonstrated, using the last three metals as well as with Na and Ba. On the other hand, metals with higher work functions, such as Cu, Au, and Pt, form ohmic contact with semiconducting polyacetylene.

Tani *et al.* (1979) have deposited polyacetylene film either just thermally isomerized or lightly doped with AsF$_5$ on a sapphire substrate and formed Schottky barriers by evaporation of a semitransparent In layer on the one side and Au on the other side for ohmic contact. The $I-V$ characteristic is shown in Fig. 12.11 (Grant *et al.*, 1981). The conductivity of polyacetylene was 8.4×10^{-3} (Ω cm)$^{-1}$ and the back-to-forward ratio was 140. For the purpose of comparison, the back-to-forward resistance ratio of an Au:n-Si Schottky diode is typically 10^{10}. The semilog plot of I versus V (Fig. 12.12) in the forward bias region shows a nearly exponential relationship according to Eq. (12.13). From the slope one finds $n = 1.98 \equiv (q/k_B T)\, \partial V/\partial(\ln I)$. At a forward bias potential greater than 0.3 V (see intercept on the V axis in Fig. 12.11 and deviation from the exponential relation in Fig. 12.12), the $I-V$

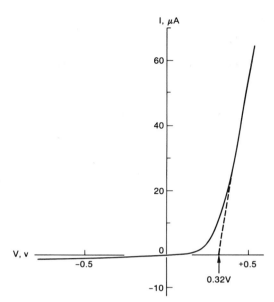

Fig. 12.11 *I–V* characteristic of a lightly doped $[CH(AsF_5)_y]_x$:In. [After Grant *et al.* (1981).]

characteristic becomes linear, indicating that the device resistance is now dominated by the roughly 3-kΩ bulk resistance of polyacetylene. Below 0.3 V, the linear part of the semilog plot (Fig. 12.12) yields $I_s = 5.4 \times 10^{-6}$ A cm^{-2}. Taking A^* to be 120 A K^{-2} cm^{-2}, one finds the value of ϕ_b at room temperature to be 0.74 V.

Capacitance measurements can provide useful information concerning the Schottky barrier and depletion layer. In the abrupt junction model, where the charge density within the depletion layer is assumed to be uniform, the square of the reciprocal of the differential junction capacitance varies linearly with reverse bias voltage V_{re} according to the relation

$$C^{-2} = 2(V_{bi} - V_{re})/(q\epsilon N), \tag{12.15}$$

where ϵ is the dielectric constant and N the impurity density. The $C–V$ characteristic of a lightly AsF$_5$-doped (CH)$_x$:In junction is shown in Fig. 12.13. The intercept gives the built-in potential $V_{bi} \approx 0.9$ eV, and the slope gives $N = 2.2 \times 10^{18}$ cm^{-3}. Together with the conductivity value of 10^{-4} (Ω cm)$^{-1}$, a hole mobility of only 3×10^{-4} cm^2 (V sec)$^{-1}$ is obtained. Also, the difference between the work function ϕ_s and the gap energy E_G leads to a value of 3.0 eV for the electron affinity X_s. The relation between these quantities, which describe the junction profile for a low-work-function

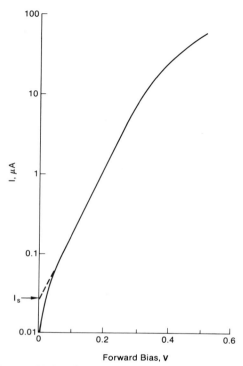

Fig. 12.12   log *I* versus *V* plot of the forward bias region of Fig. 12.11. [After Grant *et al.* (1981).]

metal to *p*-type polyacetylene for the ideal condition of no intervening surface states and under zero bias, is shown in Fig. 12.14.

The *C–V* relation was found by Tani *et al.* (1979) to exhibit frequency dependence. Grant and Krounbi (1980) cautioned that the complex impedance of the polyacetylene Schottky junction could display a spurious frequency dependence because of the high series resistance introduced by the polymer in the measurement of junction capacitance by ac measurements. If the frequency-dependent *C–V* results are interpreted literally, then the change of slopes would imply a frequency dependence for the carrier concentration. Furthermore, the frequency strongly affects the intercept and thus the built-in potential. Such behavior would indicate the presence of deep acceptor levels coexisting in the forbidden gap with shallow levels.

Grant *et al.* (1981) have also measured the *I–V* and *C–V* properties of undoped and oxygen-doped *trans*-polyacetylene and found hole concentrations of 6×10^{16} and 8.9×10^{16} cm^{-3}, respectively, and corresponding

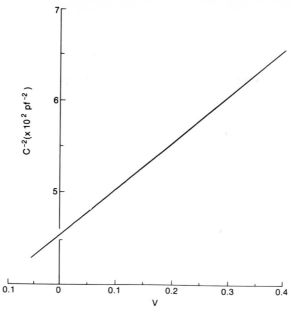

Fig. 12.13 $C-V$ characteristic of the sample in Fig. 12.11. [After Grant *et al.* (1981).]

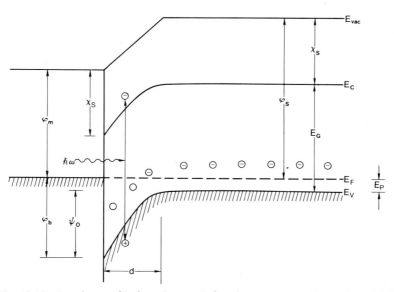

Fig. 12.14 Junction profile for a low work function to p-type polyacetylene Schottky contact under zero bias with no intervening surface states. [After Grant *et al.* (1981).]

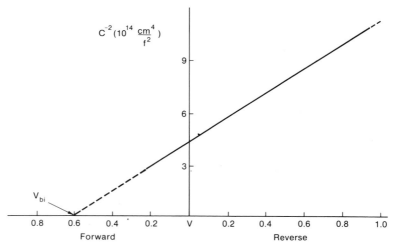

Fig. 12.15 $C-V$ characteristic of Al: *trans*-(CH)$_x$ Schottky diode. [After Weinberger *et al.* (1982).]

mobility values of 3×10^{-6} and 2×10^{-5} cm^2 (V sec)$^{-1}$. The estimates for carrier concentration in undoped polyacetylene by heterojunction (Ozaki *et al.*, 1980) is about 2×10^{18} cm^{-3}. Similar values were obtained by other techniques. The undoped *trans*-polyacetylene in the study of Grant *et al.* (1981) seems to have the unusually low conductivity of 3×10^{-8} (Ω cm)$^{-1}$, whereas most workers found it to be $\sim 10^{-5}$ (Ω cm)$^{-1}$. This raises the question of whether the polymer was either incompletely isomerized or heated too long in the thermal isomerization. However, see below about measurements that support the low-hole concentration.

Al: *trans*-(CH)$_x$ and Pb: *trans*-(CH)$_x$ junctions were studied by Weinberger *et al.* (1982). Figure 12.15 presents the $C-V$ characteristic of the former. The plot gives $V_{bi} \approx 0.6$ eV and $N = 3.3 \times 10^{16}$ cm^{-3}. Thus the carrier concentration is in agreement with that given by Grant *et al.* (1981). Apparently, the $C-V$ technique measures a lower carrier concentration than the other techniques.

Polyacetylene doped by dipping the film in HCl–methanol solution was used for the *p*-type semiconductor to form a Schottky junction with Al by Tsukamoto *et al.* (1981). The log I versus V plot is linear from 0 to 0.8 V forward bias, and the diode quality factor ranges from 3 to 4. The authors estimated a value of $V_{bi} = 1.1$ eV from the extrapolation of the straight portion of the $I-V$ curve. This value is about twice that of the undoped *trans*-(CH)$_x$: Al junction by Weinberger *et al.* (1982).

12.2.4 Deviation for Schottky Behaviors

Although the metal:p-type $(CH)_x$ and the semiconductor:metallic $[CH(AsF_5)_y]_x$ form blocking junctions that obey the simple thermionic model relations of Eqs. (12.13) and (12.14), there are indications that these devices are not simple Schottky diodes. For instance, in the case of a lightly doped $[CH(AsF_5)_y]_x$:In junction, with the work function of In being $\phi_{In} = 4.1$ eV, the barrier height is $\phi_b = E_G - (\phi_{In} - X_s) = 0.4$ eV. Thus this value is different from that obtained from an $I-V$ plot, which is 0.72 eV. Moreover, both values of ϕ_b are smaller than V_{bi}, as determined from $C-V$ measurements. According to the thermionic model (Fig. 12.14), ϕ_b should be larger than ϵV_{bi}.

The theoretical diode quality factor is $n = 1$. However, the values obtained from the $I-V$ measurements are 1.98 or greater. Departures from the theoretical $n = 1$ are often caused by extensive recombinations in the depletion region. Under such conditions, the thermionic emission model cannot be used to analyze for the barrier height. Even in those Schottky junctions where $n \approx 1$, it is often found that ϕ_b is independent of the metal used for the contact. The properties of Schottky barriers are strongly influenced by the surface states that affect the position of the Fermi level vis-à-vis the metal and semiconductor band edges.

It had been pointed out above that the forward-to-back resistance ratios of polyacetylene Schottky junctions are many orders of magnitude smaller than those of inorganic semiconductor–metal contacts. Furthermore, it has been noticed by Grant *et al.* (1981) that there was a gradual decrease in back resistance in time in measurements under reverse bias. The junctions lost much of their rectifying property after an hour or longer. This decrease in resistance was more pronounced, the higher the doping level. After returning to zero bias for about the same length of time, normal junction properties were recovered. A plausible explanation for this behavior is that negative charge is leaking into the depletion region as a function of time under reverse bias. A possible source of negative charge is the weakly bound dopant anion.

According to Eq. (12.13), there should be a rapid increase in the slope of $\ln I$ versus V with decreasing temperature along with a sharp decrease in I_s. This was found to be not the case by Weinberger *et al.* (1982) for the Pb:*trans*-$(CH)_x$ junction. The linear region of the $\ln I$ versus V curve is clearly in evidence (Fig. 12.16). However, n deviates greatly from unity and the slope of the plots, $d(\ln I)/dV$ is virtually independent of temperature. According to the value $\phi_b = 0.36$ eV determined from $C-V$ data and Eq.

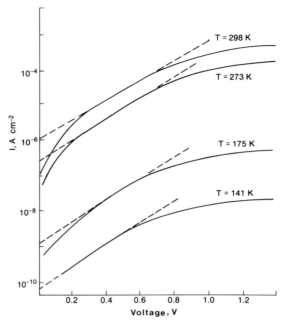

Fig. 12.16 *I–V* characteristics for Pb:*trans*-(CH)$_x$ Schottky diode at various temperatures. [After Weinberger *et al.*(1982).]

(12.14), there should be a pronounced temperature effect. Furthermore, the
front-to-back rectification ratio decreased measurably with decreasing temperature.

The carrier mobilities as estimated from *C–V* and conductivity data are
very low. From elementary considerations based on band theory one would
expect the hole mobility to be of the order of 1 cm² (V sec)⁻¹ and independent of concentration. Park *et al.* (1980) and other workers who used direct
analysis of conductivity data as a function of dopant concentration concluded that the mobility is concentration depenent.

There are many possible reasons for the deviation from simple
thermionic emission model for Schottky diodes containing polyacetylene. A
few of these are listed below.

(1) In some disordered semiconductors, such as amorphous and polycrystalline solicon, anomalously low and concentration-dependent mobilities were observed. These systems have high density of localized states in the
forbidden gap that can trap or scatter carriers at low dopant concentration.
As carrier concentration increases, these localized states fill up until none

are left to trap the carriers. Consequently, additional carriers are able to occupy the more extended states, resulting in a sudden increase in carrier mobility.

(2) There may be tunneling by means of field or thermionic-field emission. Local field anomalies due to edge effects and the fibrous morphology of polyacetylene could be responsible.

(3) The net mobility could be an average over various inhomogeneous dopant distributions in doped polyacetylenes. If the polymer fibrils were doped more at the exterior than in the interior, then the measured mobility would be the average of the conducting outer sheath and the insulating interior.

(4) The mechanism of conduction in lightly doped or undoped polyacetylene may be by means of intersoliton hopping, which would be very different from the usual mechanisms of conduction in an amorphous semiconductor, such as variable-range hopping.

Very unusual behaviors were observed by Tsukamoto *et al.* (1982) for the Al–HCl-doped $(CH)_x$–Au device. They made the device by forming the Schottky barrier on the dull side of the polyacetylene film (**D**) and the shiny surface of the film (**S**) and found different I–V characteristics, as shown in Figs. 12.17 and 12.18, respectively. The saturation current of **D** is 10^{-8} A

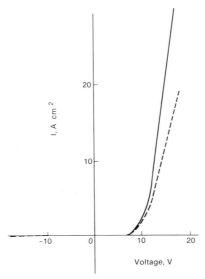

Fig. 12.17 I–V characteristics of Al–HCl-doped $(CH)_x$ Schottky cell: (———) device **D**; (---) device **S** with forward bias. [After Tsukamoto *et al.* (1982).]

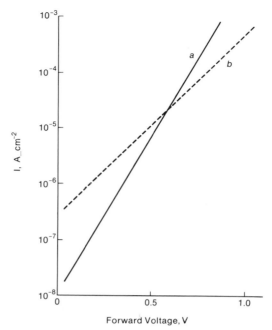

Fig. 12.18   log I versus V plot for the data in Fig. 12. [After Tsukamoto *et al.* (1982).]

cm^{-2}; it is about 10^{-7} A cm^{-2} for **S**. The diode quality factor range is 3–4 and the **D** device has a smaller n.

The authors also studied the $I-V$ characteristics, using a reverse bias. The results are shown in Fig. 12.19. The plot of log I_{re} versus $(V_{re} + V_D - k_B T/e)^{1/4}$ is linear for device **D** but not for **S**. Conversely, the plot of log I_{re} versus V_{re} is linear for device **S** but not for **D**. V_D is the diffusion voltage and is assumed to be 1 V. Tsukamoto *et al.* (1982) suggested that there may be barrier lowering due to image force in device **D** but tunneling occurring through thin barrier in device **S**.

There were significant differences between the two surfaces of polyacetylene in the $C-V$ characteristics of these diodes. The data obtained at 1 kHz are shown in Fig. 12.20. The C^{-2} versus reverse bias plot is linear with a large slope for device **D**, giving a small impurity concentration of 7.4 × 10^{16} cm^{-3}. However, the plot bends over on the side of forward bias. On the other hand, in the plot for device **S**, C^{-2} is virtually independent of bias. The very small slope gave a large impurity concentration of 1.4 × 10^{18} cm^{-3}. It almost appears as though there is no intercept on the V axis and thus a very large or indefinite built-in voltage.

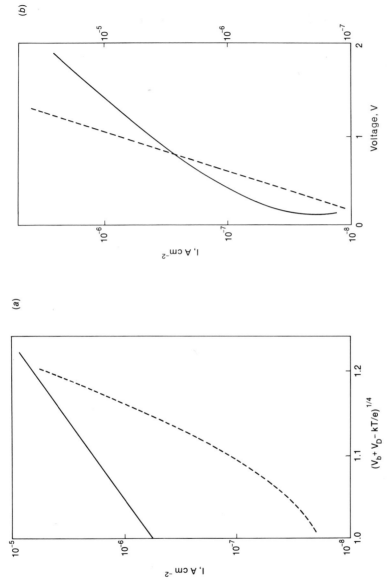

Fig. 12.19 The I_{re}–V characteristics of Al–HCl-doped $(CH)_x$ Schottky cell under reverse bias: (——) device **D**; (----) device **S**. (a) plots of $\log I_{re}$ versus $(V_{re} + V_D - kT/e)^{1/4}$; (b) plots for $\log I$ versus V_{re}. [After Tsukamoto et al. (1982).]

The authors claimed that reflection X ray showed the shiny surface side of the polyacetylene to be less than 10% crystalline, whereas the dull surface has the usual 90% crystallinity. This was used as justification that the shiny surface is in an almost amorphous state, which results in the formation of more defects acting as trapping or recombination centers.

The above results are very unusual. The shiny side of the polyacetylene film is next to the reactor wall so that the growth of the polymer fibril is constrained and becomes flattened. On the other side, the fibrils were able to grow freely, but the surface is rich in voids. However, by merely rubbing the surface it becomes shiny as the fibrils are compacted and the voids covered up. If the shiny side is as low in crystallinity as claimed, then this amorphous layer must be so thin that it does not reduce the average crystallinity of 80% of the film, as measured by X-ray diffraction.

Some physical and chemical complications can be suggested. A vacuum-deposited layer of Al may be very different for the more regular shiny surface than for the irregular dull surface. The latter may not be a continuous film and/or it may be of uneven thickness. Because the shiny side is attached to the reactor wall, it may be that Tsukamoto *et al.* (1981, 1982) did not wash it thoroughly enough. Any inorganic catalyst residue will be transformed to semiconducting TiO_2 during thermal isomerization. If this TiO_2 is amorphous, as is likely, it will tend to give halos in reflection X-ray measurements. Finally, the nature of the HCl dopant is unknown and may contribute carriers or traps. It seems necessary that the above systems be studied further.

12.2.5   Heterojunctions

Heterojunctions have been made from *p*-type *trans*-polyacetylene and *n*-type ZnS and CdS. The devices behave somewhat differently depending on whether the polymer is grown on the inorganic semiconductor surface or the $p-n$ junction is made by pressure contacts. Figure 12.21 shows rectifying behavior characteristic of a $p-n$ junction diode, thus confirming as-grown polyacetylene film to be *p*-type. The $C-V$ data are given in Fig. 12.22 for CdS with two levels of doping. The intercept gives a very large V_{bi} of 1.9 eV and the slope is 6.7×10^{13} cm^4 (F^2 V)$^{-1}$ for 10-ppm doped CdS. For the 50-ppm doped CdS, $V_{bi} = 1.8$ eV and the slope was found to be 1.2×10^{13} cm^4 (V P^2)$^{-1}$. The slope corresponds to

$$\frac{d(C^{-2})}{dV} = 2\frac{\epsilon_p N_A + \epsilon_n N_D}{q N_A N_D \epsilon_p \epsilon_n},$$

(12.16)

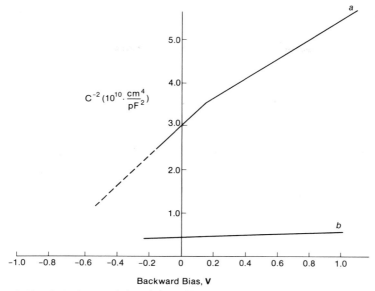

Fig. 12.20 $C-V$ characteristics of Al–HCl-doped (CH)$_x$ Schottky cell. (——) device **D**; (---) device **S**. [After Tsukamoto *et al.* (1982).]

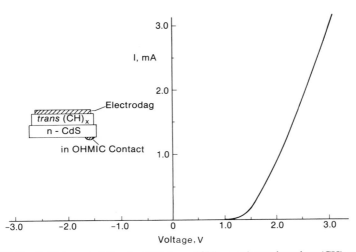

Fig. 12.21 $I-V$ characteristic of *n*-CdS : *trans*-(CH)$_x$ *p–n* heterojunction; (CH)$_x$ polymerized directly on CdS surface. [After Ozaki *et al.* (1980).]

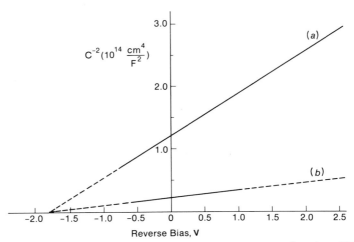

Fig. 12.22   *C–V* characteristics of *n*-CdS : *trans*-(CH)$_x$ *p–n* heterojunction; (CH)$_x$ polymerized directly on CdS surface: (a) CdS doped to 10 ppm; (b) CdS doped to 50 ppm. [After Ozaki *et al.* (1980).]

where $\epsilon_p(\epsilon_n)$ is the dielectric constant of the *p*-type (*n*-type) semiconductor and N_A (N_D) is the acceptor (donor) dopant concentration. For polyacetylene, $\epsilon_p = (\frac{1}{3}\epsilon_\parallel + \frac{2}{3}\epsilon_\perp) \simeq 6$, for CdS, $\epsilon_n = \epsilon_0$ (ϵ_0 is the vacuum permittivity). The acceptor concentration becomes

$$N_A = \frac{2\epsilon_n N_D}{[(d(C)^{-2}/dV)N_D\epsilon_n - 2]\epsilon_p}. \tag{12.17}$$

Using experimental values of slope of the 10-ppm doped CdS in Fig. 12.18, we find $N_A = 1.6 \times 10^{18}$ cm^{-3} for the polyacetylene films.

At an impurity concentration of 50 ppm, CdS may no longer be characterized as a nondegenerate semiconductor with the dielectric constant of the intrinsic material. Furthermore, $\epsilon_n N_D \gg \epsilon_p N_A$, and the heterojunction Eq. (12.16) is reduced to the Schottky formula

$$N_A = 2\left[q\epsilon_p \frac{d(C)^{-2}}{dV} \right]^{-1}. \tag{12.18}$$

The experimental data in Fig. 12.18 give $N_A = 2 \times 10^{18}$ cm^{-3}. The values of N_A obtained here are in agreement with the results of analysis of the thermopower data, of conductivity versus dopant concentration results, and are as expected for the intersoliton model. The values are about two orders

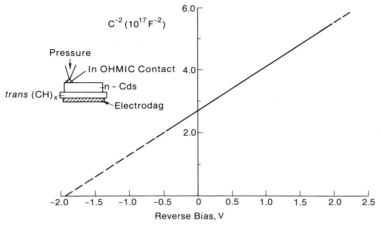

Fig. 12.23 $C-V$ characteristic of a pressure contact n-CdS : $trans$-(CH)$_x$ heterojunction. [After Ozaki *et al.* (1980).]

of magnitude greater than those obtained from the $C-V$ data of metal : (CH)$_x$ Schottky junctions.

The $C-V$ characteristic of a heterojunction produced by a pressure contact between n-CdS and a freestanding *trans*-(CH)$_x$ film, shown in Fig. 12.23, gave the name 1.94 eV for the built-in potential. However, the slope is different from that of a grown polyacetylene $p-n$ junction. This was attributed to an unusually good and consequently large contact area for the mechanically soft polyacetylene.

12.3 SOLAR CELLS

The photovoltaic responses seem to differ with different junction devices.

12.3.1 Schottky Diodes

Tani *et al.* (1980) reported the phototransport effects of a device with light incident on a sapphire substrate through a blocking metal layer and

Wavelength, μm

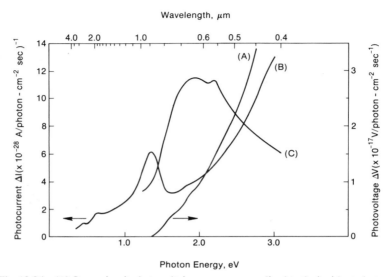

Photon Energy, eV

Fig. 12.24   (A) Open-circuit photovoltaic response normalized to the incident photon-flux density and the transmittance of the In-blocking contact with sapphire substrate. (B) Photo-conductivity response similarly normalized. (C) Absorbance of undoped *trans*-(CH)$_x$ at 298 K. [After Tani *et al.* (1980).]

then to the polyacetylene film backed by an ohmic metal. The open-circuit photovoltaic response was shown in Fig. 12.24. The conductivity of the polyacetylene lightly doped with AsF$_5$ was 5.7×10^{-3} $(\Omega \text{ cm})^{-1}$ ($y \approx 6 \times 10^{-3}$). The thickness of the depletion region (d in Fig. 12.14) was estimated at 10–60 nm from C–V measurements. The photoresponse has a definite threshold at 1.48 eV, which grows with increasing incident photon energy. The absorption spectrum of undoped *trans*-polyacetylene, also included in the figure, shows close correspondence between the absorption edge and onset of photovoltage. The latter was relatively insensitive to the doping level. Therefore, 1.48 eV represents the single-particle band gap in *trans*-polyacetylene. An important fact is that the photovoltage (PV) threshold is well below the maximum in absorption.

   For low light intensities, the PV response should vary linearly with the density of the free-electron–hole pairs generated by the absorption coefficient. Recent measurements indicate that the PV peaks at 3 eV and decreases thereafter.

   Typical photocurrents near the threshold were 9×10^{-12} Å. The dark current background was 1.2×10^{-6} Å. The photocurrent–dark current ratio was $\sim 8 \times 10^{-6}$. With the particular geometry of the device, the total photocurrent I_{pc} is

$$I_{pc} = 2\mu\tau e V p W/L, \tag{12.19}$$

where μ is the carrier mobility, τ the inverse trapping rate, V the voltage drop across the sample, p the incident photon-flux density in photons (cm^2 sec)$^{-1}$, W the width of the illuminated area, and L its length between the electrodes. The quantum yield is defined by

$$G = I_{pc}/(eNWL) = 2\mu\tau V/L^2. \tag{12.20}$$

The light source was calibrated at 8×10^{15} photons (cm^2 sec^{-1}), we find $G \approx 3 \times 10^{-8}$ carrier per absorbed photon. The quantity $\tau\mu$ is 2×10^{-10} cm^2 V^{-1}, which is to be compared with 10^{-5} for CdS and unity for Si.

The photoconductivity response displays three distinct features (Fig. 12.24b). There is a peak at 1.35 eV, which is absent in the PV spectrum, there is the steep increase in the high-energy region similar to the PV response, and the background is relatively large, which seems to be proportional to the incident photon energy after normalization to the incident photon flux but is not related to the dark dc. The last was attributed to the thermal modulation of the sample resistance by the incident light, which reaches about 0.01 W cm^{-2} in the near infrared region. The mean temperature rise is estimated at 3×10^{-5} K. Correction of the thermal effect yields a threshold of about 1.5 eV, as it should be. The peak at 1.35 eV was suggested by Tani *et al.* (1980) to represent dissociation of a weakly bound exciton. The efficiency of the diode is quite low, being roughly 0.002% under 1 sun illumination.

Weinberger *et al.* (1981) evaporated a thin Al layer (6% transmission) onto a free standing *trans*-polyacetylene film and ohmic contact was made with Electrodag. The device was immediately transferred to a glass chamber evacuated to less than 10^{-3} torr. The I–V characteristics of a 1-cm^2 Al:(CH)$_x$ Schottky junction in the dark and light are shown in Fig. 12.25. The dark I–V curve shows good rectifying behavior. In the photovoltaic response spectrum (Fig. 12.26) there is a knee in the range 1.5–1.8 eV which may be associated with the optical creation of electron–hole pairs by excitation across the polyacetylene band gap. The region between 1.0 and 1.5 eV is exponential. The exponential edge and Urbach tail below 1.0 eV may be disorder-induced band tailing. But more probably it is due to excitation into midgap states generating soliton–antisoliton pairs for which the threshold energy is $2E_G/\pi$. The dependence of photocurrent on photon energy about E_G may be due to the probability of the photoexcited electron–hole pairs escaping their mutual Coulombic attraction and increasing with the energy imparted to them by the exciting photon.

For the device whose response spectrum is shown in Fig. 12.22, the

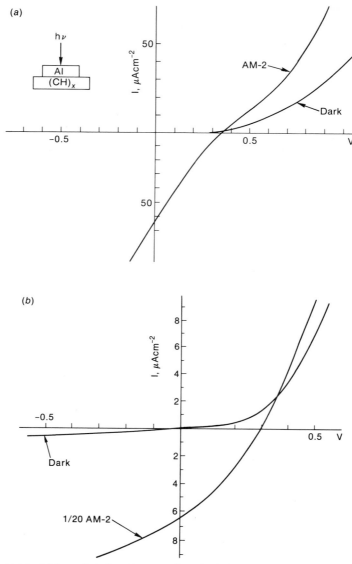

Fig. 12.25   Light and dark $I–V$ curves for Al:(CH)$_x$ Schottky cell: (a) AM-2 (70 mW cm^{-2}); (b) 0.05 AM-2 (3.5 mW cm^{-2}). [After Weinberger *et al.* (1981).]

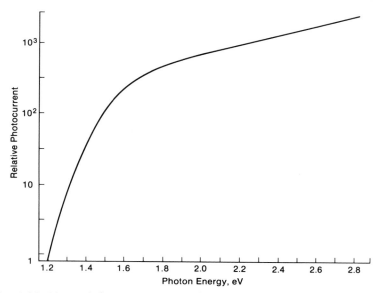

Fig. 12.26   Photovoltaic response spectrum for Al:(CH)$_x$ Schottky cell. [After Weinberger *et al.* (1981).]

absolute quantum efficiency defined as I_{sc} per photon was measured as 10% at 1.95 eV, using He–Ne light of 0.2 mW cm^{-2} intensity. The illuminated $I–V$ curves were shown in Fig. 12.21 under the simulated solar irradiation AM-2 (70 mW cm^{-2}). A short-circuit current of 63 μA cm^{-2}, a V_{oc} of 0.31 V, and a maximum power point $V_{mp}I_{mp} = 0.21\ V_{oc}I_{sc}$ were obtained. In terms of the light actually reaching the polyacetylene depletion layer, the internal conversion efficiency was estimated to be 0.1%. At 0.05 AM^{-2}, $I_{sc} = 6\ \mu$A cm^{-2}, $V_{oc} = 0.3$ V, fill factor $= 0.34$, and an enhanced internal conversion efficiency of 0.3% was obtained. The dependence of I_{sc} with light intensity is shown in Fig. 12.27.

For photovoltaic studies, Tsukamoto *et al.* (1982) concentrated on the diodes with Al in contact with the dull side of the polyacetylene film, i.e., device **D**. The relative collection efficiency Q as a function of photon energy is shown in Fig. 12.28; the photon flux is $\sim 10^{14}$ (cm^2 sec)$^{-1}$. It decreases rapidly below 1.4 eV and tends to be saturated above 2.5 eV. The absolute collection efficiency at 2 eV is estimated to be about 4%.

The photocurrent was increased by reverse bias, as observed in other organic solar cells. The increase is wavelength dependent. At 0.7 V of bias, the photocurrent was increased tenfold for 630-nm light; it is a sixfold increase for 450-nm radiation (Fig. 12.29). The variations of I_{sc} and V_{oc} with incident light intensity are shown in Fig. 12.30. I_{sc} was found to vary with

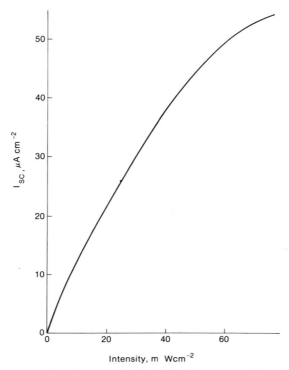

Fig. 12.27 I_{sc} versus light intensity for Al:(CH)$_x$ Schottky cell. [After Weinberger *et al.* (1981).]

(intensity)$^{0.78}$. At an incident light intensity of 7 mW cm^{-2}, $V_{oc} \cong 0.32$ V, $I_{sc} \cong 35$ μA cm^{-2}, and a fill factor $\cong 0.26$ was obtained. The value of V_{oc} is in agreement with the results of Weinberger *et al.* (1982); however, the I_{sc} is about three times greater at comparable light intensity.

The results of Fig. 12.30 suggest that the relation for conventional solar cells

$$V_{oc} = n'k_B T[e \ln(I_{sc}/I_0)] \qquad (12.21)$$

also applies to the polyacetylene photodiodes. The value of n' was found to be 1.4 from a plot of I_{sc} versus V_{oc}. The value is much smaller than the value of $n \approx 3-4$ obtained from the slope of the $I-V$ plot. The difference may result from the holes flowing toward bias while electrons are dominant carriers in the short-circuit current.

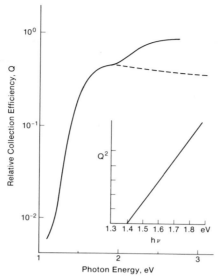

Fig. 12.28   Relative collection efficiency, Q, of Al–HCl-doped $(CH)_x$ solar cell. The dashed line is the absorption spectrum normalized at 1.8 eV. The inset shows a plot of Q^2 versus $h\nu - E_G$. [After Tsukamoto *et al.* (1982).]

12.3.2   Heterojunction

The relative collection efficiency for the *trans*-$(CH)_x$:n-CdS (10 ppm) $p-n$ heterojunction is shown in Fig. 12.31. It appears to be far less efficient than the Al–HCl-doped $(CH)_x$ shown in Fig. 12.28. Ozaki *et al.* (1980) suggested that the photocurrent may be limited to a conduction-band spike at the interface due to the differences in electron affinity of CdS and polyacetylene. The spike, illustrated in the inset of Fig. 12.31, would pose a barrier to collection of photogenerated minority carrier (electrons) from $(CH)_x$ to CdS. Alternatively, the relatively low Q may indicate primary exciton formation.

If the collected current results only from carriers generated near the junction and carried to the interface by the internal electric field, then I_{sc} will be proportional to $E\mu\tau$ or d, whichever is larger, where E is the electric field, μ the mobility and τ is the minority-carrier lifetime. Then there should be an increase in photoresponse by reverse biasing the junction. Figure 12.32 shows that at $V_{re} = 2$ V, and the photocurrent begins to increase for light of 625-nm wavelength. A maximum of about a threefold increase occurs for

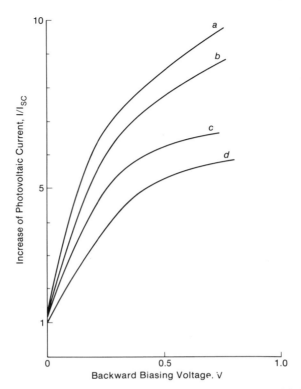

Fig. 12.29   Variation of ratio of monochromatic photocurrent to I_{sc} with backward bias voltage: (a) 450 nm; (b) 500 nm; (c) 560 nm; (d) 630 nm. [After Tsukamoto *et al.* (1982).]

$\lambda \approx 543$ nm. However, there is very little effect for $V_{re} = 1$ V. This is in contrast to the large increase of up to sixfold with $V_{re} = 0.7$ V for the Al–HCl-doped $(CH)_x$ solar cell.

12.3.3   Limitations

The polyacetylene solar cell devices suffer from instability toward oxygen. For instance, exposure of the Al : *trans*-$(CH)_x$ photodiodes causes large reductions in photocurrent and fill factor (Table 12.5). On the other hand, the dark diode parameters R_s and V_{bi} were unaffected by oxygen. The reduced collection efficiency was attributed to the release of oxygen at the

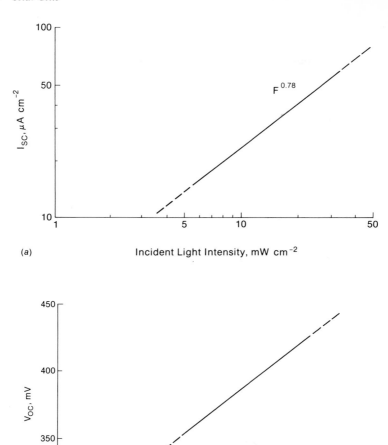

(a)

(b)

Fig. 12.30   Characteristics of Al–HCl-doped (CH)$_x$ solar cell. (a) log I_{sc} versus log intensity; (b) V_{oc} versus log intensity. [After Tsukamoto *et al.* (1982).]

photocarrier recombination center. Tsukamoto *et al.* (1982) found that by keeping their diode under flowing N$_2$, there was a rapid initial decrease of photocurrent. But after two days the I_{sc} had dropped 40% and it was stabilized for many days.

The source of the low conversion efficiency and the saturation of the

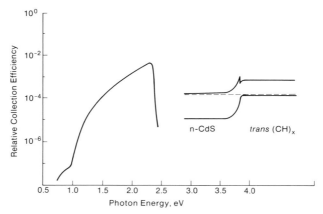

Fig. 12.31 Photoresponse of *trans*-(CH)$_x$:n-CdS (10 ppm) p–n heterojunction. [After Ozaki *et al.* (1980).]

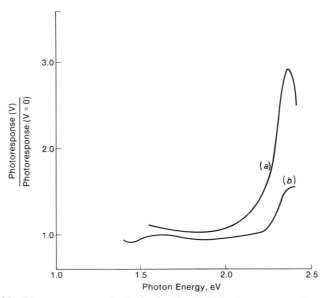

Fig. 12.32 Photoresponse to the device in Fig. 12.31 in the presence of reverse bias: (a) $V_{re} = -1$ V; (b) $V_{re} = -2$ V. [After Ozaki *et al.* (1980).]

TABLE 12.5

Effect of Oxygen on Al:*trans*-(CH)$_x$ Junction Parameters

Oxygen pressure (torr)	R_s (kΩ)	V_{bi} (V)	AM-2 I_{sc} (μA)	0.05 AM-2 fill factor
0	3.4	0.98	63	0.34
2	2.8	1.06	4	0.18
30	2.2	1.03	0.8	—

conversion efficiency at appreciable light levels must be identified and overcome if commercially viable polyacetylene-based photovoltaic collectors are to be developed. Ozaki *et al.* (1980) have studied deep trapping states in polyacetylene. Figure 12.33 shows the photoresponse spectrum of a *trans*-(CH)$_x$:*n*-CdS (10 ppm) before and after exposing the junction to radiation of 2.0 eV. In the initial curve the junction was irradiated with light

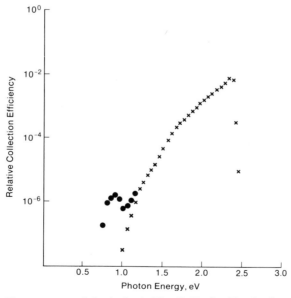

Fig. 12.33 Photoresponse of the device in Fig. 12.31 after illumination with $\hbar\omega \geqslant$ (●) 1.4 eV, (×) after illumination with $\hbar\omega = 0.9$ eV light for 20 min. [After Ozaki *et al.* (1980).]

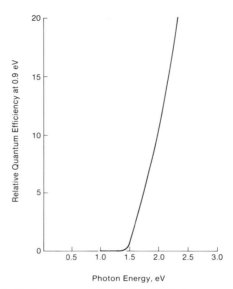

Fig. 12.34 Threshold photon energy for the 0.9-eV photoresponse peak. [After Ozaki *et al.* (1980).]

of $\geqslant 1$ eV. A response peak at 0.9 eV is developed. Exposure of this sample to 0.9-eV radiation bleaches the photoresponse at that energy. To obtain the threshold energy to form the 0.9-eV photoresponse peak, the junction is irradiated with photon energy > 1.0 eV, and the photoresponse at 0.9 eV is recorded. Figure 12.34 shows that the threshold is at 1.45 ± 0.05 eV. The 0.9-eV photoresponse peak decays very slowly in the dark (Fig. 12.35a) with a characteristic dark decay time of about 4.5 hr. Under illumination with 3.6×10^{15} photons $(cm^2 \ sec)^{-1}$ of 1390-nm wavelength, the decay time was about 40 sec. These results indicate that there is a metastable deep trapping in or near the junction. The traps are efficiently filled by pumping electron–hole pairs with light of energy $\geqslant E_G$. It seems that the deep traps are the soliton midgap states that are intrinsic to polyacetylene. There seems to be no way to ameliorate this barrier.

Generally, in low-efficiency solar cells, high-series resistance R_s is the limiting factor for the conversion efficiency. R_s can be evaluated from the amount of current that can be drawn externally with a large shunt resistance from the photogenerated current. The relationship under short-circuit conditions is

$$\ln\{[(I_p - I_{sc})/I_0] + 1\} = (l/k_B T)R_s I_{sc}, \qquad (12.22)$$

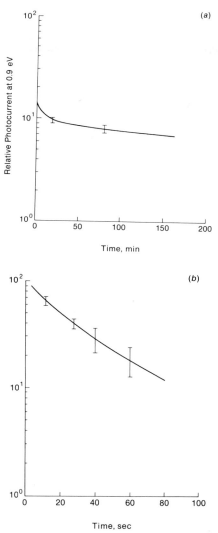

Fig. 12.35   Decay of the 0.9-eV photoresponse peak: (a) in the dark; (b) in the presence of 1290-nm light. [After Ozaki *et al.* (1980).]

where I_p is the photogenerated current. The typical R_s for a polyacetylene diode is about several ohms from the $I-V$ characteristics in the dark. This series resistance is not so high as to reduce I_{sc} significantly. So long as $\exp(e/k_B T)R_s I_{sc} \cong 1$, series resistance is not a limiting factor.

References

Akaishi, T., Miyasaka, K., Ishikawa, K., Shirakawa, H., and Ideda, S. (1980). *J. Polym. Sci. Polym. Phys. Ed.* **18,** 745.

Aldissi, M., Schué, F., Giral, L., Rolland, M. (1982b). *Polymer* **23,** 246.

Allen, W. N., Decorpo, J. J., Saalfeld, F. E., Wyatt, J. R., and Weber, D. C. (1979). *Synth. Met.* **1,** 371 (1979).

Anderson, P. N. (1958). *Phys. Rev.* **109,** 1492.

Anderson, L. R., Pez, G. P., and Hsu, S. L. (1978). *J. Chem. Soc. Chem. Commun.* p. 1066.

Audenaert, M., Gusman, G., and Deltour, R. (1981). *Phys. Rev. B* **24,** 7380.

Austin, I. G., and Mott, N. F. (1969). *Adv. Phys.* **18,** 41.

Ball, R., Su, W. P., and Schrieffer, J. R. (1982). *Int. Conf. Phys. Chem. Conducting Polym., Les Arc, France, December.*

Bartlett, N., Biagioni, R. N., McQuillan, R. W., Robertson, A. S., and Thompson, A. C. (1978). *J. Chem. Soc. Chem. Commun.* 200.

Baughman, R. H., Hsu, S. L., Pez, G. P., and Signorelli, A. J. (1978). *J. Chem. Phys.* **68,** 5405.

Baughman, R. H., Shacklette, L. W., Chance, R. R., Ivory, D. M., Miller, G. G., Preziosi, A. F., and Lahav, M. (1980). *Bull. Am. Phys. Soc.* **25,** 314.

Bawn, C. E. H., and Symcox, R. (1959). *J. Polym. Sci.* **34,** 139.

Bechgaard, K., Jacobsen, C. S., Morlensen, K., Pederson, H. J., and Thorup, N. (1980). *Solid State Commun.* **33,** 1119.

Belinsky, A. J. (1976). *Contemp. Phys.* **17,** 331.

Benoit, C., Rolland, M., Aldissi, M., Rossi, A., Cadene, M., and Bernier, P. (1981). *Phys. Status Solidi* **68,** 209.

Berlin, A. A., and Cherkashin, M. I. (1971). *Vysokomol. Soedin. Ser. A* **13,** 2298.

Bernier, P., Rolland, M., Linaya, C., and Disi, M. (1980). *Polymer* **21,** 7.

Bernier, P., Schué, F., Sledz, J., Rolland, M., and Giral, L. (1981a). *Chem. Scr.* **17,** 151.

Bernier, P., Linaya, C., Rolland, M., and Aldissi, M. (1981b). *J. Phys. (Paris) Lett.* **42,** L-295.

Blanchet, G. B., Fincher, C. R., Chung, J. C., and Heeger, A. J. (1983). *Phys. Rev. Lett.* **50,** 1938.

Bloor, D., Ando D. J., Preston, F. H., Stevens, G. C. (1974). *Chem. Phys. Lett.* **24,** 407.

Blumberg, M., Boss, C. R., and Chien, J. C. W. (1965). *J. Appl. Polym. Sci.* **9,** 3837.
Brédas, J. L. (1979). Ph.D. Thesis, Facultés Universitaires de Namur, Belgium.
Brédas, J. L., Chance, R. R., Silbey, R. (1981). *J. Phys. Chem.* **85,** 756.
Brédas, J. L., Chance, R. R., and Silbey, R. (1982). *Phys. Rev. B* **26,** 5843.
Brenig, W., Dohler, G., and Heyszenau, H. (1973). *Philos. Mag.* **27,** 1093.
Buravov, L. I., and Shchegolev, I. F. (1971). *Prib. Tekh. Eksp.* **2,** 171.
Bryant, G. W., and Glick, A. J. (1982). *Phys. Rev. B* **26,** 5855.
Campbell, D. K., and Bishop, A. R. (1981). *Phys. Rev. B* **24,** 4859.
Cao, Y., and Qian, R. (1982). *Makromol. Chem. Rapid Commun.* **3,** 687.
Chance, R. R., Shacklette, L. W., Miller, G. G., Ivory, D. M., Sowa, J. M., Eisenbaumer, R. L., and Baughman, R. H. (1980). *J. Chem. Soc. Chem. Commun.* p. 349.
Chiang, C. K. (1981). *Polymer* **22,** 1454.
Chiang, C. K., Fincher, C. R., Jr., Park, Y. W., Heeger, A. J., Shirakawa, H., Louis, E. J., Gau, S. C., and MacDiarmid, A. G. (1977). *Phys. Rev. Lett.* **39,** 1098.
Chiang, C. K., Gau, S. C., Fincher, C. R., Jr., Park, Y. W., MacDiarmid, A. G., and Heeger, A. J. (1978a). *Appl. Phys. Lett.* **33,** 18.
Chiang, C. K., Park, Y. W., Heeger, A. J., Shirakawa, H., Louis, E. J., and MacDiarmid, A. G. (1978b). *J. Chem. Phys.* **69,** 5098.
Chiang, C. K., Druy, M. A., Gau, S. C., Heeger, A. J., Louis, E. J., MacDiarmid, A. G., Park, Y. W., and Shirakawa, H. (1978c). *J. Am. Chem. Soc.* **100,** 1013.
Chien, J. C. W. (1959). *J. Am. Chem. Soc.* **81,** 86.
Chien, J. C. W. (1963a). *J. Polym. Sci. Polym. Chem. Ed.* **1,** 425.
Chien, J. C. W. (1963b). *J. Polym. Sci. Polym. Chem. Ed.* **1,** 1939.
Chien, J. C. W. (1968). *J. Polym. Sci. Polym. Chem. Ed.* **6,** 375.
Chien, J. C. W., ed. (1975). "Coordination Polymerization." Academic Press, New York.
Chien, J. C. W. (1979a). *J. Polym. Sci. Polym. Chem. Ed.* **17,** 2555.
Chien, J. C. W. (1979b). *Polym. News* **6,** 52.
Chien, J. C. W. (1980). *In* "Preparation and Properties of Stereoregular Polymers" (R. W Lenz and F. Ciardelli, eds.), pp. 113–130. Reidel, Boston, Massachusetts.
Chien, J. C. W. (1981). *J. Polym. Sci. Polym. Lett. Ed.* **19,** 249.
Chien, J. C. W. (1982). *In* "Macromolecules" (H. Benoit and P. Rempp, eds.), p. 233. Pergamon, New York.
Chien, J. C. W., and Babu, G. N. (1983). Unpublished results.
Chien, J. C. W., and Boss, C. R. (1966). *J. Polym. Sci. A1* **4,** 1543.
Chien, J. C. W., and Boss, C. R. (1967a). *J. Polym. Sci. Polym. Chem. Ed.* **5,** 1683.
Chien, J. C. W., and Boss, C. R. (1967b). *J. Polym. Sci. Polym. Chem. Ed.* **5,** 3091.
Chien, J. C. W., and Boss, C. R. (1967c). *J. Am. Chem. Soc.* **89,** 571.
Chien, J. C. W., and Boss, C. R. (1972). *J. Polym. Sci. Polym. Chem. Ed.* **10,** 1579.
Chien, J. C. W., and Carlini, C. (1983). Unpublished results.
Chien, J. C. W., and Connor, W. P. (1968). *J. Am. Chem. Soc.* **90,** 100.
Chien, J. C. W., and Hsieh, J. T. T. (1976). *J. Polym. Sci. Polym. Chem. Ed.* **14,** 1915.
Chien, J. C. W., and Kiang, J. K. Y. (1978). *In* "Stabilization of Degradation of Polymers" (D. L. Allara and W. L. Hawkins, eds.), pp. 175–197. *Adv. Chem. Ser., No.* 169. Am. Chem. Soc., Washington, D.C.
Chien, J. C. W., and Kiang, J. K. Y. (1979a). *Eur. Polym. J.* **15,** 1059.
Chien, J. C. W., and Kiang, J. K. Y. (1979b). *Macromolecules* **12,** 1077.
Chien, J. C. W., and Kiang, J. K. Y. (1979c). *Macromol. Chem.* **181,** 47.
Chien, J. C. W., and Kiang, J. K. Y. (1980). *Macromolecules* **13,** 280.
Chien, J. C. W., and Schen, M. A. (1984). Unpublished results.
Chien, J. C. W., and Wang, D. S. T. (1975). *Macromolecules* **8,** 920.

Chien, J. C. W., and Westhead, E. W. (1971). *Biochemistry* **10**, 3198.

Chien, J. C. W., and Wu, J. C. (1982a). *J. Polym. Sci. Polym. Chem. Ed.* **20**, 2445.

Chien, J. C. W., and Wu, J. C. (1982b). *J. Polym. Sci. Polym. Chem. Ed.* **20**, 2461.

Chien, J. C. W., and Yang, X. (1983a). *J. Polym. Sci. Polym. Lett Ed.* (to be published).

Chien, J. C. W., and Yang, X. (1983b). *Macrmolecules* (to be published).

Chien, J. C. W., Vandenberg, E. J., and Jabloner, H. (1968). *J. Polym. Sci. Polym. Chem. Ed.* **6**, 381.

Chien, J. C. W., Karasz, F. E., Wnek, G. E., MacDiarmid, A. G., and Heeger, A. J. (1980a). *J. Polym. Sci. Polym. Chem. Ed.* **18**, 45.

Chien, J. C. W., Karasz, F. E., and Wnek, G. E. (1980b). *Nature (London)* **285**, 390.

Chien, J. C. W., Wnek, G. E., Karasz, F. E., and Hirsch, J. A. (1981). *Macromolecules* **14**, 479.

Chien, J. C. W., Karasz, F. E., and Shimamura, K. (1982a). *J. Polym. Sci. Polym. Lett. Ed.* **20**, 97.

Chien, J. C. W., Wnek, G. E., Karasz, F. E., Warakomski, J. M., Dickinson, L. C., Heeger, A. J., and MacDiarmid, A. G. (1982b). *Macromolecules* **15**, 614.

Chien, J. C. W., Uden, P. C., and Fan, J. L. (1982c). *J. Polym. Sci. Polym. Chem. Ed.* **20**, 2159.

Chien, J. C. W., Wu, J. C., Kuo, C. I. (1982d). *J. Polym. Sci. Polym. Chem. Ed.* **20**, 2019.

Chien, J. C. W., Karasz, F. E., and Shimamura, K. (1982e). *Macromolecules* **15**, 1012.

Chien, J. C. W., Capistran, J. D., Dickinson, L. C., Karasz, F. E., and Schen, M. A. (1982f). *J. Polym. Sci. Polym. Lett. Ed.* **21**, 93.

Chien, J. C. W., Yamashita, Y., Hirsch, J. A., Fan, J. L., Schen, M. A., Karasz, F. E. (1982g). *Nature (London)* **299**, 608.

Chien, J. C. W., Karasz, F. E., and Shimamura, K. (1982h). *Makromol. Chem. Rapid Commun.* **3**, 655.

Chien, J. C. W., Wu, J. C., and Kuo, C. I. (1982i). *J. Polym. Sci. Polym. Chem. Ed.* **20**, 2019.

Chien, J. C. W., Wu, J. C., and Kuo, C. I. (1983a). *J. Polym. Sci. Polym. Chem. Ed.* **21**, 725.

Chien, J. C. W., Wu, J. C., and Kao, C. I. (1983b). *J. Polym. Sci. Polym. Chem. Ed.* **21**, 737.

Chien, J. C. W., Karasz, F. E., Schen, M. A., and Yamashita, Y. (1983c). *Makromol. Chem. Rapid Commun.* **4**, 5.

Chien, J. C. W., Yang, X., and Dickinson, L. C. (1983d). *Macromolecules* (to be published).

Chien, J. C. W., Karasz, F. E., Schen, M. A., and Hirsch, J. A. (1983e). *Macromolecules* (to be published).

Chien, J. C. W., Warakomski, J. M., Karasz, F. E., Chia, W. L., and Lillya, C. P. (1983f). *Phys. Rev. B* (to be published).

Chien, J. C. W., Schen, M. A., and Karasz, F. E. (1983g). *J. Polym. Sci. Polym. Chem. Ed.* (to be published).

Chung, T. C., Feldblum, A., Heeger, A. J., and MacDiarmid, A. G. (1981). *J. Chem. Phys.* **74**, 5504.

Chung, T. C., MacDiarmid, A. G., Feldblum, A., Heeger, A. J. (1982). *J. Polym. Sci. Polym. Lett. Ed.* **20**, 427.

Clarke, T. C., and Street, G. B. (1979). *Synth. Met.* **1**, 119.

Clarke, T. C., Geiss, R. H., Kwak, J. F., and Street, G. B. (1978). *J. Chem. Soc. Chem. Commun.* p. 489.

Clarke, T. C., Geiss, R. H., Gill, W. D., Grant, V. M., Morawitz, H., and Street, G. B. (1979). *Synth. Met.* **1**, 21.

Clarke, T. C., Krounbi, M. T., Lee, V. Y., and Street, G. B. (1981). *J. Chem. Soc. Chem. Commun.* p. 384.

Clarke, T. C., Scott, J. C., and Street, G. B. (1983). *IBM J.* **27**, 313

Cohen, M. H., Fritzsche, H., and Ovshinsky, S. R. (1969). *Phys. Rev. Lett.* **22**, 1065.

Colson, R., and Nagels, P. (1980). *J. Non-Cryst. Solids* **35**, 129.

Coulson, C. A. (1938). *Proc. R. Soc. London Ser. A* **164**, 383.

Daniels, W. E. (1964). *J. Org. Chem.* **29**, 2936.

Davis, E. A., and Mott, N. F. (1970). *Philos. Mag.* **22**, 903.

Dawes, D. H., and Winkler, C. A. (1964). *J. Polym. Sci. A* **2**, 3029.

Deits, W., Cukor, P., Rubner, M., and Jopson, H. (1981). *J. Electron. Mater.* **10**, 683.

Devreux, F., Döry, I., Mihàly, L., Pekker, S., Jánossy, A., and Kértesz, M. (1981a). *J. Polym. Sci. Polym. Chem. Ed.* **19**, 743.

Devreux, F., Genoud, F., Holczer, K., Nechtschein, M., and Travers, J. M. (1981b). *Mol. Cryst. Liq. Cryst.* **77**, 97.

Druy, M. A., Tsang, C. H., Brown, N., Heeger, A. J., and MacDiarmid, A. G. (1980). *J. Polym. Sci. Polym. Phys. Ed.* **18**, 429.

Dugay, M., and Rouston, J. (1979). *Lect. Notes Phys.* **95**, 199.

D'yachkovskii, F. S., Yarovitskii, P. A., and Bystrov, V. F. (1964). *Vysokomol. Soedin.* **6**, 659.

Dyson, F. J. (1955). *Phys. Rev.* **98**, 349.

Dzhabiev, T. S., Dyachkovskii, F. S., and Shilov, A. Ye (1971). *Vysokomol. Soedin. Ser. A* **13**, 2474.

Enkelmann, V., Müller, W., and Wegner, G. (1979). *Synth. Met.* **1**, 185.

Enkelmann, V., Lieser, G., Müller, W., and Wegner, G. (1981a). *Chem. Scr.* **17**, 141.

Enkelmann, V., Lieser, G., Monkenbusch, M., Müller, W., and Wegner, G. (1981b). *Mol. Cryst. Liq. Cryst.* **77**, 111.

Enkelmann, V., Monkenbusch, M., and Wegner, G. (1982). *Polymer* **23**, 1581.

Epstein, A. J., Gibson, H. W., Chaikin, P. M., Clark, W. G., and Grüner, G. (1980). *Phys. Rev. Lett.* **45**, 1730.

Epstein, A. J., Rommelmann, H., Druy, M. A., Heeger, A. J., and MacDiarmid, A. G. (1981a). *Solid State Commun.* **38**, 683.

Epstein, A. J., Gibson, H. W., Chaikin, P. M., Clark, W. G., and Grüner, G. (1981b). *Chem. Scr.* **17**, 135.

Epstein, A. J., Rommelmann, H., Abkowitz, M., and Gibson, H. W. (1981c). *Phys. Rev. Lett.* **47**, 1549.

Epstein, A. J., Rommelmann, H., Abkowitz, M., and Gibson, H. W. (1981d). *Mol. Cryst. Liq. Cryst.* **77**, 81.

Epstein, A. J., Rommelmann, H., Bigelow, R., Gibson, H. W., Hoffman, D. M., and Tanner, D. B. (1983). *J. Phys. (Paris).* (to be published).

Ermin, D. (1973). *In* "Electronic and Structural Properties of Amorphous Semiconductors" (P. G. LeCumber and J. Mort, eds.), p. 261. Academic Press, New York.

Etemad, S., Pron, A., Heeger, A. J., MacDiarmid, A. G., Mele, E. J., and Rice M. J. (1981a). *Phys. Rev. B* **23**, 5137.

Etemad, S., Mitani, T., Ozaki, M., Chung, T. C., Heeger, A. J., and MacDiarmid, A. G. (1981b). *Solid State Commun.* **40**, 75.

Etemad, S., Ozaki, M., Heeger, A. J., and MacDiarmid, A. G. (1981c). *Chem. Scr.* **17**, 159.

Fan, J. L., and Chien, J. C. W. (1983). *J. Polym. Sci. Polym. Chem. Ed.* (to be published).

Feher, G., and Kip, A. F. (1955). *Phys. Rev.* **98**, 337.

Feldblum, A., Park, Y. W., Heeger, A. J., MacDiarmid, A. G., Wnek, G. E., Karasz, F. E., and Chien, J. C. W. (1981). *J. Polym. Sci. Polym. Phys. Ed.* **19**, 173.

Feldblum, A., Kaufman, J. H., Etemad, S., Heeger, A. J., Chung, T. C., and MacDiarmid, A. G. (1982a). *Phys. Rev. B* **26**(2), 815.

Feldblum, A., Heeger, A. J., Chung, T. C., and MacDiarmid, A. G. (1982b). *J. Chem. Phys.* **77**, 5114.

Ferraris, J. P., Webb, A. W., Weber, D. C., Fox, W. B., Carpenter, E. R., and Brant, P. (1980). *Solid State Commun.* **35**, 15.

Fincher, C. R., Jr., Peebles, D. L., Heeger, A. J., Druy, M. A., Matsumura, Y., MacDiarmid, A. G., Shirakawa, H., and Ikeda, S. (1978). *Solid State Commun.* **27**, 489.

Fincher, C. R., Jr., Ozaki, M., Tanaka, M., Peebles, D. L., Lauchlau, L., Heeger, A. J., and MacDiarmid, A. G. (1979a). *Phys. Rev. B* **20**, 1589.

Fincher, C. R., Jr., Ozaki, M., Heeger, A. J., and MacDiarmid, A. G. (1979b). *Phys. Rev. B* **19**, 4140.

Fincher, C. R., Jr., Chen, C. E., Heeger, A. J., MacDiarmid, A. G., and Hastings, J. B. (1982). *Phys. Rev. Lett.* **48**, 100.

Fitchen, D. B. (1982). *Mol. Cryst. Liq. Cryst.* **83**, 95.

Flood, J. D., Ehrenfreund, E., Heeger, A. J., and MacDiarmid, A. G. (1982). *Solid State Commun.* **44**, 1055.

Francois, B., Bernard, M., and Andre, J. J. (1981a). *J. Chem. Phys.* **75**, 4142.

Francois, B., Mermilliod, N., and Zuppiroli, L. (1981b). *Synth. Met.* **4**, 131.

Fritzsche, H. (1971). *Solid State Commun.* **9**, 1813.

Galvin, M. E., and Wnek, G. E. (1982). *Polym. Commun.* **23**, 795.

Gammel, J. T., and Krumhansl, J. A. (1981). *Phys. Rev. B* **24**, 1035.

Garito, A. F., and Heeger, A. J. (1974). *Acc. Chem. Res.* **7**, 232.

Gau, S. C., Milliken, J., Pron, A., MacDiarmid, A. G., and Heeger, A. J. (1979). *J. Chem. Soc. Chem. Commun.* p. 662.

Gibson, H. W., and Pochan, J. M. (1982). *Macromolecules* **15**, 242.

Gibson, H. W., Bailey, F. C., Epstein, A. J., Rommelmann, H., and Pochan, J. M. (1980). *J. Chem. Soc. Chem. Commun.* p. 426.

Gibson, H. W., Pochan, J. M., and Kaplan, S. (1981). *J. Am. Chem. Soc.* **103**, 4619.

Goldberg, I. B., Crowe, H. R., Newman, P. R., Heeger, A. J., and MacDiarmid, A. G. (1979). *J. Chem. Phys.* **70**, 1132.

Gooding, E. P. (1976). *Chem. Soc. Rev.* **5**, 95.

Gooding, R., Lillya, C. P., and Chien, J. C. W. (1983). *J. Chem. Soc. Chem. Commun.* p. 151.

Gould, C. M., Bates, D. M., Bozler, H. M., Heeger, A. J., Druy, M. A., and MacDiarmid, A. G. (1981). *Phys. Rev. B* **23**, 6820.

Gourley, K. D., Lillya, C. P., Reynolds, J. R., Karasz, F. E., and Chien, J. C. W. (1983). *Macromolecules* (to be published).

Grant, P. M., and Batra, I. P. (1979a). *Synth. Met.* **1**, 193.

Grant, P. M., and Batra, I. P. (1979b). *Solid State Commun.* **29**, 225.

Grant, P. M., and Krounbi, M. (1980). *Solid State Commun.* **36**, 291.

Grant, P. M., Tani, T., Gill, W. D., Krounbi, M., and Clarke, T. C. (1982). *J. Appl. Phys.* **52**, 869.

Grant, P. M., *et al.* (1982).

Guckelsberger, K., Rödhammer, P., Gmelin, E., Peo, M., Menke, K., Hocker, J., Roth, S., and Dransfeld, K. (1981). *Z. Phys. B* **43**, 189.

Guttman, J. Y., and Guillet, J. E. (1970). *Polym. Prepr. Am. Chem. Soc. Div. Polym. Chem.* **30**(1), 177.

Haberditzl, W. (1968). "Magnetochemie." Akademie Verlag, Berlin.

Haberkorn, H., Naarmann, H., Peuzien, K., Schlag, J., and Simak, P. (1982). *Synth. Met.* **5**, 51.

Hankin, A. G., and North, A. M. (1967). *Trans. Faraday Soc.* **63**, 1525.

Harada, I., Tasumi, M., Shirakawa, H., and Ikeda, S. (1978). *Chem. Lett.* p. 1411.

Hartman, R. D., and Pohl, H. A. (1968). *J. Polym. Sci. Polym. Chem. Ed.* **6**, 1116.

Hatano, M. (1962). *Kogyo Kagaku Zasshi* **65**, 723.

Hatano, M., Kambara, S., and Okamoto, S. (1961). *J. Polym. Sci.* **51**, 526.

Heeger, A. J., and MacDiarmid, A. G. (1978). "Quasi One-Dimensional Conductors II" (S. Barisic, A. Bjelis, J. R. Cooper, and B. Leontic, eds.), p. 361. Springer-Verlag, Berlin and New York.

Hennessy, M. J., McElwee, C. D., and Richards, P. M. (1973). *Phys. Rev. B* **7**, 930.

Hirai, H., Hiraki, K., Noguchi, I., and Mukishima, S. (1972). *J. Polym. Sci. Polym. Chem. Ed.* **8**, 147.

Hiraki, K., Kaneko, S., and Hirai, H. (1972). *J. Polym. Sci. Polym. Lett. Ed.* **10**, 199.

Holczer, K., Devreux, F., Nechtschein, M., and Travers, J. P. (1981a). *Solid State Commun.* **39**, 881.

Holczer, K., Boucher, J. P., Devreux, F., and Nechtschein, M. (1981b). *Phys. Rev. B* **23**, 1051.

Holob, G. M., Ehrlich, P., and Allendoerfer, R. D. (1972). *Macromolecules* **5**, 569.

Horovitz, B. (1981). *Chem. Scr.* **17**, 127.

Horovitz, B. (1982). *Solid State Commun.* **41**, 729.

Horsch, P. (1981). *Phys. Rev. B* **24**, 7351.

Hsu, S. L., Signorelli, A. J., Pez, G. P., and Baughman, R. H. (1978). *J. Chem. Phys.* **69**, 106.

Ikeda, S. (1967). *Kogyo Kagaku Zasshi* **70**, 1880.

Ikeda, S., and Tanaka, A. (1966). *J. Polym. Sci. Polym. Lett. Ed.* **4**, 605.

Ikeda, S., and Tamaki, A. (1966). *Int. Symp. Macromol. Chem. Tokyo Kyoto* I-24.

Ikehata, S., Kaufer, J., Woerner, T., Pron, A., Druy, M. A., Sivak, A., Heeger, A. J., and MacDiarmid, A. G. (1980). *Phys. Rev. Lett.* **45**, 1123.

Ikehata, S., Druy, M., Woerner, T., Heeger, A. J., and MacDiarmid, A. G. (1981). *Solid State Commun.* **39**, 1239.

Ikemoto, I., Sakairi, M., Tsutsumi, T., Kuroda, H., Harada, I., Tasumi, M., Shirakawa, H., and Ikeda, S. (1979). *Chem. Lett.* p. 1189.

Ikemoto, I., Cao, Y., Yamada, M., Kuroda, H., Harada, I., Shirakawa, H., and Ikeda, S. (1982). *Bull. Chem. Soc. Jpn.* **55**, 721.

Imhoff, E. A., Fitchen, D. B., and Stahlbush, R. E. (1982). *Solid State Commun.* **44**, 329.

Inagaki, F., Tasumi, M., and Miyazawa, T. (1974). *J. Mol. Spectrosc.* **50**, 286.

Inagaki, F., Tasumi, M., and Miyazawa, T. (1975). *J. Raman Spectrosc.* **3**, 335.

Ingram, P., and Schindler, A. (1968). *Makromol. Chem.* **111**, 267.

Ito, T., Shirakawa, H., and Ikeda, S. (1974). *J. Polym. Sci. Polym. Chem. Ed.* **12**, 11.

Ito, T., Shirakawa, H., and Ikeda, S. (1975). *J. Polym. Sci. Polym. Chem. Ed.* **13**, 1943.

Ivory, D. M., Miller, G. G., Sowa, J. M., Shacklette, L. W., Chance, R. R., and Baughman, R. H. (1979). *J. Chem. Phys.* **71**, 1506.

Jerome, D., Mazaud, A., Ribault, M., and Bechgaard, K. (1980). *J. Phys. Lett.* **41**, L95.

Kaindl, G., Wortmann, G., Roth, S., and Menke, K. (1982). *Solid State Commun.* **41**, 75.

Kambara, S., Hatano, M., and Hosoe, T. (1962). *Kogyo Kagaku Zasshi* **65**, 720.

Kanazawa, K. K., Diaz, A. F., Geiss, R. H., Gill, W. D., Kwak, J. F., Logan, J. A., Rabolt, J. F., and Street, G. B. (1979). *J. Chem. Soc. Chem. Commun.* p. 854.

Kanazawa, K. K., Diaz, A. F., Gill, W. D., Grant, P. M., Street, G. B., Gardini, G. P., and Kwak, J. F. (1980). *Synth. Met.* **1**, 329.

Kaneto, K., Maxfield, M., Nairns, D. P., MacDiarmid, A. G., and Heeger, A. J. (1982). *J. Chem. Soc. Faraday Trans. 1* **78**, 3417.

Karasz, F. E., Chien, J. C. W., Galkiewicz, R., Wnek, G. E., Heeger, A. J., and MacDiarmid, A. G. (1979). *Nature (London)* **282**, 236.

Karpfen, A., and Höller, R. (1981). *Solid State Commun.* **37**, 179.

Karpfen, A., and Petkov, J. (1979a). *Solid State Commun.* **29**, 251.

Karpfen, A., and Petkov, J. (1979b). *Theor. Chim. Acta* **53**, 65.

Kaufman, J. H., Mele, E. J., Heeger, A. J., Kaner, R., and MacDiarmid, A. G. (1983). *J. Electrochem. Soc.* **130**, 571.

Kiang, J. K. Y., Uden, P. C., and Chien, J. C. W. (1980). *Polym. Degrad. Stab.* **2**, 113.

Kiess, H., Meyer, W., Baeriswyl, D., and Harbeke, G. (1980). *J. Electron. Mater.* **9**, 763.

Kivelson, S. (1981a). *Mol. Cryst. Liq Cryst.* **77**, 65.

Kivelson, S. (1981b). *Phys. Rev. Lett.* **46**, 1344.

Kivelson, S. (1982). *Phys. Rev. B* **25**, 3798.

Kivelson, S., Lee, T. K., Lin-Liu, Y. R., Peschel, I., and Yu, L. (1982). *Phys. Rev. B* **25**, 4173.

Kletter, M. J., Woerner, T., Pron, A., MacDiarmid, A. G., Heeger, A. J., and Park, Y. W. (1980). *J. Chem. Soc. Chem. Commun.* p. 426.

Koide, N., Iimura, K., and Takeda, M. (1967). *J. Chem. Soc. Jpn. Ind. Chem. Sect.* **70**, 1224.

Kuhn, H. (1948). *Helv. Chim. Acta* **31**, 144.

Kurkijarvi, J. (1973). *Phys. Rev. B* **8**, 922.

Kuzmany, H. (1980). *Phys. Status Solidi* **97**, 521.

Kuzmany, H. (1981). *Chem. Scr.* **17**, 155.

Kwak, J. F., Gill, W. D., Greene, R. L., Seeger, K., Clarke, T. C., and Street, G. B. (1979a). *Synth. Met.* **1**, 213.

Kwak, J. F., Clarke, T. C., Greene, R. L., and Street, G. B. (1979b). *Solid State Commun.* **31**, 355.

Lauchlan, L., Etemad, S., Chung, T. C., Heeger, A. J., and MacDiarmid, A. G. (1981). *Phys. Rev. B* **24**, 3701.

Lefrant, S., Lichtmann, L. S., Temkin, H., Fitchen, D. B., Miller, P. C., Whitwell, G. E., and Burlitch, J. M. (1979). *Solid State Commun.* **29**, 191.

Lefrant, S., Rzepka, E., Bernier, P., Rolland, M., and Aldissi, M. (1980). *Polymer* **21**, 1235.

Lefrant, S., Faulques, E., Lauchlan, L., Kletter, M. J., and Etemad, S. (1982). *Mol. Cryst. Liq. Cryst.* **83**, 1149.

Lennard-Jones, J. E. (1937). *Proc. R. Soc. (London) Ser. A* **158**, 280.

Li, G. C. G., Kidwell, D. A., Brown, D. W., Hall, E. E., Wnek, G. E. (1982). *J. Polym. Sci. Polym. Chem. Ed.* **21**, 301.

Lichtmann, L. S. (1981). Ph.D. Thesis, Cornell Univ., Ithaca, New York.

Lichtmann, L. S., and Fitchen, D. B. (1979). *Synth. Met.* **1**, 139.

Lichtmann, L. S., and Fitchen, D. B. (1981). *Bull. Am. Phys. Soc.* **26**, 485.

Lichtmann, L. S., Sarhangi, A., and Fitchen, D. B. (1980). *Solid State Commun.* **36**, 869.

Lichtmann, L. S., Sarhangi, A., and Fitchen, D. B. (1981). *Chem. Scr.* **17**, 149.

Lieser, G., Wegner, G., Müller, W., and Enkelmann, V. (1980a). *Makromol. Chem. Rapid Commun.* **1**, 621.

Lieser, G., Wegner, G., Müller, W., Enkelmann, V., Meyer, W. H. (1980b). *Makromol. Chem. Rapid Commun.* **1**, 627.

Lieser, G., Monkenbusch, M., Enkelmann, V., and Wegner, G. (1981). *Mol. Cryst. Liq. Cryst.* **77**, 169.

Lin-Liu, Y. R. (1982). *Phys. Rev. B* **25**, 5540.

Longuet-Higgins, H. C., and Salem, L. (1959). *Proc. R. Soc. (London) Ser. A* **25**, 172.

Louboutin, J. P., and Beniere, F. (1982). *J. Phys. Chem. Solids* **43**, 233.

Luttinger, L. B. (1960). *Chem. Ind. (London)* p. 1135.

Luttinger, L. B. (1962). *J. Org. Chem.* **27**, 1591.

McConnell, H., Heller, C., Cole, T., and Fessenden, R. (1960). *J. Am. Chem. Soc.* **82**, 766.

MacDiarmid, A. G., and Heeger, A. J. (1979). *Synth. Met.* **1**, 101.

MacDiarmid, A. G. (1982). Personal communication.

Mackley, M. R. (1975). *Colloid Polym. Sci.* **213**, 373.

MacInnes, D., Jr., Druy, M. A., Nigrey, P. J., Nairns, D. P., MacDiarmid, A. G., and Heeger, A. J. (1981). *J. Chem. Soc. Chem. Commun.* p. 317.

McQuillan, B., Street, G. B., and Clarke, T. C. (1982). *J. Electron. Mater.* **11**, 471.

Maisin, F., Gusman, G., and Deltour, R. (1981). *Solid State Commun.* **39**, 505.

Maki, K., and Nakahara, M. (1981). *Phys. Rev. B.* **23**, 5005.

Maricq, M. M., Waugh, J. S., MacDiarmid, A. G., Shirakawa, H., and Heeger, A. J. (1978). *J. Am. Chem. Soc.* **100**, 7729.

Matsuyama, T., Sakai, H., Yamaoda, H., Maeda, Y., and Shirakawa, H. (1981). *Solid State Commun.* **40**, 563.

Mele, E. J. (1982). *Solid State Commun.* **44**, 827.

Mele, E. J., and Rice, M. J. (1980a). *Solid State Commun.* **34**, 339.

Mele, E. J., and Rice, M. J. (1980b). *Phys. Rev. Lett.* **45**, 926.

Mele, E. J., and Rice, M. J. (1981). *Phys. Rev. B* **23**, 5397.

Merriwether, L. S. (1961). *J. Org. Chem.* **26**, 5163.

Meuer, B., Spegt, P., Weill, G., Mathis, C., and Francois, B. (1982). *Solid State Commun.* **44**, 201.

Meyer, W. H. (1981a). *Synth. Met.* **4**, 81.

Meyer, W. H. (1981b). *Mol. Cryst. Liq. Cryst.* **77**, 137.

Mihàly, L., Pekker, S., and Jánossy, A. (1979a). *Synth. Met.* **1**, 349.

Mihàly, G., Vanscó, G., Pekker, S., and Jánossy, A. (1979b). *Synth. Met.* **1**, 357.

Monkenbusch, M., Morra, B. S., and Wegner, G. (1982). *Makromol. Chem. Rapid Commun.* **3**, 69.

Montaner, A., Galtier, M., Benoit, C., and Aldissi, M. (1981a). *Solid State Commun.* **39**, 99.

Montaner, A., Galtier, M., Benoit, C., and Aldissi, M. (1981b). *Phys. Status Solidi* **66**, 267.

Mortenson, K., Thewalt, M. L. W., and Tomkiewicz, Y. (1980). *Phys. Rev. Lett.* **45**, 490.

Moses, D., Denenstein, A., Pron, A., Heeger, A. J., and MacDiarmid, A. G. (1980). *Solid State Commun.* **36**, 219.

Moses, D., Chen, J., Denenstein, A., Kaveh, M., Chung, T. C., Heeger, A. J., and MacDiarmid, A. G. (1981). *Solid State Commun.* **40**, 1007.

Moses, D., Feldblum, A., Denenstein, A., Chung, T. C., Heeger, A. J., and MacDiarmid, A. G. (1982a). *Mol. Cryst. Liq. Cryst.* **83**, 87.

Moses, D., Denenstein, A., Chen, J., Heeger, A. J., McAndrew, P., Woerner, T., MacDiarmid, A. G., and Park, Y. W. (1982b). *Phys. Rev. B.* **25**, 7652.

Moses, D., Denenstein, A., Chen, J., Heeger, A. J., McAndrew, P., Woerner, T., MacDiarmid, A. G., and Park, Y. W. (1982c). *Phys. Rev. B.* **25**, 7652.

Mott, N. F. (1970). *Philos. Mag.* **22**, 7.

Nagels, P. (1976). *In* "Linear and Nonliner Electron Transport in Solids" (J. T. Devreese and V. E. Van Doren, eds.), p. 435. Plenum, New York.

Nagels, P. Callerts, R., Denayer, M., and DeConinck, R. (1970). *J. Non-Cryst. Solids* **4**, 295.

Nakahara, M., and Maki, K. (1981). *Phys. Rev. B* **24**, 1045.

Nakano, T., and Fukuyama, H. (1980). *J. Phys. Soc. Jpn.* **49**, 1679.

Natta, G., Mazzanti, G., and Corradini, P. (1958). *Alti Accad. Naz. Lincei Rend. Sci. Fis. Mat. Nat.* **25**, 2.

Natta, G., Poori, L., and Stoppa, G. (1964). *Makromol. Chem.* **77**, 114.

Nechtschein, M., Devreux, F., Greene, R. L., Clarke, T. C., and Street, G. B. (1980). *Phys. Rev. Lett.* **44**, 356.

Nechtschein, M., Devreux, F., Genoud, F., Guglielmi, M., and Holczer, K. (1983). *Phys. Rev. B* **27**, 61.

Newman, P. R., Ewbank, M. D., Mauthe, C. D., Winkle, M. R., and Smolyncki, W. D. (1981). *Solid State Commun.* **40**, 975.

Nigrey, P. J., MacDiarmid, A. G., and Heeger, A. J. (1979). *J. Chem. Soc. Chem. Commun.* p. 594.

Nigrey, P. J., MacInnes, D., Jr., Nairns, D. P., MacDiarmid, A. G., and Heeger, A. J. (1981a). *J. Electrochem. Soc.* **128**, 1651.

Nigrey, P. J., MacInnes, D., Jr., Nairns, D. P., MacDiarmid, A. G., and Heeger, A. J. (1981b). *In* "Conductive Polymers" (R. D. Seymour, ed.), p. 227. Plenum, New York.

Ovchinnikov, A. A., Ukrainski, I. I., and Kventsel, G. V. (1978). *Sov. Phys. Usp.* **15**, 575.

Ozaki, M., Peebles, D. L., Weinberger, B. R., Chiang, C. K., Gau, S. C., Heeger, A. J., and MacDiarmid, A. G., (1979). *Appl. Phys. Lett.* **35**, 83.

Ozaki, M., Peebles, D. L., Weinberger, B. R., Heeger, A. J., and MacDiarmid, A. G. (1980). *J. Appl. Phys.* **51**, 4252.

Park, Y. W., Druy, M. A., Chiang, C. K., MacDiarmid, A. G., Heeger, A. J., Shirakawa, H., and Ikeda, S. (1979a). *J. Polym. Sci. Polym. Lett. Ed.* **17**, 195.

Park, Y. W., Denenstein, A., Chiang, C. K., Heeger, A. J., and MacDiarmid, A. G. (1979b). *Solid State Commun.* **29**, 745.

Park, Y. W., Heeger, A. J., Druy, M. A., and MacDiarmid, A. G. (1980). *J. Chem. Phys.* **73**, 946.

Peierls, R. E. (1955). "Quantum Theory of Solids," Chapter 5. Oxford Univ. Press, London and New York.

Pennings, A. J., Zwijenenburg, A., and Lageveen, R. (1973). *Koll. Z. Z. Polym.* **251**, 500.

Peo, M., Förster, H., Menke, K., Hocker, J. A., Roth, S., and Dransfield, K. (1981a). *Solid State Commun.* **38**, 467.

Peo, M., Roth, S., and Hocker, J. (1981b). *Chem. Scr.* **17**, 133.

Peo, M., Forster, H., Menke, K., Hocker, J., Gardner, J. A., Roth, S., and Dransfeld, K. (1981a). *Solid State Commun.* **38**, 467.

Peo, M., Roth, S., and Hocker, J. (1981b). *Chem. Scr.* **17**, 133.

Perlstein, J. H. (1977). *Angew. Chem. Int. Ed. Engl.* **16**, 519.

Pez, G. P. (1976). *J. Am. Chem. Soc.* **98**, 8072.

Phillipp, A., Mayr, W., and Seeger, K. (1982). *Solid State Commun.* **43**, 857.

Pochan, J. M., Gibson, H. W., and Bailey, F. C. (1980). *J. Polym. Sci. Polym. Chem. Ed.* **18**, 447.

Pochan, J. M., Pochan, D. F., Rommelmann, H., and Gibson, H. W. (1981). *Macromolecules* **14**, 110.

Pople, J. A., and Walmsley, S. H. (1962). *Mol. Phys.* **5**, 15.

Pron, A., Bernier, P., Rolland, M., Lefrant, S., Aldissi, M., Pachdi, F., and MacDiarmid, A. G. (1981). *J. Mater. Sci.* **7**, 305.

Rabolt, J. F., Clarke, T. C., and Street, G. B. (1979). *J. Chem. Phys.* **71**, 4614.

Rabolt, J. F., Clarke, T. C., Kanazawa, K. K., Reynolds, J. R., Street, G. B. (1980). *J. Chem. Soc. Chem. Commun.* p. 347.

Rabolt, J. F., Clarke, T. C., and Street, G. B. (1982). *J. Chem. Phys.* **76**, 5781.

Rachdi, F., Bernier, P., Billaud, D., Pron, A., and Przyluski, J. (1981). *Polymer* **22**, 1606.

Redfield, D. (1975). *Adv. Phys.* **24**, 463.

Reikel, C., Hässlin, H. W., Menke, K., and Roth, S. (1982). *J. Chem. Phys.* **77**, 4624.

Reynolds, J. R., Chien, J. C. W., Karasz, F. E., Lillya, C. P., and Curran, D. J. (1982). *J. Chem. Soc. Chem. Commun.* p. 1358.

Rice, M. J. (1979). *Phys. Lett. A* **71**, 152.

Rice, M. J., and Mele, E. J. (1980). *Solid State Commun.* **35**, 487.

Rice, M. J., and Mele, E. J. (1981). *Chem. Scr.* **17**, 121.

Rice, M. J., and Timonen, J. (1979). *Phys. Lett. A* **73**, 378.

Roberts, S., and von Hipple, A. (1946). *J. Appl. Phys.* **17**, 610.

Robin, P., Pouget, J. P., Comes, R., Gibsn, H. W., and Epstein, A. J. (1983). *Phys. Rev. B* **27**, 3938.

Rolland, M., Bernier, P., Lefrant, S., and Aldissi, M. (1980a). *Polymer* **21**, 1111.

Rolland, M., Bernier, P., Disi, M., Linaya, C., Sledz, J., Schué, F., Fabre, J. M., and Giral, L. (1980b). *J. Phys. Lett.* **41**, L-165.

Rolland, M., Bernier, R., and Aldissi, M. (1980c). *Phys. Status Solidi* **62**, K5.

Rolland, M., Lefrant, S., Aldissi, M., Bernier, P., Rzepka, E., and Schué, F. (1981a). *J. Electron. Mater.* **10**, 619.

Rolland, M., Cadène, M., Bresse, J. F., Rossi, A., Riviere, D., Aldissi, M., Benoit, C., and Bernier, P. (1981b). *Mater. Res. Bull.* **16**, 1045.

Rommelmann, H., Fernquist, R., Gibson, H. W., Epstein, A. J., Druy, M. A., and Woerner, T. (1981). *Mol. Cryst. Liq. Cryst.* **77**, 177.

Ruland, W. (1961). *Acta Crystallogr.* **14**, 1180.

Salaneck, W. R., Thomas, H. R., Duke, C. B., Plummer, E. W., Heeger, A. J., and MacDiarmid, A. G. (1979a). *Synth. Met.* **1**, 133.

Salaneck, W. R., Thomas, H. R., Duke, C. B., Paton, A., Plummer, E. W., Heeger, A. J., and MacDiarmid, A. G. (1979b). *J. Chem. Phys.* **71**, 2044.

Salaneck, W. R., Thomas, H. R., Bigelow, R. W., Duke, C. B., Plummer, E. W., Heeger, A. J., and MacDiarmid, A. G. (1980). *J. Chem. Phys.* **72**, 3674.

Schmeal, W. R., and Street, J. R. (1971). *AIChE J.* **17**, 1188.

Schügerl, F. B., and Kuzmany, H. (1981). *J. Chem. Phys.* **74**, 953.

Schwoerer, M., Lauterbach, U., Müller, W., and Wegner, G. (1980). *Chem. Phys. Lett.* **69**(2), 359.

Shacklette, L. W., Chance, R. R., Ivory, D. M., Miller, G. G., and Baughman, R. H. (1979). *Synth. Met.* **1**, 307.

Shante, V. K. S., and Varma, C. M. (1973). *Phys. Rev. B* **8**, 4885.

Sheng, P. (1980). *Phys. Rev. B* **21**, 2180.

Sheng, P., Sichel, E. K., and Gittleman, J. I. (1978). *Phys. Rev. Lett.* **40**, 1197.

Shimamura, K., Karasz, F. E., Hirsch, J. A., and Chien, J. C. W. (1981). *Makromol. Chem. Rapid Commun.* **2**, 473.

Shimamura, K., Karasz, F. E., Chien, J. C. W., and Hirsch, J. A. (1982). *Makromol. Chem. Rapid Commun.* **3**, 269.

Shirakawa, H., and Ikeda, S. (1971). *Polym. J.* **2**, 231.

Shirakawa, H., and Ikeda, S. (1974). *J. Polym. Sci. Polym. Chem. Ed.* **12**, 929.

Shirakawa, H., and Ikeda, S. (1979). *Synth. Met.* **1**, 175.

Shirakawa, H., Ito, T., and Ikeda, S. (1973). *Polym. J.* **4**, 460.

Shirakawa, H., Louis, E. J., MacDiarmid, A. G., Chiang, C. K., and Heeger, A. J. (1977). *J. Chem. Soc. Chem. Commun.* p. 578.

Shirakawa, H., Ito, T., and Ikeda, S. (1978). *Makromol. Chem.* **179**, 1565.

Shirakawa, H., Sato, M., Hamano, A., Kawakami, S., Soga, K., and Ikeda, S. (1980a). *Macromolecules* **13**, 457.

Shirakawa, H., Hamano, A., Kawakani, S., Sato, M., Soya, K., and Ikeda, S. (1980b). *Z. Phys. Chem.* **120**, 5235.

Shirakawa, H., Ikeda, S., Aizawa, M., Yoshitake, J., and Suzuki, S. (1981). *Synth. Met* **4**, 43.

Shiren, N. S., Tomkiewicz, Y., Kazyaka, T. G., Taranko, A. R., Thomann, H., Dalton, L., and Clarke, T. C. (1982). *Solid State Commun.* **44**, 1157.

Sichel, E. K., Gittleman, J. I., and Sheng, P. (1978). *Phys. Rev. B* **18**, 5712.

Sichel, E. K., Knowles, M., Rubner, M., and Georges, J., Jr. (1982). *Phys. Rev. B* **25**, 5574.

Simionescu, C., Dumitrescu, Sv., Negulescu, I., Percec, V., Grigoras, M., Diaconu, I., Leanca, M., and Goras, L. (1974). *Vysokomol. Soedin. Ser. A* **16**, 790.

Singh, D., and Merrill, R. P. (1971). *Macromolecules* **4**, 599.

Slater, J. C. (1934). *Phys. Rev.* **45**, 794.

Soga, K., Kobayashi, Y., Ikeda, S., and Kawakami, S. (1982). *J. Chem. Soc. Chem. Commun.* p. 931.

Springett, B. E. (1973). *Phys. Rev. Lett.* **31**, 1463.

Su, W. P., and Schrieffer, J. R. (1980). *Proc. Natl. Acad. Sci. U.S.A.* **77**, 5626.

Su, W. P., Schrieffer, J. R., and Heeger, A. J. (1979). *Phys. Rev. Lett.* **42**, 1698.

Su, W. P., Schrieffer, J. R., and Heeger, A. J. (1980). *Phys. Rev. B* **22**, 2099.

Suzuki, N., Ozaki, M., Etemad, S., Heeger, A. J., and MacDiarmid, A. G. (1980). *Phys. Rev. Lett.* **45,** 1209.

Takayama, H., Lin-Liu, Y. R., and Maki, K. (1980). *Phys. Rev. B* **21,** 2388.

Takeda, M., Iimura, K., Nozawa, Y., Hisatome, M., and Koide, N. (1968). *J. Polym. Sci. C* **23,** 741.

Tanaka, M., Watanabe, A., and Tanaka, J. (1980). *Bull. Chem. Soc. Jpn.* **53,** 645.

Tanaka, M., Watanabe, A., and Tanaka, J. (1981). *Chem. Scr.* **17,** 131.

Tani, T., Gill, W. D., Grant, P. M., Clarke, T. C., and Street, G. B. (1979). *Synth. Met.* **1,** 301.

Tani, T., Grant, P. M., Gill, W. D., Street, G. B., and Clarke, T. C. (1980). *Solid State Commun.* **33,** 499.

Tanner, D. B., Jacobson, C. S., Garito, A. F., and Heeger, A. J. (1976). *Phys. Rev. B* **13,** 3381.

Thomas, H. R., Salaneck, W. R., Duke, C. B., Plummer, E. W., Heeger, A. J., and MacDiarmid, A. G. (1980). *Polymer* **21,** 1238.

Tomkiewicz, Y., Schultz, T. D., Broom, H. B., Clarke, T. C., and Street, G. B. (1979). *Phys. Rev. Lett.* **43,** 1532.

Tomkiewicz, Y., Schultz, T. D., Broom, H. B., Taranko, A. R., Clarke, T. C., and Street, G. B. (1981). *Phys. Rev. B* **24,** 4348.

Tomkiewicz, Y., Shiren, N. S., Schultz, T. D., Thomann, H., Dalton, L. R., Zettl, A., Gruner, G., and Clarke, T. C. (1982). *Mol. Cryst. Liq. Cryst.* **83,** 17.

Tsukamoto, J., Ohigashi, H., Matsumura, K., and Takahashi, A. (1981). *Jpn. J. Appl. Phys.* **20,** L127.

Tsukamoto, J., Ohigashi, H., Matsumura, K., and Takahashi, A. (1982). *Synth. Met.* **4,** 177.

Vanderbilt, D., and Mele, E. J. (1980). *Phys. Rev. B* **22,** 3939.

Vogel, F. L. (1977). *J. Mater. Sci.* **12,** 982.

Vogel, F. L. (1979). *In* "Molecular Metals" (W. E. Halfield, ed.), pp. 261–279. Plenum, New York.

Voronkov, M., Pukhnarevich, V., Suchchinskaya, S., Annenkova, V. Z., Annekova, V. M., and Andreeva, N. (1980). *J. Polym. Sci. Polym. Chem. Ed.* **18,** 53.

Wada, Y., and Schrieffer, J. R. (1978). *Phys. Rev. B* **18,** 3897.

Walling, C. (1957). "Free Radicals in Solution," pp. 397–466. Wiley, New York.

Weber, D. C., Holtzclaw, J. R., Pron, A. B., Brant, P., Wyatt, J. R., Decorpo, J. J., and Saalfeld, F. D. (1981). *Int. J. Mass Spectrom. Ion Phys.* **40,** 101.

Wegner, G. (1969). *Z. Naturforschteil B* **25,** 824.

Wegner, G. (1981). *Angew. Chem. Int. Ed. Engl.* **20,** 361.

Weinberger, B. R., Kaufer, J., Heeger, A. J., Pron, A., and MacDiarmid, A. G. (1979). *Phys. Rev. B* **20,** 223.

Weinberger, B. R., Ehrenfreund, E., Pron, A., Heeger, A. J., and MacDiarmid, A. G. (1980). *J. Chem. Phys.* **72,** 4749.

Weinberger, B. R., Gau, S. C., and Kiss, Z. (1981). *Appl. Phys. Lett.* **38,** 555.

Weinberger, B. R., Akhtar, M., and Gau, S. C. (1982). *Synth. Met.* **4,** 187.

Wnek, G. E., Chien, J. C. W., Karasz, F. E., Druy, M. A., Park, Y. W., MacDiarmid, A. G., and Heeger, A. J. (1979a). *J. Polym. Sci. Polym. Lett. Ed.* **17,** 779.

Wnek, G. E., Chien, J. C. W., Karasz, F. E., and Lillya, C. P. (1979b). *Polymer* **20,** 1441.

Wnek, G. E., Capistran, J. D., Chien, J. C. W., Dickinson, L. C., Gable, R., Gooding, R., Gourley, K., Karasz, F. E., Lillya, C. P., and Yao, K. D. (1981). *In* "Conductive Polymers" (R. B. Seymour, ed.), pp. 183–208. Plenum, New York.

Woerner, T., MacDiarmid, A. G., and Heeger, A. J. (1982). *J. Polym. Sci. Polym. Lett. Ed.* **20,** 305.

Woon, P. S., and Farona, M. F. (1974). *J. Polym. Sci. Polym. Chem. Ed.* **12,** 1749.

Wristlers, J. (1973). *J. Polym. Sci. Polym. Phys. Ed.* **11,** 1601.

Yamabe, T., Tanaka, K., Terame, H., Fukui, H., Imamura, A., Shirakawa, H., and Ikeda, S. (1979). *Solid State Commun.* **29**, 329.

Yamabe, T., Akagi, K., Shirakawa, H., Ohzeki, K., and Fukui, K. (1981). *Chem. Scr.* **17**, 157.

Yamabe, T., Akagi, K., Ohzeki, K., Fukui, K., and Shirakawa, H. (1982). *J. Phys. Chem. Solids* **43**, 577.

Yamamoto, T., Sanechika, K., and Yamamoto, A. (1980). *J. Polym. Sci. Polym. Phys. Ed.* **18**, 9.

Yamaoka, H., Sakai, H., Matsuyina, T., Moeda, Y., and Shimamura, H. (1981). *Polym. Prepr. Jpn.* **30**, 1444.

Yen, S. P. S., Somoano, R., Khanna, S. K., Rembaum, A. (1980). *Solid State Commun.* **36**, 339.

Zannoni, G., and Zerbi, G. (1982). *Chem. Phys. Lett.* **87**, 50.

Zwijnenburg, A., and Pennings, A. J. (1976). *Colloid Polym. Sci.* **254**, 866.

Index